LIPID METABOLISM IN MAMMALS · 2

Edited by

Fred Snyder
Oak Ridge Associated Universities
Oak Ridge, Tennessee

Plenum Press · New York and London

Library of Congress Cataloging in Publication Data

Main entry under title:

Lipid metabolism in mammals.

(Monographs in Lipid Research)
Includes bibliographies and index.
1. Lipid metabolism. 2. Mammals–Physiology. I. Snyder, Fred, 1931- II. Series

QP751.L548 599'.01'9247 77-913

ISBN 978-1-4684-0867-6 ISBN 978-1-4684-0865-2 (eBook)
DOI 10.1007/978-1-4684-0865-2

softcover reprint of the hardcover 1st edition 1977

A Division of Plenum Publishing Corporation
227 West 17th Street, New York, N.Y. 10011

LIPID METABOLISM IN MAMMALS · 2

MONOGRAPHS IN LIPID RESEARCH

David Kritchevsky, Series Editor
Wistar Institute
Philadelphia, Pennsylvania

FUNGAL LIPID BIOCHEMISTRY
By John D. Weete

LIPID METABOLISM IN MAMMALS, Volume 1
Edited by Fred Snyder

LIPID METABOLISM IN MAMMALS, Volume 2
Edited by Fred Snyder

Contributors

John M. Bailey — Biochemistry Department, George Washington Medical School, Washington, D.C.

R. M. Broekhuyse — Department of Ophthalmology, University of Nijmegen, Nijmegen, The Netherlands

John G. Coniglio — Department of Biochemistry, Vanderbilt University, Nashville, Tennessee

F. J. M. Daemen — Department of Biochemistry, University of Nijmegen, Nijmegen, The Netherlands

R. R. Dils — Department of Physiology and Biochemistry, University of Reading, Whiteknights, Reading, United Kingdom

Thomas R. Dirksen — Departments of Oral Biology and Cell and Molecular Biology, Schools of Dentistry and Medicine, Medical College of Georgia, Augusta, Georgia

M. F. Frosolono — Pulmonary Division of the Department of Pediatrics and Department of Pathology, Albert Einstein College of Medicine, Bronx, New York

M. R. Grigor — Department of Biochemistry, University of Otago, Dunedin, New Zealand

Clyde G. Huggins — Department of Biochemistry, University of South Alabama College of Medicine, Mobile, Alabama

Ten-ching Lee Medical and Health Sciences Division, Oak Ridge Associated Universities, Oak Ridge, Tennessee

C. A. Pasternak Department of Biochemistry, St. George's Hospital Medical School, London, England

Charles O. Rock Medical and Health Sciences Division, Oak Ridge Associated Universities, Oak Ridge, Tennessee

Fred Snyder Medical and Health Sciences Division, Oak Ridge Associated Universities, Oak Ridge, Tennessee

Jen-sie Tou Department of Biochemistry, Tulane University School of Medicine, New Orleans, Louisiana

Keizo Waku Laboratory of Medical Research Institute, Tokyo Medical and Dental University, Tokyo, Japan

Preface

During the past decade we have witnessed a vast expansion in our knowledge of lipid metabolism, especially for mammalian tissues. One obvious conclusion arising from these studies is that no single overall scheme of lipid metabolism can be classed as distinctly characteristic of all mammalian organs. Although certain synthetic and degradative lipid pathways are similar in a variety of organs, I have been impressed by the notable exceptions. I was motivated to organize this work on *Lipid Metabolism in Mammals* because of the lack of a single reference source containing a comparative organ approach to lipid metabolism in mammals that emphasizes the uniqueness of pathways in the various organs of the body. Because of the escalation in lipid research, I also feel strongly that there is an urgent need for an updated concise account of this field.

The group of authors for the chapters in the two volumes of *Lipid Metabolism in Mammals* were selected for their expertise and personal experience with the lipid metabolism of the organs or blood constituents that are the subjects of the chapters. Sufficient leeway has been given each author to approach the subject matter from a personal viewpoint. However, the overall direction of each chapter has been slanted to emphasize the similarities and differences in lipid metabolism among organ systems. The introductory chapter on general pathways provides a convenient reference to illustrations of specific reaction sequences that are well established and that occur in a number of organs.

The topics covered in *Lipid Metabolism in Mammals* are pertinent to scientists and graduate students engaged in studies of lipids in biomedical research. Furthermore, the books should be valuable supplements to graduate courses in biochemistry.

My special thanks go to Dorothy Vola and Kathy Spence for their excellent assistance in this endeavor. I am indebted to all contributors for a job well done.

Fred Snyder

Oak Ridge, Tennessee

The shorthand notation 16:0, 18:0, 18:1, 18:2, etc., used throughout the text represents chain length : number of double bonds.

Contents

Gonadal Tissue

JOHN G. CONIGLIO

Mammary Glands

R. R. DILS

The Eye

R. M. BROEKHUYSE AND F. J. M. DAEMEN

Skeletal Muscle

KEIZO WAKU

Skin

M. R. GRIGOR

Calcified Tissues

Thomas R. Dirksen

Cancer Cells

TEN-CHING LEE AND FRED SNYDER

Harderian Gland

CHARLES O. ROCK

Cultured Cells

JOHN M. BAILEY

Lipid Changes in Membranes during Growth and Development

C. A. PASTERNAK

Contents of Volume 1

Lung

M. F. FROSOLONO

I. Introduction

Very early experiments with radioactively labeled precursors led to the recognition that the mammalian lung has an extremely active lipid metabolism (Popjak and Beeckmans, 1950). The recent exponential increase of information concerning the complexities of pulmonary biochemistry, especially the many studies dealing with lipid biosynthesis and turnover, has reinforced the concept that the lung is more than an organ passively engaged in respiratory gas exchange. However, the prime physiological function of the lung is this exchange of respiratory gases, a process which requires structurally intact and open alveoli. The maintenance of alveolar structural integrity is dependent upon the reduction of surface tension forces in the acellular alveolar lining layer by the pulmonary surfactant system (Scarpelli, 1968a,b). Much of the metabolic potential of the lung is seemingly directed toward the biosynthesis, elaboration, and secretion of lipid constituents of the pulmonary surfactant system.

The major compositional and functional constituent of the pulmonary surfactant system is a unique phospholipid, dipalmitoyl phosphatidylcholine (dipalmitoyl-PC), which is found in high concentration in the mammalian lung. This chapter will attempt to survey pulmonary lipid metabolism as a whole; however, it must be realized that dipalmitoyl-PC is often, with some justification, considered to be a major, but certainly not exclusive, focus of lung lipid metabolism. Dipalmitoyl-PC synthesis probably accounts for the major portion of the characteristically high rate of pulmonary lipid biosynthesis (Naimark, 1971).

M. F. FROSOLONO • Pulmonary Division of the Department of Pediatrics and Department of Pathology, Albert Einstein College of Medicine, Bronx, New York 10461.

Differentiation of the lung is on a strict developmental timetable. There is a close correlation of response to fetal hormonal systems, ultrastructural cytodifferentiation, acquisition of biochemical capabilities, and appearance of surface activity, all associated with extrauterine survival (Blackburn *et al.*, 1972, 1973). Consequently, pulmonary lipid metabolism must be considered in relation to different stages of development and hormonal influences. The relationship between extrauterine survival and appearance of surface forces, presumably the result of maturation of phospholipid biosynthesis, continues to be an active area of pulmonary research. This subject has a profound clinical relevance since prediction of fetal lung maturation and, hence, ability of the fetal lung to undergo a successful transition from a liquid- to a gas-filled environment, is an important aspect of neonatology. On the other hand, the interaction of hormonal systems with pulmonary lipid metabolism, both as a function of gestational age and in the adult, is only now beginning to be fully explored.

The lung is a very difficult organ in which to study metabolic processes because of its cellular diversity. Over 40 different cell types are present in the lung (Bertalanffy, 1964; Sorokin, 1970). We can reasonably assume that each cell type does not contribute equally to metabolic processes. As will be discussed later, some specific cell species, the type II epithelial cells, on the basis of both direct experimental and circumstantial evidence, are thought to be major sites of pulmonary lipid metabolism. Unfortunately, until now, most of this evidence has not been quantitative or direct in the biochemical sense since it has been based in large part upon light- and electron-microscopic radioautography. Specific quantitative information is dependent upon the isolation and study of homogeneous populations of pulmonary parenchymal cells, an area which is just now yielding significant information (Kikkawa and Yoneda, 1974; Kikkawa *et al.*, 1975). Additionally, the extensive extracellular alveolar lining may have specific capabilities in certain areas of lipid metabolism (Scarpelli, 1967; Scarpelli *et al.*, 1975), a concept which has not yet been fully explored. Most of the information in this chapter must, therefore, be presented in terms of the lung as a total organ of great cellular diversity with only relatively little knowledge available concerning the lipid biochemistry of specific cell types.

II. Fatty Acid Metabolism

A. Biosynthesis

The major emphasis in the study of pulmonary fatty acid biosynthesis has been directed toward mechanisms for palmitate production. This is due to the requirement of palmitate for dipalmitoyl-PC synthesis. The lung will actively remove palmitate from the circulation and incorporate it into dipalmitoyl-PC and other complex lipids (Thomas and Rhoades, 1970; Frosolono *et al.*, 1971; Pawlowski *et al.*, 1971; Naimark, 1971; Young and Tierney, 1972). In at least one report, pulmonary arterial–venous differences in palmitate concentration were demonstrated in fasting and hyperlipemic states (Condorelli *et al.*, 1972). This reinforces the role of pulmonary lipid metabolism as related to digestion since the lungs present the first capillary barrier to the lipids that are absorbed from the digestive system.

From a consideration of dipalmitoyl-PC turnover alone, and not including other lipids, Naimark (1971) suggested that only 25% of the palmitate flux in the lung was used in dipalmitoyl-PC synthesis. The remaining palmitate requirement must then be accounted for by *de novo* fatty acid synthesis or acyl exchange between other pulmonary and plasma lipids.

By 1950 it was known that [1-^{14}C]acetate was incorporated predominantly into phospholipids in the intact lung (Popjak and Beeckmans, 1950). Studies with lung slices indicated that palmitate and myristate accounted for 87% and 12%, respectively, of the [1-^{14}C]acetate incorporated into PC fatty acids (Chida and Adams, 1967). For some time, however, it was thought that the lung might differ significantly from other tissues in the way it synthesizes palmitate. Tombropoulos (1964) examined the biosynthesis of fatty acids from [1-^{14}C]acetate in rat pulmonary subcellular fractions and found that the mitochondria were the only significant organelle for incorporation of this precursor into long-chain fatty acids. Very little activity was found in the cytosol, which, as discussed in the introductory chapter (Volume 1), is usually associated with the *de novo* biosynthesis of palmitate through fatty acid synthetase and acetyl-CoA carboxylase. Even though Tombropoulos (1964) pointed out that his procedure would not distinguish between *de novo* biosynthesis and chain elongation, the more typical mitochondrial mechanism, the lack of demonstrable cytoplasmic activity caused considerable uncertainty regarding pulmonary palmitate biosynthesis.

Subsequently, in a very well designed and detailed study, Schiller and Bensch (1971) resolved this apparent discrepancy with the demonstration that rabbit lung cytosol does indeed contain an active system for the *de novo* biosynthesis of long-chain fatty acids, presumably palmitate in large part. Both fatty acid synthetase and acetyl-CoA carboxylase were found in the 95,000 × *g* supernatant. Lung mitochondria, in common with other tissues, synthesized fatty acids predominantly by elongation as outlined in the introductory chapter (Volume 1). These studies indicated that while a *de novo* mitochondrial pathway utilizing malonyl-CoA may also be present, it is of minor importance in the lung. Fatty acid biosynthesis through chain elongation was also present in lung microsomes and, similarly to other tissues, this organelle fraction incorporated a greater percentage of malonyl-CoA than acetyl-CoA into fatty acids.

With regard to the actual biosynthesis of palmitate by the cytoplasmic *de novo* pathway, the experiments of Schiller and Bensch (1971), which were based on the decarboxylation technique, gave a very strong indication but did not conclusively demonstrate that this fatty acid was indeed the major product. Gas–liquid chromatography of the fatty acids produced by the cytoplasmic fatty acid synthetase did reveal that palmitate was indeed the major product (Gross and Warshaw, 1974). Further, these latter investigators found that the pulmonary activity of enzymes related to *de novo* fatty acid biosynthesis was at adult levels by gestational day 23 in the fetal rabbit. This suggests that the capacity for production of palmitate occurs well in advance of the time when the lung is able to produce sufficient surface-active lipids, i.e., dipalmitoyl-PC, for adaptation to extrauterine life at 30 days of gestation.

Although the results of Schiller and Bensch (1971) more rigidly defined and markedly increased our understanding of pulmonary fatty acid biosynthesis and placed the mechanism in proper perspective with other tissues, the apparent discrepancy between their findings and the earlier failure of Tombropoulos (1964) to demonstrate cytoplasmic *de novo* synthesis is intriguing. Leaving aside species differences and the possible deactivation of either fatty acid synthetase or acetyl-CoA carboxylase during the preparative procedure of Tombropoulos, the discrepancy may be related to the fact that Schiller and Bensch, in part, used [1-^{14}C]acetyl-CoA as a precursor, whereas [1-^{14}C]acetate was used in the earlier studies. This could indicate that the lung cytosol lacks the capacity to synthesize acetyl-CoA from acetate. From the discussions in the introductory chapter, it would therefore be of interest to determine if the lung cytosol contains the ATP-dependent citrate cleavage enzyme by which acetyl-CoA is regenerated after its transport as citrate from mitochondria, the original site of acetyl-CoA synthesis.

Frosolono *et al.* (1970) found an unusual unsaturated 18-carbon fatty acid in the PC associated with a surface active macromolecular fraction isolated from dog lung. Most of this fatty acid was esterified on the glycerol 2-position. King *et al.* (1973), who also worked with the PC derived from a dog lung surfactant-related preparation, showed that this fatty acid is *cis*-5-octadecenic and further suggested that it might serve as a marker for the pulmonary surfactant system. At present there is no definitive information concerning either the organelle site or mechanism by which *cis*-5-octadecenic acid may be synthesized in the dog lung.

B. *Oxidation*

Most of the emphasis in pulmonary fatty acid metabolism has been concerned with the biosynthesis of these compounds. However, as pointed out by Tierney (1974), using the results of Salisbury-Murphy *et al.* (1966), Wolfe *et al.* (1970), and Wang and Meng (1972), there is now considerable evidence that lung tissue can readily oxidize long-chain fatty acids. Presumably this takes place by the β-oxidation process described in the introductory chapter.

III. Complex Lipid Metabolism

Studies of pulmonary lipid composition and distribution imply that the lung has the capacity for biosynthesis of many different complex lipid classes (White, 1973). Until very recently, the major emphasis in the study of lung lipid metabolism has been directed toward elucidation of mechanisms for the biosynthesis of PC and, especially, dipalmitoyl-PC. This is understandable in view of the previously mentioned contribution of dipalmitoyl-PC to maintenance of alveolar structural integrity and lung function. Because of this emphasis, it is convenient to organize most of the material in the following section relative to the biosynthesis of dipalmitoyl-PC. This should not be construed as indicating that other complex lipids are not important factors in lung metabolism and structure.

A. PC and Dipalmitoyl-PC Biosynthesis

For some time it was thought that pulmonary PC biosynthesis might take place by two major pathways usually referred to as the *de novo* and transmethylation sequences. Formation of PC through the successive methylation of PE, utilizing S-adenosyl methionine as the methyl donor, was proposed to be an important mechanism, especially during gestation and the early neonatal period (Gluck *et al.*, 1972). The possible role of the transmethylation pathway created a considerable stir of interest in the pulmonary biochemistry field in recent years. However, the preponderance of evidence now indicates that lung PC biosynthesis takes place primarily through the *de novo* or CDP-choline pathway as described in the introductory chapter (Weinhold, 1968; Akino *et al.*, 1971; Epstein and Farrell, 1975).

Assuming that palmitate and other acyl residues are readily available, there are four major aspects involved in *de novo* PC biosynthesis: (1) formation of phosphatidic acid (PA); (2) production of the 1,2-diacylglyceride (DG) from PA; (3) systems for incorporation of the choline base onto the glycerol backbone; and (4) mechanisms that may be involved in determining the specific acyl group distribution patterns of PC and dipalmitoyl-PC. This is a convenient but admittedly artificial division.

1. *Biosynthesis of PA*

Reference to Fig. 5 of the introductory chapter indicates that PA is a common intermediate arising from glucose and glycerol. Results from several *in vivo* and *in vitro* experiments indicate that glycerol is readily incorporated into the backbone of lung PC, presumably through *sn*-glycerol-3-phosphate (*sn*-glycerol-3-P) (Mims and Zee, 1971; Mims and Kotas, 1973; Moriya and Kanoh, 1974; Frosolono *et al.,* 1975). The latter intermediate may arise either directly from glycerol by the action of glycerokinase or from dihydroxyacetone phosphate, an intermediate of glycolysis.

Although it is an observable fact that radioactive glycerol, administered *in vivo,* is incorporated into pulmonary PC, the exact role of this compound as a direct circulating precursor has been questioned since glycerokinase was reported to be absent in mammalian lung (Wieland and Suyter, 1957). This could imply an extrapulmonary phosphorylation of glycerol to *sn*-glycerol-3-P before utilization in the lung. Mims and Zee (1971), using lung slices from newborn rabbits, showed that glycerol was incorporated preferentially into the PC fraction. These results with slices would seem to mitigate against the necessity of an extrapulmonary source of glycerokinase, leading to the inference that this enzyme may be present at least in neonatal lung. This may be significant because circulating glycerol is known to rise rapidly after birth (Persson and Gentz, 1966) when gluconeogenesis is low and may provide a readily available source for incorporation into lung PC, as suggested by Mims and Zee (1971). Further supportive, but not conclusive, evidence of a precursor role for circulating glycerol and, by inference, the presence of glycerokinase in the lung, was presented by Mims and Kotas (1973). When neonatal rabbit lung slices were incubated in the presence of equimolar concentrations of [^{3}H]glycerol and [^{32}P], both isotopes were incorporated into PC on a 1:1 molar basis. Apparently, however, glycerokinase has not yet been conclusively demonstrated in adult lung.

O'Neil and Tierney (1974) have pointed out the difficulty involved in detecting predicted arterial–venous differences in glucose concentration in order to estimate glucose utilization by the lung *in vivo*. This is due to the fact that observed errors in sampling and analysis are about seven times the arterial–venous differences for *in vivo* glucose concentrations, which might be predicted from values obtained *in vitro*. However, several studies have shown that radioactive glucose is rapidly incorporated into lung PC, predominantly labeling the glycerol moiety (Gassenheimer *et al.,* 1972), although up to 40% of the incorporated glucose can be found in the esterified fatty acids of phosphoglycerides (Scholz and Rhoades, 1971). Glucose, therefore, provides a ready source of *sn*-glycerol-3-P.

Definitive information is not yet available concerning the relative contributions to PA biosynthesis of *sn*-glycerol-3-P, through lyso-PAs, and dihydroxyacetone phosphate (via acyldihydroxyacetone-P), by passing *sn*-glycerol-3-P (introductory chapter, Fig. 5). The recent results of Snyder and Malone (1975) do suggest that the latter pathway may be operative in the lung, in addition to that through *sn*-glycerol-3-P.

Once *sn*-glycerol-3-P and/or dihydroxyacetone-P are available, these would serve as receptor substrates for acyltransferase systems which would catalyze the formation of PA using various acyl-CoA derivatives as donor substrates. The role that these acyltransferases, which operate at the first stage of glycerophosphatide biosynthesis, may play in conferring the specific fatty acid distribution pattern of PA and, subsequently PC and dipalmitoyl-PC, will be dealt with later. Hendry and Possmayer (1974) demonstrated that there is an active and stable *sn*-glycerol-3-P:acyl-CoA acyltransferase (*sn*-glycerol-3-P acyltransferase, EC 2.3.1.15) present in rabbit lung endoplasmic reticulum, in keeping with the concept that this organelle is the site of the majority of glycerolipid biosynthesis in mammalian tissues. In their assay system, acyl-CoA donor substrates were produced from free fatty acids by the endogenous acyl-CoA synthetase in the microsomes rather than by adding exogenous acyl-CoAs. It was noted that acyl-CoAs generated *in situ* support a more rapid rate of PA synthesis in nonpulmonary tissues (Jamdar and Fallon, 1973; Sanchez de Jimenez and Cleland, 1969; Brindley, 1973). The major product of the overall reaction was PA, which suggests that, while the microsomes may have some activity for the acylation of lyso-PA, this intermediate presumably does not accumulate to a significant degree. The previously referred to work of Snyder and Malone (1975) indicates a possibly important role for acyltransferases in the production of PA utilizing dihydroxyacetone phosphate and acyldihydroxyacetone-P as receptors.

2. *Production of DG from PA*

DG species are the actual receptor molecules for the transfer of phosphorylcholine to form PC in the *de novo* pathway. Therefore, phosphatidic acid phosphohydrolase (PAP'ase, EC 3.1.3.4.), which catalyzes the formation of DG from PA, occupies a central position in this scheme as well as in the *de novo* pathway for synthesis of other complex lipids. Meban (1972) demonstrated PAP'ase activity in the perilamellar surface of lamellar bodies in hamster type II cells using an electron-microscopic histochemical technique. Spitzer *et al.* (1975), working with lamellar bodies isolated from pig lung, have shown that these organelles contain PAP'ase at very high enzymatic specific activity. Their calculations indicate that as much as 40% of the activity in type II cells may be associated with the lamellar body system. The presence of a very active PAP'ase in lamellar bodies is an indication that these organelles are actively involved in PC biosynthesis and do not function only as phospholipid storage, transfer, and secretory vesicles. This is of interest since PAP'ase is a key regulatory enzyme in lipid biosynthesis (Lamb and Fallon, 1974). Spitzer *et al.* (1975) suggest that PAP'ase activity in the lamellar body membrane may be metabolically responsible for biosynthesis of dipalmitoyl-PC.

3. *Incorporation of Choline*

Conversion of choline to the phosphoglyceride *de novo* involves mediation of three sequential enzymatic reactions: (a) phosphorylation of the free base by choline kinase (ATP:choline phosphotransferase, EC 2.7.1.32); (b) conversion of choline phosphate to CDP-choline by choline phosphate cytidyl transferase (CTP:choline phosphate cytidyl transferase, EC 2.7.7.15); and (c) transfer of phosphorylcholine to the DG by choline phosphotransferase (CDP-choline:1,2-diglyceride choline phosphotransferase, EC 2.7.8.1). Artom (1968) assayed specifically for these three enzymes in fetal rat lung and found that they reached full activity at about the time of birth. Choline kinase, the first enzyme in this pathway, has been demonstrated in homogenates from human newborn lungs and some of its properties described (Zachman, 1971). Choline kinase and choline phosphotransferase, assayed in homogenates of fetal and neonatal rat lungs, exhibited a dramatic increase in specific activity 3 days before birth, at 19 days gestation (Farrell *et al.,* 1974). The increases in specific activities were correlated with a significant elevation of PC synthesis during the same gestational period. These authors point out that choline kinase and choline phosphotransferase exhibit many features characteristic of key regulatory steps in metabolic sequences and speculated that they may play an important role in the initiation of enhanced pulmonary PC synthesis near birth. This would also imply that these enzymes might be important factors in regulation of PC synthesis in the adult lung, although Spitzer *et al.* (1975) have proposed that PAP'ase may be a more important regulatory enzyme.

4. *Determination of Acyl Group Distribution*

Dipalmitoyl-PC, in terms of positional distribution of its fatty acyl residues, is both symmetrical and highly nonrandom. This is the direct opposite of the situation that exists in most other tissues, where the phospholipids are highly asymmetric with a preponderance of saturated residues on the glycerol-1 position and unsaturated residues on the 2-position. Also, while the distribution of fatty acids at each position may well be characteristic for each tissue, the degree of randomness is typically much greater in nonpulmonary tissues where there is a variety of saturated residues on the 1-position and unsaturated moieties on the 2-position.

The putative role of dipalmitoyl-PC in the maintenance of alveolar structural integrity and its known high concentration in lung suggest specific mechanisms for its biosynthesis in pulmonary tissue. These pathways might be assumed to operate in a significantly different manner in lung than in other tissues. Consequently, an important aspect of pulmonary biochemistry is the determination of which point or points in the metabolic pathways leading to the synthesis of PC may be responsible for imparting the unique fatty acid distribution of dipalmitoyl-PC.

There are at least two major mechanisms by which dipalmitoyl-PC could be synthesized in the lung: (a) The initial acylations of *sn*-glycerol-3-P may be highly nonrandom, thereby producing large amounts of dipalmitoyl-PA whose acyl distribution pattern would be maintained through subsequent steps to the PC level. Dipalmitoyl-PC would, then, be produced as a consequence of the substrate and positional specificity of *sn*-glycerol-3-P acyltransferases. (b) The initial acylations of *sn*-glycerol-3-P may produce PA and, subsequently, PC species with varying degrees of randomness and asymmetry. The nonspecific fatty acid distribution pattern of these *de novo* produced PC species would then be modified to give dipalmitoyl-PC through the interaction of other enzymatic reacylation systems.

a. De Novo. Some reports have suggested that the asymmetrical distribution of saturated and unsaturated fatty acids at the 1- and 2-positions of nonpulmonary phospholipids may be introduced during the synthesis of PA (Possmayer *et al.*, 1969; Akesson *et al.*, 1970). Since the fatty acid distribution of dipalmitoyl-PC is symmetrical, as opposed to asymmetrical for the majority of nonpulmonary phospholipids, it is not immediately apparent that lung *sn*-glycerol-3-P acyltransferase will be the determining factor in the synthesis of dipalmitoyl-PC. If, however, a significant portion of dipalmitoyl-PC is produced by the *de novo* pathway, one would expect that this enzyme system would show some, if not considerable, preference for palmitoyl-CoA as a donor substrate at each position of glycerolphosphate. This might lead to production of significant amounts of dipalmitoyl-PA and, hence, dipalmitoyl-PC. Hendry and Possmayer (1974) found, in their *in vitro* studies, that the relative rates of acyl group incorporation into PA were linoleate > palmitate, oleate, stearate > myristate > laurate > linolenate, arachidonate. Phospholipase A_2 degradation of the PAs produced revealed that palmitic and stearic acids were incorporated at both the glycerol-1 and -2 positions, with slightly higher percentages of each found at the latter position. These *in vitro* data concerning relative incorporation rates and positional distribution of palmitate into PA certainly support the concept that dipalmitoyl-PA can be a product of microsomal *sn*-glycerol-3-P acyltransferase in rabbit lungs.

Several lines of investigation focusing on incorporation of radioactive precursors and turnover of labeled pulmonary lipids with whole animals and lung slices support the concept that at least some dipalmitoyl-PC can be produced by the *de novo* mechanism. It must be kept in mind, however, that many of these same studies also suggest that dipalmitoyl-PC may not be produced solely or even predominantly by this pathway. Thus, Vereyken *et al.* (1972) studied the *in vivo* incorporation of labeled glycerol into various molecular species of PC and concluded that, while the *de novo* pathway may be responsible for synthesis of some disaturated species, it contributed primarily to linoleic-containing PCs and also represented an important mechanism for production of tetraenoic and monoenoic classes. The marked specificity of linolenoyl-CoA as a donor substrate with the *sn*-glycerol-3-P acyltransferase system has been noted both *in vitro* (Hendry and Possmayer, 1974) and *in vivo* Vereyken *et al.* (1972). *In vitro* experiments with rat lung slices utilizing radioactive glycerol, palmitate, and choline also led Akino *et al.* (1971) to the interpretation that the *de novo* pathway was primarily responsible for the biosynthesis of dienoic and other unsaturated PC species. Snyder *et al.* (1973) studied the biosynthesis of surfactant-related PC in homogenates of urethan-induced pulmonary adenomas, which are thought to be derived from type II cells (Brooks, 1968). They also concluded, on the basis of unequal radioactive palmitate incorporation at the glycerol-1 and -2 positions and by using calcium to inhibit *de novo* synthesis that dipalmitoyl-PC cannot be produced primarily by the *de novo* mechanism.

These results were confirmed and extended by Moriya and Kanoh (1974) in an elegant *in vivo* study of the turnover of rat lung DG and PC. Saturated species accounted for both the highest percentage incorporation and radiochemical specific activity in the total DG class 5 min after injection of labeled glycerol. However, the specific activities of the dienoic and trienoic PC species were higher than those of the disaturated PC. Calculated turnover rates showed that dipalmitoyl-PC accounted for only 17% of PC formed by the *de novo* mechanism. When [^{3}H]palmitate was injected, 60% of the radioactivity was distributed in disaturated PA and DG species in 2 min and this labeling pattern was reflected in the synthesis of PC. The distribution of radioactivity in the 1- and 2-positions of glycerolipids showed that the ratio of labeled palmitate residues was almost 1:1 in the disaturated DG species, supporting the *de novo* pathway, but that the 2-position of dipalmitoyl-PC contained some three times the radioactivity, as did the 1-position, indicating that subsequent modification of preformed PC species plays an important role in the production of dipalmitoyl-PC. Although disaturated DG species were found to be most active in the incorporation of labeled glycerol and palmitate, no marked differences were found in the calculated turnover rate constants of PC species synthesized *de novo*. This result was interpreted by Moriya and Kanoh (1974) to indicate that choline phosphotransferase does not exhibit substrate selectivity toward DG species, in contrast to unpublished observations referred to by Hendry and Possmayer (1974). Additionally, further experiments by Possmayer (personal communication) indicated that rabbit lung microsomal choline phosphotransferase probably does not exhibit substrate preference for disaturated-DG.

b. Reacylation Mechanisms. From the above discussion, it is reasonable to conclude that dipalmitoyl-PC can be produced by the *de novo* pathway. It is by no means certain that this mechanism represents the origin of all, or even most, of the disaturated-PC in the lung. Thus, an important question which is not yet resolved concerns the degree to which the *de novo* pathway may contribute to the synthesis of dipalmitoyl-PC. Most of the evidence does suggest that the major PC products of *de novo* synthesis are linoleate and other unsaturated species and that dipalmitoyl-PC accounts for only a relatively small fraction of the total production of this pathway. These considerations, coupled with the known high concentration of dipalmitoyl-PC in the lung, suggest that mechanisms other than *de novo* biosynthesis are responsible for the major production of disaturated PCs. Generally, the more important of these alternative mechanisms invoke reaction of lyso-PCs, which serve as receptor substrates, with acylating enzyme systems having high degrees of donor substrate specificity for palmitoyl and other saturated acyl groups.

There is no evidence to indicate that lyso-PC, as opposed to PC, is a major biosynthetic product of pulmonary metabolism. Consequently, phospholipase A_2 hydrolysis of *de novo* produced unsaturated PC species is an integral feature of these proposed mechanisms. As will be discussed later, phospholipase A_2 activity has been found in the lung (Franson and Waite, 1973; Franson *et al.*, 1973; Garcia *et al.*, 1975). Presumably, that fraction of the activity which is not associated with alveolar macrophages can be considered another instance where a degradative enzyme actually fulfills a synthetic function, much in the same way as does PAP'ase in the *de novo* pathway.

Two major reacylation mechanisms that may be important in the biosynthesis of pulmonary disaturated-PC species have been proposed, lyso-PC acyltransferases and transacylases.

i. Lyso-PC Acyltransferases. These enzymes are more properly designated 1-acyl-2-lyso-PC (or 1-lyso-2-acyl-PC):acyl-CoA acyltransferases (EC.3.1.22) to distinguish them from the acyltransferases that utilize *sn*-glycerol-3-P and lyso-PA as receptor substrates. Although acyltransferases active with lyso-PC (Frosolono *et al.*, 1971) and *sn*-glycerol-3-P (Hendry and Possmayer, 1974) have been found in lung microsomes, no information is presently available that would allow us to determine if these two activities reside in separate enzymes or are the result of the same enzyme with broad substrate specificity.

Lands (1958) and Webster (1965) were the first to report that there was significant lyso-PC acyltransferase activity in lung. The first detailed studies of pulmonary lyso-PC acyltransferases were provided by Frosolono *et al.* (1971). This dog lung microsomal system was studied by determining the reactivities of various acyl-CoA derivatives with 1-lyso-2-acyl-PC and 1-acyl-2-lyso-PC under uniform conditions. In contrast to results obtained with liver microsomes and other nonpulmonary tissues where acylation with saturated acyl-CoAs at the 2-lyso-PC position is very slow in comparison to the 1-lyso position, the palmitoyl-CoA derivative had equal and marked reactivity for both lyso positions with lung microsomes. The stearoyl-CoA derivative was also reactive with both lyso positions although the specific activity with the palmitoyl-CoA was approximately twice that of the stearoyl-CoA at each position. The results with palmitoyl-CoA were interpreted to indicate that lyso-PC acyltransferase positional specificity may be an important determinant of dipalmitoyl-PC synthesis in the lung. It was further speculated that a phospholipase A-lyso-PC acyltransferase cycle, operative at the PC level, may be an important mechanism to ensure that sufficient amounts of dipalmitoyl-PC are available for use in the biosynthesis of surface active materials.

A complicating factor in the findings of Frosolono *et al.* (1971) was that although both the unsaturated palmitoleoyl- and oleoyl-CoAs were more active with the 2-lyso than with the 1-lyso-position, as might be expected, palmitoleoyl-CoA was slightly more active at the 1-position than was palmitoyl-CoA. Oleoyl-CoA, however, was not very active towards the 1-lyso-position, but exhibited considerable specificity for the 2-position. Since the physical state of the acyl-CoA derivatives undoubtedly influences their reactivities toward each position, we may infer that in its physical properties palmitoleoyl-CoA resembles shorter-chain saturated analogs and thus is active with 1- and 2-lyso-PC. An alternative interpretation is that lyso-PC acyltransferase in lung microsomes does not have the same degree of positive positional specificity as found in other tissues.

Minimally, the findings of Frosolono *et al.* (1971) lead to the inference that, if lyso-PCs are available as receptor substrates, pulmonary lyso-PC acyltransferases do have the capacity to transfer palmitoyl residues equally well to both the 1- and 2-positions, as might be expected on the basis of the large amount of dipalmitoyl-PC in the lung. Interestingly, these studies also indicated that 1-acyl-2-lyso-PE and 2-lyso-choline plasmalogen (1-alk-1-enyl-*sn*-glycero-3-phosphorylcholine) were poor receptor substrates for the acyltransferase system.

Vereyken *et al.* (1972) studied the microsomal lyso-PC acyltransferase system in rat lungs. They both confirmed and extended the findings of Frosolono *et al.* (1971). Experiments with 1-palmitoyl-2-lyso-PC showed that uptake of palmitate into the 2-position was almost equal to that of oleate and linoleate. More information was obtained by following the incorporation of mixtures of palmitate and oleate or palmitate and linoleate. In agreement with their findings, both *in vivo* and *in vitro*, oleate and palmitate incorporated equally well into the 2-position from mixtures of the donor substrates. A similar tendency was observed with palmitate and linoleate, but a small preference was seen for linoleate. This suggests that the activity and preference of the system for palmitate residues is real and not merely a reflection of *in vitro* availability of palmitate since, when offered a choice between palmitate and unsaturated acyl derivatives (the normal nonpulmonary constituent at the 2-position), palmitate was esterified equally well.

More detailed kinetic studies (Hasegawa-Sasaki and Ohno, 1975) suggest that there may be at least two distinct activities: (a) 1-acyl-2-lyso-PC:palmitoyl-CoA acyltransferase, which has low values for V_{max}, activation energy, and K_m with both donor and receptor substrates and (b) 1-acyl-2-lyso-PC:arachidonoyl-CoA acyltransferase, with markedly higher values for K_m, V_{max}, and activation energy. The first activity was inhibited by the addition of other acyl-CoAs as well as by palmitoyl-CoA at higher concentrations. The second activity had a high substrate specificity for arachidonoyl-CoA, was not affected by palmitoyl-CoA or oleoyl-CoA, and was only slightly inhibited by linoleoyl-CoA.

Tansey and Frosolono (1975) investigated some of the kinetic characteristics of rabbit lung microsomal 1-acyl-2-lyso-PC acyltransferase. Earlier studies (Frosolono *et al.*, 1970, 1971) indicated that oleate was the major unsaturated acyl residue present in lung PC. Consequently, the kinetic parameters exhibited by this residue were contrasted with palmitate. The K_m for palmitoyl-CoA was almost five times greater than that for oleoyl-CoA. V_{max} for oleoyl-CoA was twice that of palmitoyl-CoA. The V_{max} values obtained in this study, only for activity towards the 2-lyso position, were in agreement with standard velocities obtained in the dog lung microsomal system (Frosolono *et al.*, 1971). It was pointed out that the lyso-PC acyltransferases studied were derived from total lung microsomes and, hence, were a reflection of these systems from many different pneumocyte species. When these studies were extended to long-term cultured cells derived from cell isolates enriched with type II pneumocytes, presumed to be a major cellular source of pulmonary surfactant and dipalmitoyl-PC (Kikkawa *et al.*, 1975; Chevalier and Collet, 1972), palmitoyl-CoA gave a significantly higher V_{max} than oleoyl-CoA while the K_m values were equivalent. Tansey and Frosolono (1975), on the basis of these results, suggested that the lyso-PC acyltransferase system in these cells has a proclivity for producing PC species with palmitate on the glycerol-2 position. Since K_m for both acyl-CoA derivatives were equivalent, V_{max} is most likely the more important criterion. If these long-term cultured cells are relevant to type II cells *in situ,* then the lyso-PC acyltransferase may well be a very important factor in the biosynthesis of dipalmitoyl-PC in the lung *in vivo.* These studies (Tansey and Frosolono, 1975), in general, support the conclusions reached by Snyder and Malone (1975), who investigated lyso-PC and other acyltransferases in urethan-induced type II cell adenomas in mouse lung.

On the other hand, the higher K_m, V_{max}, and activation energies found with the arachidonyl-CoA activity in rat lung microsomes, as compared to the palmitoyl-CoA, led Hasegawa-Sasaki and Ohno (1975) to the interpretation that lyso-PC acyltransferases do not play an important role in the synthesis of dipalmitoyl-PC but, rather, are involved in the production of PC species containing unsaturated fatty acids. However, if one takes the general interpretation that, among a number of possible substrates, the one with the lowest K_m is considered to be the preferred substrate of the enzyme system, these results could indicate that palmitoyl-CoA, with a markedly lower K_m than arachidonoyl-CoA, may be the preferred substrate. The lower V_{max} seen with palmitoyl-CoA could be interpreted to mean that, even though the reaction with arachidonate is significantly faster, the much larger available concentration of palmitate in the lungs, as opposed to arachidonate (White, 1973), enables the system to produce dipalmitoyl-PC.

Snyder and Malone (1975) have pointed out that one must be extremely cautious in interpreting results obtained for lyso-PC acyltransferase donor substrate preferences since the specificity varies with the concentration of acyl-CoA used. In their studies, the acyl-CoA derivatives were generated *in situ.* When the ratio of palmitate or arachidonate to 2-lyso-PC was kept constant at 4, increased concentrations of both substrates caused a preferential utilization of palmitic acid instead of arachidonate. At lower substrate concentrations, the relative incorporation rates at the 2-position were approximately the same for both fatty acids.

It should be kept in mind, in consideration of the *in vitro* specificities of the lyso-PC acyltransferase system as related to the biosynthesis of pulmonary PC, especially dipalmitoyl-PC, that the results from the several studies above may actually be a reflection of loss of acyl group and positional specificity resulting from tissue homogenization and organelle disruption. Mitigating against this is the fact that the *in vitro* specificities apparently do bear some positive correlations with results reflective of the *in vivo* state (Frosolono *et al.,* 1971; Vereyken *et al.,* 1972).

ii. Transacylases. 1-Acyl-2-lyso-PC:1-acyl-2-lyso-PC acyltransferase, or transacylase, which catalyzes acyl group transfer from one molecule of 1-acyl-2-lyso-PC to another, thereby forming PC and *sn*-glycerol-3-phosphorylcholine, was suggested to be an important factor in the biosynthesis of disaturated-PC from studies dealing with incorporation of radioactive precursors into lipids of rat lung slices (Akino *et al.,* 1971, 1972). When palmitate was the precursor, an extremely high radiochemical specific activity was found in disaturated-PC species, as might be expected, but the incorporation was much higher on the 2-position of PC (60%) than on the 1-position (40%). This finding in itself indicated that the *de novo* pathway probably could not account for all of the disaturated-PC synthesis in the lung. Both acyl positions would be equally labeled if the *de novo* pathway were totally responsible for dipalmitoyl-PC synthesis. When lyso-PC labeled with [^{3}H]glycerol and [^{14}C]choline was the precursor, the ratio of ^{14}C to ^{3}H was approximately 1 in PC, indicating that the intact *sn*-glycerol-3-phosphorylcholine moiety was incorporated. In this case, the highest radiochemical specific activity was found in tetraenoic and hexaenoic PC species and it was felt that the *sn*-glycerol-3-P acyltransferase system exhibited a higher specificity for unsaturated acyl-CoA derivatives, thereby functioning primarily in the biosynthesis of unsaturated PC species. The ratio of ^{14}C to ^{3}H activity was approximately 2 in saturated PC species and half that in polyenoic species utilizing lyso-PC labeled with [^{3}H]glycerol and [^{14}C]palmitate at the 1-position. The ratio of 2 in the disaturated species was taken as direct evidence that transacylation between two 1-palmitoyl-2-lyso-PC species was a key factor in the biosynthesis of disaturated-PC. This interpretation was not completely unambiguous since considerably more [^{14}C]palmitate was found on the 1-position (60%) of disaturated PC than on the 2-position (40%).

When 2-lyso-PC labeled with [^{14}C]palmitate at the glycerol-1 position and [^{3}H]glycerol was injected into rats, there was a preferential incorporation into tetraenoic PC species (Akino *et al.,* 1972). This unsaturated PC had a ^{14}C/^{3}H ratio of 0.78 and presumably was synthesized as a result of the acyltransferase pathway (Frosolono *et al.,* 1971). That lyso-PC that was incorporated into saturated + monoenoic species gave PC that had a ^{14}C/^{3}H ratio of 1.91 or nearly 2. The inference was that dipalmitoyl-PC was synthesized by transacylation. Preliminary turnover studies indicated that, in spite of the fact that dipalmitoyl-PC could be formed from exogenous lyso-PC, the majority of this species was probably not derived from plasma lyso-PC.

Actual transacylase activity was demonstrated in rat lung cytosol or soluble fraction, i.e., 105,000 × *g* supernatant (Van Den Bosch *et al.,* 1965) and subsequently studied in some detail by Abe *et al.* (1972). There was no requirement for ATP, CoA, or palmitoyl-CoA. A mixed molecular species preparation of 1-acyl-2-lyso-PC, biosynthetically labeled with [^{3}H]glycerol and [^{14}C]palmitate at the glycerol 1-position, gave rise to a PC with a ^{14}C/^{3}H ratio of 2. Interestingly, 70% of the [^{14}C]palmitate was found at the glycerol 2-position and only 30% at the 1-position. It was speculated, therefore, that 1-palmitoyl-2-lyso-PC is a better donor than receptor substrate. Formation of PC from 1-stearoyl-2-lyso-PC was approximately one-half that with the palmitoyl analog but stearate was incorporated almost equally well into the 1- and 2-positions.

Hallman and Raivio (1974), working with rabbit lung, confirmed many of the above findings concerning the transacylation pathway. After incubation of lung slices with 2-lyso-PC doubly labeled with [^{3}H]glycerol and [^{14}C]palmitate at the glycerol 1-position, disaturated-PC species exhibited a higher ^{14}C/^{3}H ratio than did the precursor or PC containing unsaturated linkages. As with the previous findings (Akino *et al.*, 1971, 1972; Abe *et al.*, 1972), more than half of the ^{14}C radioactivity was associated with the 2-position of disaturated-PC. The addition of excess unlabeled palmitate to the incubation medium did not markedly change the distribution of label between positions 1 and 2 of the PC. This was interpreted to mean that transacylation between two 1-acyl-2-lyso-PC molecules, rather than direct acylation of 2-lyso-PC, was operative. It was pointed out, however, that their experiments did not exclude the possibility that some disaturated-PC may be derived by direct acylation of lyso-PC.

A marked lysophospholipase activity seems to be associated with the transacylase activity in rat lung supernatant (Ohta and Hasegawa, 1972; Ohta *et al.*, 1972). Subsequently, both activities were copurified about 100-fold and found to have the same K_m value for 2-lyso-PC and the same pH-activity curve (Abe *et al.*, 1974). Stability and behavior toward a number of chemicals and treatments were identical or very similar for both enzymic activities. It was pointed out that even though pancreatic lysophospholipase does not have transacylase activity (Van den Bosch *et al.*, 1973), this does not preclude the possibility that the lung enzyme may have a significant role in the synthesis of PC by transacylation. Various other hydrolytic enzymes such as glycosidases (Nisizawa and Hashimoto, 1970), nucleases (Lehman, 1963), and peptidases (Fruton, 1963) have been reported to have some degree of transferease activity in addition to their generally much more potent hydrolytic activities. In their studies directed toward purification of the lysophospholipase and transacylase, Abe *et al.* (1974) reported that, at all stages, the specific activity, i.e., units of enzyme activity/mg protein, of the hydrolase was 30 times greater than that of the transferase. This may indicate that the hydrolytic function is the more important role of this enzyme system and may reinforce the earlier suggestion that the contribution of the transacylation mechanism, stated to be detectable *in vitro* only under energy-poor conditions (introductory chapter), is not a significant contributor to the production of disaturated-PC species *in vivo* (Vereyken *et al.,* 1972). In this context, it should also be kept in mind that the endoplasmic reticulum, the site of *de novo* phospholipid synthesis and acyltransferase activity, is generally considered to be quantitatively much more important in complex lipid (i.e., as opposed to fatty acid) biosynthesis rather than the cytosol, the site of the transacylase activity.

c. Summary. In assessing the relative contributions of the *de novo* lyso-PC acyltransferase and transacylase systems to the biosynthesis of pulmonary DPL, the previously referred to studies of Moriya and Kanoh (1974) are of particular interest. At various intervals after the injection of [^{3}H]palmitate, rat lung DG and PC were isolated and subfractionated into molecular species according to the degree of acyl group saturation. Although disaturated-DG species accounted for only 10.6 mole% of the total DG class, over one-half of the [^{3}H]palmitate incorporated into total DG went into the disaturated subclass. The incorporation of radioactive palmitate into the 1- and 2-positions of these subclasses was determined. At the earlier postinjection periods, palmitate was distributed almost equally between the 1- and 2-positions of the disaturated DG, 47.2% and 52.8%, respectively, at 2 min. However, in DPL at the same time, 76.4% of the incorporated [^{3}H]palmitate was on the 2-position. This rose to 82.1% by 5 min postinjection.

Earlier, Frosolono *et al.* (1970), in their work describing a surface active macromolecular fraction from dog lung, observed that, after *in vivo* injection of radioactive palmitate, the glycerol-2 position of PC isolated from this fraction had a significantly higher incorporation of the label than did the 1-position.

The interpretation of these studies (Moriya and Kanoh, 1974) is that *de novo* synthesis cannot account for the finding that the 2-position of DPL contained much more radioactive palmitate than the 1-position and, consequently, this pathway from *sn*-glycerol-3-P is not responsible for the majority of DPL produced in the lung. If it were, the 1- and 2-positions of DPL should have the same ratio of palmitate incorporation as found at least in the disaturated-DG subclass, i.e., ~ 1:1. As pointed out in the introductory chapter (Vol. 1), PA and DG may also arise in the *de novo* pathway through dihydroxyacetone phosphate. In Fig. 5, reaction 4, of that chapter, it can be seen that an acyltransferase could place a palmitate residue on the primary hydroxyl of dihydroxyacetone phosphate and, following reduction with NADPH, 1-acyl-2-lyso-glycerol-3-P would be produced. Another acylation, step 3a, would then give PA, as in the pathway from *sn*-glycerol-3-P. If palmitoyl-CoA from step 3a came from a different pool from that in step 4, the resultant PA and dipalmitoyl-PC produced from it via the DG could have a higher specific activity in the 2-position. However, this *de novo* pathway also seems to be ruled out since the salient observation is that, in the disaturated-DG class, the ratio of incorporated radioactivity at each position was 1:1.

These considerations provide reasonable evidence that a reacylation mechanism may be operable that functions to produce dipalmitoyl-PC; however, Moriya and Kanoh (1974) suggest that neither lyso-PC acyltransferase nor transacylase can explain these findings. If 1-[^{3}H]palmitoyl-2-lyso-PC, produced as a result of *de novo* biosynthesis and phospholipase A_2 hydrolysis, serves as the substrate-receptor for a transacylase, the resultant DPL should have equal radiochemical specific activity at both the 1- and 2-positions. As this was not the case, the interpretation was that a transacylase does not function significantly in the production of DPL. The conclusion of Moriya and Kanoh (1974) regarding the lyso-PC acyltransferase is not as clear cut. The fact that dipalmitoyl-PC comprised only about 17% of the PC formed by the *de novo* pathway may indicate the existence of multiple palmitoyl-CoA pools in the lung that are not used in the same way for *de novo* synthesis and for reacylation. If that portion of the palmitoyl-CoA pool available for reacylation has a higher radiochemical specific activity than that used for *de novo* synthesis, then this could explain the higher specific activity of palmitate found at the 2-position of DPL. A higher radiochemical specific activity on the 2-position would then result by reacylation with [^{3}H]palmitoyl-CoA, which itself has a higher specific activity than that palmitate already present on the 1-position.

It is clear, however, at this time that the exact mechanisms and their relationships by which pulmonary dipalmitoyl-PC is synthesized have not yet been clearly elucidated. This is one of the most intriguing problems in lung lipid metabolism and will certainly be a focus of research in this field for the immediate future.

B. *Transmethylation of PE to PC*

For some time it was thought that PC might be formed from PE via three successive methylations at the free amino moiety utilizing S-adenosyl methionine as the methyl donor (Gluck *et al.,* 1967a,b, 1970). An exciting aspect of this work was the interpretation that the transmethylation mechanism might be significant in the lungs of neonates born before full biochemical development of the *de novo* pathway for PC, and presumably, dipalmitoyl-PC synthesis. Animals born prematurely in the later phases of gestation often develop the Respiratory Distress Syndrome, which, from many lines of evidence, has been associated with a lack of pulmonary surfactant activity (Avery and Mead, 1959; Scarpelli, 1968c). The transmethylation pathway was thought to be responsible for the biosynthesis of 1-palmitoyl-2-myristoyl-PC, which would then function as a surfactant species until *de novo* production of dipalmitoyl-PC became fully developed

(Gluck *et al.*, 1970). Thus, premature neonates born during the period of greatest risk of respiratory distress syndrome might be aided in survival if the transmethylation pathway were developed to such an extent that sufficient amounts of 1-palmitoyl-2-myristoyl-PC could be produced to provide surface tension reducing forces in the alveolar lining layer. These experiments and observations from Gluck's laboratory seemed to indicate that considerable PC biosynthesis might take place via this route. The difficulty with assigning a significant role to the transmethylation pathway was that, as pointed out earlier, the majority of other experimental evidence indicates that the *de novo* sequence, in connection with a reacylation mechanism, is the primary means by which pulmonary PC and dipalmitoyl-PC are synthesized (Weinhold, 1968; Akino *et al.*, 1971; Epstein and Farrell, 1975).

The presumed presence of the transmethylation pathway in mammalian lungs rested upon the identification of phosphatidyldimethylethanolamine (PDME), one of the postulated intermediates between PE and PC. The identification of PDME and, hence, the importance of the transmethylation sequence, was based primarily upon the mobility of this compound on TLC plates to a position intermediate between PC and PE (Morgan *et al.*, 1965; Morgan, 1971; Gluck *et al.*, 1972). Pfleger *et al.* (1972) pointed out that the material thought to be PDME was in fact phosphatidylglycerol (PG), which they found accounted for 10% of surfactant phospholipid phosphorus isolated from beagle dog lungs. No appreciable quantities of PDME were found. The same conclusions were subsequently reached by Rooney *et al.* (1974) from studies of rat, rabbit, and human lungs and by Godinez *et al.* (1975) with isolated perfused rat lung preparations. Hallman and Gluck (1974) have recently indicated that transmethylation may not be a significant factor in pulmonary PC biosynthesis and have, instead, supported the identification of PG.

As a result of these findings it appears that the transmethylation pathway, if it exists at all in the lung, is of relatively little importance. However, some further considerations should be kept in mind before completely ruling out the possible relevance of the transmethylation sequence. Tombropoulos (1973), studying palmitate and *sn*-glycerol-3-P incorporation into hamster lung subcellular fractions, found that the highest *de novo* incorporation of the former occurred with PE. Earlier, Scarpelli (1967) presented evidence, based on incorporation of palmitate into lipids of fetal lung and fetal pulmonary fluid, which indicated that PE is rapidly labeled and may serve as an important precursor of PC. If PE palmitate residues are incorporated into PC, transmethylation could be an obvious means. The accumulation of PDME or other intermediates is not a necessary prerequisite for identification of the transmethylation pathway, since these intermediates might be short-lived and transient. Alternatively, transfer of palmitate from PE to PC might also take place through deacylation of PE and reacylation of lyso-PC with the palmitate originally released from PE. In any event, the work of Gluck and his associates (1967a,b, 1970, 1972), even though it may have been predicated upon misidentification of PDME, is still of interest because it points to a possible significant role for PG in pulmonary lipid metabolism.

C. *Metabolism of PG*

Henderson and Pfleger (1972) showed that PG, isolated from dog lung surfactant, was a potent surface tension lowering agent and suggested that it could contribute to the characteristic surface behavior of alveolar lining layer constituents. When equimolar quantities of PG and PC are cosonicated, inverted vesicles are formed whose outer sufaces contain twice as many PG as PC molecules (Michaelson *et al.*, 1973). Henderson *et al.* (1974), on the basis of nuclear magnetic resonance studies, found that PG possesses a greater hydrogen bonding capacity than PC. The pertinent observations concerning PG have been summarized by Godinez *et al.* (1975), who suggested, in light of their own and previous findings, that PG may play a role in stabilizing the surfactant lipoprotein complex after its secretion or in its storage form in the lamellar bodies of the type II cell.

As discussed in section III.B, it now seems reasonable to conclude that the compound once identified as PDME is in fact PG. PG may be considered distinctive in lung both because of its relatively high concentration in that organ and because it incorporates phospholipid precursors at an initial rate greater than any other pulmonary phospholipid (Godinez *et al.*, 1975). Previously, Weinhold and Villee (1965) found that a relatively non-polar phospholipid, which in retrospect probably was PG, incorporated ^{32}P at a faster rate than PC in rat lungs. Isolated rat lungs were perfused with [2-^{14}C]lactate and it was found that PG accounted for 15% of the lactate incorporated into the total phospholipids and the relative specific activity of PG was five times greater than PC (Godinez *et al.*, 1975). When the lungs were perfused with [1-^{14}C]acetate and [U-^{14}C]glucose, PG also had a specific activity higher than PC.

Sanders and Longmore (1975) found that the IB surfactant fraction, as defined by Frosolono *et al.* (1970), from isolated perfused rat lung tissue accounted for approximately 35% of the total PG. Sanders and Longmore (1975) also studied the incorporation of radioactive glucose, palmitate, and acetate into the IB and residual fractions of their isolated rat lung preparation. Perfusion with [U-^{14}C]glucose, [9,10-^{3}H]palmitate, and [1-^{14}C]acetate yielded, in each case, a higher specific activity for residual PC and PG than for these components in the surfactant fraction. With glucose and acetate, the specific activity of the PG constituent in both residual and surfactant fractions was markedly higher than the PC components except for the surfactant preparation with acetate. In the latter case, PG and PC specific activities were essentially equal. On the other hand, perfusion with palmitate gave higher specific activities for residual and surfactant PC than for PG. It was noted that those precursors which yield palmitic acid by *de novo* synthesis give rise to PG with a higher specific activity than that found for PC with incorporation of exogenous palmitate. Another intriguing aspect of these studies was that palmitate esterification to the 1- and 2-positions of the glycerol portion of PG and PC occurs at different rates and is dependent upon the precursor source of palmitate.

Although the data of Sanders and Longmore (1975) are certainly not conclusive, especially since only a single time period, 2 hr, was investigated, it is tempting to speculate that PG may serve as a precursor, at least in part, for PC in the surfactant fraction. This may be intimated from the observation that perfusion with radioactive glucose and acetate did give surfactant and residual PG with a higher specific activity than PC. The exact nature of the precursor–product relationship, if indeed it exists, is unclear because of the previously referred to results with palmitate perfusion, i.e., PC had a higher specific activity than PG in both fractions. A further complication is that comparison of relative specific activities of palmitate at the *sn*-1 and *sn*-2 positions of PG and PC in the surfactant and residual fractions, after perfusion with each precursor, does not yield data that are readily interpretable in terms of a PG–PC precursor–product relationship. This may be a reflection of the way the data are presented. DiAugustine (1971) earlier suggested that PG should be considered as an important surfactant constituent in its own light rather than as a precursor to PC. In any event, these experiments provide a basis and frame of reference for investigation of the metabolic relationship which may exist between PG and PC.

Hallman and Gluck (1974) have studied the subcellular localization of PG biosynthesis in rat lung by incubating various organelle fractions with *sn*-[U-^{14}C]glycerol-3-P and CDP-diglyceride. Enzymatic specific activity, i.e., cpm incorporated from *sn*-[U-^{14}C]glycerol-3-P per mg of organelle protein, indicated significant synthesis in both mitochondria and microsomes. The latter had the highest specific activity. Very little PG synthesis was detected in lamellar bodies and it was speculated that this may have been due to contamination by microsomes. Interestingly, Hallman and Gluck (1974), utilizing a "pulse-chase" technique in their *in vitro* system found that previously synthesized microsomal PG could be transferred either to lamellar bodies or to alveolar wash fractions. Mitochondria were not able to effect this transfer of PG to surfactant type material, thereby leading to the speculation that PG synthesis in these organelles is primarily directed toward cardiolipid production. They postulated that lamellar bodies contain a protein component with high affinity binding to PG. This would, then, explain the difference between the site of PG synthesis (microsomes) and location (lamellar bodies) of the majority of the compound.

D. *Degradative Processes*

To date, the major emphasis of most studies in pulmonary lipid metabolism has been directed toward synthetic rather than degradative aspects. Even though elucidation of pathways responsible for biosynthesis of dipalmitoyl-PC is a major focus of pulmonary biochemistry, considerably less is known about turnover and degradation of this important compound. The same is generally true for other lung lipids.

The presence of phospholipase A in rat lung has previously been reported (Robertson and Lands, 1962; Gallai-Hatchard and Thompson, 1965), but these studies were not detailed. The data for lung were presented as part of surveys for phospholipase A in various organ systems. The findings of Kyei-Aboagye *et al.* (1973) that rabbit lung homogenates catalyze a Ca^{2+}-dependent release of [^{3}H]oleate from [^{3}H]oleate containing PC can be interpreted to mean that a phospholipase A was present. Ohta and Hasegawa (1972) demonstrated a Ca^{2+}-independent phospholipase A activity in rat lung and found that one-half to three-fourths of the total activity was in the $105{,}000 \times g$ supernatant fraction. Their data suggested that phospholipase A_2 activity was higher than phospholipase A_1. Under their experimental conditions, no significant difference was found in the extent of hydrolysis between PC and PE; that is, both phospholipids appeared to be equally good substrates. A more detailed study of the supernatant phospholipase A_2 showed that the activity was optimal at pH 6.5 and was strongly inhibited by sodium dexoycholate, other detergents, bovine serum albumin, and F- (Ohta *et al.*, 1972). In their work demonstrating that the same enzyme was responsible for lysophospholipase and transacylase activities in rat lung supernatant, Abe *et al.* (1974) found that phospholipase A_2 was distinct and completely separable from these activities.

Recently, Garcia *et al.* (1975) presented the first direct evidence for a highly active, Ca^{2+}-dependent phospholipase A_2 in rat lung microsomes. Mitochondria also exhibited phospholipase A_2 with a specific activity significantly larger than the microsomal. Marker enzyme studies, however, indicated that the microsomal activity was not a result of contamination with mitochondria. Incubation of microsomes and mitochondria with 1-[^{3}H]palmitoyl-2-[^{14}C]linoleoyl-PC gave rise to stoichiometric amounts of 1-[^{3}H]palmitoyl-2-lyso-PE and [^{14}C]linoleate. Significant amounts of [^{3}H]palmitate or 1-lyso-2-[^{14}C]linoleoyl-PE did not accumulate. These results appear conclusively to demonstrate a phospholipase A_2 in the two organelle fractions. The microsomal phospholipase A_2 was absolutely dependent upon Ca^{2+}. The pH optimum was 9.0. Some microsomal lysophospholipase activity was found but no lipase, using tri-[1-^{14}C]oleoylglycerol as substrate, was observed.

The report by Garcia *et al.* (1975) is significant since it provides direct evidence that a phospholipase A_2, postulated to be an integral component of a deacylation–reacylation cycle involved in biosynthesis of pulmonary PC and dipalmitoyl-PC (Frosolono *et al.*, 1971), is present in lung microsomes. Microsomes are generally considered to be responsible for the majority of the *de novo* biosynthesis of complex lipids (introductory chapter), and acyltransferases, both those involved in *de novo* and reacylation pathways, are present in these organelles (Hendry and Possmeyer, 1974; Frosolono *et al.*, 1971). Thus, the microsomal phospholipase A_2 is present in the anatomical location where a reacylation process would be presumed to function in concert with the hydrolytic activity.

In this context, it is probably more proper to consider that the lung microsomal phospholipase A_2 may be primarily involved in the synthesis of dipalmitoyl-PC, as well as other phospholipids, rather than degradative in purpose. This would be analogous to PAP'ase, which, although a hydrolytic enzyme, provides a product that is a key substrate in the biosynthesis of lipids. It is interesting to speculate that pulmonary microsomal phospholipase A_2 may exhibit marked differences in activity according to the fatty acid composition of PC presented as a substrate. The enzyme might be more active with PC species containing an unsaturated acyl moiety at the *sn*-2 position than with disaturated PCs. The former, as discussed earlier, quite likely arise from the *de novo* PC synthetic pathway. If such substrate preference, as determined by comparative kinetic studies with various PC species (Frosolono *et al.*, 1973), is subsequently found to be an important feature of the microsomal phospholipase A_2 system, this, in conjunction with the specificities previously observed with the microsomal lyso-PC acyltransferases (Frosolono *et al.*, 1971; Vereyken *et al.*, 1972; Snyder and Malone, 1975), would provide a very potent mechanism for biosynthesis of dipalmitoyl-PC.

It should be pointed out that, while the work of Garcia *et al.* (1975) is important, one aspect remains to be clarified. Their work was carried out utilizing PE as the substrate for phospholipase A_2. While it is tempting and, perhaps, reasonable to assume that this enzyme, like that reported by Ohta and Hasegawa (1972) in rat lung supernatant, will react in a similar fashion with PC as with PE, this has not yet been proven and must remain an assumption.

Alveolar macrophages, known to be very active in phagocytosis (Myrvik *et al.*, 1961a,b), contain vacuoles that appear to be lipid-rich and bear a strong resemblance to myelin figures. There has been at least an assumption that these cells, because of their anatomical location in the alveolus, are active in digestion of PC, dipalmitoyl-PC, and other constituents of the alveolar lining layer (Scarpelli, 1968d; Weibel, 1967). Alveolar macrophages do contain several phospholipase A activities which can be differentiated on the basis of pH optima, Ca^{2+} dependence or inhibition, and intracellular distribution (Franson *et al.*, 1973). A phospholipase A_2 with an acid pH optimum is associated with the lysosomal fraction. Some phospholipase A_1 activity is also found in these organelles. Although there was considerable loss of activity as a consequence of the fractionation procedure, phospholipase A with an alkaline pH optimum followed the same distribution pattern seen with the microsomal marker, NADPH:cytochrome *c* reductase. It was speculated that the phospholipase As with an alkaline pH optimum, not primarily associated with lysosomes, may play a role in the turnover of endogenous phospholipid, especially in relation to metabolism of macrophage membranes. These same studies revealed that a very potent lysophospholipase activity is present in alveolar macrophage homogenates.

Injection of rabbits with heat-killed bacille Calmette Guérin (BCG) results in an accumulation of activated macrophages in the alveolar spaces (Myrvik *et al.*, 1962) and increases in their hydrolytic enzyme activities such as lysozyme (Leake and Myrivik, 1968), and acid phosphatase and β-glucuronidase (Franson and Waite, 1973). However, the specific activity of phospholipase A was decreased in BCG-stimulated alveolar macrophages (Franson and Waite, 1973). The distribution of phospholipase A and lysosomal marker enzymes resulting from sucrose density gradient centrifugation indicated that the phospholipase A active at pH 4.0 was of lysosomal origin. This was in agreement with this group's earlier findings (Franson *et al.*, 1973). Both phospholipase A_1 and A_2 had optimal activity in the presence of EDTA at pH 7.4 when 1-acyl-2-[^{14}C]linoleoyl-PE was the substrate. The amount of [^{14}C]linoleic acid released was two to three times greater than the amount of 1-lyso-[^{14}C]linoleoyl-PE released. The first would be indicative of phospholipase A_2 activity, the latter, phospholipase A_1. Franson and Waite (1973) suggest these findings might be due to a phospholipase A_2, which is more active than a phospholipase A_1 or to the concerted action of a phospholipase A_1 and lysophospholipase. Further studies with 1-[^{3}H]palmitoyl-2-[^{14}C]linoleoyl-PE confirmed the presence of both phospholipase A_1 and A_2 since 1-lyso-2-[^{14}C]linoleoyl-PE and 1-[^{3}H]palmitoyl-2-lyso-PE were produced; however, because 70–90% of both lyso-PEs were further hydrolyzed, lysophospholipase activity was also confirmed.

An interesting facet of alveolar macrophage lipid metabolism concerns the relationship of the phospholipase activities, which one would perhaps expect to be a reflection of phagocytosis, and the ability of these cells to convert lyso-PC to PC (Elsbach, 1968). Indeed, alveolar macrophages, as shown through incorporation of radioactive precursors, can synthesize dipalmitoyl-PC (Mason *et al.*, 1972a). Unpublished experiments with alveolar macrophage homogenates (Frosolono and Tansey) indicate that acyltransferase activity is present in these cells. The phospholipase A activities in alveolar macrophages, especially those associated with microsomes (Franson *et al.*, 1973), may more properly be considered part of a synthetic pathway leading to dipalmitoyl-PC by a reacylation mechanism. Since membrane proliferation and conservation are integral features of phagocytosis (Bennet, 1956; Zucker-Franklin and Hirsch, 1964), synthetic capacity of alveolar macrophages for PC and dipalmitoyl-PC may be related to the biosynthesis of intracellular and plasma membranes resulting from phagocytic activity. In this respect, lysosomal phospholipase A activities would then be associated with degradation of internalized lipid material as a consequence of phagocytosis.

Scarpelli *et al.* (1971) showed that fetal pulmonary fluid contains surface active phospholipids that are products of alveolar cell metabolism and secretion. Constituents of fetal pulmonary fluid presumably become part of the acellular alveolar lining when air breathing begins. Thus, fetal pulmonary fluid can be considered the analog of the acellular alveolar lining in the air-breathing animal. The biological half-life of pulmonary PC has been reported to be 12–14 hr and 17.5 hr for PC in the acellular alveolar lining (Thomas and Rhoades, 1970; and Tierney *et al.*, 1967). However, Scarpelli *et al.* (1975) found a much shorter half-life, 15–57 min, for sheep PC in fetal pulmonary fluid when protein complexed [U-^{14}C]PC was injected into the fetal pulmonary fluid. The results indicated that the PC was metabolized to lyso-PC and FA within the fluid itself. Interestingly, FA derived from PC in the fetal pulmonary fluid entered the pulmonary circulation and thereby established a pulmonary arteriovenous FA gradient. These results suggested to Scarpelli *et al.* (1975) that phospholipase A activity may be present in the fetal pulmonary fluid or, perhaps more likely, at the surface of alveolar epithelial cells. It was considered unlikely that, with this fetal lamb system, alveolar macrophages played a significant role in degradation of PC in the fetal pulmonary fluid, since these cells are very scarce, if present at all, during gestation (Kikkawa, as reported in Scarpelli *et al.*, 1975). As suggested by Scarpelli, if these studies with the fetal pulmonary fluid can be directly extrapolated to the adult lung, phospholipase A systems may be present either in the alveolar lining layer itself, exclusive of alveolar macrophages, or as constituents of epithelial cell plasma membranes in a situation analogous to those reported in the plasmalemma of liver cells (Newkirk and Waite, 1971).

IV. Cellular Sites of Lipid Metabolism

The great cellular heterogeneity and complexity of the lung has, in many respects, made it very difficult to determine which biochemical events may be a principal feature of a given pneumocyte species. Not only does the lung contain a large number of cell types, some 40 different species having been identified (Bertalanffy, 1964; Sorokin, 1970), but the lung consists of several distinct anatomical regions, i.e., trachea, bronchioles, alveoli, interstium, large blood vessels, capillaries, etc. It is not unreasonable to assume that different cell types in each of the regions may participate in, or emphasize, different aspects of lipid metabolism.

The cellular diversity of the lung has afforded an excellent opportunity for interaction between several scientific disciplines in exploring features of pulmonary metabolism, and this approach has been a feature of studies in lung lipid metabolism. The principal example has been concerned with the elucidation of the biochemistry, especially biosynthesis, elaboration, and secretion, of the pulmonary surfactant system and the role played therein by type II alveolar epithelial cells. An important correlation was made relatively early in this field when it was found that the morphological differentiation of type II cells, especially the appearance of the lamellar body system in the last phases of gestation, was paralleled by increased phospholipid biosynthesis and the demonstrable appearance of surface active forces in lung minces (Avery and Mead, 1959; Kikkawa *et al.*, 1965, 1968; Scarpelli, 1968c,d,e). The work of Goldenberg *et al.* (1969) and Chevalier and Collet (1972) has been important in describing the intracellular morphological events associated with synthesis, "packaging," and secretion of the phospholipid-rich surfactant system. Thus, electron microscopy, biochemistry, physiology, and clinical experience have all contributed to the hypothesis that the type II cell is a primary focus of phospholipid metabolism in the lung.

Since the type II cell lamellar body system is directly involved in elaboration and, perhaps, synthesis, of phospholipids, these organelles have been a center of interest in lung lipid biochemistry. A number of procedures have been developed for the isolation of lamellar bodies (Hoffman, 1972; Page-Roberts, 1972; Williams *et al.*, 1971; Valdivia, 1973; Gil and Reiss, 1973; DiAugustine, 1974; Littman *et al.*, 1974; Spitzer *et al.*, 1975). Although some investigators have emphasized the possibility that these organelles, because of their content of acid phosphatase and other hydrolytic enzymes (DiAugustine, 1974), may be degenerative lysosomes, others have found evidence that they may be capable of phospholipid biosynthesis. Thus, Spitzer *et al.* (1975), who estimated that as much as 40% of the PAP'ase in type II cells was associated with the lamellar bodies, suggested that PC stored within these organelles may be synthesized at their perilamellar surface. Adamson and Bowden (1973) had previously concluded, on the basis of electron microscopic autoradiographic experiments that the lamellar body perilamellar membrane was an active site of surfactant lipid biosynthesis. Lyso-PC acyltransferase activity has also been found in isolated type II cell lamellar bodies (Engle and Longmore, 1975).

Although the lamellar body system may be involved to a presently undetermined extent in lipid biosynthesis and/or, perhaps, PC acyl group rearrangement, it is most likely that the endoplasmic reticulum is the major site of complex lipid biosynthesis. Pertinent references in support of this concept have been given in previous sections dealing with the biosynthesis of PC, dipalmitoyl-PC, and PG. Also, as discussed earlier, the bulk of *de novo* palmitate synthesis is associated with the cytoplasmic fraction.

Kikkawa and Yoneda (1974) have developed a procedure for the isolation of homogenous populations of well-preserved type II cells from rats. This procedure was subsequently modified and extended to the isolation of rabbit type II cells (Kikkawa *et al.,* 1975). It is difficult to overestimate the importance of this work because, for the first time, pulmonary biochemists have access to sufficient numbers of these cells for detailed studies of their lipid metabolism and other parameters.

Studies by Kikkawa's group (Kikkawa and Yoneda, 1974; Kikkawa *et al.,* 1975) have been instrumental in showing that the *sn*-glycerol-3-phosphocholine backbone of PC in type II cells is synthesized by the *de novo* sequence and not by transmethylation. Additionally, their results can be interpreted as confirmation of the concept that dipalmitoyl-PC may be synthesized, at least in part, by a reacylation mechanism. These findings are supported by the recent work of Spitzer *et al.* (1975), who apparently used a different procedure for isolation of type II cells. Obviously, these studies with isolated type II cells represent an initial effort in what promises to be a most productive area of pulmonary lipid metabolism.

Niden (1967), utilizing electron microscopic autoradiograms, suggested that the nonciliated bronchiolar Clara cell was a site of surfactant synthesis. His premise was based in part on the fact that [^{3}H]palmitate and [^{3}H]acetate accumulated in these cells. Etherton and Conning (1971) found a very early incorporation of labeled palmitate in Clara cells, again by electron microscopic autoradiography. These latter findings were extended (Etherton *et al.,* 1973) with the observation that labeling of Clara cells with palmitate exceeded that of all other cell species within 3 min of administration of the isotope. Only 1 hr later did type II cell lamellar bodies become heavily labeled. These investigators suggested that palmitic acid is synthesized in specialized mitochondria of the nonsecretory region of the Clara cell and is then incorporated into dipalmitoyl-PC in the membranous apex of the cell. Secretion into the bronchiolar lining layer was thought to take place possibly by an apocrine mechanism. Although not addressed specifically to the question of the cellular origin of lipid consituents of the surfactant system, the findings of Kuhn *et al.* (1974) supported exocytosis as a more probable mechanism of Clara cell secretion.

Petrik and Collet (1974) studied the participation of Clara cells in the biosynthesis of pulmonary surfactant, also from the perspective of electron microscopic autoradiographs. They found that [^{3}H]choline, regarded as a specific precursor of dipalmitoyl-PC, was not incorporated into these cells. [^{3}H]Acetate was, however, actively incorporated and arguments were marshaled to suggest that, while these cells possibly do not synthesize dipalmitoyl-PC, they may participate in the synthesis of cholesterol destined for incorporation into the hypophase of the acellular alveolar lining layer.

The lack of incorporation of tritiated choline into Clara cells (Petrik and Collet, 1974), in contrast to its incorporation into type II cells (Chevalier and Collet, 1972), is very strong evidence that the former cell type is not involved as a major site of dipalmitoyl-PC synthesis. In this context, results of labeling studies with radioactive palmitate (Etherton *et al.*, 1973) can be considered less definitive since this compound might reasonably be expected to be a less specific marker for dipalmitoyl-PC biosynthesis. That is, palmitate could be utilized in the synthesis of a number of lipid classes, whereas choline would be directed primarily to PC and, perhaps, sphingomyelin synthesis. The electron microscopic autoradiogram results with Clara cells, however, do provide information that these cells may indeed have a very active lipid metabolism. Obviously, a procedure for isolation of homogeneous Clara cell populations would offer an important and, perhaps, definitive answer to the role of these cells in pulmonary lipid metabolism.

V. Developmental and Control Aspects

A general observation seen with a number of mammalian species is that pulmonary surfactant production, which can be correlated with certain aspects of cellular differentiation, increased phospholipid biosynthesis, and other parameters, is turned on during the last phases of gestation (Scarpelli, 1968c,e). The previously mentioned involvement of the type II epithelial cell in surfactant synthesis and phospholipid metabolism offers an excellent example of the integration of morphological, biochemical, and physiological events that coincide to prepare the fetus for air breathing and extrauterine survival. An excellent summary of the evidence linking the lamellar body system of type II cells to surfactant production has been collected by Blackburn *et al.* (1972). An important clinical observation is that premature fetal delivery is associated with RDS if birth takes place prior to development of biochemical systems giving rise to production of surface active material, notably dipalmitoyl-PC, in the lung. It is a well-established tenet of neonatology that RDS is associated with deficient phospholipid production as a consequence of prematurity (Avery and Mead, 1959; Chu *et al.*, 1967; Adams and Enhorning, 1966; Baum *et al.*, 1971). Thus, the fetal lung appears to be on a rather strict developmental timetable. Since phospholipids, especially dipalmitoyl-PC, and other lipids are integral functional and structural constituents of the pulmonary surfactant system (Frosolono *et al.*, 1970; King and Clements, 1972; King, 1974), biochemical development, as exemplified by changes in lipid metabolism, is an important feature of pulmonary maturation during the neonatal period. Interestingly, lung development apparently continues for a considerable period after birth (Reid, 1967; Bartlett, 1970).

Recently, elucidation of factors which initiate or influence synthesis of phospholipids destined for incorporation into the surfactant system and release of the latter into the alveolar spaces has been one of the most exciting areas of pulmonary research. The clinical implications and potential benefits are obvious and profound. Even though this research is in its earliest stages, it is generally accepted that neurohumoral agents and hormones are important in the processes (Blackburn *et al.*, 1972, 1973).

Acetyl choline, pilocarpine, and adrenaline stimulate the release of surfactant (Goldenberg *et al.*, 1967, 1969; Wang *et al.*, 1971). Vagotomy gives rise to a respiratory syndrome characterized by atelectasis, apparently in response to a deficiency in surface active phospholipids (Klaus *et al.*, 1962; Goldenberg *et al.*, 1967). It has been suggested, therefore, that surfactant synthesis and release may be regulated by a vagal system (Bensch *et al.*, 1964; Goldenberg *et al.*, 1967, 1969). An increasing accumulation of evidence suggests that ACTH, cortisone, hydrocortisone, and thyroxine promote functional maturity of the lungs by stimulating surfactant production (DeLemos *et al.*, 1970; Kotas and Avery, 1971; Kikkawa *et al.*, 1971; Redding *et al.*, 1972; Wu *et al.*, 1973).

As summarized by Blackburn *et al.* (1973) several lines of evidence have shown that the fetal pituitary begins to function well in advance of birth with the result that the trophic hormones of this gland bring the adrenal, thyroid, and testis in the fetus under its control. Blackburn *et al.* (1972) carried out a series of experiments in which the pituitary–adrenal–thyroid axes were interrupted by decapitation of fetal rats in utero on the 16th day of gestation. Cell proliferation was not thought to be affected, but pulmonary cytodifferentiation was retarded. The developing pneumocytes retained large cytoplasmic accummulations of glycogen with relatively few organelles. There was a decrease in type II cell lamellar bodies, and minces of these lungs showed abnormally high surface tension values when tested by standard methods. These authors interpreted their findings to indicate that the interruption of the pituitary–adrenal axis results in impaired glycogen utilization and phospholipid biosynthesis. Since decapitation of fetal rats did not prevent the formation of normal-appearing lamellar bodies, but only produced a decrease in their numbers, it was suggested that an intact pituitary–adrenal axis is not necessary for induction of enzymes involved in phospholipid synthesis. Rather, phospholipid production probably did not occur at a normal rate or was qualitatively abnormal. The failure of glycogen to disappear from type II cells was taken as an indication that enzymes necessary for its utilization were absent or deficient. In a subsequent report, Blackburn *et al.* (1973) found that quantities of lipid, phospholipid, and PC in lung were decreased in the decapitated fetuses, but that pulmonary hypercellularity was also an important feature. Phospholipid fatty acids were less saturated than those of the normal fetal lung. This would explain, it was felt, the abnormal surface activity seen in lung minces from decapitated animals.

Present evidence indicates that the *de novo* or CDP-choline pathway for PC biosynthesis reaches full activity at birth as reflected by increases in levels of choline kinase, choline phosphate cytidyl transferase, and choline phosphotransferase (Artom, 1968; Farrell *et al.*, 1974). Schultz *et al.* (1974) found that the activity of PAP'ase in fetal rabbit lung increases fourfold in the last seven days of gestation and preceded the increase in PC concentration by 24 hr. In the same study, no increase was seen with choline phosphotransferase, and these authors therefore postulated that PAP'ase activity may function in the regulation of PC biosynthesis in the developing lung by increasing the availability of substrates for PC production. Since the *de novo* pathway is thought to be primarily responsible for biosynthesis of PC species containing unsaturated acyl groups (see previous sections), the close association between increased surface activity and dipalmitoyl-PC biosynthesis in the fetal lung near birth suggests that the enzymes involved in the reacylation mechanism should also increase near term. Presently, very little information is available on this point except for a report by Rooney *et al.* (1975) that cortisol administration to rabbit fetuses at 24 days gestation gave no demonstrable increases in lyso-PC acyltransferase at 27 days. However, a marked increase in glycerolphosphate phosphatidyltransferase activity, leading to PG synthesis, was seen. Full term in the rabbit is 30 days.

Fetal steroids promote type II cell ultrastructural and, presumably, biochemical maturation (Kotas and Avery, 1971; Kikkawa *et al.*, 1971; Wang *et al.*, 1971; Smith *et al.*, 1974) and 9-flurorednisolone, a potent glucocorticoid, has been shown to induce choline phosphotransferase in the fetal rabbit lung (Farrell and Zachman, 1973). As discussed earlier, this enzyme can be considered to operate at a key step in the biosynthetic pathway leading to PC and, ultimately, to dipalmitoyl-PC.

The fetal lung contains specific cytoplasmic receptors which bind glucocorticosteroids and then migrate to the nucleus (Ballard and Ballard, 1972; Giannopoulos *et al.*, 1972; Toft and Chytil, 1973). This supports the suggestion (Blackburn *et al.*, 1973) that cortisol deficiency in the fetal lung is associated with pneumocyte proliferation and deficient choline phosphotransferase mRNA transcription.

Although not without some possible complications (Sutnick *et al.*,

1970 and Kotas *et al.*, 1974), antepartum glucocorticoid treatment may be effective as a prophylactic treatment of RDS in premature human infants (Liggins and Howie, 1972). A complicating factor in our understanding of the role glucocorticosteroids may play in fetal lung maturation, especially regarding induction of specific enzyme systems, is that Rooney *et al.* (1975), in contrast to the results with 9-fluoroprednisolone (Farrell and Zachman, 1973), found no increase in choline phosphotransferase activity in fetal rabbit lung after treatment with cortisol.

Thyroxine has also been reported to accelerate maturation of the fetal lung as evidenced by an increase in the number of lamellar bodies and early disappearance of glycogen in type II cells (Wu *et al.*, 1973). In adult animals, thyroxine administration leads to an enlargement of type II cells and increases both in lamellar bodies and yield of a specific surface active fraction obtained by lung lavage (Redding *et al.*, 1972). These findings also support the suggestion that thyroxine may be an important regulator of dipalmitoyl-PC metabolism, although Mason *et al.* (1972b) reported no increase in its concentration in lungs from rats treated with the hormone.

Barrett *et al.* (1974) found that adenylate cyclase of lung homogenates from fetal and neonatal rabbits is markedly higher than in adults. Epinephrine stimulated the activity at all days studied (21, 24, 27, 30, and 31 days postconception). NaF depressed adenylate cyclase activity at 21 and 24 days, but stimulated it afterwards. Glucagon stimulated the activity only after birth. These results were interpreted to indicate that epinephrine stimulation of adenylate cyclase activity occurs earlier in development and to a greater degree through a separate receptor system than does glucagon stimulation. This led to the speculation that adenyl cyclase may be important in regulation of pulmonary phospholipid biosynthesis in fetal animals through the utilization of glycogen.

ACKNOWLEDGMENTS

Dr. Frosolono is the recipient of a Research Career Development Award, National Institutes of Health No. 1-K04-HL-00105; and his research is supported in part by the Alveolar Biology Program Project grants No. HL-16137 and HL-19010, National Heart and Lung Institute, National Institutes of Health.

References

Abe, M., Akino, T., and Ohno, K. 1972. The formation of lecithin from lysolecithin in rat lung supernatant. *Biochim. Biophys. Acta* **280**:275–280.

Abe, M., Ohno, K., and Sato, R. 1974. Possible identity of lysolecithin acyl-hydrolase with lysolecithin-lysolecithin acyl-transferase in rat-lung soluble fraction. *Biochim. Biophys. Acta* **369**:361–370.

Adams, F. H., and Enhörning, G. 1966. Surface properties of lung extracts. I.A dynamic alveolar model. *Acta Physio. Scand.* **68**:23–27.

Adamson, I. Y. R., and Bowden, D. H. 1973. The intracellular site of surfactant synthesis. Autoradiographic studies on murine and avian lung explants. *Exp. Mol. Pathol.* **18**:112–124.

Akesson, B., Elovson, J., and Arvidson, G. 1970. Initial incorporation into rat liver glycerolipids of intraportally injected [^{3}H] glycerol. *Biochim. Biophys. Acta* **210**:15–27.

Akino, T., Abe, M., and Arai, T. 1971. Studies on the biosynthetic pathways of molecular species of lecithin by rat lung slices. *Biochim. Biophys. Acta* **248**:274–281.

Akino, T., Yamazaki, I., and Abe, M. 1972. Metabolic fate of lysolecithin injected into rats. *Tohoku J. Exp. Med.* **108**:133–139.

Artom, C. 1968. Enzymes for the synthesis of lecithins from choline in tissues of developing rats. *Fed. Proc.* **27**:457.

Avery, M. E., and Mead, J. 1959. Surface properties in relation to atelectasis and hyaline membrane disease. *Am. J. Dis. Child.* **97**:517–523.

Ballard, P. L., and Ballard, R. A. 1972. Glucorcorticoid receptors and the role of glucocorticoids in fetal lung development. *Proc. Natl. Acad. Sci. USA* **69**:2668–2672.

Barrett, C., Sevanian, A., and Kaplan, S. A. 1974. Adenylate cyclase activity in immature rabbit lung. *Pediat. Res.* **8**:244–247.

Bartlett, D., Jr. 1970. Postnatal growth of the mammalian lung: Influence of low and high oxygen tensions. *Respir. Physiol.* **9**:58–64.

Baum, M., Benzer, H., Lempert, J., Regele, H., Stuhlinger, W., and Tolle, W. 1971. The surface tension properties of the lungs of newborn babies. Investigations from autopsies on stillborn and premature infants and babies dying from hyaline membrane disease 'respiration distress' syndrome. *Respiration* **28**:409–429.

Bennet, H. S. 1956. The concepts of membrane flow and membrane vesiculation as mechanisms for active transport and ion pumping. *J. Biophys. Biochem. Cytol.* **2**:99–103.

Bensch, K., Schaefer, K., and Avery, M. E. 1964. Granular pneumocytes: Electron microscopic evidence of their exocrine function. *Science* **145**:1318–1319.

Bertalanffy, F. D. 1964. Respiratory tissue: structure, histopathology, cytodynamics. Part I. Review and basic cytomorphology. *Int. Rev. Cytol.* **16**:233–328.

Blackburn, W. R., Travers, H., and Potter, D. M. 1972. The role of the pituitary–adrenal–thyroid axes in lung differentiation. I. Studies of the cytology and physical properties of anencephalic fetal rat lung. *Lab. Invest.* **26**:306–318.

Blackburn, W. R., Kelly, J. S., Dickman, P. S., Travers, T., Lopata, M. A., and Rhoades, R. A. 1973. The role of the pituitary–adrenal–thyroid axes in lung differentiation. II. Biochemical studies of developing lung in anencephalic fetal rats. *Lab. Invest.* **28**:352–360.

Brindley, D. N. 1973. The relationship between palmitoyl-coenzyme A synthetase activity and esterification of *sn*-glycerol 3-phosphate by the microsomal fraction of guinea-pig intestinal mucosa. *Biochem. J.* **132**:707–715.

Brooks, R. E. 1968. Pulmonary adenoma of strain A mice: An electron microscopic study. *J. Natl. Canc. Inst.* **41**:719–742.

Chevalier, G., and Collet, A. J. 1972. In vivo incorporation of choline-^{3}H, leucine-^{3}H, and galactose-^{3}H in alveolar type II pneumocytes in relation to surfactant synthesis. A quantitative radioautographic study in mouse by electron microscopy. *Anat. Rec.* **174**:289–310.

Chida, N., and Adams, F. H. 1967. Incorporation of acetate into fatty acids and lecithin by lung slices from fetal and newborn lambs. *J. Lipid Res.* **8**:335–341.

Chu, J., Clements, J. A., Cotton, E. K., Klaus, M. H., Sweet, A. Y., and Tooley. 1967. Neonatal pulmonary ischemia. Part I: Clinical and physiological studies. *Pediatrics* **40**:709–782.

Condorelli, S., Lombardi, D., Gusman, R., and Pisano, L. 1972. Pulmonary A–V differences of lipids and free fatty acids in relation to fasting and to high fat meals. *Clin. Chim. Acta* **38**:141–146.

DeLemos, R. A., Shermeta, D. W., Knelson, J. H., Kotas, R., and Avery, M. E. 1970. Acceleration of appearance of pulmonary surfactant in the fetal lamb by administration of corticosteroids. *Am. Rev. Respir. Dis.* **102**:459–461.

DiAugustine, R. P. 1971. Lung phospholipids. I. In vivo studies of the incorporation of ^{32}P, (methyl-^{14}C) choline, 1-^{14}C-palmitic acid and 1-^{14}C-oleic acid into phosphatidylethanolamine, phosphatidyl-*n,n*-dimethylethanolamine and phosphatidylcholine. *Biochem. Biophys. Res. Commun.* **43**:311–317.

DiAugustine, R. P. 1974. Lung concentric laminar organelle. Hydrolase activity and compositional analysis. *J. Biol. Chem.* **249**:584–593.

Elsbach, P. 1968. Increased synthesis of phospholipid during phagocytosis. *J. Clin. Invest.* **47**:2217–2229.

Engle, M. J., and Longmore, W. J. 1975. Phospholipid composition and acyl transferase activity of lamellar bodies isolated from rat lung. *Fed. Proc.* **34**:633.

Epstein, M. F., and Farrell, P. M. 1975. The choline incorporation pathway: primary mechanism for *de novo* lecithin synthesis in fetal primate lung. *Pediat. Res.* **9**:658–665.

Etherton, J. E., and Conning, D. M. 1971. Early incorporation of labeled palmitate into mouse lung. *Experientia* **27**:554–555.

Etherton, J. E., Conning, D. M., and Corrin, B. 1973. Autoradiographical and morphological evidence for apocrine secretion of dipalmitoyl lecithin in the terminal bronchiole of mouse lung. *Am. J. Anat.* **138**:11–36.

Farrell, P. M., and Zachman, R. D. 1973. Induction of choline phosphotransferase and lecithin synthesis in the fetal lung by corticosteroids. *Science* **179**:297–298.

Farrell, P. M., Lundgren, D. W., and Adams, A. J. 1974. Choline kinase and choline phosphotransferase in developing fetal rat lung. *Biochem. Biophys. Res. Commun.* **57**:696–701.

Franson, R. C., and Waite, M. 1973. Lysosomal phospholipases A_1 and A_2 of normal and Bacillus Calmette Guerin-induced alveolar macrophages. *J. Cell Biol.* **56**:621–627.

Franson, R., Beckerdite, S., Wang, P., Waite, M., and Elsbach, P. 1973. Some properties of phospholipases of alveolar macrophages. *Biochim. Biophys. Acta* **296**:365–373.

Frosolono, M. F., Charms, B. L., Pawlowski, R., and Slivka, S. 1970. Isolation, characterization, and surface chemistry of a surface-active fraction from dog lung. *J. Lipid Res.* **11**:439–457.

Frosolono, M. F., Slivka, S., and Charms, B. L. 1971. Acyl transferase activities in dog lung microsomes. *J. Lipid Res.* **12**:96–103.

Frosolono, M. F., Pawlowski, R. P., Charms, B. L., Corbusier, C., Abrams, M., and Jones, J. 1973. Lung surface-active fraction as a model system for macromolecular ultrastructural studies with *Crotalus atrox* venom. *J. Lipid Res.* **14**:110–120.

Frosolono, M. F., Pawlowski, R., and Charms, B. L. 1975. Relationship between intra- and extracellular surface-active fractions from rat lungs. *Chest* **67**:16s–19s.

Fruton, J. S. 1963. The Proteins, p. 189. Academic Press, New York.

Gallai-Hatchard, J. J., and Thompson, R. H. S. 1965. Phospholipase-A activity of mammalian tissues. *Biochim. Biophys. Acta* **98**:128–136.

Garcia, A., Newkirk, J. D., and Mavis, R. D. 1975. Lung surfactant synthesis: A Ca^{++}-dependent microsomal phospholipase A_2 in the lung. *Biochem. Biophys. Res. Commun.* **64**:128–135.

Gassenheimer, L., Rhoades, R. A., and Scholz, R. W. 1972. *in vivo* incorporation of ^{14}C-1-palmitate and ^{3}H-U-glucose into lung lecithin. *Respir. Physiol.* **15**:268–275.

Giannopoulos, G., Mulay, S., and Solomon, S. 1972. Cortisol receptors in rabbit fetal lung. *Biochem. Biophys. Res. Commun.* **47**:411–418.

Gil, J., and Reiss, O. K. 1973. Isolation and characterization of lamellar bodies and tubular myelin from rat lung homogenates. *J. Cell Biol.* **58**:152–171.

Gluck, L., Motoyama, E. K., Smits, H. L., and Kulovich, M. V. 1967a. The biochemical development of surface activity in mammalian lung. I. The surface-active phospholipids; the separation and distribution of surface-active lecithin in the lung of the developing rabbit fetus. *Pediat. Res.* **1**:237–246.

Gluck, L., Sribney, M., and Kulovich, M. V. 1967b. The biochemical development of surface activity in mammalian lung. II. The biosynthesis of phospholipids in the lung of the developing rabbit fetus and newborn. *Pediat. Res.* **1**:247–265.

Gluck, L., Landowne, R. A., and Kulovich, M. V. 1970. Biochemical development of surface activity in mammalian lung. III. Structural changes in lung lecithin during development of the rabbit fetus and newborn. *Pediat. Res.* **4**:352–364.

Gluck, L., Kulovich, M. V., Eidelman, A. I., Cordero, L., and Khazin, A. F. 1972. Biochemical development of surface activity in mammalian lung. IV. Pulmonary lecithin synthesis in the human fetus and newborn and etiology of the respiratory distress syndrome. *Pediat. Res.* **6**:81–99.

Godinez, R. I., Sanders, R. L., and Longmore, W. J. 1975. Phosphatidylglycerol in rat lung. I. Identification as a metabolically active phospholipid in isolated rat lung. *Biochemistry* **14**:830–834.

Goldenberg, V. E., Buckingham, S., and Sommers, S. C. 1967. Pulmonary alveolar lesions in vagotomized rats. *Lab. Invest.* **16**:693–705.

Goldenberg, V. E., Buckingham, S., and Sommers, S. C. 1969. Pilocarpine stimulation of granular pneumocyte. *Lab. Invest.* **20**:147–158.

Gross, I., and Warshaw, J. B. 1974. Enzyme activities related to fatty acid synthesis in developing mammalian lung. *Pediat. Res.* **8**:193–199.

Hallman, M., and Gluck, L. 1974. Phosphatidly glycerol in lung surfactant: I. Synthesis in rat lung microsomes. *Biochem. Biophys. Res. Commun.* **60**:1–7.

Hallman, M., and Raivio, K. 1974. Studies on the biosynthesis of disaturated lecithin of the lung: The importance of the lysolecithin pathway. *Pediat. Res.* **8**:874–879.

Hasegawa-Sasaki, H., and Ohno, K. 1975. Acyltransferase activities in rat lung microsomes. *Biochim. Biophys. Acta* **380**:486–495.

Henderson, R. F., and Pfleger, R. C. 1972. Surface tension studies of phosphatidyl glycerol isolated from lungs of beagle dogs. *Lipids* **7**:492–494.

Henderson, T. O., Glonek, T., and Myers, T. C. 1974. Phosphorus-^{31} nuclear magnetic resonance spectroscopy of phospholipids. *Biochemistry* **13**:623–628.

Hendry, A. T., and Possmayer, F. 1974. Pulmonary phospholipid biosynthesis. Properties of a stable microsomal glycerophosphate acyltransferase preparation from rabbit lung. *Biochim. Biophys. Acta* **369**:156–172.

Hoffman, L. 1972. Isolation of inclusion bodies from rabbit lung parenchyma. *J. Cell Phys.* **79**:65–72.

Jamdar, S. C., and Fallon, H. J. 1973. Glycerolipid biosynthesis in rat adipose tissue. I. Properties and distribution of glycerophosphate acyltransferase and effect of divalent cations on neutral lipid formation. *J. Lipid Res.* **14**:509–516.

Kikkawa, Y., and Yoneda, K. 1974. The type II epithelial cell of the lung. I. Method of isolation. *Lab. Invest.* **30**:76–84.

Kikkawa, Y., Motoyama, E. K. and Cook, C. D. 1965. The ultrastructure of the lungs of lambs: the relation of osmiophilic inclusions and alveolar lining layer to fetal maturation and experimentally produced respiratory distress. *Am. J. Pathol.* **47**:877–903.

Kikkawa, Y., Motoyama, E. K., and Gluck L. 1968. Study of the lungs of fetal and newborn rabbits. *Am. J. Pathol.* **52**:177–209.

Kikkawa, Y., Kaibara, M., Motoyama, E. K., Orzalesi, M., and Cook, C. D. 1971. Morphologic development of fetal rabbit lung and its acceleration with cortisol. *Am. J. Pathol.* **54**:423–442.

Kikkawa, Y., Yoneda, K., Smith, F., Packard, B., and Suzuki, K. 1975. The type II epithelial cells of the lung. II. Chemical composition and phospholipid synthesis. *Lab. Invest.* **32**:295–302.

King, R. J. 1974. The surfactant system of the lung. *Fed. Proc.* **33**:2238–2247.

King, R. J., and Clements, J. A. 1972. Surface active materials from dog lung. II. Composition and physiological correlations. *Am. J. Phyisol.* **223**:715–726.

King, R. J., Ruch, J., and Clements, J. A. 1973. An uncommon phosphatidylcholine specific for surface-active material in canine lung. *J. Appl. Physiol.* **35**:778–781.

Klaus, M. H., Reiss, O. K., Tooley, W. H., Piel, C., and Clements, J. A. 1962. Alveolar epithelial cell mitochondria as source of the surface-active lung lining. *Science* **137**:750–751.

Kotas, R. V., and Avery, M. E. 1971. Accelerated appearance of pulmonary surfactant in the fetal rabbit. *J. Appl. Physiol.* **30**:358–361.

Kotas, R. V., Mims, L. C., and Hart, L. K. 1974. Reversible inhibition of lung cell number after glucocorticoid injection into fetal rabbits to enhance surfactant appearance. *Pediatrics* **53**:358–361.

Kuhn, C., Callaway, L. A., and Askin, F. B. 1974. The formation of granules in the bronchiolar Clara cells of the rat. *J. Ultrastruct. Res.* **49**:387–400.

Kyei-Aboagye, K., Rubinstein, D., and Beck, J. C. 1973. Biosynthesis of dipalmitoyl-lecithin by the rabbit lung. *Can. J. Biochem.* **51**:1581–1587.

Lamb, R. G., and Fallon, H. J. 1974. Glycerolipid formation from *sn*-glycerol-3-phosphate by rat liver cell fractions. The role of phosphatidate phosphohydrolase. *Biochim. Biophys. Acta* **348**:166–178.

Lands, W. E. M. 1958. Metabolism of glycerolipids: A comparison of lecithin and triglyceride synthesis. *J. Biol. Chem.* **231**:883–888.

Leake, E. S., and Myrvik, Q. N. 1968. Changes in morphology and in lysozyme content of free alveolar cells after the intravenous injection of killed BCG in oil. *J. Reticuloendothel. Soc.* **5**:33–53.

Lehman, I. R. 1963. Progress in Nucleic Acid Research. p. 115. Academic Press, New York.

Liggins, G. C., and Howie, R. N. 1972. A controlled trial of antepartum glucocorticoid treatment for prevention of the respiratory distress syndrome in premature infants. *Pediatrics* **50**:515–525.

Littman, J., Kress, Y., Frosolono, M. F., Rosenbaum, R. M., Colacicco, G., and Scarpelli, E. M. 1974. A morphological and biochemical characterization of the lamellar inclusion bodies from rabbit alveolar pneumocytes. *Fed. Proc.* **33**:345.

Mason, R. J., Huber, G., and Vaughan, M. 1972a. Synthesis of dipalmitoyl lecithin by alveolar macrophages. *J. Clin. Invest.* **51**:68–73.

Mason, R. J., Manganiello, V., Vaughan, M. 1972b. Effect of thyroxine on the disaturated lecithin content of lung. *Am. Rev. Respir. Dis.* **106**:767–768.

Meban, C. 1972. Localization of phosphatidic acid phosphatase activity in granular pneumocytes. *J. Cell Biol.* **53**:249–252.

Michaelson, D. M., Horwitz, A. F., and Klein, M. P. 1973. Transbilayer asymmetry and surface homogeneity of mixed phospholipids in cosonicated vesicles. *Biochemistry* **12**:2637–2645.

Mims, L. C., and Kotas, R. V. 1973. Glycerol as a phosphatidyl choline precursor in the developing mammalian lung. *Biol. Neonate* **22**:436–443.

Mims, L. C., and Zee, P. 1971. Utilization of glycerol by the newborn mammalian lung. *Biol. Neonate* **18**:356–362.

Morgan, T. E. 1971. Biosynthesis of pulmonary surface-active lipid. *Arch. Int. Med.* **127**:401–407.

Morgan, T. E., Finley, T. N., and Fialkow, H. 1965. Comparison of the composition and surface activity of "alveolar" and whole lung lipids in the dog. *Biochim. Biophys. Acta* **106**:403–413.

Moriya, T., and Kanoh, H. 1974. In vivo studies on the de novo synthesis of molecular species of rat lung lecithins. *Tohoku J. Exp. Med.* **112**:241–256.

Myrvik, Q. N., Leake, E. S., and Fariss, B. 1961a. Studies on pulmonary alveolar macrophages from the normal rabbit: a technique to procure them in a high state of purity. *J. Immunol.* **86**:128–132.

Myrvik, Q. N., Leake, E. S., and Fariss, B. 1961b. Lysozyme content of alveolar and peritoneal macrophages from the rabbit. *J. Immunol.* **86**:133–136.

Myrvik, Q. N., Leake, E. S., and Oshima, S. 1962. A study of macrophages and epithelioid-like cells from granulomatous (BCG-induced) lungs of rabbits. *J. Immunol.* **89**:745–751.

Naimark, A. 1971. Cellular dynamics and lipid metabolism in the lung. *Fed. Proc.* **32**:1967–1971.

Newkirk, J. D., and Waite, M. 1971. Identification of phospholipase A_1 in plasma membranes of rat liver. *Biochim. Biophys. Acta* **225**:224–233.

Niden, A. H. 1967. Bronchiolar and large alveolar cell in pulmonary phospholipid metabolism. *Science* **158**:1323–1324.

Nisizawa, K., and Hashimoto, Y. 1970. The Carbohydrate Chemistry and Biochemistry, Vol. IIA, pp.241–300. Academic Press, New York.

Ohta, M,, and Hasegawa, H. 1972. Phospholipase A activity in rat lung. *Tohoku J. Exp. Med.* **108**:85–94.

Ohta, M., Hasegawa, H., and Ohno, K. 1972. Calcium independent phospholipase A_2 activity in rat lung supernatant. *Biochim. Biophys. Acta* **280**:552–558.

O'Neil, J. J., and Tierney, D. F. 1974. Rat lung metabolism: Glucose utilization by isolated perfused lungs and tissue slices. *Am. J. Physiol.* **226**:867–873.

Page-Roberts, B. A. 1972. Preparation and partial characterization of a lamellar body fraction from rat lung. *Biochim. Biophys. Acta* **260**:334–338.

Pawlowski, R., Frosolono, M. F., Charms, B. L., and Przybylski, R. 1971. Intra- and extracellular compartmentalization of the surface-active fraction in dog lung. *J. Lipid Res.* **12**:538–544.

Persson, B., and Gentz, J. 1966. The pattern of blood lipids, glycerol and ketone bodies during the neonatal period, infancy and childhood. *Acta Paediatr. Uppsala* **55**:353–362.

Petrik, P., and Collet, A. 1974. Quantitative electron microscopic autoradiography of in vivo incorporation of ^{3}H-choline, ^{3}H-leucine, ^{3}H-acetate and ^{3}H-galactose in non-ciliated bronchilar (clara) cells of mice. *Am. J. Anat.* **139**:519–534.

Pfleger, R. C., Henderson, R. F., and Waide, J. 1972. Phosphatidyl glycerol—a major component of pulmonary surfactant. *Chem. Phys. Lipids* **9**:51–68.

Popjak, G., and Beeckmans, M. L. 1950. Extrahepatic lipid synthesis. *Biochem. J.* **47**:233–238.

Possmayer, F., Scherphof, G. L., Dubbelman, T. M. A. R., Van Golde, L. M. G., and Van Deenen, L. L. M. 1969. Positional specificity of saturated and unsaturated fatty acids in phosphatidic acid from rat liver. *Biochim. Biophys. Acta* **176**:95–110.

Redding, R. A., Douglas, W. H., and Stein, M. 1972. Thyroid hormone influence upon lung surfactant metabolism. *Science* **175**:994–996.

Reid, L. 1967. The embryology of the lung, pp. 109–124. *In* V. S. DeReuck and R. Porter (eds.). Ciba Foundation Symposium, Development of the Lung. Little, Brown, and Co. Boston.

Robertson, A. F., and Lands, W. E. M. 1962. Positional specifities in phospholipid hydrolysis. *Biochemistry* **1**:804–810.

Rooney, S. A., Canavan, P. M., and Motoyama, E. K. 1974. The identification of phosphatidylglycerol in the rat, rabbit, monkey, and human lung. *Biochim. Biophys. Acta* **360**:56–67.

Rooney, S. A., Gross, I., Gassenheimer, L. N., and Motoyama, E. K. 1975. Stimulation of glycerolphosphate phosphatidyltransferase activity in fetal rabbit lung by cortisol administration. *Biochim. Biophys. Acta* **398**:433–441.

Salisbury-Murphy, S., Rubinstein, D., and Beck, J. C. 1966. Lipid metabolism in lung slices. *Am. J. Physiol.* **211**:988–992.

Sanchez de Jimenez, E., and Cleland, W. W. 1969. Studies of the microsomal acylation of L-glycerol-3-phosphate. I. The specificity of the rat brain enzyme. *Biochim. Biophys. Acta.* **176**:685–691.

Sanders, R. L., and Longmore, W. J. 1975. Phosphatidylglycerol in rat lung. II. Comparison of occurrence, composition, and metabolism in surfactant and residual lung fractions. *Biochemistry* **14**:835–840.

Scarpelli, E. M. 1967. The lung, tracheal fluid, and lipid metabolism of the fetus. *Pediatrics* **40**:951–947.

Scarpelli, 1968a. Introduction, pp. 1–7. *In* The Surfactant System of the Lung. Lea and Febiger, Philadelphia.

Scarpelli, 1968b. Pulmonary mechanics and surfactants, pp. 9–29. *In* The Surfactant System of the Lung. Lea and Febiger, Philadelphia.

Scarpelli, 1968c. Fetal physiology and surfactants, pp. 153–175. *In* The Surfactant System of the Lung. Lea and Febiger, Philadelphia.

Scarpelli, 1968d. Morphology and histochemistry, pp. 31–52. *In* The Surfactant System of the Lung. Lea and Febiger, Philadelphia.

Scarpelli, 1968e. Respiratory distress syndrome of the newborn, pp. 177–199. *In* The Surfactant System of the Lung. Lea and Febiger, Philadelphia.

Scarpelli, E. M., Colacicco, G., and Chang, S. J. 1971. Significance of methods for isolation and characterization of pulmonary surfactants. *Respir. Physiol.* **12**:179–198.

Scarpelli, E. M., Condorelli, S., Colacicco, G., and Cosmi, E. V. 1975. Lamb fetal pulmonary fluid. II. Fate of phosphatidylcholine. *Pediat. Res.* **9**:195–201.

Schiller, H., and Bensch, K. 1971. De novo fatty acid synthesis and elongation of fatty acids by subcellular fractions of lung. *J. Lipid Res.* **12**:248–255.

Scholz, R. W., and Rhoades, R. A. 1971. Lipid metabolism by rat lung in vitro. Effect of starvation and re-feeding on utilization of (U-^{14}C) glucose by lung slices. *Biochem. J.* **124**:257–264.

Schultz, F. M., Jimenez, J. M., MacDonald, P. C., and Johnston, J. M. 1974. Fetal lung maturation. I. Phosphatidic acid phosphohydrolase in rabbit lung. *Gynecol. Invest.* **5**:222–229.

Smith, B. T., Torday, J. S., and Giroud, C. J. P. 1974. Evidence for different gestation-dependent effects of cortisol on cultured fetal lung cells. *J. Clin. Invest.* **53**:1518–1526.

Snyder, F., and Malone, B. 1975. Acyltransferases and the biosynthesis of pulmonary surfactant lipid in adenoma alveolar type II cells. *Biochem. Biophys. Res. Comm.* **66**:914–919.

Snyder, C., Malone, B., Nettesheim, P., and Snyder, F. 1973. Urethan-induced pulmonary adenoma as a tool for the study of surfactant biosynthesis. *Cancer. Res.* **33**:2437–2443.

Sorokin, S. P. 1970. The cells of the lungs, pp. 3–43. *In* P. Nettesheim, M. G. Hanna, Jr., and J. W. Deathevage, Jr. (eds.). Morphology of Experimental Respiratory Carcinogenesis, USAEC Symposium Series 21, CONF-700501.

Spitzer, H. L., Rice, J. M., MacDonald, P. C., and Johnston, J. M. 1975. Phospholipid biosynthesis in lung lamellar bodies. *Biochem. Biophys. Res. Commun.* **66**:17–23.

Sutnick, A. I., Cronlund, M. M., and Ambacher, S. B. 1970. Cortisone therapy and pulmonary surfactant. *Res. Commun. Chem. Pathol. Pharmacol.* **1**:57–66.

Tansey, F. A., and Frosolono, M. F. 1975. Role of 1-acyl-2-lyso-phosphatidylcholine acyl transferase in the biosynthesis of pulmonary phosphatidylcholine. *Biochem. Biophys. Res. Commun.* **67**:1560–1566.

Thomas, T., and Rhoades, R. A. 1970. Incorporation of palmitate-1-^{14}C into lung tissue and "alveolar" lecithin. *Am. J. Physiol.* **219**:1535–1538.

Tierney, D. F. 1974. Lung metabolism and biochemistry. *Annu. Rev. Physiol.* **36**:209–231.

Tierney, D. F., Clements, J. A., and Trahan, H. J. 1967. Rates of replacement of lecithins and alveolar instability in rat lungs. *Am. J. Physiol.* **213**:671–676.

Toft, D., and Chytil, F. 1973. Receptors for glucocorticoids in lung tissue. *Arch. Biochem. Biophys.* **157**:464–469.

Tombropoulos, E. G. 1964. Fatty acid synthesis by subcellular fractions of lung tissue. *Science* **146**:1180–1181.

Tombropoulos, E. G. 1973. Palmitate incorporation into lipids by lung subcellular fractions. *Arch. Biochem. Biophys.* **158**:911–918.

Valdivia, E. 1973. Isolation and identification of pulmonary lamellar bodies from guinea pigs. *Prep. Biochem.* **3**:19–30.

Van den Bosch, H., Bonte, H. A., and Van Deenan, L. L. M. 1965. On the anabolism of lysolecithin. *Biochim. Biophys. Acta* **98**:648–651.

Van den Bosch, H., Aarsman, A. J., de Jong, J. G. N., and Van Deenen, L. L. M. 1973. Studies on lysophospholipases. I. Purification and some properties of lysophospholipase from beef pancreas. *Biochim. Biophys. Acta* **296**:94–104.

Vereyken, J. M., Montfoort, A., and Van Golde, L. M. G 1972. Some studies on the biosynthesis of the molecular species of phosphatidylcholine from rat lung and phosphatidylcholine and phosphatidylethanolamine from rat liver. *Biochim. Biophys. Acta* **260**:70–81.

Wang, M. C., and Meng, H. C. 1972. Lipid synthesis by rat lung in vitro. *Lipids* **7**:207–211.

Wang, N. S., Kotas, R. V., Avery, M. E., and Thurlbeck, W. M. 1971. Accelerated appearance of osmiophilic bodies in fetal lungs following steroid injection. *J. Appl. Physiol.* **30**:362–365.

Webster, G. R. 1965. The acylation of lysophosphatides with long-chain fatty acids by rat brain and other tissues. *Biochim. Biophys. Acta* **98**:512–519.

Weibel, E. R. 1967. Discussion, pp. 170–176. *In* A. V. A. de Reuck and R. Porter (eds.). Development of the Lung, Churchill, London.

Weinhold, P. A. 1968. Biosynthesis of phosphatidyl choline during prenatal development of the rat lung. *J. Lipid Res.* **9**:262–266.

Weinhold, P. A., and Villee, C. A. 1965. Phospholipid metabolism in the liver and lung of rats during development. *Biochim. Biophys. Acta* **106**:540–550.

White, D. A. 1973. Phospholipid composition of mammalian tissues, pp. 441–482. *In* G. B. Ansell, R. M. C. Dawson, and J. N. Hawthorne (eds.). Form and Function of Phospholipids, Elsevier, Amsterdam.

Wieland, O., and Suyter, M. 1957. Glycerokinase: Isolierung and Eigenschaften des Enzyms. *Biochem. Z.* **329**:320–331.

Williams, C. H., Vail, W. J., Harris, R. A., Green, D. E., and Valdivia, E. 1971. The isolation and characterization of the lamellar body of bovine lung. *Prep. Biochem.* **1**:37–45.

Wolfe, B. M. J., Anhalt, B., Beck, J. C., and Rubinstein, D. 1970. Lipid metabolism in rabbit lungs. *Can. J. Biochem.* **48**:170–177.

Wu, B., Kikkawa, Y., Orzalesi, M. M., Motoyama, E. K., Kaibara, M., Zigas, C. J., and Cook, C. D. 1973. The effect of thyroxine on the maturation of fetal rabbit lungs. *Biol. Neonate* **22**:161–168.

Young, S. L., and Tierney, D. F. 1972. Dipalmitoyl lecithin secretion and metabolism by the rat lung. *Am. J. Physiol.* **222**:1539–1544.

Zachman, R. D. 1971. The enzymes of lecithin bio-synthesis in human newborn lungs. *Biol. Neonate* **19**:211–219.

Zucker-Franklin, D., and Hirsch, J. G. 1964. Electron microscope studies on the degranulation of rabbit peritoneal leukocytes during phagocytosis. *J. Exp. Med.* **120**:569–576.

Kidney

JEN-SIE TOU AND CLYDE G. HUGGINS

I. Introduction

Kidney, like other organs of the mammals, has its distinct pattern of lipid metabolism, but it also shares the similarity in most of the metabolic pathways involved in lipid synthesis and degradation. In this chapter, we review the metabolism of free fatty acids, phospholipids, cholesterol, prostaglandins, sphingolipids, and vitamin D_3 in this organ. There are several unique features of lipid metabolism in kidney: (1) Cytochrome P-450 in kidney microsomes appears to be involved specifically in omega and penultimate hydroxylation of fatty acids (kidney microsomes exhibit low hydroxylation activities when various drugs or steroids are tested as substrates); (2) the rapid labeling of polyphosphoinositides by $[^{32}P]P_i$ *in vivo* followed by a rapid decline of the radioactivity has not been observed in other organs examined; (3) kidney is the primary site of the metabolism of the circulating D-mevalonate, most of which is incorporated into cholesterol precursors and only a small amount of which is incorporated into cholesterol; (4) in normal human subjects digalactosylceramide is present only in kidney and intestine in significant amounts; and (5) kidney is considered as an endocrine organ because of its involvement in the metabolism of vitamin D_3.

The distribution of lipids in kidney is not homogeneous. A regional distribution of all lipids has been reported. In addition, the metabolic activities of the cortex and medulla of the kidney are different, e.g., prostaglandin synthesis is predominantly localized in the medulla while its

JEN-SIE TOU • Department of Biochemistry, Tulane University School of Medicine, New Orleans, Louisiana. CLYDE G. HUGGINS • Department of Biochemistry, University of South Alabama College of Medicine, Mobile, Alabama.

catabolism is primarily in the cortex. Furthermore, the metabolism of all lipids in kidney may be influenced by age, sex, and species of the mammals, though only glycosphingolipid metabolism is known to be affected by these factors. It is thus important to study lipid metabolism in different anatomical regions of the kidney in terms of age, sex, and species of the mammals.

II. Free Fatty Acids

Free fatty acids were found in all zones of the rat kidney (Hohenegger and Novak, 1969; Muller *et al.*, 1972). The highest levels were found in the cortex (0.216 g/100 g wet tissue), somewhat lower in the outer (0.135 g/100 g wet tissue), and lowest in the inner (0.078 g/100 g wet tissue) medulla (Muller *et al.*, 1972). The respiratory quotient of dog kidney cortex (the greatest part of the kidney) is 0.75 (Bainbridge and Evans, 1914; Dickens and Simer, 1930; Shorr *et al.*, 1930), indicating that oxidation of fatty acids is the major source of energy for this organ, whereas the contribution of glucose and amino acids as energy sources is negligible. Kidneys perfused 24 hr without added oleate lost 35% of their total lipid content and 27% of their phospholipids (Huang *et al.*, 1971). Addition of serum albumin-bound oleate to the perfusate prevented the loss of phospholipids to 8%.

The kidney can utilize fatty acids *in vitro* (Quastel and Wheatly, 1933; Geyer *et al.*, 1949; Geyer and Cunningham, 1950) and *in vivo* (Gold and Spitzer, 1964). It oxidizes fatty acids mainly by β-oxidation to produce acetyl-CoA, which is then converted to CO_2 through the Krebs cycle. It is known that acyl-CoA, not free fatty acid, is the substrate for β-oxidation. Fatty acids enter the cell and are converted to the CoA ester extramitochondrially. Carnitine (β-hydroxy-α-trimethyl-ammonium butyrate) long-chain fatty acyltransferase converts the acyl-CoA to an acyl-carnitine in the outer mitochondrial compartment (Brdiczka *et al.*, 1969). The acyl-carnitine then penetrates to the inner mitochondrial compartment where the fatty acid moiety is again esterified with CoA and is available for β-oxidation. In kidney, both long-chain carnitine acyltransferase, carnitine palmitoyltransferase, and short-chain carnitine acyltransferase, carnitine acetyltransferase, are exclusively mitochondrial enzymes, whereas in liver carnitine acetyltransferase was also found in peroxisomes where the specific activity of carnitine acetyltransferase was two- to threefold higher than that in the mitochondrial fraction (Markwell *et al.*, 1973).

The oxidation of fatty acids provides a major source of energy for the reabsorptive work of the renal cortex. Kleinman *et al.* (1973) reported that 4-pentenoic acid, an inhibitor of fatty acid oxidation, impaired sodium and bicarbonate (possibly potassium) reabsorption by dog kidney. Hohenegger *et al.* (1973) found that the oxidation rates of fatty acids were the highest in the cortex and outer medulla, the lowest in the inner medulla. In a sodium-free incubation medium, palmitate and acetate oxidation fell equally to about 50% in all zones. Depression of oxygen consumption under this condition was in the same range. These results show that fatty acids may provide energy for sodium transport in all zones of the kidney.

Fatty acids can also be metabolized to their ω-hydroxyl derivatives by kidney microsomes (Ichihara *et al.*, 1969). An activated fatty acid substrate is not involved. The fatty acid hydroxylation activity has been attributed to the CO-binding hemoprotein, cytochrome P-450, present in these particles. Unlike liver microsomes, kidney microsomes show low or no hydroxylation activities when various drugs or steroids are tested as substrates (Jakobsson *et al.*, 1970). The reduced CO-bound hemoprotein of kidney cortex microsomes absorbs light maximally at 453–454 nm, as compared to the maximum at 450–451 nm of the liver cytochrome (Ichihara *et al.*, 1969). Ichihara *et al.* (1971), using Triton X-100, solubilized an enzyme system catalyzing the ω-hydroxylation of medium-chain fatty acids from porcine kidney cortex microsomes, and resolved the solubilized system into two fractions: CO-binding hemoprotein (cytochrome P-452) and NADPH-cytochrome *c* reductase. Recently Ellin *et al.* (1972) have presented evidence for the hydroxylation of lauric acid to produce ω- and (ω–1)-hydroxyl derivatives by rat kidney cortex microsomes in the presence of NADPH and molecular oxygen. 12-Hydroxylaurate and 11-hydroxylaurate are formed at the ratio of approximately 2:1; NADH is only half as efficient as NADPH. The formation of the hydroxylated products from laurate is inhibited by carbon monoxide and the inhibition is reversed by irradiation of 450 nm wavelength. Furthermore, addition of laurate to suspensions of microsomes produces a type I spectral change, indicative of the formation of a cytochrome P-450–laurate complex.

The rate of laurate hydroxylation in the isolated kidney cortex microsomes or the distribution of the ω- and (ω–1)-hydroxyl products was not affected by treatment of rats with repeated doses of phenobarbital or 3,4-benzpyrene (Ellin and Orrenius, 1971). However, starvation for 48 hr caused an enhanced rate of laurate hydroxylation, as well as increases in the concentration of cytochrome P-450 and in the magnitude of the type I spectral change produced by the addition of an excess of laurate to the isolated kidney cortex microsomes (Ellin and Orrenius, 1971). These results together with the substrate specificity of cytochrome P-450 in kidney suggest that cytochrome P-450 of rat kidney cortex microsomes may be specialized for fatty acid hydroxylation.

Recently, Sasame *et al.* (1974) presented evidence that cytochrome b_5 is also involved in omega and penultimate hydroxylation of laurate by rat liver and kidney microsomes. These investigators showed that antibody against NADPH-cytochrome *c* reductase inhibited the NADPH-dependent omega and penultimate hydroxylation of lauric acid by microsomes from kidney cortex and liver of rats, but did not inhibit the NADH-dependent hydroxylation of lauric acid. By contrast, an antibody against cytochrome b_5 inhibited both the NADH- and the NADPH-dependent hydroxylation of lauric acid by these microsomal preparations. These findings suggest that NADPH-cytochrome *c* reductase mediates the NADPH-dependent hydroxylation of lauric acid but not its NADH-dependent hydroxylation, whereas cytochrome b_5 plays a role in both the NADPH- and NADH-dependent hydroxylation of the fatty acid.

Fatty acid synthesis in kidney tissue has not received intensive investigation under the same experimental conditions as in liver (see the chapter Liver), but kidney has been found capable of synthesizing fatty acids from glucose and acetate *in vitro* (Chernick *et al.*, 1950; Medes *et al.*, 1952). Furthermore, evidence indicates that the enzymatic pathway for fatty acid synthesis in kidney is similar to that established for liver (Ganguly, 1960). Masoro and Porter (1965) reported that kidney slices synthesize fatty acids from [1-^{14}C]acetate at a 10–20% slower rate than do liver slices.

The control of fatty acid synthesis in kidney tissue appears to be different from that in liver. Kidney tissue from diabetic rats incorporated 2–3 times as much acetate into fatty acids as kidney tissue from normal animals, while the labeling of fatty acids by [1-^{14}C]acetate was depressed in the diabetic livers (Burns and Elwood, 1969).

III. Phospholipids

A. Composition

Phospholipids are the major lipid constituents of both cortex and medulla of the kidney. The quantity of each class of phospholipids extracted from kidney may reflect to some extent the methods of extraction, separation, and analysis, but a close similarity of kidney phospholipid (except the phosphoinositides) values has been found in whole kidneys of several species (Scott *et al.*, 1967; Rouser *et al.*, 1969) and has recently been reviewed by White (1973). Although a similarity in the fatty acid patterns of total phospholipids from the kidneys of mouse, rat, and rabbit was reported (Veerkamp *et al.*, 1962), the molecular species of phosphatidylethanolamine from dog and pig kidney have recently been found to be different (Yeung and Kuksis, 1974). The total diacyl ethanolamine phosphatides of the dog kidney contained about 1.5 times more palmitate than stearate, whereas the diacyl derivatives of the ethanolamine phosphatides of pig kidney possessed about 2.5 times as much stearate as palmitate. Ethanolamine phospholipid from dog kidney contained about equal proportions of diacyl and alkenylacyl derivatives, whereas that from pig kidney contained twice as much alkenylacyl as diacyl derivatives.

Morgan *et al.* (1963) analyzed phospholipid classes in the cortex and medulla of rabbit kidney by column chromatography on silicic acid. They found that the cortex contained more total lipids (3.32 g/100 g wet weight of tissue) and phospholipids (88% of total lipids) than the medulla (2.08 g/100 g wet weight of tissue are total lipids, and 68.4% of the total lipids are phospholipids). The prodominant phospholipids in the medulla and the cortex are phosphatidylcholine (30.1% of total lipid P for cortex, 40.7% for medulla) and phosphatidylethanolamine (21.6% of total lipid P for cortex, 25.0% for medulla). Most of the plasmalogens are present in the ethanolamine phospholipid fraction of both cortex and medulla of rabbit kidney. Plasmalogen components are also identified in small amounts in serine, inositol, and choline phospholipids (Morgan *et al.,* 1963). The content of alk-1-enyl and alkyl groups in the phospholipids from mammalian kidneys has been reviewed by Horrocks (1972).

Morgan *et al.* (1963) also determined the positional distribution of fatty acid composition of rabbit kidney phospholipids and triglycerides. Unsaturated fatty acids were found predominantly in the 2-position, and saturated fatty acids in the 1-position (and 3-position in triglycerides). Studies by Hagen (1971) on the compositional specificity of the major diacyl phosphatides, plasmalogens, and triglycerides of whole kidney of the pig showed that the triglycerides are esterified predominantly with saturated fatty acids in the 2-position while the phosphatides have predominantly unsaturated fatty acids in this position. Such complete "inversion" of the structure of triglycerides and phospholipids is not seen in rabbit kidney (Morgan *et al.,* 1963). These data suggest that an active deacylation–reacylation process may control the positional distribution of fatty acids in pig kidney phospholipids.

The intracellular distribution profile of kidney phospholipids has been examined in several mammals. Getz *et al.* (1968) fractionated sheep kidney into nuclear, mitochondrial, microsomal, and supernatant fractions. Among these fractions, microsomes are the richest in phospholipids (296 μmoles/g dry weight of tissue fraction), whereas the mitochondria and nuclei contain 214 μmoles and 124 μmoles of phospholipids/g dry weight of tissue fraction, respectively. The cell sap has much less phospholipids. Phosphatidylcholine and phosphatidylethanolamine are the major phospholipids in both mitochondria and microsomes. Microsomes contain more sphingomyelin than mitochondria. Fleischer *et al.* (1967) reported that pure mitochondria from bovine renal cortex contain three major phospholipid classes: phosphatidylcholine, phosphatidylethanolamine, and cardiolipin. Phosphatidylinositol is a minor component (about 3% of the total phospholipid). A trace (0.1% or less) of phosphatidylserine was found in the mitochondrial preparation. The phospholipid composition of plasma membranes, studied in cultured baby hamster kidney cells (Renkonen *et al.*, 1972), was found different from that of endoplasmic reticulum. The plasma membrane phospholipids of these cells contained less phosphatidylcholine, but more phosphatidic acid, phosphatidylserine, and sphingomyelin than the endoplasmic reticulum. However, recently Zambrano *et al.* (1975) have shown that the content of phosphatidic acid, phosphatidylserine, and sphingomyelin in the plasma membrane fraction from rat kidney is similar to that in the rough microsomes. Zambrano *et al.* (1975) also reported that the Golgi apparatus of rat kidney has the highest content of phosphatidylcholine and phosphatidylinositol on a protein basis, among mitochondria, rough microsomes, and plasma membranes.

Although the major phospholipids in kidney tissue of various species are quantitatively similar (White, 1973), different values of phosphoinositides (phosphatidylinositol, diphosphoinositide, and triphosphoinositide) have been reported as summarized in Table 1. The phosphoinositides are minor phospholipid constituents of the kidney and they were first isolated by Huggins and Cohn (1959) from kidney cortex of various species. The quantitative variation of these lipids in kidney could be attributed to the differences in methods of extraction.

B. *Metabolism*

1. *The Incorporation of Isotopes into Phospholipids*

Radioactive tracers have been used to study the metabolism of phospholipids *in vitro* and *in vivo*. When [2-^{3}H]inositol was injected to rats, the greatest incorporation into chloroform–methanol soluble fraction per mg of tissue was found in the kidney (Agranoff *et al.*, 1958). In 1964, Andrade and Huggins showed that when slices of pig kidney were incubated in the presence of [^{32}P]P_i, most of the radioactivity in the phosphoinositides could be recovered from diphosphoinositide and triphosphoinositide. The specific activity of the phosphoinositides increased progressively with the degree of phosphorylation, indicating a rapid turnover of the phosphomonoester groups of the polyphosphoinositides. The incorporation of [^{32}P]P_i into the phosphoinositides by rat kidney has also been studied *in vivo* (Tou *et al.*, 1972). The time-course of the labeling of the phosphoinositides with [^{32}P]P_i emphasizes the rapid metabolism of the phosphomonoester groups of the polyphosphoinositides in rat kidney. Both diphosphoinositide and triphosphoinositide exhibited an accelerated metabolic rate in that the rates of appearance and disappearance of the isotope into the components were similar to the metabolic rates of inorganic phosphate and acid-labile phos-

TABLE 1
The Phosphoinositide Content of Rat Kidney (μmoles/g wet tissue)

Inositide	Wagner *et al.* (1963)	Dawson and Eichberg (1965)	Dittmer and Douglas (1969)	Tou *et al.* (1972)	Hauser and Eichberg (1973)
Phosphatidylinositol	2.72	—	2.62	1.35	—
Diphosphoinositide	0.05	0.05–0.09	0.04	0.153	0.067
Triphosphoinositide	0.03	0.04–0.07	0.03	0.168	0.130

phate of the nucleotides. Phosphatidylinositol did not show this metabolic pattern (Fig. 1). The rapid incorporation and rapid decline of [^{32}P]P_i in diphosphoinositide and triphosphoinositide were not observed in heart and submaxillary glands of the rat (Fig. 2). The physiological significance of the rapid metabolism of the phosphomonoesters of diphosphoinositides and triphosphoinositides in kidney tissue is not known. Since polyphosphoinositides have been found to be associated with membrane fraction in kidney (Flint, 1967), it is possible that these molecules participate in some important functional component of the tubule membranes during the secretion

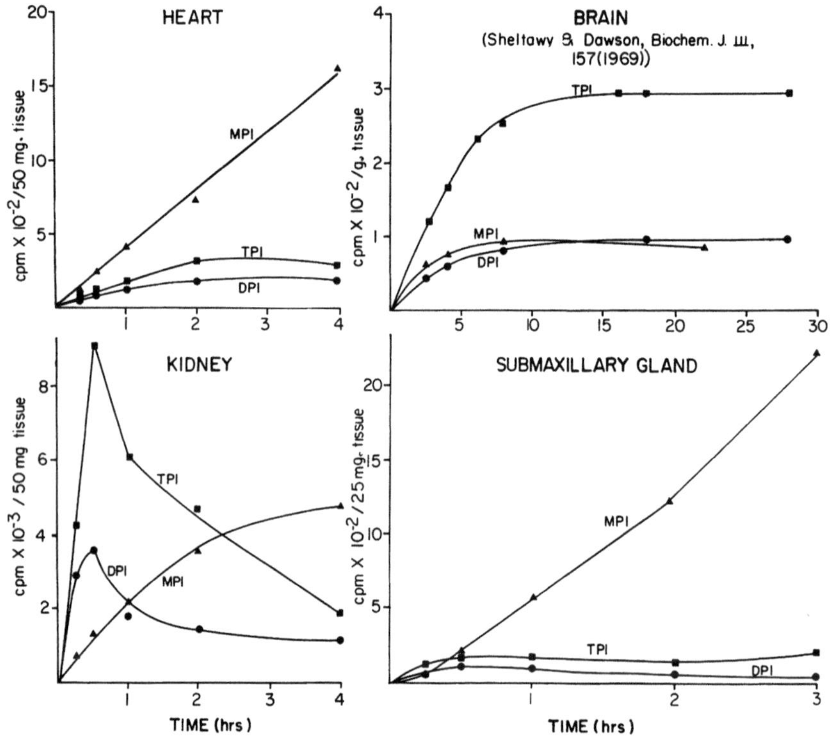

FIGURE 1. Time-course of labeling of phosphoinositides in kidney, heart, and submaxillary gland of the rat. Carrier-free ^{32}P (0.5 mCi) was injected intraperitoneally to each rat. The lipids were separated by silicic acid-impregnated glass fiber paper chromatography. Each time point represents the average data obtained from four rats. MPI = monophosphoinositide or phosphatidylinositol; DPI = diphosphoinositide; TPI = triphosphoinositide. The data for hen brain from Sheltawy and Dawson (1969) were calculated from the specific activity (cpm/μg P) at each time interval multiplied by the total amount (μg P) of each inositide per g of brain. MPI = 24.5 μg; DPI = 17.6 μg; TPI = 42.1 g P. Reproduced from Tou *et al.* (1973) by permission of the copyright owner.

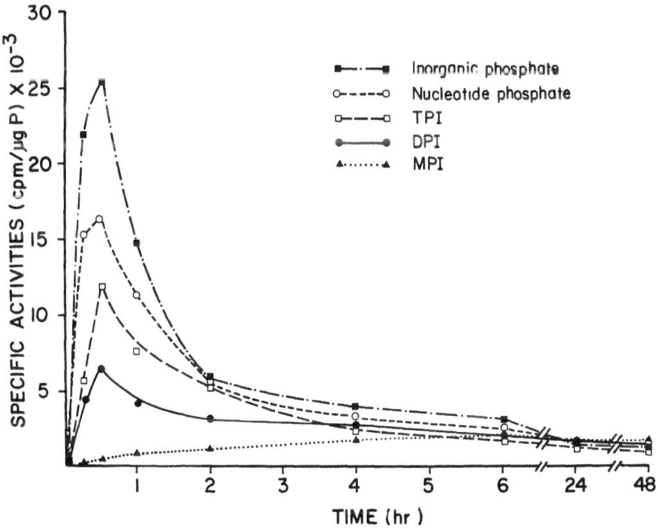

FIGURE 2. Time-course of the specific activity of inorganic phosphate, nucleotide phosphate, and phosphoinositides of kidney. Carrier-free ^{32}P (0.5 mCi) was given to each rat. Reproduced from Tou *et al.* (1973) by permission of the copyright owner.

and reabsorption of solutes from the lumen of the tubule. Evidence is not now available to resolve this important role of the inositides in the functional characteristics of the tubular membrane.

Tinker and Hanahan (1966) studied the metabolism of phospholipids in rabbit kidney by incubating the cortex slices with [^{32}P]P$_i$, [6-^{14}C]glucose, [3-^{14}C]serine, [1-^{14}C]palmitic acid, or [1-^{14}C]linoleic acid, and determined the pattern of labeling of individual phospholipids isolated by silicic acid column chromatography. These investigators found that, with the exception of cardiolipin and sphingomyelin, all the phospholipids of rabbit renal cortex slices are metabolically active. This was also observed in rat kidney *in vivo,* when [^{32}P]P$_i$ was injected intraperitoneally to the animal (Soula *et al.,* 1972). When [^{32}P]P$_i$ was incubated with renal cortex slices of the rabbit, the specific activity of phosphatidylinositol at all time intervals was the highest among the isolated phospholipids (Tinker and Hanahan, 1966). The labeling of diphosphoinositide and triphosphoinositide by [^{32}P]P$_i$ was not examined in these experiments. [^{32}P]P$_i$-labeled phosphatidylcholine had the second highest specific activity. Much lesser amount of radioactivity appeared in the phosphate moiety of phosphatidylethanolamine; labeling of cardiolipin and sphingomyelin was negligible.

All phospholipids, except cardiolipin, were rapidly labeled by [6-^{14}C]glucose in slices of kidney cortex (Tinker and Hanahan, 1966). The glycerol moiety of phosphatidylcholine was labeled from [6-^{14}C]glucose. The time-course of [6-^{14}C]glucose incorporation into phosphatidylcholine and that of [^{32}P]P_i were parallel, suggesting that the glycerophosphate moiety of this lipid "turns over" as a unit. Hauser (1963) injected rats intraperitoneally with [6-^{14}C]glucose or sodium [2-^{14}C]pyruvate and found that phosphatidylinositol and phosphatidylglycerol in kidney became labeled at 1.5 hr after the injection.

Although the phosphate moiety of phosphatidylserine is metabolically stable, [3-^{14}C]serine was actively incorporated into phosphatidylserine via an energy-independent exchange reaction (Tinker and Hanahan, 1966). A lesser amount of [3-^{14}C]serine was incorporated into the ethanolamine of phosphatidylethanolamine via a direct decarboxylation of phosphatidylserine. The low incorporation of [^{32}P]P_i and [6-^{14}C]glucose into phosphatidylethanolamine as compared to phosphatidylcholine suggests that the pathway involving *N*-methylation of phosphatidylethanolamine is not an important route leading to the biosynthesis of phosphatidylcholine.

In addition to the studies on the labeling of the phosphorus, glycerol, and the base moieties, Tinker and Hanahan (1966) also examined the metabolism of the acyl groups of the phosphatides in rabbit kidney cortex slices. A rapid incorporation of both [1-^{14}C]linoleic acid and [1-^{14}C]palmitic acid into total lipids was observed. After 2 hr incubation, 29.1% of the administered [1-^{14}C]linoleic acid and 12.2% of the administered [1-^{14}C]palmitic acid were recovered in the total lipids of the cortex slices. Slightly over half of the incorporated radioactivities were found in the neutral lipids, the bulk being in triglycerides. Phosphatidylcholine comprised most of the radioactivity in the phospholipid fraction. The specific activities of 1- and 2-esterified linoleic acid of phosphatidylcholine were about the same. However, the specific activity of 2-esterified palmitic acid was much higher than that of the 1-esterified palmitic acid, indicating the palmitic acid in the 2-position "turned over" much faster than that in the 1-position. Tinker and Hanahan (1966) suggested that the turnover of 2-esterfied palmitic acid in phosphatidylcholine may be involved in the oxidation of this fatty acid by renal cortex tissue.

A metabolic heterogeneity of phospholipids in terms of the two acyl groups has been noted. Wurster and Copenhaver (1965) employed reductive ozonolysis to determine [^{32}P]P_i incorporation into different species of phosphatidylcholine in kidney slices of the rabbit. The highest [^{32}P]P_i incorporation was found in the disaturated phosphatidylcholine and the lowest was found among the 1-saturated-2-unsaturated phosphatidylcholines.

The turnover of the glycerol moiety of phosphatidylcholine and phosphatidylethanolamine has been studied in BHK cells, a permanent line of baby hamster kidney cells (Gallaher *et al.*, 1973). When BHK-21 cells were pulse-labeled with [2-^{3}H]glycerol for 1 hr, followed by chase periods of 0–22 hr, a half-life of 2.0–2.5 hr was found for both phospholipid classes.

2. *Physiological Significance of Phospholipid Turnover in Kidney*

Phospholipid turnover was speculated to be connected with active cation transport in kidney tissue (Wurster and Copenhaver, 1965), but in the presence of 10^{-4} ouabain, which would strongly inhibit cation transport, the turnover of lipid phosphorus in renal cortex slices was increased in a nonspecific manner (Tinker and Hanahan, 1966; Bartlett and Bossart, 1969). Tinker *et al.* (1963) reported that diphenylhydantoin increased the rate of uptake of potassium but had no effect on the amount or rate of [^{32}P]P_i incorporation into phospholipids by renal cortex slices. Furthermore, Nelson and Cornatzer (1964) found no evidence for the participation of phospholipid turnover in sodium transport by the kidney *in vivo*. However, recently, Wilson *et al.* (1973) have shown that medullary phospholipids are doubled in rats after 11 days on the low-potassium diet and continued increase occurs with longer depletion. The relative proportion of different medullary phospholipid species was not altered by the low-potassium diet. The elevation of phospholipid levels returned to base level following the feeding of a normal diet to the rats. Kidney medulla slices from potassium-depleted rats were found to incorporate more [^{32}P]P_i and [2-^{3}H]glycerol into phospholipids per tissue weight than the medulla slices from control animals. These observations led Wilson *et al.* (1973) to conclude that the increased phospholipids in kidney medulla from rats on K-depleted diet was attributed to an increased synthesis of phospholipids.

An increased phospholipid synthesis has been observed in the remaining compensatory kidney of the mouse after uninephrectomy (Toback *et al.*, 1974). The rate of [^{14}C]choline incorporation into phospholipids in renal cortical slices obtained from the contralateral kidney was increased by 37%, 5 min after uninephrectomy. It was increased 24%, 3 hr after uninephrectomy *in vivo*. The increased phospholipid synthesis appears to be an early event during the compensatory growth, since neither the protein content nor the rate of protein synthesis was increased until 14 hr following uninephrectomy (Coe and Korty, 1967).

3. *Enzymes Involved in Phospholipid Metabolism in Kidney*

Only limited enzymes involved in phospholipid metabolism in kidney have been examined. In 1962, Robertson and Lands detected the activities of phospholipase A_1 and A_2 in rat kidney homogenates. Later Gallai-Hatchard and Thompson (1965) reported the presence of phospholipase A_2 activity in a glycerol-treated extract of acetone powder from rat kidney.

The activity of phospholipase A_2 may be important in regulating the renin–angiotensin system, since lysophospholipid has been speculated as a renin inhibitor (Smeby and Bumpus, 1971). A recent finding by Antonello *et al.* (1973) has shown that lysophospholipid but not phospholipid from hog kidney lowered the blood pressure in rats with renal hypertension *in vivo*. However, the structure of the lysophospholipid has not been well identified.

Some of the enzymes involved in the synthesis and hydrolysis of the phosphoinositides have been studied in kidney tissue. Agranoff *et al.* (1958) described an enzyme system present in the insoluble residue of guinea-pig kidney mitochondria that catalyzes the incorporation of inositol into inositol phosphatide in the presence of magnesium ion and cytidine diphosphate-choline or cytidine-5′-phosphate. Cytidine-5′-phosphate was less effective than cytidine diphosphate-choline. Furthermore, the incorporation of inositol into the lipid fraction was stimulated by the addition of phosphatidic acid, but not by diglyceride. In the absence of inositol, a cytidine lipid, which is soluble in chloroform–methanol mixtures and thought to be cytidine diphosphate-diglyceride, was formed during the incubation. Based on these observations, Agranoff *et al.* (1958) proposed the participation of cytidine diphosphate-diglyceride in the synthesis of phosphatidylinositol.

A phosphatidylinositol kinase catalyzing the phosphorylation of phosphatidylinositol by $[\gamma^{-32}P]$ATP to form diphosphoinositide was demonstrated in rat kidney homogenates (Colodzin and Kennedy, 1965; Tou *et al.*, 1968). The enzyme was localized in the microsomal fraction from rat kidney cortex (Tou *et al.*, 1969). It was dependent on added magnesium ion for activity and was stimulated by nonionic detergents, Cutscum, or Triton X-100. Harwood and Hawthorne (1969) reported that the distribution of phosphatidylinositol kinase activity among the kidney subcellular fractions resembled that of $(Na^+ + K^+)$-ATPase, with most of the activity in the nuclear and microsomal fractions; only the microsomal enzyme was stimulated by Cutscum.

Triphosphoinositide was shown to be biosynthesized via the phosphorylation of diphosphoinositide by $[\gamma^{-32}P]$ATP catalyzed by diphosphoinositide kinase in rat kidney cortex (Tou *et al.*, 1970). The enzyme was predominantly localized in the plasma membrane fraction. Diphosphoinositide kinase required added magnesium ions for activity. In the presence of detergents the formation of triphosphoinositide was impaired.

The hydrolysis of phosphatidylinositol in kidney tissue has not been studied, but kidney cortex and medulla have been found to be rich sources of the enzyme D-myoinositol 1:2-cyclic phosphate 2-phosphohydrolase (Dawson and Clarke, 1972). This enzyme catalyzes the hydrolysis of D-myoinositol 1:2-cyclic phosphate, which is an intermediary in the enzymic degradation of phosphatidylinositol (Dawson *et al.* 1971) to produce D-myoinositol 1-phosphate. Furthermore, Dawson and Clarke (1972) have shown that the kidney enzyme is located in the brush borders of the proximal tubules.

The hydrolysis of polyphosphoinositides has been examined in rat kidney cortex. T. -C. Lee and Huggins (1968a,b) reported triphosphoinositide phosphomonoesterase to be a microsomal enzyme in rat kidney cortex; a 50-fold purification of the enzyme was achieved. The partially purified enzyme was activated by magnesium or manganese ions and was inhibited by calcium ions. It hydrolyzed triphosphoinositide to yield phosphatidylinositol and inorganic phosphate; diphosphoinositide was found to be formed as a transient intermediate in the dephosphorylation process. Kidney was found to have the highest triphosphoinositide phosphomonoesterase activity; brain, spleen, and lung were intermediate, and liver and heart were the least active. The hydrolysis of triphosphoinositide by a phosphodiesterase was demonstrated in rat kidney cortex (Tou *et al.*, 1973). Subcellular fractionation studies revealed that the enzyme was predominantly in the supernatant fraction. After acid precipitation and ammonium sulfate fractionation, the soluble enzyme was free from triphosphoinositide phosphomonoesterase activity. Although the partially purified enzyme did not require divalent cations for activity, it was strongly inhibited by EDTA (0.1 mM). In the absence of EDTA, added magnesium or calcium ions depressed the enzyme activity. The enzyme hydrolyzed both diphosphoinositide and triphosphoinositide with the formation of 1,2-diglyceride and organic phosphate.

Very few studies have been performed on the regulation of phospholipid metabolism in kidney tissue. Wheldrake (1972) reported a 16% decrease in the phospholipid content of rat kidneys 7 days after adrenalectomy in young female rats. Adrenalectomy decreased the incorporation of $[^{32}P]P_i$ and $[^{14}C]$choline into kidney phosphatidylcholine, and the injection of aldosterone, cortisol, and corticosterone could not restore phosphatidylcholine synthesis, whereas deoxycorticosterone was only slightly effective.

It seems that cyclic nucleotides may play a role in the regulation of phospholipid metabolism in kidney. Baricos (1972) has shown that dibutyryl cyclic 3′,5′-adenosine monophosphate (1 mM), caffeine (3 mM), theophylline (3 mM), and prostaglandin E_1 (10 μg/ml) all inhibited $[^{32}P]P_i$ incorporation into the phosphoinositides in renal cortex slices of the rabbit.

IV. *Cholesterol*

Cholesterol biosynthesis by kidney is less active than that of liver, gut, and skin (Srere *et al.*, 1950); however, kidney has been shown to be the primary site of the metabolism of circulating D-mevalonate (Elwood and

Van Bruggen, 1961; Duncan and Best, 1963; Edgren and Hellström, 1972; Gans *et al.*, 1973; Hellström *et al.*, 1973), which is a unique and a more direct precursor of squalene and cholesterol than is acetate (Bloch, 1959). Elwood and Van Bruggen (1961) reported that an average of 64% of the total body nonsaponifiable lipid radioactivity was recovered in the kidneys 2 hr after an intraperitoneal injection of DL-[2-^{14}C]mevalonate into the rat. In rabbits and mice, the incorporation of DL-[2-^{14}C]mevalonate into nonsaponifiable lipids of the kidneys was found to be greater than that of the liver (Edgren and Hellström, 1972). The pattern of distribution of labeled nonsaponifiable lipids differed markedly in these two organs. In rabbit liver, 80% of the nonsaponifiable label was present as a C_{27} sterol, presumably cholesterol, approximately 2 hr after an intravenous injection of DL-[2-^{14}C]mevalonate. During the same time period, most of the labeled nonsaponifiable lipids in the kidneys was found in squalene (13%) and lanosterol (34%); only 9% of this radioactivity was recovered in cholesterol (Edgren and Hellström, 1972). A similar finding was demonstrated in rat kidney. Hellström *et al.* (1973) showed that rat kidneys contained 13% of the radioactivity of the injected DL-[2-^{14}C]mevalonate as nonsaponifiable lipids, while only 1.5% of the administered ^{14}C was recovered in the nonsaponifiable lipids of the liver 30 min after intravenous administration of DL-[2-^{14}C]mevalonate. After subfractionation of the ^{14}C-labeled nonsaponifiable lipids from the kidney by thin-layer chromatography, 65% of the radioactivity was found to be incorporated into squalene, 21% in lanosterol, and only 10% in cholesterol. On the contrary, in the liver, cholesterol was found to be the chief product of the injected mevalonate (Hellström *et al.*, 1973).

In vitro studies by Bojesen *et al.* (1973) and Bojesen and Roepstorff (1973) have shown that renal papillae have an active sterologenesis when [2-^{14}C]acetate or [2-^{14}C]mevalonate is used as precursor. Nearly all the sterol radioactivity could be recovered from lanosterol (4,4,14α-trimethyl-5α-cholesta-8,24-dien-3β-ol), desmosterol (cholesta-5,24-dien-3β-ol), lathosterol (5α-cholest-7-3β-ol), and 5α-cholesta-7,24-dien-3β-ol; only a very small amount of labeled cholesterol could be detected in the sterol fraction.

The role of kidney in sterol metabolism was further substantiated by the effect of nephrectomy on the incorporation of DL-[2-^{14}C]mevalonate into nonsaponifiable lipids in the liver and blood. Cuzzopoli *et al.* (1972) demonstrated that in sham-operated rats the kidneys had the highest capacity *in vivo* to incorporate DL-[2-^{14}C]mevalonate into nonsaponifiable lipids with squalene and lanosterol as the major labeled lipid fraction. In nephrectomized rats, there was a marked increase of the labeled nonsaponifiable lipids of the liver and blood.

A direct enzymatic synthesis of cholesterol from squalene, lanosterol, and desmosterol has recently been studied in the microsomes from rat kidney (Johnson and Shah, 1974). The 105,000*g* supernatant fraction of kidney homogenates was found to contain two noncatalytic proteins that enhance the synthesis of cholesterol by kidney microsomes. One of these proteins is heat labile and is required for the cyclization of squalene to lanosterol by kidney microsomes, and the other is heat stable and participates in cholesterol synthesis from lanosterol and desmosterol. A detailed analysis of these two activators has not been performed.

The mechanism that regulates cholesterol synthesis in kidney tissue is not fully understood. Dietschy and Siperstein (1967) reported that the level of total digitonin-precipitable sterols and cholesterol in rat kidneys was not sensitive to cholesterol feeding and fasting. In contrast the incorporation of [2-^{14}C]acetate into total digitonin-precipitable sterols and cholesterol by liver slices from rats fed a cholesterol diet for 6 weeks was depressed about 100-fold. Hepatic cholesterologenesis was reduced nearly 11-fold after the rats were fasted for 48 hr. Dietschy and Wilson (1968) further reported that the incorporation of [2-^{14}C]acetate into digitonin-precipitable sterols by kidney slices from squirrel monkeys fed a cholesterol diet for 1 week was reduced 50% of that in the control animals. Under identical conditions, a 25-fold decrease in the labeling of digitonin-precipitable sterols by [2-^{14}C]acetate in liver slices was observed. It should be pointed out that the data of Dietschy and Siperstein (1967) indicated that the radioactivity in cholesterol fractions constituted 79% of that in the digitonin-precipitable sterol in rat kidney. However, recent studies by Edgren and Hellström (1972) and Bojesen *et al.* (1973) have shown that only small amounts of [2-^{14}C]acetate incorporated in the digitonin-precipitable sterols can be detected as cholesterol.

Cholesterol can be metabolized into cholesterol sulfate by kidney *in vitro* and *in vivo*. Rice *et al.* (1968) reported that rat kidney incorporated 10 times more radioactive sulfate into lipids than the liver did, 21 hr after a subcutaneous injection of [^{35}S]sulfate. Cholesterol sulfate was found to be the major labeled acidic lipid and cerebroside sulfate was the minor labeled lipid. Rice *et al.* (1968) also demonstrated the formation of cholesterol sulfate *in vitro*. The 105,000*g* supernatant fraction of rat kidney incorporated [^{35}S]sulfate into endogenous and exogenous cholesterol. In addition, Hochberg *et al.* (1974) have studied the uptake of cholesterol sulfate by rat kidney *in vivo*. One hour after an intravenous injection of ammonium [^{3}H]cholesterol sulfate, it was found that about one-half of the ^{3}H taken from the blood by the kidney appeared as the conjugate. The finding indicates that kidney has the ability to take up cholesterol sulfate from the blood and also has the ability to cleave the conjugate.

V. Prostaglandins

Several comprehensive review articles of prostaglandin metabolism and its biological functions in kidney have recently been published (McGiff and Itskovitz, 1973; McGiff *et al.*, 1974; Zins, 1975). In this chapter, however, we should like to mention briefly the general properties, biosynthesis, and catabolism of renal prostaglandins, and cite the subsequent papers that were not covered in previous reviews.

A. General Properties

The prostaglandins are basically 20-carbon monocarboxylic fatty acids containing a five-membered ring. Interest in renal prostaglandins began in 1965 when J. B. Lee *et al.* isolated three biologically active acidic lipids from rabbit renal medulla. These acidic lipids were identified as prostaglandin (PG) E_2, $PGF_2\alpha$, and PGA_2 (J. B. Lee, 1967; J. B. Lee *et al.*, 1967, Crowshaw and Szlyk, 1970). Their structures are shown in Fig. 3.

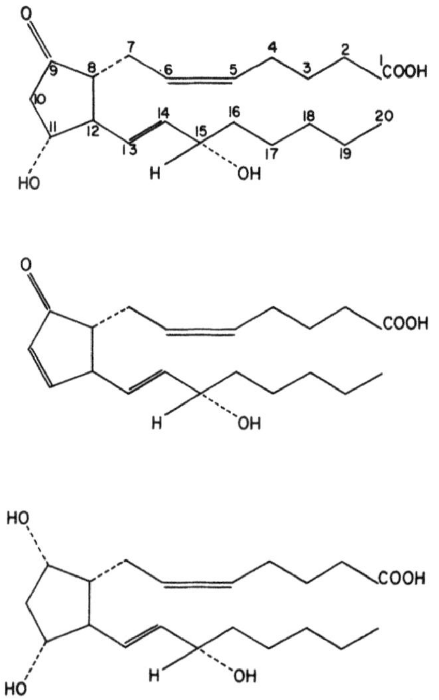

PGE_2 11α,15α - Dihydroxy - 9 - oxo - 5 *cis*, 13 *trans* - prostadienoic Acid

PGA_2 15α - Hydroxy - 9 - oxo - 5 *cis*, 10,13 *trans* - prostadienoic Acid

$PGF_{2\alpha}$ 9α, 11α, 15α - Trihydroxy - 5 - *cis* 13 *trans* - prostadienoic Acid

FIGURE 3. Structures of three renal prostaglandins.

PGE_2 is characterized by having a keto group at position 9, a hydroxyl at position 11, and two double bonds in its side chain. $PGF_2\alpha$ has hydroxyls at positions 9 and 11 and two double bonds in its side chain. PGA_2 is also called medullin (J. B. Lee *et al.*, 1965). It is the dehydrated product of PGE_2 and absorbs ultraviolet light at 217 mμ, hence it is also called PGE_2-217.

Renal prostaglandins, unlike other lipids, are localized primarily in the supernatant fraction of renal medulla, with a small portion associated with microsomes, mitochondria, and lipid droplets (Anggård *et al.*, 1972). PGE_2 is the major renal PG (Daniels *et al.*, 1967). It dilates renal medullary blood vessels and thereby regulates blood flow to the inner cortex. The participation of $PGF_2\alpha$ in intrarenal action has not been reported (McGiff *et al.*, 1974).

B. *Biosynthesis and Metabolism*

The biosynthesis of prostaglandins from arachidonic acid in kidney tissue was first demonstrated by Hamberg (1969) after incubation of tritiated arachidonic acid with the homogenates from rabbit renal medulla; most of the radioactivity (22–47%) was recovered as PGE_2 and a small portion (5–7%) was found in $PGF_2\alpha$. Only trace amounts of the radioactivity was identified as PGA_2, most of which was considered to be formed nonenzymatically from PGE_2 during the extraction procedure. It remains to be established whether PGA_2 originates in the kidney.

In kidney tissue, most of the arachidonic acid exists in the phospholipid fraction (Morgan *et al.*, 1963; Crowshaw, 1973). It would have to be cleaved by phospholipase A_2 in order to serve as a substrate for prostaglandin synthesis. Kunze and Vogt (1971) proposed that the phospholipase A_2 activity may be a rate-limiting step in the regulation of prostaglandin synthesis. In kidney, this hypothesis was supported by the observation of Kalisker and Dyer (1972) on an inhibition of prostaglandin release from slices of rabbit medulla incubated in a low calcium medium, and an increase of prostaglandin release from medulla slices exposed to cobra venom phospholipase A_2. Recently Danon *et al.* (1975) have reported that pancreatic lipase enhances prostaglandin synthesis by rat renal papilla *in vitro,* indicating that the arachidonic acid released from papillary triglycerides is utilized for prostaglandin synthesis.

After the arachidonic acid is released from phospholipids or triglycerides, it is rapidly converted to PGE_2 and $PGF_2\alpha$ by the enzyme complex, prostaglandin synthetase (Hamberg, 1969). Two endoperoxide intermediates, 15-hydroperoxy-9α, 11α-peroxidoprosta-5, 12-dienoic acid (PGG_2), and 15-hydroxy-9α,11α-peroxidoprosta-5,13-dienoic acid (PGH_2) were identified when arachidonic acid was incubated with the microsomes from vesicular glands of sheep (Hamberg *et al.*, 1974). However, PGG_2 and PGH_2 have not yet been identified in the renal system. The formation of the endoperoxide intermediates may be important in controlling the level of prostaglandins. Aspirin and indomethacin have been found to inhibit prostaglandin synthesis by preventing the formation of these intermediates (Hamberg *et al.*, 1974).

Prostaglandin synthetase is predominantly localized in the endoplasmic reticulum of the renal medulla and has been reviewed by McGiff *et al.* (1974) and Zins (1975). Catecholamines, angiotensins I and II, vasopressin, and bradykinin promote prostaglandin synthesis by the kidney, and these studies have been reviewed by Zins (1975). Dibutyryl cyclic AMP (10^{-2} M) was reported to reduce the release of prostaglandins from slices of rabbit renal medulla *in vitro* (Kalisker and Dyer, 1972), but the mechanism of inhibition remains to be studied. Recently, Danon *et al.* (1975) using an isotope dilution technique have shown that angiotensin II stimulates the formation of both PGE_2 and $PGF_2\alpha$ in rat renal papilla by increasing the availability of arachidonic acid. These investigators speculate that angiotensin II may activate a specific acyl hydrolase in the kidney to make arachidonic acid available for prostaglandin synthesis.

The catabolism of prostaglandins by the kidney from swine and rabbit takes place primarily in the cortex and has been reviewed by McGiff *et al.* (1974) and Zins (1975). Prostaglandins in the renal cortex are first metabolized by NAD^+-dependent 15-hydroxyprostaglandin dehydrogenase in the cell cytoplasm, and then by NADH-dependent prostaglandin Δ^{13}-reductase. These two enzymes have also been demonstrated in the soluble fraction of rat kidney cortex (Pace-Asciak and Domazet, 1975). The activities of 15-hydroxyprostaglandin dehydrogenase and prostaglandin Δ^{13}-reductase in rat kidney homogenates vary with animal development (Pace-Asciak, 1975b); both enzymes reached maximal activity 19 days postnatally, followed by a decline to adult values by 40 days after birth. Recently S.-C. Lee *et al.* (1975) identified two types of 15-hydroxyprostaglandin dehydrogenase in swine kidney. Type I is NAD^+-dependent and inhibited by NADH, but not by NADPH; type II is $NADP^+$-dependent and inhibited by NADPH, but not by NADH. The cortex was found to contain eleven times as much NAD^+-dependent 15-hydroxyprostaglandin dehydrogenase activity as the medulla, while about twice as much $NADP^+$-dependent dehydrogenase activity was found in the medulla than in the cortex. Both types of dehydrogenase have been partially purified from the particle-free supernatant fraction of cortex and medulla from swine kidneys.

PGE_2 in the kidney can be metabolized to $PGE_2\alpha$. S.-C. Lee and Levine (1974) reported the presence of a cytoplasmic NADPH-dependent PGE_2 9-keto-reductase, which converts the 9-keto group of PGE_2 to form

$PGF_{2}\alpha$ in monkey kidney homogenate. The enzyme utilized NADPH much more efficiently than NADH. S.-C. Lee *et al.* (1975) have recently shown that PGE_2-9-keto-reductase is equally distributed in the swine kidney cortex and medulla. On the other hand, $PGF_{2}\alpha$ can be metabolized to the PGE_2 metabolite by adult rat kidney. Pace-Asciak and Miller (1974) incubating $PGF_{2}\alpha$ with rat kidney homogenates identified 15-keto-13,14-dihydro-PGE_2 as the major product. The conversion of F-type to E-type prostaglandin was found to take place at the 15-keto-13,14-dihydro-$PGF_{2}\alpha$ step and was catalyzed by 9-hydroxyprostaglandin dehydrogenase (Pace-Asciak, 1975a). The enzyme is located primarily in the cortex of rat kidney with most of its activity in the high-speed supernatant fraction (Pace-Asciak and Domazet, 1975); the dehydrogenase is absent in newborn rat kidney.

A probable scheme for prostaglandin biosynthesis and metabolism by kidney is shown in Fig. 4. Prostaglandins are not stored intrarenally but are

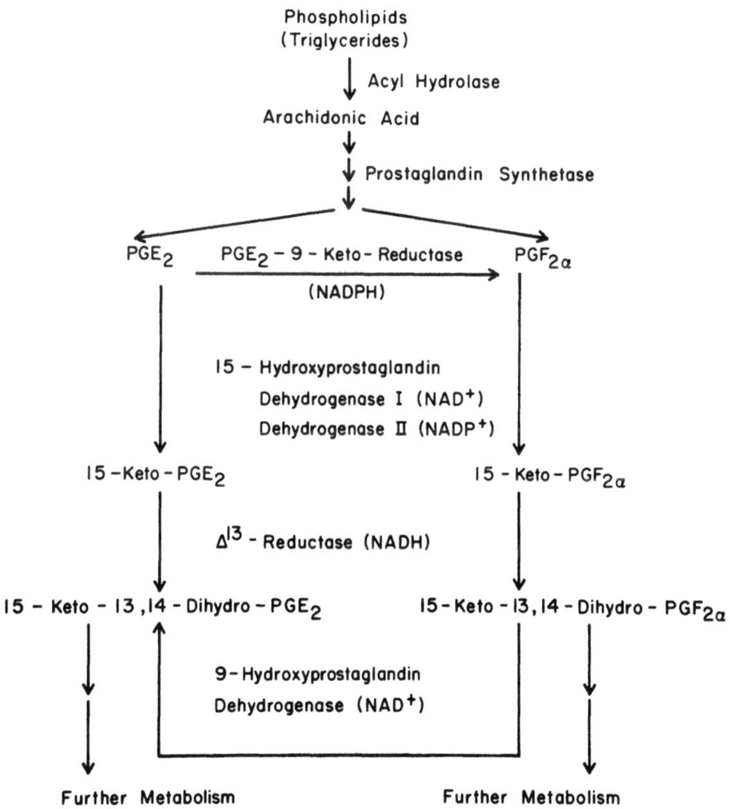

FIGURE 4. Probable scheme for renal prostaglandin biosynthesis and metabolism.

released into the extracellular compartment after their formation (Crowshaw, 1973). It appears that under physiological conditions the availability of NADH, NAD^+, NADPH, and $NADP^+$ in the kidney plays an important role in controlling the metabolism of renal prostaglandins.

VI. Sphingolipids

Interest in sphingolipid metabolism has increased rapidly during the last decade, because normal glycosphingolipid metabolism is altered in the kidney in the presence of several lipid-storage diseases. However, most of the investigations on sphingolipids in kidneys were focused on qualitative and quantitative analyses and structure eludications. The metabolic pathways of these lipids in kidneys are incompletely known. In this chapter we shall separately discuss long-chain bases, ceramide, sphingomyelin, neutral glycosphingolipids, sulfatides, and gangliosides.

A. Long-Chain Bases

Kidneys from humans (Karlsson, 1966; Mårtensson, 1966a; Karlsson and Mårtensson, 1968; Karlsson *et al.*, 1968a), cattle (Karlsson and Steen, 1968; Carter and Hirschberg, 1968), and rats (Carter and Hirschberg, 1968) contain three major long-chain bases: (1) sphingosine or 4*t*-sphingenine (D-erythro-1, 3-dihydroxy-2-amino-4-*trans*-octadecene); (2) dihydrosphingosine or sphinganine (D-*erythro*-1, 3-dihydroxy-2-aminooctadecane); and (3) phytosphingosine or 4D-hydroxysphinganine (D-ribo-1,3,4-trihydroxy-2-aminooctadecane).

Phytosphingosine was shown to make up half of the base fraction of human kidney cerebrosides, but to be practically absent from brain cerebrosides (Karlsson, 1964). In sphingomyelin from bovine kidney, the three major bases were found in an approximate ratio of 20:2:3 (Karlsson *et al.*, 1973c). Although Carter and Hirschberg (1968) reported the absence of detectable phytosphingosine in the sphingomyelin fraction from bovine and rat kidneys, Karlsson and Steen (1968) demonstrated the existence of this long-chain base in sphingomyelin of bovine kidney. In sphingomyelin from bovine kidney (Karlsson and Steen, 1968) and in the Forssman hapten from horse kidney (Karlsson *et al.*, 1974), it has been shown that phytosphingosine preferentially combines with longer-chain fatty acids (C_{24}). A small amount (13.4% of total long-chain base) of branched-chain sphingosine was also found in sphingomyelin from bovine kidney (Carter and Hirscherg, 1968).

The biosynthesis and degradation of sphingosine bases have been studied in yeast, brain, and liver of the rat and have been recently reviewed by Morell and Braun (1972). However, they have not been examined in kidneys of the mammals. Assmann and Stoffel (1972) using radioactive phytosphingosine studied its fate in rats following oral and intravenous administration. Phytosphingosine was found to be incorporated particularly into sphingomyelin and cerebrosides of the intestinal wall, liver, and kidney. The activity was found primarily in phytosphingosine and a small portion in pentadecanoic acid, a catabolite of phytosphingosine, but not in sphingosine and dihydrosphingosine. These results led Assmann and Stoffel (1972) to conclude that phytosphingosine, present in the sphingolipids of the mammalian cell, originates from dietary sources. However, a recent finding of Karlsson *et al.* (1973c) and Morrison (1973) on the very similar paraffin chain distributions of dihydroxy and triphydroxy bases suggests that they may be synthesized in bovine kidney from a common pool of fatty acids.

B. Ceramide

Ceramide plays a central role in sphingolipid metabolism. It is the basic lipid moiety of the sphingolipids and may be the precursor of sphingomyelin, cerebrosides, sulfatides, and gangliosides (Morell and Braun, 1972). Free ceramide is present in considerable amounts in kidney. In bovine kidney, they decreased in concentration from cortex (1.0 mg/g dry wt) to papilla (0.5 mg/g dry wt) (Karlsson *et al.*, 1973b). Using high performance liquid chromatography, Sugita *et al.*, (1974) analyzed the ceramide level in human tissues, and in serum and urine. Kidney tissue contained the highest level of free ceramide among lung, stomach, cerebellum, and heart from normal subjects and patients with Fabry's disease. Normal whole kidney in humans contained approximately 95 μg/g wet tissue of total ceramide, predominantly nonhydroxy fatty acid-containing ceramide. In kidneys obtained from individuals with Fabry's disease, the total ceramide content amounted to 515 μg/g wet tissue. Only trace amounts of hydroxy fatty acid-containing ceramide were detected in normal human kidneys, whereas large amounts were found in kidneys from patients with Fabry's disease (Sugita *et al.*, 1973; 1974). The hydroxy fatty acids from these kidneys consist mostly of longer-chain hydroxy fatty acids ($C_{22:0}$ and above) predominantly $C_{22:0}$. The major nonhydroxy fatty acid constituent of ceramide isolated from normal and diseased (Fabry) kidney is $C_{24:1}$ (Sugita *et al.*, 1974).

Detailed studies on the biosynthesis and degradation of ceramide in kidney tissue have not been reported. It has only been shown that ceramide is hydrolyzed in kidney by a ceramidase to form long-chain base and free fatty acid. The kidney from patients with Fabry's disease has very low activity of this enzyme (Sugita *et al.*, 1972).

C. Sphingomyelin

Sphingomyelin represents the major sphingolipid fraction from bovine kidney, where it is 14.4 mg/g dry weight (Karlsson *et al.*, 1973c). The concentration of sphingomyelin is decreased from cortex to papilla. Morgan *et al.* (1963) found that the sphingomyelin isolated from the cortex of rabbit kidney is markedly saturated while the sphingomyelin in the medulla has more unsaturated fatty acids. Recently Soula *et al.* (1972) have reported the existence of two kinds of sphingomyelin in both cortex and medulla of human kidney. These two sphingomyelins differ in their fatty acid composition—the ratio of saturated fatty acids to unsaturated fatty acids is 1.5 in the sphingomyelin I and 10 in the sphingomyelin II.

Employing $[^{32}P]P_i$ as a tracer, Tinker and Hanahan (1966) found that the phosphate moiety in sphingomyelin is metabolically inactive in rabbit kidney cortices. In studies of the enzymatic formation of sphingomyelin in mouse kidney Ullman and Radin (1974) determined that the phosphorylcholine moiety in 16:0 sphingomyelin was derived from phosphatidyl [1, 2-^{14}C]choline. No cofactors were required for the transfer of phosphorylcholine from phosphatidylcholine to 16:0 ceramide. Kidney microsomes had the highest transferase activity among lung, liver, spleen, and heart (in decreasing order of activity), but activity was not detected in brain. Enzymatic hydrolysis of sphingomyelin by kidney tissue was reported by Weinreb *et al.* (1968). Kidney lysosomes contained the highest specific activity of sphingomyelinase among the subcellular fractions.

D. Neutral Glycosphingolipids

Five neutral glycosphingolipid fractions have been found in kidney tissue: (1) monohexosylceramide (cerebroside), (2) dihexosylceramide, (3) trihexosylceramide, (4) tetrahexosylceramide (aminoglycolipid, or globoside), and (5) pentahexosylceramide (Forssman hapten). The first four glycosphingolipids have been isolated from kidneys of humans (Mårtensson, 1963b; Makita, 1964; Mårtensson, 1966a) and several mouse strains (Adams and Gray, 1968). Forssman hapten has been isolated from horse kidney (Makita *et al.*, 1966) and canine kidney (Sung *et al.*, 1973). The distribution of glycosphingolipids appears to be specific for different species and organs. In rat kidney, the main neutral glycosphingolipid is monohexosylceramide (glucosylceramide and galactosylceramide); dihexosylceramide, trihexosylceramide, and tetrahexosylceramide are smaller fractions (Kawanami, 1968). Tetrahexosylceramide has not been detected in bovine kidney (Karlsson *et al.*, 1973b). In human kidney, tetrahexosylceramide is the dominating neutral glycosphingolipid (Mårtensson, 1966a), while in brain tissue galactosylceramide is essentially the only neutral glycosphingolipid (Svennerholm, 1964). Forssman hapten is a minor constituent of glycosphingolipids in canine kidney (0.008 μmoles/g wet wt), but it is the major constituent in canine intestine (0.110 μmoles/g wet wt) (Sung *et al.*, 1973).

The quantity of neutral glycosphingolipids in human kidney varies with age (Mårtensson, 1966a). A higher concentration (4.85 mg/g of dry wt) of these lipids was found in the juvenile group (10–17 years) than in senile and middle aged material. However, the relative distribution of the neutral glycosphingolipids is unchanged. Furthermore, sex also influences the distribution of these lipids. For instance, a substantial amount of dihexosylceramide is in the kidney of male mice and testosterone-treated female mice, but only trace amounts in the kidneys of female mice (Coles and Gray, 1970; Coles *et al.*, 1970; Gray, 1971).

All the neutral glycosphingolipids from human kidney share a similar pattern of long-chain base and fatty acid composition (except glucosylceramide) (Mårtensson, 1966a). They all contain C_{16}- to C_{20}-sphingosines, with C_{18}-sphingosines as the major fraction. The fatty acid composition of the neutral glycosphingolipids shows a characteristic common pattern with $C_{22:0}$, $C_{24:0}$, and $C_{24:1}$ as the predominant acids. This may indicate that they are metabolically related to each other. Tetrahexosylceramide and trihexosylceramide have an almost identical fatty acid composition. However, the fatty acid composition of glucosylceramide is distinctly different from that of trihexosyl- and tetrahexosylceramides in that the former contains a much higher content of hydroxy fatty acids (Mårtensson, 1966a).

1. Monohexosylceramide (Cerebroside)

Both glucosyl(1 $\xrightarrow{\beta}$ 1)ceramide and galactosyl(1 $\xrightarrow{\beta}$ 1)ceramide have been found in kidneys of human (Mårtensson, 1963b; Makita, 1964; Mårtensson, 1966a), mice (Adams and Gray, 1968), rats (Kawanami, 1968), and cattle (Karlsson *et al.*, 1973b). In bovine kidney they exist in different anatomical regions. Galactosylceramide is concentrated in the medulla (1.3 mg/g dry tissue) and glucosylceramide in the papilla (1.4 mg/g dry tissue). Whole bovine kidney contains twice as much glucosylceramide as galactosylceramide, with an average concentration of 1.2 mg/g dry weight.

Kidney monohexosylceramide is very heterogeneous with respect to its *N*-acyl fatty acids and long-chain bases. Bovine monohexosylceramide contains both saturated and unsaturated hydroxy and nonhydroxy fatty acids with 16–22 carbons, and dihydroxy and trihydroxy long-chain bases (Karlsson *et al.*, 1973a). Of the total monohexosylceramide, 67% are glucosylceramide with phytosphingosine and 2-hydroxydocosanoic acid, and 23% are galactosylceramide with sphingosine and 2-hydroxyhexadeconoic acid.

The synthesis of galactosylceramide in mouse kidney is catalyzed by an enzyme, ceramide: UDP galactosyltransferase (Constantino-Ceccarini and Morell, 1973) as shown in reaction (1):

$$\text{ceramide} + \text{UDP-galactose} \rightarrow \text{galactosylceramide} + \text{UDP} \qquad (1)$$

The kidney enzyme in the particulate fraction showed a marked specificity for the galactosylation of ceramide with 2-hydroxy fatty acids as compared to ceramide with nonhydroxy fatty acids (Coles and Gray, 1970; Constantino-Ceccarini and Morell, 1973).

Constantino-Ceccarini and Morell (1973) have shown that the synthesis of glucosylceramide in kidney is similar to that of galactosylceramide. When UDP-glucose was substituted for UDP-galactose, the product was glucosylceramide. Both 2-hydroxy fatty acid- and nonhydroxy fatty acid-ceramide were effective acceptors for glucose incorporation.

The levels of the two enzymes in mouse kidney are sex dependent and vary markedly during development (Constantino-Ceccarini and Morell, 1973). Galactosylceramide synthesis activity was increased fivefold in preparations from male kidneys of mice between 20 and 30 days of age, while the enzyme activity in kidneys of female mice was decreased slowly from 10 to 70 days of age. Glucosylceramide synthesis in both male and female kidneys of mice was increased during the development (10–64 days of age), but the enzyme activity in female kidney was lower than that in male kidney.

Esselman *et al.* (1973) reported that the concentration of glucosylceramide was increased, while that of galactosylceramide was decreased in allograft-rejected canine kidney. This observation presents another example of the difference in the biosynthetic activity of galactosylceramide and glucosylceramide.

An alternative biosynthetic pathway that may lead to the formation of galactosylceramide is galactosylation of a long-chain base to form psychosine followed by acylation according to reactions (2) and (3):

$$\text{long-chain base} + \text{UDP-galactose} \rightarrow \text{psychosine} + \text{UDP} \qquad (2)$$

$$\text{psychosine} + \text{acyl-CoA} \rightarrow \text{galactosylceramide} + \text{CoA} \qquad (3)$$

This pathway [reactions (2) and (3)] has not been examined in kidney tissue, but Miyatake and Suzuki (1972) have recently reported an extremely low activity of psychosine galactosidase in kidney of patients with globoid cell leukodystrophy (Krabbe's disease), a genetically determined, rapidly fatal neurological disorder of infants. The activity of psychosine galactosidase in kidneys from normal controls is ten times that of kidneys from patients with globoid cell leukodystrophy. The diseased kidney is also deficient of galactosylceramide β-galactosidase (Suzuki, 1971; Austin *et al.*, 1970), but not glucosylceramide β-galactosidase (Austin *et al.*, 1970). Despite the enzymatic block for the degradation of galactosylceramide, both glucosylceramide and galactosylceramide increase in the pathological kidney, and the ratio of these two compounds remained essentially unchanged (Suzuki, 1971). An elevated level of both glucosylceramide and galactosylceramide was also found in kidney of patients with lactosylceramidosis (Dawson, 1972).

Both glucosylceramide and galactosylceramide can be hydrolyzed by the respective β-galactosidase in rat kidney to produce ceramide and hexose. Both enzymes are localized in the lysosomes (Weinreb *et al.*, 1968).

2. *Dihexosylceramide*

Dihexosylceramide in human kidney is a mixture of galactosyl(1 $\xrightarrow{\alpha}$ 4)-galactosyl(1 $\xrightarrow{\beta}$ 1)ceramide and galactosyl(1 $\xrightarrow{\beta}$ 4)glucosyl(1 $\xrightarrow{\beta}$ 1)ceramide (lactosylceramide) (Makita and Yamakawa, 1964; Mårtensson, 1966a). Digalactosylceramide is the predominant dihexosylceramide in kidneys of several mouse strains (Adams and Gray, 1968). In humans, it is present only in kidney (Sweeley and Klionsky, 1963; Miyatake, 1969; Schibannoff *et al.*, 1969) and intestine (Dawson, 1972) in significant amounts. Digalactosylceramide and trihexosylceramide accumulate in kidney from patients with Fabry's disease (Sweeley and Klionsky, 1963, 1966). They have also been found in the spinal ganglia of a patient with Fabry's disease (Miyatake and Ariga, 1972). Its anomeric structure has only recently been elucidated through the contribution of Li *et al.* (1972) by using purified enzymes. These investigators employed fig α-galactosidase to liberate the terminal galactose of digalactosylceramide and jack bean β-galactosidase to liberate the penultimate galactose residue after previously incubating with α-galactosidase. These data led Li *et al.* (1972) to conclude that the anomeric configuration of the terminal galactosyl residue in digalactosylceramide is α and that of the penultimate galactose is β. Thus the structure of digalactosylceramide is galactosyl(1 $\xrightarrow{\alpha}$ 4)galactosyl(1 $\xrightarrow{\beta}$ 1)ceramide.

Lactosylceramide is biosynthesized from UDP-galactose and glucosylceramide in kidney homogenates from rat (Hildebrand and Hauser, 1969) and C57/BL mouse (Coles and Gray, 1970). The synthesis of lactosylceramide was depressed in kidneys of mice bearing BP8 ascites tumor (Hay and Gray, 1970). Kidney homogenates from C57/BL mouse also catalyzed the synthesis of digalactosylceramide in the presence of UDP-galactose and galactosylceramide (Coles and Gray, 1970). Recently, Mårtensson *et al.*, (1974) characterized a galactosyltransferase in rat kidney that catalyzes the transfer of galactose from UDP-galactose to galactosylceramide. Unlike the glucosylceramide galactosyltransferase, the galactosylceramide galactosyltransferase was detected only in rat kidney microsomes, not in brain, spleen, liver, or lung. The enzyme attaches the terminal galactose in an α-linkage; it requires manganese ions for activity and shows a pH optimum of 6.0. The concentration of digalactosylceramide was found to be decreased in rejected canine kidney, while that of lactosylceramide increased (Esselman *et al.*, 1973). An elevated level of lactosylceramide was also found in kidneys from a patient with lactosylceramidosis (Dawson, 1972).

Lactosylceramide β galactosidase was demonstrated in kidney tissue (Sandhoff *et al.*, 1964). The enzyme hydrolyzes lactosylceramide producing glucosylceramide and galactose.

3. *Trihexosylceramide*

Trihexosylceramide is a minor neutral glycosphingolipid in rat (Kawanami, 1968) and bovine kidneys (Karlsson *et al.*, 1973b), but in normal human kidney, the concentration of trihexosylceramide is only second to aminoglycolipid (Mårtensson, 1966a). Trihexosylceramide and digalactosylceramide accumulate in the kidney (Sweeley and Klionsky, 1963, 1966) and spinal ganglion (Miyatake and Ariga, 1972) of patients with Fabry's disease. The concentration of trihexosylceramide was found to be increased in rejected canine kidney (Esselman *el at.*, 1973) and the synthesis of trihexosylceramide is impaired by polyoma transformation of baby hamster kidney cells (Hammarström and Bjursell, 1973).

In 1964, Makita and Yamakawa described the chemical structure of trihexosylceramides from normal human kidney as galactosyl(1 → 4)-galactosyl(1 → 4)glucosyl(1 → 1)ceramide. In 1971, Li and Li using fig α-galactosidase and jack bean β-galactosidase determined the anomeric linkages of the carbohydrate moiety in trihexosylceramides from the kidney of patient with Fabry's disease and normal human kidney. They found that trihexosylceramides from both sources contain terminal α-galactosyl and penultimate β-galactosyl linkages. From the nuclear magnetic resonance spectra of the intact molecules, optical rotatory data, and enzymatic data (partially purified α-galactosidase from green coffee beans), Clarke *et al.* (1971) also demonstrated the presence of a terminal α-D-galactopyranosyl residue in trihexosylceramides from kidneys of normal subjects and patients with Fabry's disease.

The synthesis of trihexosylceramide was demonstrated in rat kidney homogenates (Hildebrand and Hauser, 1969) and rat kidney microsomes (Mårtensson *et al.*, 1974) according to reaction (4):

Lactosylceramide + UDP-galactose → trihexosylceramide + UDP (4)

Kinetic studies showed this galactosyltransferase to be different from that catalyzing the transfer of galactose from UDP-galactose to galactosylceramide. The specific activities of both enzymes were found to decrease with age, and the total activities per animal increased with age (Mårtensson *et al.*, 1974). Employing the method of permethylation, Stoffyn *et al.* (1974) have shown that the synthesis of trihexosylceramide *in vitro* from lactosylceramide and UDP-galactose by rat kidney microsomes produces a mixture of two isomers. In one of them the galactose molecules are linked 1 → 3 and in the other 1 → 4.

Trihexosylceramide can be hydrolyzed by a ceramide trihexosidase in rat kidney homogenates to form lactosylceramide and galactose (Brady *et al.*, 1967). A 16,072-fold purified ceramide trihexosidase from human kidney has been achieved by using affinity chromatography (Mapes and Sweeley, 1973). Kano and Yamakawa (1974a) using isoelectric focusing demonstrated the presence of a heat-stable α-galactosidase and a heat-labile α-galactosidase in normal human kidney. Both forms of α-galactosidase exhibit ceramide trihexosidase activity. Kano and Yamakawa (1974b) have further demonstrated that there is a greater reduction in ceramide trihexosidase activity in the kidney than in the liver of patients with Fabry's disease. On isolectric focusing, α-galactosidase from the kidney and liver of patients with Fabry's disease appeared as a single component corresponding to the heat-stable fraction.

4. *Tetrahexosylceramide (Aminoglycolipid, Cytolipin K, or Globoside)*

Tetrahexosylceramide has been isolated from kidneys of humans (Makita *et al.*, 1964; Rapport *et al.*, 1964), rats (Kawanami, 1968), and mice (Adams and Gray, 1968). It is the major glycosphingolipid in human kidney (1.5–2.5mg/g dry tissue)(Mårtensson, 1966a). Rapport *et al.* (1964) isolated tetrahexosylceramide from normal human kidney as a cellular hapten and named it "cytolipin K." The tetrahexosylceramide isolated from human erythrocytes and pig erythrocytes was termed "globoside" (Yamakawa and Suzuki, 1952; Yamakawa *et al.*, 1965).

Siddiqui *et al.* (1972), using the methods of methylation, periodate oxidation, and sequential degradation by β-*N*-acetylhexosaminidase, α-galactosidase, and β-galactosidase demonstrated two isomers of tetrahexysylceramides in rat kidney. The two isomers have identical carbohydrate sequences and anomeric structures. The major part of them is shown in structure (a), and the minor part is shown in structure (b), which has galactosyl ($1 \xrightarrow{\alpha} 4$) instead of galactosyl ($1 \xrightarrow{\alpha} 3$) at the penultimate residue:

(a) *N*-acetylgalactosamine($1 \xrightarrow{\beta} 3$)galactosyl($1 \xrightarrow{\alpha} 3$)galactosyl($1 \xrightarrow{\beta} 4$)-glucosyl($1 \xrightarrow{\beta} 1$)ceramide
(b) *N*-acetylgalactosamine($1 \xrightarrow{\beta} 3$)galactosyl($1 \xrightarrow{\alpha} 4$)galactosyl($1 \xrightarrow{\beta} 4$)-glucosyl($1 \xrightarrow{\beta} 1$)ceramide

Assuming that tetrahexosylceramide can be metabolized to form trihexosylceramide after removal of the terminal hexosamine, the recent finding by Stoffyn *et al.* (1974) of two isomers of trihexosylceramide synthesized by rat kidney microsomes may correlate with the existence of the two isomers of tetrahexosylceramides in rat kidney.

The synthesis of tetrahexosylceramide by kidney tissue has not yet been studied. Hydrolysis of the terminal amino sugar from tritium-labeled globoside was observed in the supernatant fraction from normal human kidney homogenates in 0.1 M citrate phosphate buffer, pH 4.5 (Sandhoff *et al.*, 1968a). This enzyme, β-*N*-acetylhexosaminadase, was found to be deficient in kidney of Tay-Sachs disease with visceral involvement, when either tritium-labeled globoside or *p*-nitrophenyl-β-D-acetylhexosaminide was used as substrate (Sandhoff *et al.*, 1968b; Sandhoff, 1969).

5. *Pentahexosylceramide (Forssman Hapten)*

In 1911, Forssman first demonstrated the presence of a heterophile antigen in sheep red blood cells and in various organs of guinea pig. Makita *et al.* (1966) proposed the structure of the Forssman hapten from horse kidney as an anomer of a globoside, i.e., a terminal α compared with a β linkage in *N*-acetylgalactosaminyl($1 \rightarrow 3$)galactosyl($1 \rightarrow 4$)galactosyl-($1 \rightarrow 4$)glucosylceramide. Sung *et al.* (1973) isolated from canine kidney a Forssman hapten that was a pentahexosylceramide. The concentration of Forssman hapten was increased in an allograft-rejected canine kidney (Esselman *et al.*, 1973). The structure of this pentahexosylceramide was *N*-acetylgalactosaminyl($1 \xrightarrow{\alpha} 3$)*N*-acetylgalactosaminyl($1 \xrightarrow{\beta} 3$)galactosyl-($1 \xrightarrow{\alpha} 4$)galactosyl($1 \xrightarrow{\beta} 4$)glucosyl($1 \xrightarrow{\beta} 1$)ceramide (Sung *et al.*, 1973) as proposed for horse spleen by Siddiqui and Hakomori (1971).

Additional evidence for the Forssman hapten from kidney tissue to be a pentahexosylceramide was obtained from a recent study of this glycolipid by Karlsson *et al.* (1974). These investigators analyzed the Forssman hapten isolated from horse kidneys by direct inlet mass spectrometry of undegraded lipid derivatives and indicated that the glycolipid was a pentaglycosylceramide with the sequence hexosamine–hexosamine–hexose–hexose–hexose–ceramide.

Recently Israel *et al.* (1974) have characterized a Forssman hapten *N*-acetyl-α-D-galactosaminidase in the high-speed supernatant fraction of the frozen-thawed and sonicated crude mitochondrial fraction from brains and kidneys of adult rats. The enzyme catalyzes the cleavage of the terminal *N*-acetyl α-galactosamine from Forssman hapten. It required taurocholate for full activity and had an optimal pH of 4.4. The highest activity of the Forssman hapten-cleaving enzyme was found in the kidney and the lowest in the heart muscle. During development, the enzyme in both brain and kidney gradually declined in specific activity as the animal matured.

E. Sulfatides

Sulfatides are sulfuric acid ester-containing glycosphingolipids. Next to nervous tissue, they occur in highest concentration in the kidney (Moser, 1972). Among the subcellular fractions of rat kidney, sulfatides are concentrated in the Golgi apparatus, though they are also found in mitochondria, rough microsomes, and plasma membranes (Zambrano *et al.*, 1975). Patients with metachromatic leukodystrophy (Austin, 1959; Malone and Stoffyn, 1966; Mårtensson *et al.*, 1966; Philippart *et al.*, 1971), a genetically determined disorder, show an accumulation of sulfatides in the nervous system, kidney, gallbladder, and liver. It also increased in kidneys of C3H mice carrying BP8 ascites tumors (Adams and Gray, 1967).

Mårtensson (1963a) isolated two different sulfatides from normal human kidney. One of them is ceramide-galactose-sulfate (galactosyl sulfatide); the other is ceramide-glucose-galactose-sulfate (lactosyl sulfatide). Galactosyl sulfatide was also isolated from rat kidney (Kawanami, 1968) and was found concentrated in the medulla of bovine kidneys (Karlsson *et al.*, 1968b, 1973b). The sulfate group is found in position 3 of the galactose moiety in galactosyl sulfatide (Makita and Yamakawa, 1964; Malone and Stoffyn, 1966) and lactosyl sulfatide (Mårtensson, 1966b; Stoffyn *et al.*, 1968).

Galactosyl sulfatide is the major sulfatide in human kidney and its concentration is about the same in juvenile (0.48 mg/g dry tissue weight) and senile (0.39 mg/g dry tissue weight) human kidneys (Mårtensson, 1966b; Mårtensson *et al.*, 1966). Lactosyl sulfatide is a minor sulfatide in human kidney. Its concentration was decreased with age, from senile, middle-aged, and juvenile kidneys 0.14, 0.18, and 0.2 mg/g dry tissue weight were isolated, respectively (Mårtensson, 1966b). In kidneys from patients with metachromatic leukodystrophy, the level of both sulfatides was increased; lactosyl sulfatide was about one-fourth the level of galactosyl sulfatide (Mårtensson *et al.*, 1966). Lactosyl sulfatide has not been detected in bovine kidney (Karlsson *et al.*, 1973b).

The fatty acid composition and long-chain base profile of kidney sulfatides resemble those of the corresponding neutral glycosphingolipids (Mårtensson, 1966b). Mårtensson *et al.* (1966) reported an increase in hydroxy acids and normal C-22 acids and a decrease in monounsaturated acids in both sulfatides in kidney from patients of late infantile metachromatic leukodystrophy when compared to normal juvenile kidney sulfatides. However, Malone and Stoffyn (1966) showed that the sulfatides accumulated in kidneys from patients with metachromatic leukodystrophy had the same fatty acid pattern as those found in normal kidney tissue.

The synthesis of galactosyl sulfatide has been demonstrated *in vivo* in rat (McKhann *et al.*, 1965; McKhann and Ho, 1967) and mouse (Sarlieve *et al.*, 1971) kidneys. Galactosyl sulfatide can be formed by the transfer of sulfate from 3′-phosphoadenosine 5′-phosphosulfate (PAPS) to galactosylceramide as shown in reaction (5):

$$\text{PAPS} + \text{galactosylceramide} \rightarrow \text{galactose-3}'\text{-sulfate-ceramide} \qquad (5)$$

The enzyme involved is PAPS:cerebroside sulfotransferase and has been found in the microsomal fraction of kidneys from rats (McKhann *et al.*, 1965; McKhann and Ho, 1967) and mice (Sarlieve *et al.*, 1971). The enzyme activity increases steadily with kidney weight and remained at a constant level in the adult (McKhann and Ho, 1967). Stoffyn *et al.* (1971) examined the structure of galactosyl sulfatide synthesized by rat microsomes *in vitro* and found the sulfate group only at C-3 of the galactose moiety. Fleischer and Zambrano (1974) demonstrated PAPS:cerebroside sulfotransferase to be localized in the Golgi apparatus in kidney cells from rats. Both hydroxy-fatty-acid- and normal-fatty-acid-containing cerebrosides are sulfated.

A direct enzymatic synthesis of lactosyl sulfatide in kidney tissue has not been studied, though it was demonstrated in rat brain (Farrell and McKhann, 1971).

The catabolism of sulfatides in kidney tissue has been studied *in vitro*. Mehl and Jatzkewitz (1964) demonstrated the hydrolysis of galactosyl sulfatide by a lysosomal sulfatase in pig kidney with the formation of galactosylceramide and sulfate. The enzyme activity is highest in kidney, and lowest in brain. A 6,000-fold purification of the enzyme was achieved. It is of interest that the activity of the purified enzyme depends on the presence of at least two higher molecular weight components, one heat labile and the other heat stable (Mehl and Jatzkewitz, 1964). The heat-labile component is deficient in metachromatic leukodystrophy kidney (Mehl and Jatzkewitz, 1965).

F. Gangliosides

Ganglioside is the generic name for a glycosphingolipid-containing sialic acid. It was originally defined by Klenk (1942) as a lipid substance predominantly located in the glial cells. It is now known that gangliosides are also present in extraneural tissues. Gangliosides have been detected in kidneys of guinea pigs (Wherrett and Cumings, 1963), humans (Makita, 1964), and rats (Kawanami, 1968). Puro *et al.* (1969) reported ganglioside values in kidneys of rat, rabbit, and pig ranging from 10 to 16 mg of lipid-bound neuraminic acid per 100 g of dry weight.

Three major gangliosides were isolated from bovine kidney (Puro, 1969). The structures of these gangliosides are

(a) *N*-acetylneuraminyl(2 $\xrightarrow{\alpha}$ 3)galactosyl(1 $\xrightarrow{\beta}$ 4)glucosyl(1 $\xrightarrow{\beta}$ 1)ceramide
(b) *N*-glycolylneuraminyl(2 $\xrightarrow{\alpha}$ 3)galactosyl(1 $\xrightarrow{\beta}$ 4)glucosyl(1 $\xrightarrow{\beta}$ 1)-ceramide
(c) *N*-acetylneuraminyl(2 $\xrightarrow{\alpha}$ 8)*N*-acetylneuraminyl(2 $\xrightarrow{\alpha}$ 3)galactosyl-(1 $\xrightarrow{\beta}$ 4)glucosyl(1 $\xrightarrow{\beta}$ 1)ceramide

The fatty acid composition and long-chain base pattern of these gangliosides have been analyzed and characterized by Puro and Keranen (1969). Kidney gangliosides contain both normal fatty acids and 2-hydroxy fatty acids. Palmitic, stearic, behenic, and lignoceric acid constitute the major normal fatty acids. The 2-hydroxy fatty acids consist predominantly of $C_{20:0}$, $C_{22:0}$, $C_{23:0}$, and $C_{24:0}$. The major long-chain bases in bovine kidney gangliosides are C_{18} trihydroxysphingosine, C_{18} normal sphingosine, C_{20} trihydroxysphingosine, and C_{18} dihydroxysphingosine. A small amount of branched-chain sphingosines was also found in kidney gangliosides (Puro and Keranen, 1969).

An enzyme, UDP-galactose : *N*-acetylgalactosaminyl-(*N*-acetylneuraminyl)-galactosyl-glucosylceramide galactosyltransferase catalyzing reaction (6) was demonstrated in rat kidney homogenate (Yip, 1972).

N-acetylgalactosaminyl-(*N*-acetylneuraminyl)-galactosyl-glucosylceramide + UDP-galactose
→ galactosyl-*N*-acetylgalactosaminyl-(*N*-acetylneuraminyl)-galactosyl-glucosylceramide (6)

In adult rat, the enzyme level was low in kidney, brain, leg muscle, and heart, but high in spleen and the large and small intestines. Tissues (liver, brain, and spleen) from 10-day-old rats were found to contain more enzyme activity than those from adult rats (Yip, 1972).

VII. Vitamin D_3

In recent years, work from several laboratories has established that kidney plays an important role in vitamin D_3 metabolism. Vitamin D_3 (cholecalciferol, the natural form present in mammals) must be metabolized to a biologically active form by kidney before it exerts its biological functions. Kidney is considered an endocrine organ because of its involvement in the function of vitamin D_3 (DeLuca, 1973, 1975). Although most of these metabolic studies were performed in avians, some were demonstrated in mammals. In this chapter we only review those studies that were carried out in mammals.

It has been shown in perfused liver and liver homogenates from the rat (Horsting and DeLuca, 1969; Ponchon *et al.*, 1969) that cholecalciferol (Fig. 5) is metabolized to 25-hydroxycholecalciferol (Fig. 5) which is the major circulating active form of vitamin D_3. Fraser and Kodicek (1970) studied the metabolism of [4-^{14}C, 1-^{3}H]25-hydroxycholecalciferol by intact and bilaterally nephrectomized rachitic rats. They found that the intact, but not the nephrectomized, rats produced a tritium-deficient metabolite—"1-oxygenated 25-hydroxycholecalciferol," indicating kidney to be the site of synthesis of this metabolite. This was soon confirmed by Gray *et al.* (1971). The new metabolite of 25-hydroxycholecalciferol produced by kidney was found predominantly in small intestine mucosa, and a smaller amount in blood plasma. Its structure was determined to be 1,25-dihydroxycholecalciferol (Fig. 5) (Holick *et al.*, 1971; Lawson *et al.*, 1971). The conformation of the 1-hydroxyl position is α (Ghazarian *et al.*, 1973). Now it has been established that 1 α,25-dihydroxycholecalciferol is the most biologically active form of vitamin D_3 in terms of stimulating intestinal calcium transport and bone calcium mobilization in rats (Boyle *et al.*, 1972; Norman and Wong, 1972), dogs (Brickman *et al.*, 1973), and humans (Brickman *et al.*, 1974), as well as elevation of serum inorganic phosphorus levels in rachitic rats (Tanaka and DeLuca, 1974). Patients with chronic renal insufficiency failed to produce 1,25-dihydroxycholecaliferol (Mawer *et al.*, 1973).

25-Hydroxycholecalciferol-1α-hydroxylase has recently been found

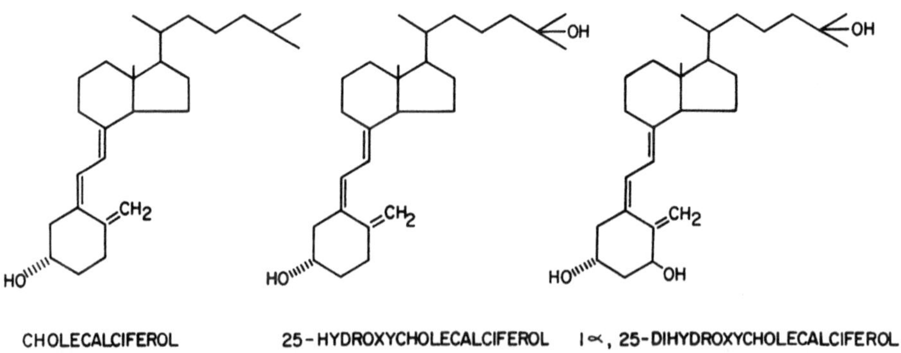

CHOLECALCIFEROL (VITAMIN D_3) 25-HYDROXYCHOLECALCIFEROL 1α, 25-DIHYDROXYCHOLECALCIFEROL

FIGURE 5. Structures of vitamin D_3 metabolites.

in kidney tissue from dogs and humans (Midgett *et al.*, 1973). In dog kidneys (Midgett *et al.*, 1973), as in chicken kidneys (Gray *et al.*, 1972), the enzyme was located in the mitochondrial fraction. In the dog, the enzymatic production of 1α,25-dihydroxycholecalciferol was found only in the kidney cortex but not in the medulla (Midgett *et al.*, 1973).

1α,25-Dihydroxycholecalciferol is now regarded as a hormone derived from vitamin D_3 (DeLuca, 1973), since it is synthesized in the kidney and carries out its function in intestine and bone to elevate serum calcium and phosphate concentrations. In addition, its biosynthesis is regulated by feedback of the products of its function—calcium and phosphorus. Boyle *et al.* (1971) demonstrated that the level of 1α,25-dihydroxycholecalciferol in these tissues was decreased 12 hr after the administration of tritiated 25-hydroxycholecalciferol, when the dietary calcium concentration was increased. On the contrary, hypocalcemia stimulated the production of 1α,25-dihydroxycholecalciferol. It is interesting that the decline of 1α,25-dihydroxycholecalciferol accompanies the appearance of a new metabolite, 24,25-diphydroxycholecalciferol. The highest concentration of this new metabolite was found in the kidney (Boyle *et al.*, 1971; Holick *et al.*, 1972). The stimulation by hypocalcemia of the conversion of 25-hydroxycholecalciferol to 1α,25-dihydroxycholecalciferol was found to be dependent on the presence of parathyroid hormone (Garabedian *et al.*, 1972). Serum phosphorus levels also play a regulatory role in 1α,25-dihydroxycholecalciferol synthesis. Tanaka and DeLuca (1973) reported that rats under hypophosphatemia produce 1α,25-dihydroxycholecalciferol from 25-hydroxycholecalciferol, a process not mediated by parathyroid hormone.

The function of 24,25-dihydroxycholecalciferol is not understood at present. Boyle *et al.* (1973) demonstrated that 24,25-dihydroxycholecalciferol has little effect on calcium mobilization from bone, though its potency in stimulating intestinal calcium transport is comparable to that of 1α,25-dihydroxycholecalciferol. Boyle *et al.* (1973) further showed that 24,25-dihydroxycholecalciferol in rat kidneys is metabolized to a more polar metabolite that is necessary for 24,25-dihydroxycholecalciferol to carry out its biological functions. Subsequent studies by Holick *et al.* (1973) demonstrated the presence of 1,24,25-trihydroxycholecalciferol in rat kidney. The trihydroxy derivative of vitamin D_3 is 60% as active as vitamin D_3 in curing rickets. It is preferentially more active in inducing intestinal calcium transport than in mobilizing calcium from bone. It is less active on a weight basis than 1α,25-dihydroxyvitamin D_3 in stimulating and sustaining intestinal calcium transport and bone calcium mineralization. However, at present it is not known whether 24,25-dihydroxycholesterol or 1α,25-dihydroxycholecalciferol is the immediate precursor of 1,24,25-trihydroxycholecalciferol.

References

Adams, E. P., and Gray, G. M. 1967. Effect of BP8 acites (sarcoma) tumors on glycolipid composition in kidneys of mice. *Nature* **216**:277–278.

Adams, E. P., and Gray, G. M. 1968. The carbohydrate structures of the neutral ceramide glycolipids in kidneys of different mouse strains with special reference to the ceramide dihexosides. *Chem. Phys. Lipids* **2**:47–155.

Agranoff, B. W., Bradley, R. M., and Brady, R. O. 1958. The enzymatic synthesis of inositol phosphatide. *J. Biol. Chem.* **233**:1077–1083.

Andrade, F., and Huggins, C. G. 1964. Myo-inositol phosphates in a phosphoinositide complex from kidney. *Biochim. Biophys. Acta* **84**:681–693.

Anggård, E. Bohman, S. O., Griffin, J. E. III., Larsson, C., and Maunsbach, A. B. 1972. Subcellular localization of the prostaglandin system in the rabbit renal papilla. *Acta Physiol. Scand.* **84**:231–246.

Antonello, A., Baggio, B., Favaro, S., Corsina, A., Todesco, S., and Borsatti, A. 1973. Effect on blood pressure of intravenous administration of a phospholipid renin preinhibitor and its active form in renal hypertensive rats. *Pflugers Arch. Eur. J. Physiol.* **341**:113–120.

Assmann, G., and Stoffel, W. 1972. On the origin of phytosphingosine (4D-hydroxysphinganine) in mammalian tissues. *Hoppe-Seyler's Z. Physiol. Chem.* **353**:971–979.

Austin, J. H. 1959. Metachromatic sulfatides in cerebral white matter and kidney. *Proc. Soc. Exp. Biol. Med.* **100**:361–364.

Austin, J. Suzuki, K., Armstrong, D., Brady, R., Bachhawat, B. K., Schlenker, J., and Stumpf, D. 1970. Studies in globoid (Krabbe) leukodystrophy (GLD) *Arch. Neurol.* **23**:502–512.

Bainbridge, F. A., and Evans, C. L. 1914. The heart, lung, kidney preparation. *J. Physiol. (London)* **48**:278–286.

Baricos, W. H. 1972. Regulation of phosphoinositide metabolism in rabbit kidney cortex slices. Ph.D. Dissertation, Tulane University.

Bartlett, P., and Bossart, J. F. 1969. Cation transport. 1. Metabolic activity of the polyphosphoinositide complex in isolated renal cortex tubules. *Henry Ford Hosp. Med. J.* **17**:247–258.

Bloch, K. 1959. Biogenesis and transformation of squalene, pp 4–16. *In* G.E.W. Wolstenholme and M. O'Connor (eds.) Ciba Foundation Symposium on the Biosynthesis of Terpenes and Sterols. J. & A. Churchill, Ltd., London.

Bojessen, I., and Roepstorff, P. 1973. Identification of C_{27}-sterols produced by rat renal inner medulla *in vitro*. *Biochim. Biophys. Acta* **316**:83–90.

Bojesen, E., Bojesen, I., and Capito, K. 1973. Sterol biosynthesis *in vitro* of rat renal inner medulla. *Biochim. Biophys. Acta* **306**:237–248.

Boyle, I. T., Gray, R. W., and DeLuca, H. F. 1971. Regulation by calcium *in vivo* synthesis of 1,25-dihydroxycholecaliciferol and 21,25-dihydroxycholecalciferol. *Proc. Natl. Acad. Sci. USA* **68**:2131–2134.

Boyle, I. T., Miravet, L., Gray, R. W., Holick, M. F., and DeLuca, H. F. 1972. The response of intestinal calcium transport to 25-hydroxy and 1, 25-dihydroxy vitamin D in nephrectomized rats. *Endocrinology* **90**:605–608.

Boyle, I. T., Omdahl, J. L., Gray, R. W., and DeLuca, H. F. 1973. The biological activity and metabolism of 24, 25-dihydroxy-vitamin D_3. *J. Biol. Chem.* **248**:4174–4180.

Brady, R. O., Gal, A. E., Bradley, R. M., and Mårtensson, E. 1967. The metabolism of creamide trihexoside. I. Purification and properties of an enzyme that cleaves the terminal galactose molecule of galactosylgalactosylglucosylceramide. *J. Biol. Chem.* **242**:1021–1026.

Brdiczka, D., Gergitz, K., and Pette, D. 1969. Localization and function of external and internal carnitine acetyltransferases in mitochondria of rat liver and pig kidney. *Eur. J. Biochem.* **11**:234–240.

Brickman, A. S., Reddy, C. R., Coburn, J. W., Passaro, E. P., Jowsey, J., and Norman, A. W. 1973. Biological action of 1,25-dihydroxy-vitamin D_3 in the rachitic dog. *Endocrinology* **92**:728–734.

Brickman, A. S., Coburn, J. W., Massry, S. G., and Norman, A. W., 1974. 1, 25-Dihydroxy-vitamin D_3 in normal man and patients with renal failure. *Ann. Intern. Med.* **80**:161–168.

Burns, B. J., and Elwood, J. C. 1969. Lipid metabolism in kidney and liver tissue from normal and diabetic rats. *Biochim. Biophys. Acta* **187**:307–318.

Carter, H. E., and Hirschberg, C. B. 1968. Phytosphingosines and branched sphingosines in kidney. *Biochemistry* **7**:2296–2300.

Chernick, S. S., Masaro, E. J., and Chalkoff, I. L. 1950. The *in vitro* conversion of C^{14}-labeled glucose to fatty acids. *Proc. Soc. Exp. Biol. Med.* **73**:348–352.

Clarke, J. T. R., Wolfe, L. S., and Perlin, A. S. 1971. Evidence for a terminal α-D-galactopyranosyl residue in galactosylgalactosylglucosylceramide from human kidney. *J. Biol. Chem.* **246**:5563–5569.

Coe, F. L., and Korty, P. R. 1967. Protein synthesis during compensatory renal hypertrophy. *Am. J. Physiol.* **213**:1585–1589.

Coles, L., and Gray, G. M. 1970. The biosynthesis of digalactosylceramide in kidney of the C57/BL mouse. *Biochem. Biophys. Res. Commun.* **38**:520–526.

Coles, L., Hay, J. B., and Gray, G. M. 1970. Factors affecting the glycosphingolipid composition of murine tissues. *J. Lipid Res.* **11**:158–163.

Colodzin, M., and Kennedy, E. P. 1965. Biosynthesis of diphosphoinositide in brain. *J. Biol. Chem.* **240**:3771–3779.

Constantino-Ceccarini, E., and Morell, P. 1973. Synthesis of galactosylceramide and glucosylceramide by mouse kidney preparations. *J. Biol. Chem.* **248**:8240–8246.

Crowshaw, K. 1973. The incorporation of (1-^{14}C) arachidonic acid into the lipids of rabbit renal slices and conversion to prostaglandins E_2 and $F_{2\alpha}$. *Prostaglandins* **3**:607–620.

Crowshaw, K., and Szlyk, J. Z. 1970. Distribution of prostaglandins in rabbit kidney. *Biochem. J.* **116**:421–424.

Cuzzopoli, M., Hellström, K., and Svensson, B. 1972. Effect of nephrectomy and cholesterol feeding on the metabolism of intravenously administered DL-mevalonate in the rat. *Metabolism* **21**:1161–1170.

Daniels, E. G., Hinman, J. W., Leach, B. E., and Muirhead, E. E. 1967. Identification of prostaglandin E_2 as the principal vasodepressor lipid of rabbit renal medulla. *Nature* **215**:1298–1299.

Danon, A., Chang, L. C. T., Sweetman, B. J., Nies, A., and Oates, J. A. 1975. Synthesis of prostaglandin E_2 as the principal vasodepressor lipid of rabbit renal medulla. *Nature*

Dawson, G. 1972. Glycosphingolipid levels in an unusual neurovisceral storage disease characterized by lactosylceramide galactosylhydrolase deficiency: lactosylceramidosis. *J. Lipid Res.* **13**:207–219.

Dawson, R. M. C., and Clarke, N. 1972. D-myoInositol 1:2-cyclic phosphate 2-phosphohydrolase *Biochem. J.* **127**:113–118.

Dawson, R. M. C., and Eichberg, J. 1965. Diphosphoinositide and triphosphoinositide in animal tissues, extraction, estimation, and changes *post mortem*. *Biochem. J.* **96**:634–643.

Dawson, R. M. C., Freinkel, N., Jangalwala, F. B., and Clarke, N. 1971. The enzymatic formation of myoinositol 1, 2-cyclic phosphate from phosphatidylinositol. *Biochem. J.* **122**:605–607.

DeLuca, H. F. 1973. The kidney as an endocrine organ for the production of 1, 25-dihydroxy-vitamin D_3, a calcium-mobilizing hormone. *N. Engl. J. Med.* **289**:359–365.

DeLuca, H. F. 1975. The kidney as an endocrine organ involved in the function of vitamin D. *Am. J. Med.* **58**:39–47.

Dickins, F., and Simer, F. 1930. The respiratory quotient, and the relationships of respiration to glycolysis. *Biochem. J.* **24**:1301–1326.

Dietschy, J. M., and Siperstein, M. D. 1967. Effect of cholesterol feeding and fasting on sterol synthesis in seventeen tissues of the rat. *J. Lipid Res.* **8**:97–104.

Dietschy, J. M., and Wilson, J. D. 1968. Cholesterol synthesis in the squirrel monkey: Relative rates of synthesis in various tissues and mechanisms of control. *J. Clin. Invest.* **47**:166–174.

Dittmer, J. C., and Douglas, M. G. 1969. Quantitative determination of phosphoinositides. *Ann. N.Y. Acad. Sci.* **165**:515–525.

Duncan, C. H., and Best, M. M. 1963. Determination of fractional disappearance rate of kidney cholesterol. *Am. J. Physiol.* **204**:505–508.

Edgren, B., and Hellström, K. 1972. Lipid biosynthesis from DL(2-^{14}C) mevalonic acid in intact mice and rabbits. *Acta Physiol. Scand.* **86**:250–256.

Ellin, A., and Orrenius, S. 1971. Studies on cytochrome P-450 of rat kidney cortex microsomes. *Chem.-Biol. Interact.* **3**:256–257.

Ellin, A., Jakobsson, S. V., Schenkman, J. B., and Orrenius, S. 1972. Cytochrome P_{450K} of rat kidney cortex microsomes: Its involvement in fatty acid ω- and (ω-1)-hydroxylation. *Arch. Biochem. Biophys.* **150**:64–71.

Elwood, J. C., and Van Bruggen, J. T. 1961. Cholesterol and mevalonate metabolism: A time course study *in vivo*. *J. Lipid Res.* **2**:344–349.

Esselman, W. J., Ackerman, J. R., and Sweely, C. C. 1973. Glycosphingolipids of membrane fractions from normal and transplanted canine kidney. *J. Biol. Chem.* **248**:7310–7317.

Farrell, D. F., and McKhann, G. M. 1971. Characterization of cerebroside sulfotransferase from rat brain. *J. Biol. Chem.* **246**:4694–4702.

Fleischer, B., and Zambrano, F. 1974. Golgi apparatus of rat kidney preparation and role in sulfatide formation. *J. Biol. Chem.* **249**:5995–6003.

Fleischer, S., Rouser, G., Fleischer, B. Casu, A., and Kritchevsky, G. 1967. Lipid composition of mitochondria from bovine heart, liver, and kidney. *J. Lipid Res.* **8**:170–180.

Flint, D. R. 1967. Studies on the structure and metabolism of phosphoinositide complex from renal cortex. Ph.D. Dissertation, Tulane University.

Forssman, J. 1911. Die Herstellung hochwertiger spezifischer Schafhamolysine ohne Verwendung von Schafblut. *Biochem. Z.* **37**:78–115.

Fraser, D. R., and Kodicek, E. 1970. Unique biosynthesis by kidney of a biologically active vitamin D metabolite. *Nature* **228**:764–766.

Gallaher, W. R., Weinstein, D. B., and Blough, H. A., 1973. Rapid turnover of principal phospholipids in BHK-21 cells. *Biochem. Biophys. Res. Commun.* **52**:1252–1256.

Gallai-Hatchard, J. J., and Thompson, R. H. S. 1965. Phospholipase-A activity of mammalian tissues. *Biochim. Biophys. Acta* **98**:128–136.

Ganguly, J. 1960. Studies on the mechanism of fatty acid synthesis. VII. Biosynthesis of fatty acids from malonyl CoA. *Biochim. Biophys. Acta* **40**:110–118.

Gans, J. H., Block, A. J., and Cater, M. R. 1973. *In vitro* incorporation of ^{14}C into liver and kidney sterols from parentally administered [2-^{14}C]D,L-mevalonic acid. *Proc. Soc. Exp. Biol. Med.* **144**:609–612.

Garabedian, M., Holick, M. F., DeLuca, H. F., and Boyle, I. T. 1972. Control of 25-hydroxycholecalciferol metabolism by parathyroid glands. *Proc. Natl. Acad. Sci. USA* **69**:1673–1767.

Getz, G. S., Bartley, W., Lurie, D., and Notton, B. M. 1968. The phospholipids of various sheep organs, rat liver and of their subcellular fractions. *Biochim. Biophys. Acta* **152**:325–329.

Geyer, R. P., and Cunningham, M. 1950. Metabolism of fatty acids *in vitro*, studied with odd and even members of the R^{14} COOH series. *J. Biol. Chem.* **184**:641–646.

Geyer, R. P., Mathews, L. W., and Stare, F. J. 1949. Metabolism of emulsified trilaurin (-C^{14}OO-) and octanoic acid (-C^{14}OO-) by rat tissue slices. *J. Biol. Chem.* **180**:1037–1045.

Ghazarian, J. G., Schnoes, H. K., and DeLuca, H. F. 1973. Mechanism of 25-hydroxycholecalciferol 1 α-hydroxylation. Incorporation of oxygen-18 into the 1α position. *Biochemistry* **12**:2555–2558.

Gold, M., and Spitzer, J. J. 1964. Metabolism of free fatty acids by myocardium and kidney. *Am. J. Physiol.* **206**:153–158.

Gray, G. M. 1971. The effect of testosterone on the biosynthesis of the neutral glycosphingolipids in the C_{57}/BL mouse kidney. *Biochim. Biophys. Acta* **239**:494–500.

Gray, R., Boyle, I., and DeLuca, H. F. 1971. Vitamin D metabolism: The role of kidney tissue. *Science* **172**:1232–1234.

Gray, R. W., Omdahl, J. L., Ghazarian, J. G., and DeLuca, H. F. 1972. 25-Hydroxycholecalciferol-1-hydroxylase. Subcellular location and properties. *J. Biol. Chem.* **247**:7528–7532.

Hagen, P.-O. 1971. Structural comparison between triglycerides and phospholipids from pig kidney. *Lipids* **6**:935–941.

Hamberg, M. 1969. Biosynthesis of prostaglandins in the renal medulla of rabbit. *FEBS Lett.* **5**:127–130.

Hamberg, M. Svensson, J., Wakabayashi, T., and Samuelsson, B. 1974. Isolation and structure of two prostaglandin endoperoxides that cause platelet aggregation. *Proc. Natl. Acad. Sci. USA* **71**:345–349.

Hammarström, S., and Bjursell, G. 1973. Glycolipid synthesis in baby-hamster-kidney fibroblasts transformed by a thermosensitive mutant of polyoma virus. *FEBS Lett.* **32**:69–72.

Harwood, J. L., and Hawthorne, J. N. 1969. The properties and subcellular distribution of phosphatidylinositol kinase in mammalian tissues. *Biochim. Biophys. Acta* **171**:75–88.

Hauser, G. 1963. The formation of free and lipid myo-inositol in the intact rat. *Biochim. Biophys. Acta* **70**:278–289.

Hauser, G., and Eichberg, J. 1973. Improved conditions for the preservation and extraction of polyphosphoinositides. *Biochim. Biophys. Acta* **362**:201–209.

Hay, J. B., and Gray, G. M. 1970. Glycosphingolipid biosynthesis in kidneys of normal C3H/He mice and of those with BP8 ascites tumors. *Biochem. Biophys. Res. Commun.* **38**:527–532.

Hellström, K. H., Siperstein, M. D., Bricker, L. A., and Luby, L. J. 1973. Studies of the *in vivo* metabolism of mevalonic acid in the normal rat. *J. Clin. Invest.* **52**:1303–1313.

Hildebrand, J., and Hauser, G. 1969. Biosynthesis of lactosylceramide and triglycosylceramide by galactosyltransferases from rat spleen. *J. Biol. Chem.* **244**:5170–5180.

Hochberg, R. B., Ladany, S., and Lieberman, S. 1974. Cholesterol sulfate: Some aspects of its biosynthesis and uptake by tissues from blood. *Endocrinology* **94**:207–213.

Hohenegger, M., and Novak, M. 1969. Freie Fettsäuren Zonen der Rattenniere. *Med. Exp. (Basel)* **19**:137–142.

Hohenegger, M., Wittmann, G., and Dahlheim, H. 1973. Oxidation of fatty acids by different zones of the rat kidney. *Pfluegers Arch. Eur. J. Physiol.* **341**:105–112.

Holick, M. F., Schnoes, H. K., and DeLuca, H. F. 1971. Identification of 1, 25-dihydroxycholecalciferol, a form of vitamin D_3 metabolically active in the intestine. *Proc. Natl. Acad. Sci. USA* **68**:803–804.

Holick, M. F., Schnoes, H. K., DeLuca, H. F., Gray, R. W., Boyle, I. T., and Suda, T. 1972. Isolation and identification of 24, 25-dihydroxycholecalciferol, a metabolite of vitamin D_3 made in the kidney. *Biochemistry* **11**:4251–4255.

Holick, M. F., Kleiner-Bossaller, A., Schnoes, H. K., Kasten, P. M., Boyle, I. T., and DeLuca, H. F. 1973. 1, 24, 25-trihydroxy-vitamin D_3. A metabolite of vitamin D_3 effective on intestine. *J. Biol. Chem.* **248**:6691–6696.

Horrocks, L. A. 1972. Content, composition, and metabolism of mammalian and avian lipids that contain ether groups, pp. 210–211. *In* F. Snyder (ed.) Ether Lipids Chemistry and Biology. Academic Press, New York.

Horsting, M., and DeLuca, H. F. 1969. *In vitro* production of 25-hydroxycholecalciferol. *Biochem. Biophys. Res. Commun.* **36**:251–256.

Huang, J. S., Downes, G. L., and Belzer, F. O. 1971. Utilization of fatty acids in perfused hypothermic dog kidney. *J. Lipid Res.* **12**:622–627.

Huggins, C. G., and Cohn, D. V. 1959. Studies concerning the composition, distribution, and turnover of phosphorus in a phosphatidopeptide fraction from mammalian tissue. *J. Biol. Chem.* **234**:257–261.

Ichihara, K., Kusunose, E., and Kusunose, M. 1969. Some properties and distribution of the ω-hydroxylation system of medium-chain fatty acids. *Biochim. Biophys. Acta* **176**:704–712.

Ichihara, K., Kusunose, E., and Kusunose, M. 1971. A fatty acid ω-hydroxylation system solubilized from porcine kidney cortex microsomes. *Biochim. Biophys. Acta* **239**:178–189.

Israel, M., Bach, G., Miyatake, T., Naiki, M., and Suzuki, K. 1974. Forssman hapten *N*-acetyl-α-D-galactosaminidase in rat brain and kidney. *J. Neurochem.* **23**:803–809.

Jakobsson, S., Thor, H., and Orrenius, S. 1970. Fatty acid inducible cytochrome P-454 of rat kidney cortex microsomes. *Biochem. Biophys. Res. Commun.* **39**:1073–1080.

Johnson, R. C. and Shah, S. N. 1974. Microsomal synthesis of cholesterol from squalene, lanosterol, and desmosterol. *Arch. Biochem. Biophys.* **164**:502–510.

Kalisker, A., and Dyer, D. C. 1972. *In vitro* release of prostaglandins from the renal medulla. *Eur. J. Pharmacol.* **19**:305–309.

Kano, I., and Yamakawa, T. 1974a. Human kidney α-galactosidase. *J. Biochem. (Tokyo)* **74**:347–354.

Kano, I., and Yamakawa, T. 1974b. The properties of α-galactosidase remaining in kidney and liver of patients with Fabry's disease. *Chem. Phys. Lipids* **13**:283–291.

Karlsson, K.-A. 1964. Studies on sphingosines, 7. The existence of C_{18}-and C_{20}-phytosphingosine in animal tissues. *Acta Chem. Scand.* **18**:2397–2398.

Karlsson, K.-A. 1966. Studies on sphingosines, 11. The chemical structure of phytosphingosine of human origin and a note on the lipid composition of the yeast *Hansenula ciferrii*. *Acta Chem. Scand.* **20**:2884–2885.

Karlsson, K.-A., and Mårtensson, E., 1968. Studies on sphingosines, XIV. On the phytosphingosine content of the major human kidney glycolipids. *Biochim. Biophys. Acta* **152**:230–233.

Karlsson, K.-A., and Steen, G. O. 1968. Studies on sphingosines XIII. The existence of phytosphingosine in bovine kidney sphingomyelins. *Biochim. Biophys. Acta* **152**:798–800.

Karlsson, K.-A., Samuelsson, B. E., and Steen, G. O. 1968a. Structure and function of sphingolipids. 1. Differences in sphingolipid long-chain base pattern between kidney cortex, medulla, and papillae. *Acta Chem. Scand.* **22**:1361–1363.

Karlsson, K.-A., Samuelsson, B. E., and Steen, G. O. 1968b. Structure and function of sphingolipid. 2. Differences in sphingolipid concentration, especially concerning sulfatides, between some regions of bovine kidney. *Acta Chem. Scand.* **22**:2723–2724.

Karlsson, K.-A., Samuelsson, B. E., and Steen, G. O., 1973a. Separation of monoglycosylceramides (cerebrosides) of bovine kidney into subgroups and characterization by mass spectrometry. *Biochim. Biophys. Acta* **306**.317–328.

Karlsson, K.-A., Samuelsson, B. E., and Steen, G. O., 1973b. The sphingolipid composition of bovine kidney cortex, medulla and papilla. Biochim. Biophys. Acta **316**:317–335.

Karlsson, K.-A., Samuelsson, B. E., and Steen, G. O., 1973c. Detailed structure of sphingomyelins and ceramides from different region of bovine kidney with special reference to long chain bases. *Biochim. Biophys. Acta* **316**:336–362.

Karlsson, K.-A., Leffler, H., and Samuelsson, B. E. 1974. Characterization of the Forssman glycolipid hapten of horse kidney by mass spectrometry. *J. Biol. Chem.* **249**:4819–4823.

Kawanami, J., 1968. Glycolipids from rat kidney. *J. Biochem. (Tokyo)* **64**:625–633.

Kleinman, J. G., Mandelbaum, J., and Levin, M. L. 1973. Renal functional effects of 4-pentenoic acid, an inhibitor of fatty acid oxidation. *Am. J. Physiol.* **224**:95–101.

Klenk, E. 1942. Über die Ganglioside, eineneue Gruppe von zuckerhältigen Gehirnlipoiden. *Hoppe-Seyler's Z. Physiol. Chem.* **273**:76–86.

Kunze, H., and Vogt, W. 1971. Significance of phospholipase A for prostaglandin formation. *Ann. N. Y. Acad. Sci.* **180**:123–125.

Lawson, D. E. M., Fraser, D. R., Kodicek, E. Morris, H. R., and Williams, D. H. 1971. Identification of 1, 25-dihydroxycholecalciferol, a new kidney hormone controlling calcium metabolism. *Nature* **230**:228–230.

Lee, J. B. 1967. Antihypertensive activity of the kidney—the renomedullary prostaglandins. *New Eng. J. Med.* **277**:1073–1079.

Lee, J. B., Covino, B. G., Takman, B. H., and Smith, E. R. 1965. Renomedullary vasodepressor substance, medullin: Isolation, chemical characterization, and physiological properties. *Circ. Res.* **17**:57–77.

Lee, J. B., Crawshaw, K., Takman, B. H., and Attrep, K. A., 1967. The identification of prostaglandins E_2, $F_2\alpha$ and A_2 from rabbit kidney medulla. *Biochem. J.* **105**:1251–1260.

Lee, S.-C., and Levine, C. 1974. Prostaglandin metabolism. 1. Cytoplasmic reduced nicotinamide adenine dinucleotide phosphate-dependent and microsomal reduced nicotinamide adenine dinucleotide-dependent prostaglandin E 9-ketoreductase activities in monkey and pigeon tissues. *J. Biol. Chem.* **249**::1369–1375.

Lee, S.-C., Pong, S.-S., Katzen, D., Wu, K.-U., and Levine, L. 1975. Distribution of prostaglandin E 9-ketoreductase and types I and II 15-hydroxyprostaglandin dehydrogenase in swine kidney medulla and cortex. *Biochemistry* **14**:142–145.

Lee, T. -C., and Huggins, C. G. 1968a. A triphosphoinositide phosphomonesterase from rat kidney. I. Isolation and general properties. *Arch. Biochem. Biophys.* **126**:206–213.

Lee, T. -C., and Huggins, C. G. 1968b. A triphosphonositide phosphomonesterase from rat kidney. II. Purification and characterization. *Arch. Biochem. Biophys.* **126**:214–220.

Li, Y.-T., and Li, S.-C. 1971. Anomeric configuration of galactose residues in ceramide trihexosides. *J. Biol. Chem.* **246**:3769–3771.

Li, Y.-T., Li, S.-C., and Dawson, G. 1972. Anomeric structure of ceramide digalactoside isolated from the kidney of a patient with Fabry's disease. *Biochim. Biophys. Acta* **260**:88–92.

Makita, A. 1964. Biochemistry of organ glycolipids. II. Isolation of human kidney glycolipids. *J. Biochem. (Tokyo)* **55**:269–276.

Makita, A., and Yamakawa, T. 1964. Biochemistry of organ glycolipids. III. The structures of human kidney cerebroside sulfuric ester, ceramide dihexoside and ceramide trihexoside. *J. Biochem. (Tokyo)* **55**:365–370.

Makita, A. M., Iwanaga, M., and Yamakawa, T. 1964. The chemical structure of human kidney globoside. *J. Biochem. (Tokyo)* **55**:202–204.

Makita, A., Suzuki, C., and Yosizawa, Z. 1966. Chemical and immunological characterization of the Forssman hapten isolated from equine organs. *J. Biochem. (Tokyo)* **60**:502–513.

Malone, M. J., and Stoffyn, P. 1966. A comparative study of brain and kidney glycolipids in metachromatic leucodystrophy. *J. Neurochem.* **13**:1037–1045.

Mapes, C. A., and Sweeley, C. C. 1973. Preparation and properties of an affinity column adsorbent for differentiation of multiple forms of α-galactosidase activity. *J. Biol. Chem.* **248**:2461–2470.

Markwell, M. A. K., McGroarty, E. J., Bieber, L. L., and Tolbert, N. E. 1973. The subcellular distribution of carnitine acyltransferases in mammalian liver and kidney. *J. Biol. Chem.* **248**:3426–3432.

Mårtensson, E. 1963a. On the sulfate containing lipids of human kidney. *Acta Chem. Scand.* **17**:1174–1176.

Mårtensson, E. 1963b. On the neutral glycolipids of human kidney. *Acta Chem. Scand.* **17**:2356–2357.

Mårtensson, E., 1966a. Neutral glycolipids of human kidney isolation, identification, and fatty acid composition. *Biochim. Biophys. Acta* **116**:296–308.

Mårtensson, E., 1966b. Sulfatides of human kidney isolation, identification, and fatty acid composition. *Biochem. Biophys. Acta* **116**:521–531.

Mårtensson, E., Percy, A., and Svennerholm, L. 1966. Kidney glycolipids in late infantile metachromatic leucodystrophy. *Acta Paediat. Scand.* **55**:1–9.

Mårtensson, E., Öhman, R., Graves, M., and Svennerholm, L. 1974. Galactosyltransferases catalyzing the formation of the galactosyl-galactosyl linkage in glycosphingolipids. *J. Biol. Chem.* **249**:4132–4137.

Masoro, E. J., and Porter, E. 1965. A comparison of fatty acid synthesis by liver and kidney. *Proc. Soc. Exp. Biol. Med.* **118**:1090–1095.

Mawer, E. B., Backhouse, J., Taylor, C. M., and Lumb, G. A. 1973. Failure of formation of 1, 25-dihydroxycholecalciferol in chronic renal insufficiency. *Lancet* **1**:626–628.

McGiff, J. C., and Itskovitz, H. D. 1973. Prostaglandins and the kidney. *Circ. Res.* **33**:479–488.

McGiff, J. C., Crowshaw, K., and Itskovitz, H. D. 1974. Prostaglandins and renal function. *Fed. Proc.* **33**:39–47.

McKhann, G. M., and Ho, W. 1967. The *in vivo* and *in vitro* synthesis of sulphatides during development. *J. Neurochem.* **14**:717–724.

McKhann, G. M., Levy, R., and Ho., W. 1965. Metabolism of sulfatides. I. The effect of galactocerebrosides on the synthesis of sulfatides. *Biochem. Biophys. Res. Commun.* **20**:109–113.

Medes, G., Thomas, A., and Weinhouse, S. 1952. Nutritional factors in fatty acid synthesis by tissue slices *in vitro*. *J. Biol. Chem.* **197**:181–191.

Mehl, E., and Jatzkewitz, H. 1964. Eine cerebrosidsulfatase aus Schweineniere. *Hoppe-Seyler's Z. Physiol. Chem.* **339**:260–276.

Mehl, E., and Jatzkewitz, H. 1965. Evidence for the genetic block in metachromatic leucodystrophy (ML). *Biochem. Biophys. Res. Commun.* **19**:407–411.

Midgett, R. J., Spielvogel, A. M., Coburn, J. W., and Norman, A. W. 1973. Studies on calciferol metabolism. VI. The renal production of biologically active form of vitamin D, 1, 25-dihydroxycholecalciferol; species, tissue and subcellular distribution. *J. Clin. Endocrinol. Metab.* **36**:1153–1161.

Miyatake, T. 1969. A study on glycolipids in Fabry's disease. *Jap. J. Exp. Med.* **39**:35–45.

Miyatake, T., and Ariga, T. 1972. Sphingolipids in the nervous system in Fabry's disease. *J. Neurochem.* **19**:1911–1916.

Miyatake, T., and Suzuki, K. 1972. Globoid cell leukodystrophy: additional deficiency of psychosine galactosidase. *Biochem. Biophys. Res. Commun.* **48**:538–543.

Morell, P., and Braun, P. 1972. Biosynthesis and metabolic degradation of sphingolipids not containing sialic acid. *J. Lipid Res.* **13**:293–310.

Morgan, T. E. Tinker, D. O., and Hanahan, D. J. 1963. Phospholipid metabolism in kidney. I. Isolation and identification of lipids of rabbit kidney. *Arch. Biochem. Biophys.* **103**:54–65.

Morrison, W. R. 1973. Long-chain bases in the sphingolipids of bovine milk and kidney, rumen bacteria, rumen protozoa, hay and concentrate. *Biochim. Biophys. Acta* **316**:98–107.

Moser, H. W. 1972. Sulfatide lipidosis: metachromatic leukodystrophy, pp. 688–729. *In* J. B. Stanburg, J. B. Wyngaarden, and D. S. Fredrickson (eds.). Metabolic Basis of Inherited Disease, 2nd ed. McGraw-Hill, New York.

Muller, M. M., Kaiser, E., and Hohenegger, M. 1972. Lipid composition of the rat kidney. *Pfluegers Arch. Eur. J. Physiol.* **335**:R43.

Nelson, D. R., and Cornatzer, W. E. 1964. Effect of digitoxin, aldosterone, and dietary sodium chloride on incorporation of inorganic ^{32}P into liver and kidney nuclear and mitochondrial phospholipids. *Proc. Soc. Exp. Biol. Med.* **116**:237–242.

Norman, A. W., and Wong, R. G. 1972. Biological activity of the vitamin D metabolite 1, 25-dihydroxycholecalciferol in chickens and rats. *J. Nutr.* **102**:1709–1718.

Pace-Asciak, C. 1975a. Prostaglandin 9-hydroxydehydrogenase activity in the adult rat kidney. *J. Biol. Chem.* **250**:2789–2794.

Pace-Asciak, C. 1975b. Activity profiles of prostaglandin 15- and 9-hydroxydehydrogenase and 13-reductase in the developing rat kidney. *J. Biol. Chem.* **250**:2795–2800.

Pace-Asciak, C., and Domazet, A. 1975. 9-Hydroxyprostaglandin dehydrogenase activity in the adult rat kidney. Regional distribution and subfractionation. *Biochim. Biophys. Acta* **380**:338–343.

Pace-Asciak, C., and Miller, D. 1974. Prostaglandins during development. II. Identification of prostaglandin 9-hydroxydehydrogenase activity in adult rat kidney homogenates. *Experientia* **30**:590–592.

Philippart, M., Sarlieve, L., Meurant, C., and Mechler, L. 1971. Human urinary sulfatides in patients with sulfatidosis (metachromatic leukodystrophy). *J. Lipid Res.* **12**:434–441.

Ponchon, G., Kennan, A. L., DeLuca, H. F. 1969. "Activation" of vitamin D by the liver. *J. Clin. Invest.* **48**:2032–2037.

Puro, K. 1969. Carbohydrate components of bovine-kidney gangliosides. *Biochim. Biphys. Acta* **189**:401–413.

Puro, K., and Keränen, A. 1969. Fatty acids and sphingosines of bovine-kidney gangliosides. *Biochim Biophys. Acta* **187**:393–400.

Puro, K., Maury, P., and Huttunen, J. K. 1969. Qualitative and quantitative patterns of gangliosides in extraneural tissues. *Biochim. Biophys. Acta* **187**:230–235.

Quastel, J. H., and Wheatly, A. H. M. 1933. Oxidation of fatty acids in the liver. *Biochem. J.* **27**:1753–1762.

Rapport, M. M., Gref, L., and Schneider, H. 1964. Immunochemical studies of organ and tumor lipids XIII. Isolation of cytolipin K, a glycosphingolipid hapten present in human kidney. *Arch. Biochem. Biophys.* **105**:431–438.

Renkonen, O., Gahmberg, C. G., Simons, K., and Kääriäinen, L. 1972. The lipids of the plasma membranes and endoplasmic reticulum from cultured baby hamster kidney cells (BHK_{21}). *Biochim. Biophys. Acta* **255**:66–78.

Rice, L. I., Rice, E. H., Spolter, L., Max W., and O'Brien, J. S. 1968. Sulfation of cholesterol in rat kidney and liver. *Arch. Biochem. Biophys.* **127**:37–42.

Robertson, A. F., and Lands, W. E. M. 1962. Positional specificities in phospholipid hydrolyses. *Biochemistry* **1**:804–810.

Rouser, G., Simon, G., and Kritchevsky, G. 1969. Species variations in phospholipid class distribution of organs: I. Kidney, liver and spleen. *Lipids* **4**(6):599–606.

Sandhoff, K. 1969. Variation of β-N-acetylhexosaminidase-pattern in Tay-Sachs disease. *FEBS Lett.* **4**:351–354.
Sandhoff, K., Pilz, H., and Jatzkewitz, H. 1964. Über den enzymatischen Abbau von *N*-Acetylneuraminsäure-freien Gangliosidresten (Ceramid-oligosacchariden). *Hoppe-Seyler's Z. Physiol. Chem.* **338**:281–285.
Sandhoff, K., Andreae, V., and Jatzkewitz, H. 1968a. Deficient hexosaminidase activity in an exceptional case of Tay-Sachs disease with additional storage of kidney globoside in visceral organs. *Pathol. Eur.* **3**:278–285.
Sandhoff, K., Andreae, V., and Jatzkewitz, H. 1968b. Deficient hexosaminidase activity in an exceptional case of Tay-Sachs disease with additional storage of kidney globoside in visceral organs. *Life Sci.* **7**:283–288.
Sarlieve, L. L., Neskovic, N. M., and Mandel, P. 1971. PAPS–cerebroside sulphotransferase activity in brain and kidney of neurological mutants. *FEBS Lett.* **19**:91–95.
Sasame, H. A., Thorgeirsson, S. S., Mitchell, J. R., and Gillete, J. O. 1974. The possible involvement of cytochrome b_5 in the oxidation of lauric acid by microsomes from kidney cortex and liver of rats. *Life Sci.* **14**:35–46.
Schibannoff, J. M., Kamoshita, S., and O'Brien, J. S. 1969. Tissue distribution of glycosphingolipids in a case of Fabry's disease. *J. Lipid Res.* **10**:515–520.
Scott, T. W., Setchell, B. P., and Bassett, J. M. 1967. Characterization and metabolism of ovine foetal lipids. *Biochem. J.* **104**:1040-1047.
Sheltawy, A., and Dawson, R. M. C. 1969. The metabolism of polyphosphoinositides in hen brain and sciatic nerve. *Biochem. J.* **111**:157-165.
Shorr, E., Loebel, R. O., and Richardson, H. B. 1930. The nature and phlorhizin diabetes. *J. Biol. Chem.* **86**:529–549.
Siddiqui, B., and Hakomori, S. 1971. A revised structure of the Forssman glycolipid hapten. *J. Biol. Chem.* **246**:5766–5769.
Siddiqui, B., Kawanami, J., Li, Y.-T., and Hakomori, S. 1972. Structures of ceramide tetrasaccharides from various sources: Uniqueness of rat kidney ceramide tetrasaccharide. *J. Lipid Res.* **13**:657–662.
Smeby, R. R., and Bumpus, F. M. 1971. Renin inhibitors, pp. 207–216. *In* J. W. Fisher (ed.). Kidney Hormones. Academic Press.
Soula, G., Souillard, C., and Douste-Blazy, L. 1972. Metabolisme des phospholipides du rein. Incorporation *in vivo* de ^{32}P-orthophosphate dans les lipides renaux du rat. *Biochmie* **54**:401–407.
Srere, P. A., Chaikoff, I. L., Treitman, S. S., and Burstein, L. S. 1950. The extrahepatic synthesis of cholesterol *J. Biol. Chem.* **182**:629–634.
Stoffyn, A., Stoffyn, P., and Mårtensson, E. 1968. Structure of kidney ceramide dihexoside sulfate. *Biochim. Biophys. Acta* **152**:353–357.
Stoffyn, P., Stoffyn, A., and Hauser, G. 1971. Structure of sulfatides biosynthesized *in vitro*. *J. Lipid Res.* **12**:318–323.
Stoffyn, A., Stoffyn, P., and Hauser, G. 1974. Structure of trihexosylceramide biosynthesized *in vitro* by rat kidney galactosyltransferase. *Biochim. Biophys. Acta* **360**:174–178.
Sugita, M., Dulaney, J. T., and Moser, H. W., 1972. Ceramidase deficiency in Fabry's disease (lipogranulomatosis). *Science* **178**:1100–1102.
Sugita, M., Connolly, P., Dulaney, J. T., and Moser, H. W. 1973. Fatty acid composition of free ceramides of kidney and cerebellum from a patient with Fabry's disease. *Lipids* **8**:401–406.
Sugita, M., Iwamori, J., Evans, J., McCluer, R. H., Dulaney, J. T., and Moser, H. W. 1974. High performance liquid chromatography of ceramides: Application to analysis in human tissues and demonstration of ceramide excess in Fabry's disease. *J. Lipid Res.* **15**:223–226.

Sung, S. J., Esselman, W. J., and Sweeley, C. C. 1973. Structure of a pentahexosylceramide (Forssman hapten) from canine intestine and kidney. *J. Biol. Chem.* **248**:6528 6533.

Suzuki, K., 1971. Renal cerebroside in globoid cell leukodystrophy (Krabbe's disease). *Lipids* **6**:433–436.

Svennerholm, L. 1964. The distribution of lipids in the human nervous system - I. Analytical procedure. Lipids of foetal and newborn brain. *J. Neurochem.* **11**:839–853.

Sweeley, C. C., and Klionsky, B. 1963. Fabry's disease: Classification as a sphingolipidosis and partial characterization of a novel glycolipid. *J. Biol. Chem.* **238**:PC3148–PC3150.

Sweeley, C. C., and Klionsky, B. 1966. Glycolipid lipidosis: Fabry's disease, pp. 618–632. *In* J. B. Stanbury, J. B. Wyngaarden, and D. S. Fredrickson (eds.) the Metabolic Basis of Inherited Disease. McGraw-Hill, New York.

Tanaka, Y., and DeLuca, H. F. 1973. The control of 25-hydroxyvitamin D metabolism by inorganic phosphorus. *Arch Biochem. Biophys.* **154**:566–574.

Tanaka, Y., and DeLuca, H. F. 1974. Role of 1, 25-dihydroxyvitamin D_3 in maintaining serum phosphorus and curing rickets. *Proc. Natl. Acad. Sci. USA* **71**:1040–1044.

Tinker, D. O., and Hanahan, D. J. 1966. Phospholipid metabolism in kidney. III. Biosynthesis of phospholipids from radioactive precursors in rabbit renal cortex slices. *Biochemistry* **5**:423–435.

Tinker, D. O., Koch, A., and Hanahan, D. J. 1963. Phospholipid metabolism in kidney II. Potasssium uptake, lipid composition and ^{32}P labelling of the phospholipids in rabbit kidney cortex slices *in vitro,* and the effects of diphenylhydantoin. *Arch. Biochem. Biophys.* **103**:66–73.

Toback, F. G., Smith, F. D., and Lowenstein, L. M. 1974. Phospholipid metabolism in the initiation of renal compensatory growth after acute reduction of renal mass. *J. Clin. Invest.* **54**:91–97.

Tou, J.-S., Hurst, M. W., and Huggins, C. G. 1968. A phosphatidylinositol kinase in rat kidney cortex. *Arch. Biochem. Biophys.* **127**:54–58.

Tou, J.-S., Hurst, M. W., and Huggins, C. G. 1969. Phosphatidylinositol in rat kidney cortex II. Subcellular distribution and kinetic properties. *Arch. Biochem. Biophys.* **131**:596–602.

Tou, J.-S., Hurst, M. W., Huggins, C. G., and Foor, W. E., 1970. Biosynthesis of triphosphoinositide in rat kidney cortex. *Arch. Biochem. Biophys.* **140**:492–502.

Tou, J.-S., Hurst, M. W., Baricos, W. H., and Huggins, C. G. 1972. The metabolism of phosphoinositides in rat kidney, *in vivo. Arch. Biochem. Biophys.* **149**:146–152.

Tou, J.-S., Hurst, M. W., Baricos, W. H., and Huggins, C. G. 1973. The hydrolysis of triphosphoinositide by a phosphodiesterase. *Arch. Biochem. Biophys.* **154**:593–600.

Ullman, M. D., and Radin, N. S., 1974. The enzymatic formation of sphingomyelin from ceramide and lecithin in mouse liver. *J. Biol. Chem.* **249**:1506–1512.

Veerkamp, J. H., Mulder, I., and Van Deenen, L. L. M. 1962. Comparison of the fatty acid composition of lipids from different animal tissues including some tumors. *Biochim. Biophys. Acta* **57**:299–309.

Wagner, H., Hölzl, J., Lissau, A., and Hörhammer, I. 1963. Papierchromatographic von Phosphatiden III. Quantitative papierchromatogrpahischer Bestimmung von Phosphatiden and Phosphatidsauren in Rattenorganen. *Biochem. Z.* **339**:34–45.

Weinreb, N. J., Brady, R. O., and Tappel, A. L. 1968. The lysosomal localization of sphingolipid hydrolases. *Biochim. Biophys. Acta* **159**:141–146.

Wheldrake, J. F., 1972. Kidney phospholipid synthesis. Effect of adrenalectomy and of aldosterone and other adrenocortical hormones in adrenalectomised rats. *Biochim. Biophys. Acta* **260**:583–592.

Wherrett, J. R., and Cumings, J. N. 1963. Detection and resolution of gangliosides in lipid extracts by thin-layer chromatography. *Biochem. J.* **87**:378–382.

White, D. A. 1973. The phospholipid composition of mammalian tissues, pp. 441–482. *In* G. B. Ansell, J. N. Hawthorne, and R. M. C. Dawson (eds.). Form and Function of Phospholipids Elsevier Scientific Publishing Company.

Wilson, H. Spargo, B., and Getz, G. S. 1973. Changes in kidney medullary phospholipid metabolism in potassium-deficient rat. *Am. J. Pathol.* **71**:295–307.

Wurster, C. F. Jr., and Copenhaver, J. H. Jr. 1965. Incorporation of (^{32}P) phosphate in rabbit kidney lecithins as a function of lecithin unsaturation. *Biochim. Biophys. Acta* **98**:351–355.

Yamakawa, T., and Suzuki, S. 1952. The chemistry of the lipids of post-hemolytic residue or stroma of erythrocytes. III. Globoside, the sugar-containing lipid of human blood stroma, *J. Biochem. (Tokyo)* **39**:393–402.

Yamakawa, T., Nishimura, S., and Kaminura, M. 1965. The chemistry of lipids of post hemolytic residue or stroma of erythrocytes. XIII. Further studies on human red cell glycolipids. *Jap. J. Exp. Med.* **35**:201–207.

Yeung, S. K. F., and Kuksis, A. 1974. Molecular species of ethanolamine phosphatides of dog and pig kidney. *Can. J. Biochem.* **52**:830–837.

Yip, M. C. M. 1972. The enzymic synthesis of gangliosides: Uridine diphosphate galactose: *N*-acetylgalactosaminyl-(*N*-acetylneuraminyl)-galactosyl-glucosylceramide galactosyltransferase in rat tissues. *Biochim. Biophys. Acta* **273**:374–379.

Zambrano, F., Fleischer, S., and Fleischer, B. 1975. Lipid composition of the Golgi apparatus of rat kidney and liver in comparison with other subcellular organelles. *Biochim. Biophys. Acta* **380**:357–369.

Zins, G. R. 1975. Renal Prostaglandins. *Am J. Med.* **58**:14–24.

Gonadal Tissue

John G. Coniglio

I. Introduction

The role of lipids in gonadal tissue is a topic of increasing interest. Progress on the understanding of this role has been hindered at least partly because of the lack of adequate information on the chemistry and metabolism of gonadal lipids. Our knowledge of this subject, limited earlier to observations made microsocopically or histochemically, has more recently advanced at an increasing rate due to the development of adequate chemical and biochemical techniques.

More information is available on the lipid chemistry and metabolism of male than of female reproductive tissue. For this reason and also because of the particular interest of the author, most of the effort in this chapter is devoted to the testis. The testis is important in the study of lipid metabolism not only because lipids have an important role in the structure and function of testicular tissue but also because in the testis there is an active lipid metabolism, and this should prove useful for the study of lipid metabolism in general. Furthermore, the testis is available for the localized administration of substrates (intratesticular injections) and results of experiments done *in vivo* can be compared with those obtained using substrates administered orally or parenterally or with experiments done *in vitro*.

This chapter is a discussion of the lipid composition and metabolism of reproductive tissue in normal animals and the changes brought about by a number of factors such as age and development, hormones, diet, drugs, etc. Although this book is concerned with lipid metabolism in mammals, data on nonmammalian animals are occasionally introduced in this chapter for reasons of comparisons or contrast. The types of lipids discussed

---

John G. Coniglio • Department of Biochemistry, Vanderbilt University, Nashville, Tennessee 37232.

include fatty acids, phospholipids, acylglycerols, free and esterified cholesterol, prostaglandins, glycolipids, and ether-linked glycerolipids, but they do not include steroids. Although the major amount of information is on animals other than man, whenever possible available human data are included because it appears that generally it will not be possible to extrapolate findings in animals to the situation in man. A comprehensive treatment of testicular lipids is available (Johnson, 1970).

II. *Major Lipid and Fatty Acid Composition*

A. *Normal Values in Adult Organs of Various Species*

1. *Testes*

a. Lipid Composition. Studies of lipid composition have been done in many species but the greatest amount of information available is on the rat. In Table 1 are given values for total lipids and several lipid classes in adult rat testes in two different studies. The data of Bieri and Prival (1965) were obtained in experiments using rats maintained on a purified diet containing 4% lard; those of Davis *et al.* (1966) were obtained using rats maintained on Purina laboratory chow. Phospholipids formed the greatest component of the total lipids, and of the phospholipids the major type was phosphatidyl choline. The following values (as percent of total phospholipids) were calculated from the data of Davis *et al.* (1966): phosphatidylcholine (44.6), phosphatidylethanolamine (25.8), phosphatidylserine (5.6), and sphingomyelin (6.3). In addition, small amounts of phosphatidylinositol, phosphatidic acid, lysophosphatidylethanolamine, and polyglycerol phosphatides were present.

TABLE 1
Lipid Composition of Testes of Adult Rats (mg/g wet wt of Tissue)

Lipid	Calculated from data of Davis *et al.* (1966)	Data of Bieri and Prival (1965)
Total lipids	24.2	18.1
Total phospholipids	15.0	13.5
Total cholesterol	—	2.0
Free cholesterol	1.8	—
Esterified cholesterol	0.8	—
Triacylglycerols	0.9	—
Diacylglycerols	0.1	—
Plasmalogens	1.8	—

TABLE 2
Lipid Composition of Human Testes Obtained at Orchidectomy[a]

Lipid class	mg/g wet wt tissue
Total lipids	24.4
Total phospholipids	14.7
Acylglycerols	2.84
Free cholesterol	1.40
Esterified cholesterol	0.56
Phospholipid classes	**% of total phospholipids**
Phosphatidylcholine[b]	48.3
Phosphatidylethanolamine[b]	22.3
Phosphatidylserine	7.42
Phosphatidylinositol	6.64
Sphingomyelin	8.28

[a]Data taken from Coniglio *et al.* (1974b).
[b]Includes the plasmalogen fraction.

Data on human testes are more difficult to obtain, but some information is available. Komatsu and Takahashi (1971) reported the following values: 22.44 mg/g, total lipid; 9.26 mg/g, total phospholipid; and 1.19 mg/g total cholesterol. Bieri and Prival (1965) gave the following results (in mg/g fresh tissue) for five humans (accidental deaths): total lipids, (15.9), total cholesterol (3.4), and phospholipid (8.9). These values may be compared with those reported by Coniglio *et al.* (1974b), shown in Table 2, for tissue obtained at orchiectomy. Differences in the reported values for total lipids and total phospholipids may be due partly to differences in analytical procedures used and partly to the tissues available. A study of lipid composition in testes removed at autopsy (hospital deaths in which the reproductive organs were not involved), published by Coniglio *et al.* (1975b), contains values similar to those reported by these investigators for the orchiectomy cases in the same age groups. It is suggested, therefore, that the use of autopsy tissue for this type of study is valid. This makes available for study an additional valuable source of material.

Lipid composition of testes of several other species was reported by Bieri and Prival (1965). These included the mouse, hamster, rabbit, dog, chicken, and guinea pig. None of these differed radically from the rat except for a very high total lipid value (34.3 mg/g) for the guinea pig.

About 45% of rat testicular phospholipids was determined to be phosphatidylcholine, 26% phosphatidylethanolamine, 5.6% phosphatidylserine, and 6.3% sphingomyelin (Davis *et al.,* 1966). Values obtained for human testicular phospholipids are similar (Table 2). A smaller amount of phosphatidylserine was found by Komatsu and Takahashi (1971), who also reported small amounts of cardiolipin.

The plasmalogen content and aldehyde compostion of the phosphoglycerides of ovine and porcine testes were reported by Neill and Masters (1975), who found that the concentrations of alk-1-enylacyl derivatives were present in minor proportions compared to diacyl analogs. Ovine testes were richer in plasmalogens (chiefly the ethanolamine type) than porcine testes. The aldehyde composition of choline plasmalogen in mature ovine testis was markedly different from other tissues (increased 16:0 and greatly decreased 18:2 chains). In ram testes phosphatidylcholine accounted for about 40% of the total phospholipids (Scott and Setchell, 1968). Only small amounts of choline plasmalogens were found.

The lipid composition of subcellular fractions from rat testis has also been described (Keenan *et al.,* 1972). Phospholipids were in greatest concentration in microsomes followed by Golgi apparatus, residual bodies, and lipid vesicles. The inverse order was observed for neutral lipids. The major phospholipids in all fractions were the choline, ethanolamine, inositol and serine phosphatides, and sphingomyelin, and the relative concentrations in the isolated fractions were similar to those in whole testis. Cardiolipin was found in whole testis and in residual bodies. The major neutral lipids were triacylglycerols, free fatty acids, and cholesterol, and they were present in all fractions.

b. Fatty Acid Composition. Fatty acid composition of rat testicular lipids has been studied in detail by many investigators with general agreement particularly for the major fatty acids. Data for this and other species, including humans, are given in Table 3. Of interest are the variations in linoleic, 20:3(*n*-6), 22:4(*n*-6), 22:5(*n*-6), and 22:6(*n*-3) acids. The concentration of arachidonic acid is the most consistent of the polyenes. Even so, this has the lowest concentration in the guinea pig, which also has the lowest total concentration of the higher polyenes (including 22:5 and 22:6) and which has the highest concentration of 18:2(*n*-6), the precursor of 22:4(*n*-6) and 22:5(*n*-6). It also has a significantly large concentration of 18:3(*n*-3), probably a reflection of the diet.

Holman and Hofstetter (1965) reported a trace of 24:4(*n*-6) in bull testis and 1.4% in boar testis as well as four other "unknown" components of retention time longer than 24:4. Another fatty acid found in small quantities in pork testis and not previously reported was 22:3(*n*-6), a structure deduced from ozonolysis and reduction methods. Using these methods 16 chemical structures were firmly established by these investigators for fatty acids occurring in beef reproductive tissue: 16:1(*n*-7), 18:1(*n*-9), 18:2(*n*-6), 18:3(*n*-3), 20:2(*n*-6), 20:3(*n*-9), 20:3(*n*-6), 20:4(*n*-6), 20:5(*n*-3), 22:3(*n*-9), 22:3(*n*-6), 22:4(*n*-6), 22:5(*n*-6), 22:5(*n*-3), 22:6(*n*-3), and 24:4(*n*-6). Two polyenoic acids of 24-carbon chain length were detected and isolated from rat

TABLE 3

Fatty Acid Composition of Testes of Various Species

Fatty acid	Rat[a]	Mouse[a]	Hamster[a]	Rabbit[a]	Dog[a]	Chicken[a]	Guinea pig[a]	Bull[b]	Boar[b]	Human[a]	Human[c]
14:0	n.r.[d]	n.r.	n.r.	n.r.	n.r.	n.r.	n.r.	1.5	1.4	n.r.	1.62
16:0	35.1	29.5	41.2	26.2	32.0	21.5	21.4	28.3	17.3	27.8	32.9
16:1(*n*-7)	1.7	1.9	t[e]	3.4	t	1.2	4.2	1.6	2.6	3.6	2.98
18:0	6.3	8.7	4.8	9.0	9.3	15.8	11.0	11.1	11.0	11.8	10.7
18:1(*n*-9)	14.7	27.2	15.9	17.7	15.1	26.0	22.7	23.9	13.4	16.4	19.6
18:2(*n*-6)	6.4	5.0	4.4	11.8	4.2	1.4	18.0	7.7	6.4	5.6	5.44
18:3(*n*-3)	0.3	0.9	t	3.1	0.4	0.2	7.9	0.3[f]	0.9[f]	0.7	n.r.
20:3(*n*-6)	0.8	1.1	0.7	4.9	2.9	0.8	1.2	1.8	4.7	6.7	4.88
20:4(*n*-6)	13.7	10.4	11.3	8.4	13.3	11.9	5.0	12.1	15.5	13.4	11.5
22:2(*n*-9)[g]	—	0.3	t	—	—	2.4	—	n.r.	n.r.	—	n.r.
22:4(*n*-6)	1.7	1.2	t	1.9	5.7	10.0	0.6	1.5	3.8	2.1	n.r.
22:5(*n*-6)	13.5	7.3	14.9	10.9	14.0	t	3.0	1.5	12.1	0.4	0.69
22:6(*n*-3)	1.1	4.9	4.4	t	1.6	t	2.4	2.6	4.4	8.5	6.35

[a]Data of Bieri and Prival (1965).
[b]Data of Holman and Hofstetter (1965).
[c]Data of Coniglio *et al.* (1974b).
[d]n.r. = Not reported.
[e]t = Trace.
[f]18:3 + 20:0.
[g]Identification tentative.

testicular tissue by Davis *et al.* (1966). One was shown to be a tetraene and the other a pentaene. The chemical characterization of these as 9,12,15,18-tetracosatetraenoic and 6,9,12,15,18-tetracosapentaenoic acids, respectively, was reported by Bridges and Coniglio (1970a). The structure of the 22:5, which is present in large amounts in adult rat testis, had been established chemically as 4,7,10,13,16-docosapentaenoic acid by Davis *et al.* (1966). In human testicular tissue significant quantities of 20:3(*n*-6), 20:4(*n*-6), and 22:6(*n*-3) are found. Smaller quantities of 22:4(*n*-6) and 22:5(*n*-3) were also reported by Grogan *et al.* (1973).

It is useful for benefit of comparison to indicate the fatty acid composition of vertebrates other than mammals. Such a comparison for testes of sunfish, bullfrog, and quail is available (McMullin *et al.*, 1968). The data for quail are similar to those given in Table 3 for the rooster except for a lower content in quail of 18:1 and larger content of 22:5 and 22:6. For the sunfish and bullfrog the interesting differences (from other animals) are the concentration of the 22-carbon polyenes and the 20:5 in the bullfrog and the large concentration of 22:6 in the sunfish.

In human tissue all phospholipids had a large percentage of saturated fatty acids and a low percentage of unsaturated fatty acids except phosphatidylethanolamine, which was found to have the following percentages of the major acids: 16:0, 30.3; 16:1, 6.8; 18:0, 15.8; 18:1, 13.3; 18:2, 2.8; 20:3, 3.8; 20:4, 9.2; 22:6, 12.6 (Komatsu and Takahashi, 1971). Phosphatidylcholine had less 20:4 and 22:6. The neutral lipids contained large amounts of unsaturated acids, including 18:1, 20:3, and 22:6.

TABLE 4
Fatty Acids of Lipid Classes of Rat Testicular Tissues (% wt)[a]

Fatty acid	Cholesteryl esters	Glyceryl ether diesters	Triacylglycerols	Total polar lipids
16:0	20.9	20.3	24.7	34.9
16:1	4.0	Trace	0.9	0.5
18:0	6.2	4.1	2.7	7.3
18:1	10.9	5.4	9.7	10.5
18:2	13.3	1.9	4.5	4.5
18:3	1.0	0.4	Trace	Trace
20:3	1.3	1.1	1.5	1.0
20:4	14.1	3.3	4.1	15.6
22:4	2.9	6.7	6.4	1.5
22:5	12.2	23.9	33.3	18.0
24:4	4.2	11.7	3.4	0.8
24:5	6.8	13.7	6.7	0.6

[a]Taken from Nakamura *et al.* (1968).

TABLE 5

Fatty Acids of Lipid Classes of Bovine and Porcine Testes (% of Total Fatty Acids)[a]

Fatty acid	Phospholipids	Diacylglycerols	Free fatty acids	Triacylglycerols	Cholesteryl esters
			Bovine testis		
16:0	25.4	20.2	32.3	39.2	17.6
18:0	8.7	4.5	7.4	6.9	13.7
18:1	14.3	6.7	17.0	21.2	15.1
18:2	6.0	3.6	6.0	6.6	4.3
20:4	20.3	5.2	8.4	3.3	2.9
22:5	4.1	4.5	3.9	4.3	3.5
22:6	8.0	8.6	6.9	1.9	Trace
24:4	0.4	22.3	0.9	Trace	26.9
			Porcine testis		
16:0	25.3	17.5	25.7	19.4	11.9
18:0	12.9	10.8	15.0	9.2	2.9
18:1	13.1	7.0	18.1	9.1	10.6
18:2	6.8	11.0	5.4	5.1	3.2
20:4	11.3	4.2	13.4	2.1	3.6
22:5	5.6	9.6	5.4	9.9	27.9
22:6	7.1	4.3	Trace	6.0	7.4
24:4	0	0	0	0	2.4

[a]Taken from Holman and Hofstetter (1965).

In rat testicular tissue the concentration of arachidonic acid is high in phospholipids and in cholesteryl esters (see Table 4), but the concentration of 22:5 is high in all fractions including triacylglycerols. However, the total amount of the fatty acids is largest in the phospholipids because it is the major lipid class in the testis. In contrast to the 22:5, in rat testis we have found a greater amount of the 24-carbon polyenes (tetraenes and pentaenes) in the neutral lipid fraction, as have Nakamura *et al.* (1968) and Carpenter (1971). In mouse testes (Coniglio *et al.* 1975a) about 3% of the total fatty acids of cholesteryl esters were 24:4 and another 3% were 24:5. About 1–1.5% of the fatty acids of acylated glycerols and of phospholipids was represented by each of these 24-carbon polyenes. Hydrogenated samples also revealed the presence of 26- and 28-carbon fatty acids, but the polyenoic acids yielding these saturated derivatives have not been isolated and characterized yet. The presence of 24-carbon fatty acids has been reported also in diacylalkylglycerols (Nakamura *et al.*, 1968).

In bovine testes (Table 5) very little 22:6 was found in triacylglycerols although significant concentrations were found in phospholipids, diacylglycerols, and free fatty acids. Linoleic acid and 22:5 were distributed in all fractions, but the major concentration of arachidonic acid was in phospholipids. Diacylglycerols and cholesteryl esters were relatively rich in 24:4. This was not observed in the analyses reported by Neill and Masters (1974a) on bovine testes, but very little was found in phospholipids and in free fatty acids and only a trace was reported in triacylglycerols. In porcine testes Holman and Hofstetter (1965) found that phospholipids and free fatty acids had large concentrations of arachidonic acid, that 22:5 was the fatty acid in highest concentration in cholesteryl esters, and that 22:6 was distributed in all fractions except free fatty acids.

The major saturated fatty acids of phosphatidylcholine, phosphatidylethanolamine, and sphingomyelin from rabbit testes were palmitic and stearic, and the main monoenoic acid oleic (Morin, 1967). Linoleic acid ranged around 5–7% (of total fatty acids) in these phospholipids. Significant quantities (4–5%) of 20:3(*n*-6) were present in all three phospholipids. The greatest amount of 22:5 and of arachidonic acid was in phosphatidylethanolamine (19% and 17%, respectively), with lesser amounts in phosphatidylcholine (13.6% and 15.5%, respectively) and the lowest amounts in sphingomyelin (12.0% and 7.8%, respectively).

In all subcellular fractions isolated from homogenates of rat testes there were high levels of unsaturated fatty acids, particularly 18:1, 20:4, and 22:5 (Keenan *et al.*, 1972). The most highly unsaturated polar lipids were phosphatidylcholine, phosphatidylethanolamine, and phosphatidylserine, whereas the lowest amount of unsaturated acids was in phosphatidylinositol.

2. *Ovaries*

a. Lipid Composition. In contrast to the abundant data available for lipid composition of testicular tissue, the reports of lipid composition of ovarian tissue are few. Data for bovine and porcine ovaries were given by Holman and Hofstetter (1965) and for guinea pigs by Sharma and Venkitasubramanian (1973). The total lipid content of beef ovary was 11.4%, of pork ovary 6.3%, and of guinea pig ovary 5.25%. The amounts of major lipid classes are given in Table 6. As with most organs, the chief constituents of pork and beef ovaries were phospholipids, but in the data reported by Sharma and Venkitasubramanian (1973) for guinea pig ovary, the concentration of triacylglycerols was slightly higher than that of phospholipids and, in addition, there was a considerable amount of mono- and of diacylglycerols. Both free and esterified cholesterol were present in significant quantities. In the ovary of the rabbit the free cholesterol concentration was 3.1 mg/g tissue and the esterified cholesterol 6.0 mg/g. In the human these values were 1.4 and 1.1, respectively (Morin, 1971).

The amounts of individual phospholipids for rabbit ovaries reported by Morin (1968) are shown in Table 7: The major phospholipids found were

TABLE 6
Lipid Composition of Bovine, Porcine, and Guinea Pig Ovarian Tissue (% of Total Lipids)

Lipid class	Bovine ovary[a]	Porcine ovary[a]	Guinea pig ovary[b]
Phospholipids	56.5	69.9	35.4
Free fatty acids	9.8	2.9	Not reported
Triacylglycerols	19.3	15.5	39.9
Cholesterol esters	3.0	4.3	6.1
Free cholesterol	Not reported	Not reported	4.4
Diacylglycerols	Not reported	Not reported	6.0
Monoacylglycerols	Not reported	Not reported	8.1

[a] Data taken from Holman and Hofstetter (1965).
[b] Data calculated from data of Sharma and Venkitasubramanian (1973).

phosphatidylcholine and phosphatidylethanolamine, the latter having a concentration only slightly lower than the former. Phosphatidylserine, sphingomyelin, phosphatidylinositol, and lysophosphatidylcholine followed in order of decreasing concentrations.

In ovarian follicles from laying hens Christie and Moore (1972) found the triacylglycerols to be the major lipid class (67% of total lipids), with the phospholipids being the only other large component (30%). Small amounts of diacylglycerols (2%) and of cholesteryl esters (1%) were also detected. Phosphatidylcholine made up the greatest proportion of the total phospholipids (nearly 80%), followed by phosphatidylethanolamine (17–18%), sphingomyelin (2%), and lysophosphatidylcholine (2%). Only traces of phosphatidylinositol were present and phosphatidylserine was not detected.

b. Fatty Acid Composition. The polyunsaturated fatty acids of ovarian lipids form the major component of the total lipid fatty acids (49.5% for

TABLE 7
Amounts of Individual Phospholipids in Rabbit Ovaries[a]

	mg phospholipid/g dry wt.
Lysophosphatidylcholine	1.5
Sphingomyelin	2.3
Phosphatidylcholine	17.5
Phosphatidylinositol	2.1
Phosphatidylserine	6.8
Phosphatidylethanolamine	14.1

[a] Data of Morin (1968).

beef ovary and 60.4% for pork ovary, Holman and Hofstetter, 1965). A summary of the fatty acids reported by these investigators is given in Table 8. Linoleic and arachidonic acids were the major polyenoic acids found in the ovaries, but considerable quantities (particularly in pork tissue) of 22:4(*n*-6) and some 22:5(*n*-3) also were present. Both beef and pork ovaries 22:4(*n*-6) was enriched mostly in the cholesteryl ester and in the triacylglycerol fractions, while 24:4(*n*-6) was present mostly in the cholesterol ester fraction of beef and exclusively in the cholesteryl ester fraction in pork ovary. A summary of the distribution of fatty acids in lipid classes of beef and pork ovaries as reported by Holman and Hofstetter (1965) is given in Table 9.

The major fatty acids of cholesteryl esters of rabbit ovarian tissue were reported by Morin (1971) to be (as % of total cholesteryl ester fatty acids) 16:0 (17.5%), 18:0 (15.7%), 18:1 (39.2%), and 18:2 (20.8%). Arachidonic acid was present in very low concentration (2.2). In humans the major fatty acids were 16:0 (14.7%), 18:0 (6.9%), 18:1 (33.4%), 18:2 (32.1%), and 20:4 (8.8%).

Morin (1968) determined the fatty acid composition of phosphatidylcholine and of phosphatidylethanolamine isolated from rabbit ovaries. His

TABLE 8
Fatty Acid Composition of Total Lipids from Bovine and Porcine Ovaries (% of Total Lipid)[a]

Fatty acid	Bovine ovary	Porcine ovary
14:0	1.5	0.1
16:0	14.6	11.4
16:1(*n*-7)	2.5	1.1
18:0	14.1	13.0
18:1(*n*-9)	15.3	14.2
18:2(*n*-6)	16.5	13.4
18:3(*n*-6)	Trace	0.2
18:3(*n*-3) } 20:0 }	1.3	0.7
20:2(*n*-6)	3.1	1.4
20:3(*n*-6)	3.7	1.9
20:4(*n*-6)	13.4	26.2
20:5(*n*-3)	Trace	0.2
22:4(*n*-6)	6.4	12.3
22:5(*n*-6)	Trace	0.8
22:5(*n*-3)	3.9	1.5
22:6(*n*-3)	Trace	0.8
24:4(*n*-6)	1.2	Trace

[a]Data taken from Holman and Hofstetter (1965).

TABLE 9
Fatty Acid Composition of Lipid Classes of Bovine and Porcine Ovaries (% of Total Fatty Acids)[a]

	Bovine ovary				Porcine ovary			
Fatty acids	P[b]	FFA[c]	TG[d]	CE[e]	P	FFA	TG	CE
14:0	1.2	1.9	1.8	4.5	0.3	0.8	0.4	0.7
16:0	14.8	11.1	10.1	15.3	17.8	13.4	15.7	8.2
16:1(*n*-7)	1.9	5.7	6.0	7.3	1.3	1.3	1.6	2.3
18:0	15.3	13.7	9.4	5.8	16.0	13.8	13.1	6.9
18:1(*n*-9)	10.3	14.5	15.3	12.4	10.9	15.7	20.3	21.6
18:2(*n*-6)	12.1	13.9	17.3	19.1	10.2	12.5	14.9	12.6
18:3(*n*-3) } 20:0 }	1.8	2.7	2.4	2.2	1.0	1.0	1.4	1.4
20:2(*n*-6)	2.4	6.1	6.9	2.7	0.8	1.7	1.6	2.1
20:3(*n*-6)	3.1	4.9	5.7	3.1	1.8	2.0	1.7	3.5
20:4(*n*-6)	13.0	11.1	7.8	6.9	22.0	19.7	12.8	17.0
20:5(*n*-3)	3.9	0.4	Trace	Trace	1.5	Trace	0	Trace
22:4(*n*-6)	5.3	6.8	9.9	10.8	6.4	13.3	15.0	15.2
22:5(*n*-6)	0.8	0.3	Trace	Trace	0.9	0.5	Trace	1.0
22:5(*n*-3)	8.0	4.1	4.0	3.2	2.3	1.4	1.4	1.3
22:6(*n*-3)	0	Trace	0	0	3.6	0.8	Trace	0
24:4(*n*-6)	0	0.4	0.8	5.8	0	0	0	3.4

[a]Data taken from Holman and Hofstetter (1965).
[b]P = Phospholipids.
[c]FFA = Free fatty acids.
[d]TG = Triacylglcyerols.
[e]CE = Cholesteryl esters.

data are given in Table 10. Major differences between these two classes appeared in the content of 16:0, 18:0, 18:2, and 20:4. No values were reported for 22:6. In view of the large amount of this polyene in phosphatidylethanolamine from most tissues, one wonders whether ovarian tissue is uniquely different or the polyene was not included in the fatty acid determination.

Only four fatty acids were reported by Christie and Moore (1972) in triacylglycerols of ovarian follicles from laying hens: 16:0 (27.7%), 18:0 (4.8%), 18:1 (48.5%), and 18:2 (13.1%). In addition to these four (in which the percentages were 35.4%, 9.1%, 26.8%, and 14.2%, respectively) the phosphatidylcholine fraction had 3.8% of 20:4, 0.8% of 22:4, and 7.1% of 22:6. The values for the same fatty acids of the phosphatidylethanolamine fraction were (in the same order): 19.3%, 21.0%, 14.2%, 7.5%, 11.9%, 1.9%, and 20.5%.

In the rat the major ovarian cholesteryl ester fatty acids were reported by Carney and Walker (1972) to be (in mole %) 16:0 (6.7), 18:0 (8.8), 18:1

TABLE 10
Major Fatty Acid Composition of Phosphatidylcholine and Phosphatidylethanolamine of Rabbit Ovaries (% of Total Phospholipids)[a]

Fatty acid	Phosphatidylcholine	Phosphatidylethanolamine
16:0	36.2	12.4
16:1	2.0	1.6
18:0	14.3	29.6
18:1	19.4	24.7
18:2	9.1	5.5
20:4	19.1	26.3

[a] Data of Morin (1968).

(12.3), 18:2 (4.3), 20:1 (4.0), 20:2(*n*-6) (2.1), 20:3(*n*-6) (2.1), 20:4(*n*-6) (11.3), 22:3(*n*-6) (1.5), 22:4(*n*-6) (21.5), 22:5(*n*-6) (8.8), 22:6(*n*-3) (8.4), and 24:4(*n*-6) (2.9). The major triacylglycerol fatty acids were: 16:0 (20.2), 16:1 (5.1), 18:0 (3.7), 18:1 (32.5), 18:2 (33.1), 20:4(*n*-6) (2.1), and 22:4(*n*-6) (1.0). In the phospholipids the composition of fatty acids was 16:0 (20.5), 18:0 (18.0), 18:1 (16.7), 18:2 (8.8), 20:4(*n*-6) (25.4), 22:4(*n*-6) (4.8), and 22:5(*n*-6) (1.3).

B. *Changes with Development*

1. *Testes*

a. Lipid Composition. Changes in lipid composition of testes of rats occur during development even though the total lipid concentration may remain relatively constant. The data in Table 11 show changes in esterified cholesterol, acylglycerols, and plasmalogens during the period 4–29 weeks of age. Decreased concentrations of sterol esters and acylglycerols during this period were accompanied by some increase in plasmalogens. The concentrations of phospholipids and of free cholesterol were relatively constant (Davis *et al.*, 1966). Oshima and Carpenter (1968) reported a slight decrease in testicular phospholipid content of adult rats compared to young rats and slightly less neutral lipid in prepubertal rats compared to older rats. Data for bovine testicular lipids as a function of age are given in Table 12. The values for the 8½-month prepartum period were similar to those given by these authors (Ahluwalia and Holman, 1966) for the 4- and 6-month prepartum periods. Total lipid concentration increased from prepartum to postpartum period, but was relatively constant through the 1-month to 12-year postpartum periods. This was also true for the concentrations of the phospholipids and triacylglycerols, although the latter increased slightly at the 10-month and 12-year periods. Concentrations of neutral lipids and of

TABLE 11

Concentration of Lipids in Testis of the Maturing Rat[a] *(μmoles/g wet wt of Tissue*[b]*)*

Age (weeks)	Total lipid (mg/g)	Sterol ester	Triacylglycerols	Free cholesterol	Diacylglycerols	Lipid phosphorus	Plasmalogens
4	26.4 ± 0.01	0.25 ± 0.02	2.9 ± 0.1	4.7 ± 1.0	0.33 ± 0.02	16.6 ± 0.3	1.80 ± 0.16
7	25.4 ± 0.01	0.16 ± 0.05	1.2 ± 0.2	4.0 ± 0.2	0.23 ± 0.02	16.6 ± 0.1	2.05 ± 0.04
13	26.1 ± 0.03	0.15 ± 0.01	1.4 ± 0.3	5.8 ± 1.0	0.21 ± 0.005	14.1 ± 2.3	2.10 ± 0.34
29	24.2 ± 0.05	0.12 ± 0.01	1.0 ± 0.1	4.6 ± 0.1	0.24 ± 0.005	16.6 ± 0.3	2.38 ± 0.07

[a]Data of Davis *et al.* (1966).
[b]Mean values ± standard error of the mean.

TABLE 12
Content of Lipids in Bovine Testes as a Function of Age[a]

Lipid class	8.5 months prepartum	Postpartum 1 month	Postpartum 10 months	Postpartum 12 years
Total lipid (% wet wt)	1.8–2.0	2.6–3.0	2.0–3.0	3.2–2.4
Phospholipid (% total lipid)	34–38	53–47	52–48	60–50
Neutral lipid (% total lipid)	66–58	50–46	55–48	46–40
Free cholesterol (% neutral lipid)	64–60	38–32	28–22	30–26
Triacylglycerol (% neutral lipid)	4–3	8–6	13–9	12–10

[a]Data of Ahluwalia and Holman (1966).

free cholesterol decreased from the prepartum to the postpartum period. The concentration of neutral lipids stayed constant in the postpartum period while that of free cholesterol decreased slightly with age. Diacylglycerols and cholesteryl esters were listed as minor components.

Neill and Masters (1974a) found increases in total bovine testicular lipids from age 2 weeks to maturity. Cholesteryl ester concentrations were highest at 4 months, but acylglycerols were highest in mature tissue. The proportions of phospholipids relative to each other remained constant with age, but the total phospholipids increased with age.

Little information is available for humans, but some data on infants are available (Coniglio *et al.*, 1975b), which allow comparison with adults. Data obtained at autopsy on three infants and on adults in two age groups are given in Table 13. A greater amount of all the lipid classes was found in

TABLE 13
Lipid Composition of Human Testes Removed at Autopsy (mg/g wet wt of Tissue)[a]

	Infants (25 hr)	Infants (58 hr)	Infants (4 months)	Adult (56–89 years)	Adult (19–51 years)
Total lipids	45.8	45.3	22.9	26.7	28.5
Total phospholipids	20.8	20.1	15.2	15.6	15.9
Acylglycerols	4.53	3.21	3.83	3.42	3.48
Free cholesterol	2.80	2.52	1.43	1.49	1.46
Esterified cholesterol	1.01	0.91	1.05	0.83	0.59

[a]Data of Coniglio *et al.* (1975b).

the testes of the two younger infants reported (25 and 58 hr old) compared to testes of a 4-month-old infant and of the adults. No outstanding differences were apparent between the 4-month specimen and the adult tissue, although the content of cholesteryl esters in the 19–51 age group was the lowest of all ages reported.

b. Fatty Acid Composition. The most dramatic change in lipid composition with development of testes has been noted in the fatty acid composition. In the rat the most significant change is an increase in 22:5(*n*-6); in the rooster it is 22:4(*n*-6); and in the human it is 22:6(*n*-3). It has been known for a long time that testes contain large amounts of polyenoic acids (Holman and Greenberg, 1953; Cole, 1956). Aaes-Jorgensen and Holman (1958) were the first to report an increase in the pentaenoic acid content of testes of rats during the age period of 4–21 weeks, but they did not elaborate on this finding. In a study of the polyunsaturated fatty acids of growing rats Kirschman and Coniglio (1961) reported a threefold increase in pentaenoic acid in testes of rats during the age period 3 weeks to 3 months. Ahluwalia and Holman (1966) reported that polyunsaturated fatty acids increased with age in all lipid classes of bovine testes, and prominent among these was the increase in 22:6(*n*-3) in the phospholipids of animals of ages 1 month to 12 years. Neill and Masters (1974a) reported that docosahexaenoic acid in bovine testes increased with maturation, but arachidonic acid decreased. In mature testes, plasmalogenic aldehydes were more saturated than in immature testes.

The increase in 22:5(*n*-6) in testes of rats from age 4 weeks to 6 months occurred in phospholipids, but not in triacylglycerols (Davis *et al.*, 1966). During this time there was also a decrease in the concentration of oleic acid in both the phospholipid and triacylglycerol fractions and of palmitic and linoleic acids in the triacylglycerol fraction (see Table 14). Parallel histological studies indicated that the lipid changes occurred at the same time as the appearance and maturation of the spermatids.

Hall and Burdett (1975) have reported that only slight changes occurred in relative amounts of individual fatty acids in testes of rats maintained on ordinary rat diet from age 1½ to 20 months in either the total lipids or in the phospholipid fraction. Progressive decreases in the concentration of essential fatty acids of rat testes had been observed by Turchetto *et al.* (1969) as the animals aged from 2 to 24 months. It should be noted that these investigators report significant quantities of 22:4, but no 22:5, and it is probable that the values given for 22:4 in reality should be those for 22:5.

The small amount of information available on human testes suggests that increases in certain polyenes also occur with development. Data obtained on three infants and on adults in two age groups (Coniglio *et al.*, 1975b) are shown in Table 15. Increases in 20:3(*n*-6) and in 22:6(*n*-3) with

TABLE 14
Major Fatty Acids of Phospholipids and of Triacylglycerols of Testes of Rats of Various Ages[a]

	Age (weeks)	Fatty acids (mg/100 g wet wt of tissue)						
		16:0	18:0	18:1	18:2	20:4	22:4	22:5
Phospholipids	4	303	62	163	28	189	11	80
	7	302	53	122	42	175	12	150
	13	243	39	82	25	150	10	174
	29	281	43	101	39	182	10	179
Triacylglycerols	4	60	9.3	85.0	46.7	4.3	2.7	14.3
	7	32	3.5	28.0	15.0	3.5	2.5	14.5
	13	33	2.0	25.7	13.3	3.7	4.0	20.0
	29	24	1.8	20.0	9.5	2.5	2.5	14.5

[a]Data of Davis *et al.* (1966).

increasing age are apparent with parallel decreases in 16:0, 18:0, and 18:1. There is only a small amount of 20:3(*n*-6) in testes of infants despite the large amount of its precursor, 18:2, and of its metabolic derivative, 20:4. The increase in 22:6 occurred apparently without any significant accumulation in testicular tissue of its precursor, 18:3(*n*-3).

In testes of chicks fed a corn-oil-supplemented basal diet, 22:4(*n*-6) increased at about week 8 and reached a maximal concentration after 15 weeks (Nugara and Edwards, 1970). At 15 weeks spermatids were observed histologically.

TABLE 15
Fatty Acid Composition of Total Lipids of Human Testes Removed at Autopsy (% of Total Fatty Acids)[a]

	Infants			Adult	
Fatty acid	(25 hr)	(58 hr)	(4 mo)	(19–53 years)	(56–89 years)
14:0	1.0	2.5	1.2	1.1	1.0
16:0	35.5	33.2	18.0	26.5	27.8
16:1	3.8	1.1	1.8	3.8	3.2
18:0	18.5	15.1	14.3	10.5	11.7
18:1	27.0	24.4	12.9	15.4	20.5
18:2	3.0	9.5	15.2	6.5	4.7
20:3	1.0	0.8	4.4	7.9	5.3
20:4	13.0	6.7	17.1	12.4	11.5
22:5	Trace	0.3	1.2	1.8	1.4
22:6	0.5	0.8	2.6	8.1	7.5

[a]Data of Coniglio *et al.* (1975b).

2. *Ovaries*

There is little available information on the lipid biochemistry of ovaries during development, although there are numerous morphological studies in which "lipid droplets" are described. Thus, Gondos and Hobel (1973) described "small lipid droplets" in the granulosa cell cytoplasm of the prefollicular human ovary of a few weeks gestation. They considered that the accumulation of lipid droplets was derived from the breakdown of ingested cellular material similar to the lipid material present in the Sertoli cell of the testis and hypothesized that the lipid material might be utilized as metabolites for the germ cells. Lipid droplets in interstitial tissue in fetuses were also described by Stegner (1973).

C. *Effect of Diet, Drugs, and Other Factors*

1. *Testes*

a. Dietary Effects. Lipid and fatty acid composition of testes is readily affected by dietary manipulations. The testes apparently accumulate lipid material during starvation (Lynch and Scott, 1951) as they apparently do in protein deficiency (Horn, 1955).

Deficiency of fat in the diet has been known since the work of Burr and Burr (1930) and of Evans *et al.* (1934) to result in changes in testes leading to sterility. Accumulation of 20:3(*n*-9) occurs readily in testes of fat-deficient rats as it does also in some other organs.

The total amount of neutral lipids is less in the testis of the rat fed an essential fatty acid deficient diet than in the normal rat, but the concentration (mg lipid per g testes) does not change significantly (Hølmer and Aaes-Jørgensen, 1969). The same is true of total phospholipids, chiefly due to decreases in phosphatidylcholine and phosphatidylethanolamine, whereas sphingomyelin actually increases slightly. The major part of the neutral lipid is triacylglycerol, but both free and esterified cholesterol increase slightly in the testis of the deficient rat.

In immature rabbits fed a purified fat-free diet for 14 weeks Ahluwalia *et al.* (1967) found decreased testicular total lipids, phospholipids, and free cholesterol, but triacylglycerols were increased. The seminiferous tubules of the testes showed extensive degeneration. Changes in testicular total fatty acids with the deficiency were increases in 16:1(*n*-7), 18:1(*n*-9), and 20:3(*n*-9) and decreases in 18:2(*n*-6), 20:4(*n*-6), 20:5(*n*-3), 22:4(*n*-6), 22:5(*n*-6), 22:5(*n*-3), and 22:6(*n*-3).

Predepletion by feeding a fat-deficient diet before weaning is effective in shortening the time required to produce essential fatty acid deficiency. This is well illustrated by the report of Bieri *et al.* (1969). They fed a fat-free diet to pups starting 11 days before weaning and observed damage to the testis in 6–9 weeks compared to more than 16 weeks required if the fat-free diet was started at weaning. Lipid analyses at the earlier time did not correlate with the histologically observed damage. Testes appeared to be particularly vulnerable to deficiencies during the developmental period. If maturation occurs, a longer period of deficiency is required. It was suggested by these authors (as well as others) that secondary effects (insufficiency of pituitary gonadotropins) are responsible for the degeneration of the testes.

The fatty acid composition of the testis responds to a marked degree to dietary fatty acid if the amount of fatty acids in question are sufficiently high in the diet fed. However, metabolic derivatives of the fed fatty acid may be accumulated rather than the original acid fed. Thus, feeding 18:2(*n*-6) increased the amount of 22:5(*n*-6) and feeding 18:3(*n*-3) increased the amount of 22:6(*n*-3) (Bieri and Prival, 1965). Certain 20- and 22-carbon polyenes, 20:5 and 22:6(*n*-3), do accumulate in testes of rats fed these compounds in diets containing cod liver oil for 9 weeks as 10% of the diet when compared to rats fed corn oil diets. The data are shown in Table 16 (Coniglio *et al.*, 1974a). The cod liver oil fed had 9.3% 20:5 and 11.6% 22:6 (% wt).

Deficiencies of certain vitamins, notably A and E, affect the testis adversely, leading to degeneration and sterility. Feeding a vitamin-A-deficient diet to rats for several weeks led to increased total lipids in testis and increased cholesterol (Bieri and Prival, 1966).

In short-term studies (5 weeks) of rats on a vitamin-A-free diet significant increases were observed in testicular total lipid, esterified cholesterol,

TABLE 16
Testicular Fatty Acids of Cod-Liver-Oil-Fed Rats[a] *(% of Total Fatty Acids)*[b]

Fatty acid	10% Cod liver oil	10% Cod liver oil + corn oil	Corn oil
16:0	40.9 ± 5.6	41.2 ± 6.3	43.9 ± 1.0
18:0	6.4 ± 1.4	6.0 ± 1.4	5.3 ± 0.7
18:1	25.5 ± 2.9	19.6 ± 3.3	16.2 ± 0.1
18:2	2.1 ± 1.0	3.8 ± 1.1	2.9 ± 0.1
20:4	9.0 ± 3.0	11.6 ± 1.7	14.1 ± 1.3
20:5	Trace	Trace	—
22:5	4.0 ± 2.0	9.8 ± 3.6	12.4 ± 0.4
22:6	5.6 ± 4.3	5.9 ± 5.6	—

[a]Data of Coniglio *et al.* (1974a).
[b]Mean ± standard error.

TABLE 17
Fatty Acid Composition of Total Phospholipids from Testicular Subcellular Fractions of Control and Retinol-Deficient Rats[a]

	Subcellular fractions					
	Mitochondria		Microsomes		Soluble fraction	
Fatty acid	Control	Deficient	Control	Deficient	Control	Deficient
			%[b]			
15:0[c]	Trace[e]	Trace	2.8 ± 1.6	2.2 ± 1.3	2.8 ± 1.5	2.5 ± 1.0
16:0	34.8 ± 1.9	34.1 ± 3.4	37.5 ± 3.5	31.4 ± 2.9	33.3 ± 1.8	27.5 ± 3.1
16:1	3.7 ± 1.7	Trace	Trace	1.4 ± 0.2	2.4 ± 0.3	2.3 ± 0.2
18:0	21.9 ± 0.8	13.4 ± 2.6[d]	6.8 ± 0.8	13.4 ± 1.7[d]	5.4 ± 2.4	9.6 ± 2.3
18:1	20.0 ± 5.9	19.6 ± 1.1	12.5 ± 1.3	17.9 ± 1.4[d]	11.4 ± 0.7	11.3 ± 0.8
18:2(*n*-6)	4.3 ± 0.8	5.6 ± 0.3	4.2 ± 0.7	5.5 ± 1.9	3.4 ± 1.5	7.9 ± 0.7[d]
20:4(*n*-6)	10.3 ± 2.0	12.2 ± 4.3	16.6 ± 1.9	18.8 ± 0.9	12.5 ± 2.0	20.3 ± 3.9[d]
22:4(*n*-6)	Trace	Trace	Trace	Trace	5.2 ± 1.3	3.0 ± 0.6
22:5(*n*-6)	9.2 ± 3.1	7.0 ± 1.8	18.3 ± 2.3	8.4 ± 3.1[d]	24.3 ± 3.9	11.8 ± 3.8[d]
%Saturated	56.7	47.5	47.1	47.0	41.5	39.6
%Unsaturated	47.5	44.4	51.6	52.0	54.0	53.6

[a] Data of Krause and Beamer (1974).
[b] % of total fatty acids in phospholipids; means ± SD of 3 animals.
[c] Represents the total of C_{15} and below.
[d] Significantly different from control, $P < 0.05$.
[e] Trace = < 1.0%.

phospholipids, and triacylglycerols (Butler *et al*, 1968). The effect of vitamin A deficiency on lipids of subcellular fractions of testes was studied by Krause and Beamer (1974). In mitochondria and microsomes of testes of retinol-deficient rats, there was increased lipid (principally phospholipids and cholesterol) associated with proteins. The fatty acid composition of phospholipids from the subcellular fractions of the deficient and control rats is shown in Table 17. Phosphatidylcholine was the major phospholipid in all fractions. In the organs of controls the phospholipid fatty acid patterns were similar for microsomes and the soluble fraction. In the deficient rats there was a decrease in 22:5 in both fractions, a decrease in 18:0 and 18:1 in the microsomes, and an increased amount of 18:2 and of 20:4 in the soluble fraction. Retinol deficiency resulted in an increased turnover of testicular phospholipids.

Bieri and Prival (1966) studied the effect of a deficiency of retinol as well as α-tocopherol and zinc on lipids of rat testes. Under these conditions, total lipids decreased in testes after 12–19 weeks on the deficient diet. Phospholipid was responsible for most of the change and testes of α-tocopherol-deficient rats had the lowest amount of phospholipids. The concentration of 20:4(*n*-6) was unchanged by retinol deficiency, but it increased to twice the normal concentration in testes of the α-tocopherol-deficient rats and moderately increased in the zinc-deficient rats. The concentration of 22:4(*n*-6) increased only in α-tocopherol deficiency while the concentration of 22:5(*n*-6) decreased to about one-third of normal in tocopherol deficiency and about two-thirds of normal in zinc- or retinol-deficient rats. Dietary selenium had no effect on testicular lipid composition either in the presence or absence of α-tocopherol. The retinol-deficient rats were given retinoic acid. After 14 weeks the amount of triacylglycerols in these testes was higher than in the controls. The absolute amounts of 20:4 and 22:5 did not decrease because of the dilutions of the triacylglycerol fatty acids that were not polyenoic. The results for lipid and fatty acid composition of testes of normal, retinol-, α-tocopherol-, and zinc-deficient animals are shown in Table 18.

The increase in 20:4(*n*-6) seen in testes of α-tocopherol-deficient rats occurred in phosphatidylcholine, phosphatidylethanolamine, and sphingomyelin (Bieri and Andrews, 1964). In the former two phosphatides, 22:4(*n*-6) increased fourfold. In total phospholipids, 22:5(*n*-6) decreased from a control value of about 20% to 2.8%. These authors suggested that there might be a block in the conversion of 20:4(*n*-6) to 22:5(*n*- 6) in vitamin E deficiency. This is in contrast to a suggestion of Bernhard *et al.* (1963) that increased synthesis of 20:4(*n*-6) in vitamin-E-deficient rats was responsible

TABLE 18
Summary of Lipid and Fatty Acid Composition of Testes from Normal, α-Tocopherol-, Retinol-, and Zinc-deficient Rat Testes[a,b]

	Normal	α-Tocopherol-deficient	Retinol-deficient[c]	Zinc-deficient
Number of rats	4	4	4	4
Age, weeks[d]	12–15	19	15	12–15
Wt. of testes (g)	3.02 ± 0.09	1.49 ± 0.30	1.41 ± 0.05	1.38 ± 0.10
Total lipid (mg/g)	18.1 ± 0.7	17.7 ± 1.4	14.0 ± 0.5	16.5 ± 1.4
Phospholipid (mg/g)	13.9 ± 0.4	8.6 ± 1.1	10.1 ± 0.2	11.9 ± 1.0
Cholesterol (mg/g)	1.8 ± 0.0	2.4 ± 0.1	2.3 ± 0.1	2.2 ± 0.1
Fatty acids				
16 ald.	1.6 ± 0.4	0.8 ± 0.2	0.7 ± 0.2	0.9 ± 0.2
16:0	33.0 ± 0.9	25.2 ± 0.5	26.7 ± 3.0	30.6 ± 2.5
16:1	2.1 ± 0.4	3.2 ± 0.4	2.0 ± 0.9	1.8 ± 0.5
18:0	5.6 ± 0.2	9.8 ± 0.7	9.5 ± 1.2	8.7 ± 0.8
18:1	20.8 ± 2.0	22.1 ± 1.9	27.0 ± 0.3	19.9 ± 2.4
18:2(*n*-6)	2.6 ± 0.1	3.4 ± 0.1	6.5 ± 0.1	2.9 ± 0.1
20:4(*n*-6)	12.9 ± 0.5	24.3 ± 1.1	14.6 ± 0.8	17.8 ± 0.6
22:4(*n*-6)	1.3 ± 0.1	3.8 ± 0.2	1.8 ± 0.2	2.1 ± 0.2
22:5(*n*-6)	17.3 ± 1.0	5.3 ± 0.9	10.1 ± 1.8	12.4 ± 0.7

[a]Data of Bieri and Prival (1966).
[b]Values are means ± SE.
[c]This diet contained retinoic acid, 20 mg/kg. The fat was 4% of cottonseed oil; the diets of the other groups contained 5% of stripped lard.
[d]Postweaning.

for the increased 20:4(n-6) seen in livers of such animals. In contrast to these findings, Carpenter (1971) reported that in α-tocopherol deficiency there is no accumulation of intermediates in the biosynthetic pathway for 22:5(n-6). However, those results are reported relative to the amount of stearic acid although the reason for this is not obvious. In those studies there appeared to be a greater amount (as weight percent) of stearic acid in testicular phospholipids of deficient compared to control rats of the same ages, but statistical evaluation was not reported. Consequently, if the amount of stearic acid was increased with the deficiency, a corresponding increase in other fatty acids would not be evident if the amount were calculated relative to the amount of stearic acid. In the data of Bieri and Prival (1966), the amount of stearic acid in testes of α-tocopherol-deficient rats was indeed higher than in the normals. Carney and Walker (1971) also found decreased levels of 22:5(n-6) and palmitic acid in testes of vitamin-E-deficient rats and slightly elevated levels of 20:4(n-6) and 22:4(n-6). Our own findings (Whorton and Coniglio, 1975a) indicate that the concentration of 22:5(n-6) is reduced somewhat in testes of vitamin-E-deficient rats although there may be no changes observed histologically. However, marked decreases in 22:5(n-6) were observed only in testes with marked degeneration. In the latter the amounts of 20:4(n-6) and 22:4(n-6) were increased slightly. No increase in either of these two polyenes was observed in vitamin-E-deficient rats in which there was no degeneration of seminiferous tubules.

In pigs given a cottenseed oil supplemented, basal pea diet vitamin E deficiency influenced markedly the concentration of 22:6 (McDowell *et al.*, 1974). Addition of vitamin E to the basal diet containing oil increased the concentration of 22:6 in the testes from an undetectable amount to 5.4% of the testicular fat. It also lowered the concentration of 22:5 (from 2.5% to 0.1%), that of 22:4 (from 6.9% to 2.6%), and of 22:3 (from 3.3% to 1.8%); it did not affect the concentration of 20:4. Tinsley *et al.* (1971) analyzed testes of vitamin-E-deficient chickens fed safflower oil and found very little change in the concentration of 22:4(n-6) (decreased from 14.5% to 12.5%), 18:1 (increased from 14–16% to 18–20%), and 18:2 (4.8% to 6.1%). They reported that the fatty acid changes were more pronounced in the smaller testes and that fertility was maintained even without an antitoxidant supplement.

b. Cryptorchidism and $CdCl_2$ Administration. Conditions or agents that cause degeneration of the seminiferous tubules result in changes in lipid and fatty acid composition of the testes. Surgical cryptorchidism, in which a testicle is removed into the abdomen, resulted in such changes in the cryptorchidic testicle, while the organ left in its normal position remained unchanged (Davis and Coniglio, 1967). The major changes included a reduced amount of 16:0 and 22:5(n-6) and an increased amount of 18:0, 18:1, and 22:4 in the cryptorchidic testis. There was a decrease in the total amount of phospholipid and a large increase in the total amount of cholesteryl esters in the cryptorchidic testis 28 days after surgery. In the rabbit, 6 days after surgery, Fleeger *et al.* (1968) found an accumulation of total lipids and triacylglycerols in both the scrotal and abdominal testes, but increased cholesterol only in the abdominal testis.

Cadmium administration also affects lipid and fatty acid composition of testes. A. D. Johnson *et al.* (1967) observed increased cholesterol that could be prevented by pretreatment with zinc. Fatty acid changes were observed by Davis and Coniglio (1967) along with severe testicular atrophy 28 days after intratesticular injection of cadmium chloride. Significant decreases were observed in the relative fatty acid concentrations of 16:0 and 22:5(*n*-6), in contrast to increases in 18:0, 18:1, and 22:4, in the cadmium-treated rat testes.

Attempts have also been made to correlate lipid and fatty acid composition with spermatogenic function by the use of antifertility drugs. Busulfan (1,4-dimethylsulfonoxy butane) is known to result in infertility in about 6 weeks due to degeneration of germinal cells. Increases in esterified cholesterol were observed by A. D. Johnson *et al.* (1967), whereas Kar *et al.* (1968) found decreases in total lipids, triacylglycerols, free and esterified cholesterol, and phospholipids in the isolated seminiferous tubules 40 days after treatment.

Investigations of fatty acid changes with the drugs, triethylene-melamine (2,4,6-*tris*-ethylenimino-s-triazine) and Win 18,441 N,N^1-*bis*(dichloroacetyl) 1,8-octamethylenediamine, were done in rats. Lipid changes were minimal or insignificant in those animals in which histological studies revealed only reductions in spermatogonia and primary spermatocytes. All other cellular components were normal, supporting the hypothesis that the polyenoic acids, particularly 22:5(*n*-6), in rat testes are associated primarily with spermatids and spermatozoa.

c. Hormonal Influences. It has been recognized for many years that testicular lipids are influenced by the hormonal condition of the animal. Gambal and Ackerman (1967) reported that the concentration of nonphospholipids increased in testes of rats that had been hypophysectomized and that administration of LH or FSH plus LH maintained them at the normal values. Changes also occurred in the concentrations of individual phospholipids (although the total concentration did not change) and a single treatment did not maintain the normal concentration of all individual phospholipids.

Nakamura *et al.* (1968) observed changes in several lipids and fatty acids of hypophysectomized rats including decreases in the amounts of triacylglycerols, cholesterol, cardiolipin, phosphatidylethanolamine, phosphatidylcholine, phosphatidylinositide, sulfatide, glycosyl glycerides, and sphingomyelin. Increased amounts (as well as percentages) of cholesteryl esters, diacylalkylglycerols, and phosphatidylserine were found. Fatty acid changes were most marked in the diacylalkylglycerols, in which 22:5 increased greatly, and in the cholesteryl ester fraction, in which there was a great increase in 22:5 and decreases in 16:0, 18:2, and 20:4.

Treatment of young rats with 0.16 mg testosterone propionate on alternate days for 16 days caused an increase in total content of testicular cholesterol with the arrest of spermatogenesis and atrophy of Leydig cells (Kar and Roy, 1955). However, treatment with 6 mg of the drug on alternate days for 16 days resulted in a decrease in total testicular cholesterol. Histological studies of the latter animals suggested accelerated spermatogenic activity.

2. *Ovaries*

a. Dietary Effects. Dietary changes in fatty acid composition of rat ovarian lipids also occur (Carney and Walker, 1972). In rats maintained on an essential fatty acid deficient diet (hydrogenated coconut oil), the major polyunsaturated fatty acid of cholesteryl esters was 22:3(*n*-9) (30.1%), followed by 20:3(*n*-9) (9.7%); a significant amount (2.8%) of 24:3(*n*-9) was also present. In ovarian triacylglycerols of the essential fatty acid deficient rats, only 1.2% and 1.0% of 20:3(*n*-9) and 22:3(*n*-9), respectively, were present, but the 18:1 content increased from 32.5% in controls to 45.4% in deficient rats and the 16:1 content increased from 5.1% to 12.7%. The major changes in ovarian phospholipid fatty acids were the increased concentrations of 20:3(*n*-9) (18.0%) and 22:3(*n*-9) (3.8%), and the decreased concentrations of 20:4(*n*-6) (from 25.4% to 12.9%) and 22:4(*n*-6) (from 4.8% to 0.6%). The concentration of monoenoic acids increased slightly in the deficient animals.

b. Hormonal Effects. It is well established that luteinizing hormone depletes cholesterol stores in the ovary and that this coincides with increased progesterone secretion. Claesson and Hillarp (1947) suggested that cholesterol serves as a precursor in ovarian steroidogenesis. Cholesteryl ester depletion occurs in the ovary of the luteinized rat (Herbst, 1967) and this lipid serves as the precursor of progesterone (Armstrong *et al.*, 1964; Major *et al.*, 1967). Cholesterol hydrolase activity in ovaries increased after administration of luteinizing hormone (Behrman and Armstrong, 1969), and Alfin-Slater and Aftergood (1971) suggested that the intracellular transfer of esterified cholesterol into mitochondria for conversion to pregnenolone was stimulated by luteinizing hormone.

Cholesterol concentration in ovaries was also affected by the oral contraceptive drug Enovid (G. D. Searle and Company, Chicago, Illinois), which is a mixture of progesterone and estrogen (Aftergood *et al.*, 1968;

Aftergood and Alfin-Slater, 1971). Cholesteryl esters were lower in ovaries of rats receiving the drug and there was a decrease in docosatetraenoic acid in the sterol esters.

Circadian variation in ovarian cholesterol content has also been observed at particular stages of the estrous cycle (Clark and Zarrow, 1967). Depletion of cholesterol from 1:30 P.M. to 8:30 P.M. was observed in Wistar rats at metaestrus, but not in Holtzman rats, whereas in diestrus, depletion occurred in both. On the day of proestrus both strains showed 50% depletion. No diurnal rhythm in mature animals was observed if they were kept in constant light and no diurnal rhythm was observed in animals that had been androgen sterilized with testosterone propionate the second day after birth.

Cholesteryl ester concentrations in ovaries decreased significantly during estrus in control rats maintained on an adequate corn oil diet, but not in rats maintained on an essential fatty acid deficient diet (Carney and Walker, 1972). Prolactin administration maintained cholesteryl ester levels and the ability of ovarian slices from hypophysectomized rats to synthesize progesterone (Behrman *et al.*, 1971).

c. Effect of Pregnancy. Pregnancy also affects the content of cholesteryl esters in ovaries. Morin (1971) found that ovaries of 6-day pregnant rabbits contained 20.9 mg/g esterified cholesterol compared to 6.0 for ovaries of nonpregnant animals. The free cholesterol content was similar in both (3.2 vs 3.1). Changes were also observed in certain fatty acids of the ovarian cholesteryl esters. A 6-day pregnancy decreased 18:0 from 15.7% to 7.0% (% of total fatty acids), 18:2 from 20.8% to 13.4%, 18:3 from 1.6% to 0.5%, and increased 18:1 from 39.2% to 54.1%. No significant changes were seen in 16:0, 16:1, and 20:4.

d. Effect of Prostaglandins. Prostaglandin $F_2\alpha$ depressed ovarian cholesteryl esters (similar to hypophysectomy) and decreased progesterone output in incubated slices (Behrman *et al.*, 1971).

III. Biosynthesis and Metabolism of Major Lipid Classes and of Fatty Acids

A. Testes

1. Normal Animal

a. In Vivo Studies. The testis is a dynamic organ with respect to lipid metabolism. Furthermore, it is an ideal system for metabolic studies since it can be used in experiments done *in vitro* or *in vivo* and where the substrate to be investigated is restricted to the organ under investigation (by means of an intratesticular injection).

The biosynthesis of fatty acids after intratesticular injection of [^{14}C]acetate has been studied by Davis and Coniglio (1966) and by Evans *et al.* (1971). The highest specific activities obtained 24 hr after injection were in 16:0, 18:0, and 18:1 in that order. Radioactivity was also observed in significant quantities in 20:4(*n*-6) and 22:5(*n*-6). In other experiments in which rats were killed 30 min after intratesticular injection of [1-^{14}C]acetate (Coniglio *et al.*, 1971), it was shown that testicular myristic and palmitic acids had been synthesized by the *de novo* pathway, while stearic acid had been synthesized at least partly by elongation of endogenous palmitic acid. The pattern of fatty acids synthesized after intratesticular injection of [1-^{14}C]acetate (Evans *et al.*, 1971) was similar in testes of rats of 30 days of age compared to rats of 6–12 months of age. However, significantly more ^{14}C was present in the fatty acids of testes of the young rats at all time periods studied. More than 70% of the ^{14}C incorporated in total lipids was in phospholipids, with about 15% in triacylglycerols and only minor amounts in cholesteryl esters and free fatty acids. No differences were observed between the two age groups.

Labeled long-chain fatty acids have also been injected intratesticularly and their metabolic fate followed. There is reasonable assurance that the metabolic products observed in testicular tissue after intratesticular injection of a labeled substrate have their origin in that testicular tissue. Thus, Bridges and Coniglio (1970b) injected only one testis of a rat with ^{14}C 22:5(*n*-6). Forty-eight hours later the total fatty acids of the injected testis contained 14.7% of the injected dose of ^{14}C, but the total fatty acids of the noninjected testis did not contain any significant amount of ^{14}C. Similar results have been obtained with other labeled fatty acids.

The metabolism of [1-^{14}C]palmitic acid was investigated at various times following intratesticular injection of the albumin complex (Coniglio *et al.*, 1972). Almost 20% of the injected dose was expired as $^{14}CO_2$ in 2 hr and about 45% in 4 hr; 47% of the ^{14}C injected was retained in the injected testis after 2 hr, 41% after 4 hr, 16% after 1 week, and only about 6% after 2 weeks. Of the ^{14}C retained in the testis after 2 hr less than 15% was in free fatty acids and by the end of 2 weeks less than 5% was present in the free fatty acid fraction. ^{14}C Activity appeared in phospholipid faster than in triacylglycerol fractions, and with time the activity in the phospholipids decreased relative to the triacylglycerols. About 80% of the ^{14}C in the phospholipids was in palmitic acid and the rest in stearic and oleic acids. By the second week, half of the [^{14}C]palmitate present in the testes had been newly synthesized from [^{14}C]acetate resulting from β-oxidation of the injected labeled palmitate.

Davis and Coniglio (1966) studied the metabolism of [1-^{14}C]linoleate complexed with albumin after intratesticular injection in rats. The peak specific activity of arachidonic acid was about three times larger than that of docosapentaenoic acid. Chronologically, the specific activities peaked in the order 20:3 → 20:4 → 22:4 → 22:5. Chemical degradation of the arachidonate showed that 81% of the ^{14}C activity was in C-3 and 14% in C-1 indicating that the arachidonate was biosynthesized by a two-carbon elongation of the administered [1-^{14}C]linoleate.

Bridges and Coniglio (1970c) extended the studies with [^{14}C]linoleate to show that 80–85% of the ^{14}C found in testicular lipids 4 hr to 14 days after injection was in phospholipids and about 10% in triacylglycerols. As soon as 4 hr after injection of the linoleate–albumin complex less than 2% of the total lipid ^{14}C was in the free fatty acid fraction. About 40% of the injected dose had been expired as [^{14}C]CO_2 in 4 hr. Significant quantities of radioactivity were found in palmitate (from two-carbon units obtained by β-oxidation) and with time ^{14}C activity predominated in the longer-chain, higher polyenes. A small but significant quantity of ^{14}C was present in the 24-carbon polyenes 2 weeks after injection.

Nakamura and Privett (1969a) also investigated the fate of intratesticularly injected [^{14}C]linoleate, but they used an emulsion of the acid made with a mixture of equal parts saline and rat serum. They observed significant labeling of diacylglycerols, phosphatidylcholine, and phosphatidylethanolamine in the early time periods. The specific activity reached a maximal value first in diacylglycerols. Differences between these results and those of Bridges and Coniglio (1970b) may very well be due to the different physical properties of the injected linoleate. Nakamura and Privett (1969b) also studied the fate of intratesticularly injected [1-^{14}C]trilinoleoylglycerol. The percentage distribution of ^{14}C remained at a constant level in the triacylglycerols for about 6 hr. ^{14}C activity increased in most phospholipids throughout the experiment (48 hr), but in phosphatidylcholine and phosphatidylethanolamine a plateau was reached in about 12 hr. The activity peaked in diacylglycerols in about 6–12 hr. ^{14}C activity in the free fatty acids decreased after the 0.5-hr sample.

The rather direct conversion of arachidonic acid to 4,7,10,13,16-docosapentaenoic acid in rat testis was shown by Davis and Coniglio (1966). Chemical degradation of docosapentaenoic acid synthesized after [1-^{14}C]arachidonate injection revealed essentially all the radioactivity in C-3, indicating direct elongation of arachidonic to docosapentaenoic acid by the testis. Highly labeled 7,10,13,16-docosatetraenoic acid was also isolated in these experiments, and it was, therefore, suggested that arachidonate was elongated to the 22-carbon tetraene, which was then desaturated to the pentaene. Conversion of both [1-^{14}C]linoleate and [1-^{14}C]arachidonate to a 24-carbon polyene was shown by Bridges and Coniglio (1970a). After [1-^{14}C]linoleate injection 6,9,12,15,18-tetracosapentaenoic acid was labeled primarily in the seventh carbon. After [1-^{14}C]arachidonate injection testicular 9,12,15,18-tetracosatetraenoate was labeled primarily in the fifth carbon. The authors suggested that the tetracosapentaenoate was formed by a two-carbon elongation of docosapentaenoic acid that in turn had been formed from arachidonic acid by elongation and desaturation. The 24-carbon tetraenoate had been formed by simple two-carbon elongation of 22:4, which in turn was derived by elongation of [1-^{14}C]20:4. It was concluded from these experiments that testicular tissue has the capacity for biosynthesis of all the higher polyenes derived from dietary linoleic acid.

Twenty hours after intratesticular injection, a larger proportion of the injected ^{14}C was found in total lipids after injection of labeled arachidonate than after labeled linoleate (Bridges and Coniglio, 1970c). Most of the ^{14}C activity 24 hr after labeled arachidonate was in phospholipids and triacylglycerols. Specific activities of arachidonate and of docosapentaenoate in triacylglycerols at each time period were greater than those of the corresponding fatty acids of phospholipids. The specific activity of arachidonate decreased and that of docosapentaenoate increased more rapidly in triacylglycerols than in phospholipids. The fate of intratesticularly injected [1-^{14}C]arachidonic acid was investigated also by Ayala *et al.* (1973). Only 1 hr after injection of the fatty acid as the albumin complex, less than 10% of the ^{14}C found in the testes was in the free fatty acid form; almost 50% was in phospholipids, about 30% in triacylglycerols, and the balance in other fractions. These authors also studied the fate of intratesticularly injected 22:4(n-6) at the same time period as the arachidonate; a little more ^{14}C was found in the free fatty acid fraction, but the major difference was in the triacylglycerols and phospholipids. About 60% of the ^{14}C activity in testicular lipids was in triacylglycerols, while only about 15% was in phospholipids.

Unavailability of labeled docosapentaenoic and docosahexaenoic acids has prevented study of these compounds as substrates for testicular metabolism except for one study by Bridges and Coniglio (1970b) in which docosapentaenoate was synthesized from [1-^{14}C]linoleate in rat testes. The disadvantage of the biosynthetic material is its relatively low specific activity. In these studies less than 40% of the injected ^{14}C was recovered in testicular lipids in 4 hr and less than 20% in 48 hr. In these time periods the amount of total lipid ^{14}C retained in the testes decreased from 38% to 19% of the administered dose. The amount in the phospholipids increased from 45% to 73%, while that in the triacylglycerols decreased from 31% to 23% of the total lipid ^{14}C. The arachidonic acid fraction isolated from these experiments had a higher specific acitivity than did the palmitate and, therefore, suggested a direct conversion of 22:5 to 20:4 rather than formation of 20:4 through a labeled two-carbon elongation of endogenous linoleate. The ^{14}C in the arachidonate fraction was found by chemical degradative studies to be located predominantly in carbon 3, indicating that the labeled arachidonate was formed primarily by direct biohydrogenation and cleavage of the injected [5-^{14}C]docosapentaenoic acid. This process, termed retroconversion, had been observed in intact animals (Verdino *et al.*, 1964; Schlenk *et al.*, 1969) and is due to partial β-oxidation in mitochondria (Kunau and Couzans, 1971).

Table 19 summarizes the incorporation of four intratesticularly injected fatty acids into phospholipids and triacylglycerols at various time periods after injection. Linoleic and arachidonic acids were largely incorporated into phospholipids, but a greater proportion of the arachidonate than the linoleate was esterified into triacylglycerols. Docosapentaenoate was esterified into triacylglycerols to a greater extent than arachidonate and in this way resembled the palmitate more than the polyenoic acids. In this regard it is useful to recall that most of the testicular 24-carbon polyenes are found in acylglycerol esters than in phospholipids.

Oleate- and linoleate-containing choline phospholipids in testes apparently are biosynthesized *de novo,* but a deacylation–acylation reaction is largely responsible for the biosynthesis of the highly unsaturated choline phospholipids. On the other hand, the various molecular species of rat testicular triacylglycerols apparently were synthesized at about the same rate and had about the same rate of turnover (Sprecher and Duffy, 1974).

Infusions of $[^{32}P]PO_4$ into the testicular artery of rams labeled phosphatidylinositol to a greater extent than other phospholipids. Only phosphatidic acid had a higher specific activity (Scott and Setchell, 1968). Triacylglycerols were the lipids primarily labeled after $[^{14}C]$glucose infusion.

b. In Vitro Studies. Studies of testicular lipid metabolism conducted with *in vitro* systems have dealt mainly with the biosynthesis and interconversion of fatty acids.

TABLE 19

Incorporation of Intratesticularly Injected $[^{14}C]$Fatty Acids into Phospholipids and Triacylglycerols

Time after injection	% of total ^{14}C recovered from thin-layer plate							
	Phospholipids				Triacylglycerols			
	16:0[a]	18:2[b]	20:4[b]	22:5[c]	16:0	18:2	20:4	22:5
4 hr	65.9	84.1	76.9	44.9	21.4	10.1	17.6	31.2
12 hr	n.d.[d]	84.1	84.5	45.3	n.d.	10.5	10.3	34.1
24 hr	n.d.	83.3	82.1	60.4	n.d.	10.2	13.6	33.2
48 hr	n.d.	n.d.	n.d.	72.2	n.d.	n.d.	n.d.	23.3
1 week	32.3	80.7	75.9	n.d.	52.2	11.4	14.5	n.d.
2 weeks	35.8	81.1	79.1	n.d.	49.6	13.9	15.3	n.d.

[a]Data of Coniglio *et al.* (1972).
[b]Data of Bridges and Coniglio (1970b).
[c]Data of Bridges and Coniglio (1970a).
[d]n.d. = Not determined.

TABLE 20
Distribution of ^{14}C (%) in Lipids of Rat Testes Incubated or Injected with [1-^{14}C]Acetate[a]

Lipid class	Testes incubated with [1-^{14}C]acetate[b]	Testes injected with [1-^{14}C]acetate[c]
Phospholipid	30	80
Cholesterol	7	3
Free fatty acids	25	nil
Triacylglycerols	30	15
Cholesterol esters	5	1

[a]Data of Evans *et al.* (1971).
[b]3-hr incubations.
[c]Rats killed 24 hr after injection.

That testicular tissue can incorporate [^{14}C]acetate into long-chain fatty acids was observed by Hall *et al.* (1963) using slices of rabbit testes. Caution must be exercised in extrapolating *in vitro* findings to the intact animal as shown in results from rats injected intratesticularly with [1-^{14}C]acetate and in slices incubated with [1-^{14}C]acetate (Evans *et al.*, 1971). The data in Table 20 show that with intratesticular injection of [1-^{14}C]acetate, most of the incorporated ^{14}C was in phospholipids and essentially all the rest in triacylglycerols. In incubated slices the incorporated ^{14}C was more evenly divided between phospholipids, triacylglycerols, and free fatty acids. Table 21 shows data on the incorporation of ^{14}C into the various

TABLE 21
Distribution of ^{14}C (%) in Total Fatty Acids of Testes of Rats Incubated or Injected with [1-^{14}C]Acetate[a]

Fatty acids	Testes incubated with [1-^{14}C]acetate[b]	Testes injected with [1-^{14}C]acetate[c]
Saturated (16:0 + 18:0)	33	88
Monoenes (16:1 + 18:1)	9	8
18-carbon polyenes	2	1
20-carbon polyenes	16	1
22-carbon polyenes	18	1
> 22-carbon	19	1

[a]Data of Evans *et al.* (1971).
[b]3-hr incubations.
[c]Rats killed 24 hr after injection.

fatty acid fractions of the total testicular lipids. In the injected testis almost all of the incorporation was into the saturated acids, with essentially all the rest in the monoenes. In the incubated slices incorporation into saturated plus monoenes accounted for only about 40% of the total; the 20-, 22-, and >22-carbon polyenes contained the rest of the incorporated activity, which was probably due to simple acetate-elongation of 18-, 20-, and 22-carbon endogenous polyenes.

In rabbit slices incubated with [^{14}C]acetate, more ^{14}C was incorporated into phosphatidylcholine than into phosphatidylethanolamine or sphingomyelin (Morin, 1967). Small but significant amounts of ^{14}C were incorporated into 8,11,14-eicosatrienoic, arachidonic, and docosapentaenoic acids.

The synthesis of palmitic acid in testis has been shown to be accomplished by cytoplasmic enzymes as is true of other organs and tissues (Whorton and Coniglio, 1974). Citrate, as well as acetate, can serve as fatty acid precursors in cell-free preparations made from testicular homogenates (Haeffner and Privett, 1971; Whorton and Coniglio, 1975b).

Testicular microsomes carry out the processes of elongation and desaturation. Ayala *et al.* (1973) demonstrated the desaturation of linoleic to 18:3(*n*-6) and 8,11,14-eicosatrienoic to arachidonic acid. Cofactors required were NADH, ATP, $MgCl_2$ and CoA. A decline in linoleic desaturation activity was observed after 3 weeks of age; no measurable desaturation of 22:4(*n*-6) to 22:5(*n*-6) was observed. Microsomes, however, were able to convert labeled linoleate to labeled docosapentaenoate, thus demonstrating elongation as well as desaturation. Labeled arachidonic acid was also elongated to 22:4, but desaturation to the pentaene was not observed. The fatty acids were incorporated mainly into phosphatidylcholine and neutral lipids, but much less labeled 22:4 entered the phospholipids than the other two substrates. The 22:4 acid was preferentially incorporated into neutral lipids. In these studies it was shown that testicular mitochondria can convert 22:4(*n*-6) to arachidonic acid by the process of retroconversion. The location of the enzymatic activity for conversion of 22:4(*n*-6) to 20:4(*n*-6) in the inner membrane of rat liver mitochondria had previously been shown by Stoffel *et al.* (1970).

Elongation of endogenous fatty acids by testicular microsomes incubated with [^{14}C]malonyl-CoA has been shown (Whorton and Coniglio, 1974). Testicular microsomes also have the ability to desaturate stearyl-CoA to oleyl-CoA (Whorton *et al.*, 1976).

Arachidonic acid is incorporated into ether-containing choline and ethanolamine phospholipids by acylation–deacylation reactions (Wykle *et al.*, 1973). These authors observed that labeled arachidonic acid was incorporated into the 2-position of [^{14}C]alkylacylglycerophosphorylcholine and alk-1-enylacylglycerophosphorylethanolamine by microsomes prepared from rat testes.

Various [^{14}C]fatty acids were incubated with slices of bovine testes taken from adult, 4-month, and 2-week old animals (Neill and Masters, 1974a). Mature testes incorporated less of each substrate than immature testes, but linoleate was always incorporated to the greatest extent. The most metabolically active lipids were phosphatidylinositol and diacylglycerols in mature testes, and the metabolism of cholesteryl esters was also significant in this age group. Maximal desaturase activity was in the 4-month-old testes. Fatty acid incorporations were not related to the endogenous fatty acid concentrations of testicular lipids.

2. *Effect of Diet, Hormones, Drugs, and Other Factors*

a. Dietary Effects. Changes in fatty acid patterns of testes of animals fed deficient diets have been described by many investigators, but actual metabolic changes in testes have not been studied extensively. There is increased synthesis of fatty acids (chiefly palmitic acid) in testes of rats fed a fat-free diet compared to those fed a fat-containing diet. The increased synthesis has been observed *in vivo,* in slices, and in the high-speed supernatant fraction (cytosol) of a rat testicular homogenate (Whorton and Coniglio, 1974). The effect apparently is on acetyl-CoA carboxylase.

In testes of rats maintained on an essential fatty-acid-deficient diet (Blank *et al.,* 1973), [^{3}H]arachidonic acid of diacylphospholipids was metabolized more rapidly than that esterified in the plasmalogens, resulting in a build-up of arachidonic acid in the plasmalogens. The authors suggested that plasmalogens may act as reservoirs for prostaglandin precursors in testes of animals deficient in essential fatty acids. Redistribution of arachidonate from ester lipid classes to ether lipids in testes of rats during essential fatty acid deficiency probably occurs by deacylation-acylation (Wykle *et al.,* 1973).

The changes observed in testicular fatty acids in vitamin E deficiency involve mainly the polyenoic acids. Thus, it has been observed that testes of deficient rats have increased concentrations of arachidonic acid and of 22:4(n-6) and decreased concentration of 22:5(n-6) (Bieri and Prival, 1966; Carney and Walker, 1971). The decreased level of 22:5(n-6) may be due to a number of factors such as (1) a metabolic block in the conversion of 22:4(n-6) to 22:5(n-6) (Bieri and Andrews, 1964); (2) the rapid peroxidation of 22:5(n-6) because of the lack of vitamin E as an antitoxidant (Witting *et al.,* 1967); (3) lack of receptor tissue (spermatids) in the degenerated testis for the biosynthesized 22:5(n-6) (Carney and Walker, 1971); or (4) increased turnover of 22:5(n-6), including the process of retroconversion (Carney and Walker, 1971). The process of retroconversion of 22:5(n-6) to 20:4(n-6) has been demonstrated in testes of normal rats (Bridges and Coniglio, 1970b). Metabolic studies using [^{14}C]linoleate (Carney and Walker, 1971) demonstrated that testes of rats on a vitamin-E-deficient diet can synthesize 22:5(*n*-6). These authors suggested that the decreased concentration of 22:5(*n*-6) could be a result of a higher turnover rate for 22:5(*n*-6) or increased rate of retroconversion in the testes of deficient animals, although an effect of peroxidation could not be ruled out. Preliminary evidence for increased retroconversion and oxidation of [^{14}C]22:5(*n*-6) in testes of vitamin-E-deficient rats is available (Coniglio *et al.*, 1974a). The fatty acid changes apparently are not seen [except for a slightly decreased level of 22:5(*n*-6)] unless tissue degeneration is histologically observable (Whorton and Coniglio, unpublished observation). The metabolism of intratesticularly injected arachidonate is similar in testes of these rats to that of controls. In testes of E-deficient rats that show degenerative changes, the fatty acid changes described were present (though not as marked as those reported by others), and there was an increase in the specific activity of 22:4(*n*-6) along with a decrease in that of 22:5(*n*-6) in triacylglycerols and phospholipids.

b. Hormonal Effects. Hormones may control lipid metabolism in testicular tissue. In hypophysectomized rats FSH stimulated the total incorporation and specific activity of ^{32}P in testicular phospholipids. Increased incorporation of [^{3}H]choline into sphingomyelin and phosphatidylcholine and of [^{3}H]ethanolamine into phosphatidylethanolamine was caused by FSH administration (Yokoe *et al.*, 1969).

The use of anti-ICSH serum decreased the incorporation of ^{32}P into testicular phospholipids and increased the concentration of phospholipids. Testosterone propionate given with anti-ICSH increased the concentration and incorporation of ^{32}P into testicular phospholipids (Gambal, 1967). Antiserum to FSH decreased the incorporation of ^{32}P into testicular phospholipids. The gonadotropins, FSH, and LH, either individually or together, stimulated the biosynthesis of polyenes in testicular tissue (Goswami and Williams, 1967). Goswami *et al.* (1968) reported that RNA isolated from testes of hypophysectomized rats treated with FSH and LH stimulated the *in vitro* incorporation of acetyl-CoA and of malonyl-CoA into polyenoic acids.

Decreased conversion of linoleate to arachidonate to docosapentaenoate in testis was observed coincident with atrophy in alloxan-diabetic rats (Peluffo *et al.*, 1970). Daily administration of 100 mg of ethyl arachidonate to alloxan-diabetic rats for 3 months did not cure testicular atrophy, although it did cure the scaly tail. The greatest rate of conversion of [^{14}C] linoleate to [^{14}C]18:3(*n*-6) occurred when the germinal tissue was developing, and it decreased with age.

Studies of [1-^{14}C]linoleate, injected intratesticularly in hypophysectomized rats and in hypophysectomized rats receiving luteinizing or follicle-stimulating hormone or both, indicated that the rate of catabolism of triacylglycerols and phospholipids was greatly increased by hypophysectomy (Nakamura and Privett, 1973). Cholesteryl esters and diacylalkylglycerols accumulated and there was an increase in the 22:5(*n*-6) content in both lipid classes. Apparently, the turnover of these compounds was impaired by hypophysectomy. The conversion of linoleic to arachidonic acid was impaired by hypophysectomy and the administration of gonadotropins partially prevented the effects of hypophysectomy. The authors suggested that some of the enzymes in the testes involved in lipid metabolism are hormone sensitive.

c. Effect of 5,8,11,14-Eicosatetraynoic Acid. This compound, used as the chemical precursor of arachidonic acid, is known to inhibit the conversion of arachidonic acid to prostaglandin (Ahern and Downing, 1970). It apparently also has an inhibitory action on the conversion of linoleate to arachidonate, since lower levels of arachidonic acid have been observed in testes as well as in other organs of animals receiving this compound compared to animals not receiving the compound (Coniglio *et al.*, 1976). Twenty-four hours after the intratesticular injection of [1-^{14}C]arachidonic acid, the testes of rats receiving the tetraynoic retained slightly more ^{14}C as arachidonic acid than did rats not receiving the compound. The former group oxidized less of the labeled substrate to $^{14}CO_2$ and converted less of the injected [^{14}C]arachidonate to 22- and 24-carbon polyenes.

B. *Ovaries*

1. *Normal Animals*

Slices of ovaries from rabbits oxidized the following substrates to [^{14}C]CO_2: [^{14}C]formate, [^{14}C]acetate, [^{14}C]choline, and [^{14}C]ethanolamine (Morin, 1969). Of the four substrates, choline was the most actively incorporated into phospholipids, perhaps because acetate and formate were more actively oxidized to carbon dioxide. [^{14}C]Formate was incorporated primarily into phosphatidylserine. [^{14}C]Ethanolamine was incorporated primarily into phosphatidylethanolamine and to a smaller extent into choline-containing phospholipids through the process of transmethylation.

In 1- and 2-hour incubations of slices of ovaries and testes of calves with [^{32}P]orthophosphate, high specific activities were obtained in phosphoinositides and phosphatidylcholine (Basinska *et al.*, 1973). Monophosphoinositide had the highest percentage incorporation of all phospholipids at the end of 1 hr and in the ovary at the end of the second hr also. Incorporation into phosphatidylcholine increased with time and was the highest in the testes at the end of 2 hr. The polyphosphoinositides also had high activity, and smaller amounts of ^{32}P also were incorporated into sphingomyelin, phosphatidylserine, phosphatidylethanolamine, cardiolipin, and phosphatidic acid.

2. *Effect of Pregnancy*

Phosphatidylcholine in ovaries increases in concentration during pregnancy and [1-^{14}C]acetate incorporation into phosphatidylcholine also increases (Morin, 1968). A marked increase occurs also in the concentration of cholesteryl esters and in the esterifying enzymatic activity during early pregnancy (Morin, 1971). The reaction is catalyzed by an ATP-CoA-dependent acyltransferase (primarily microsomal) and by a mitochondrial reaction that operates at low pH and does not require cofactors (Morin, 1973a). There is also an increase in the hydrolytic activity of the ovaries demonstrated by the hydrolysis of radioactive cholesteryl oleate by ovarian homogenates (Morin *et al.*, 1969), suggesting a response by the ovary to provide precursor free cholesterol for increased ovarian steroid hormone synthesis during pregnancy.

3. *Effect of Drugs*

Prostaglandin E_1 added to mitochondrial and microsomal preparations of rabbit ovaries inhibits the incorporation of cholesterol and of palmitate (or palmitoyl-CoA) into cholesteryl esters (Morin, 1973b). Prostaglandin $F_{2\alpha}$ had less inhibitory effect while polyphloretin phosphate was a more potent inhibitor of the esterification reaction. Cholesteryl ester hydrolase activity was not significantly affected by any of these agents. Oxidation of the [^{14}C]palmitate substrates to [^{14}C]CO_2 also was not affected.

IV. *Other Lipids*

A. *Prostaglandins*

1. *Testis*

The great abundance of polyunsaturated fatty acids in testes of animals and the increase in concentration with sexual maturity of certain derivatives of the linoleic and α-linolenic acid families lead one to consider their exact role in the testis. Since members of both of these families are the precursors of the prostaglandins it is logical to assign the biosynthesis of prostaglandins as one of the principal functions of these compounds in reproductive tissue. Prostaglandin E_2 injected into rats kept on a fat-deficient diet for 5 months resulted in improvement in their fertility from 45% to 100% compared with an improvement of 91% for methyl arachidonate (Hafiez, 1974).

Prostaglandins have been isolated and quantified in testicular tissue and it has been shown that testicular tissue is capable of synthesizing prostaglandins. In swine testes (Michael, 1973), about 1.5 μg total PGE's and about 0.7 μg total PGF's per g tissue were found. The major individual prostaglandins identified were PGE_1, PGF_2, $PGE_{1\alpha}$, and $PGF_{2\alpha}$. In rat testes tentative identification of PGE_1, PGE_2, and $PGF_{2\alpha}$ has been reported (Carpenter, 1974). The amount of $PGF_{2\alpha}$ was estimated at 0.03–0.04 μg/g and in immature testis about five times as much. Evidence for the biosynthesis of prostaglandins in rat testis injected with [1-^{14}C]linoleic acid was reported by Carpenter *et al.* (1971). The ability of microsomes isolated from testes to synthesize prostaglandins from arachidonate was demonstrated by Carpenter and Manning (1975); the primary prostaglandin synthesized was $PGF_2\alpha$. When the supernate was added to the incubation medium, there was a decrease in the amount of $PGF_{2\alpha}$ metabolites. Other reports of enzymes for inactivation of prostaglandins have been published. Thus, 15-hydroxyprostaglandin dehydrogenase and prostaglandin-13-reductase were reported in swine testes (Anggard *et al.*, 1971) and in rat testes (Nakano *et al.*, 1971).

Hypophysectomy or adrenalectomy results in a reduction in prostaglandin synthesized by minced rat testicular tissue incubated with arachidonate (Ellis *et al.*, 1972). These authors suggested that steroidogenesis, lipid peroxidation, and prostaglandin synthesis are related.

An effect of prostaglandin on testicular esterified cholesterol and on plasma testosterone was observed by Bartke *et al.* (1973). $PGF_{2\alpha}$ lowered plasma testosterone and increased the concentration of esterified cholesterol in mice testes. The use of an inhibitor of cholesterol esterase, phenylmethylsulfonylchloride, in male mice causes an increase in esterified cholesterol in testes and a decrease in plasma testosterone. It is suggested that hydrolysis of cholesteryl ester is required for the synthesis of androgenic steroids. The action of $PGF_{2\alpha}$ appears to be one of inhibition of steroidogenesis.

The prostaglandins may also act as modulators of testicular contractions. Two groups of prostaglandins have been described with respect to *in vitro* effects on autorhythmical contractions of intact rabbit testes: PGE at low concentrations reduced and at high concentrations (>36 nM) completely effaced contractions (Hargrove *et al.*, 1971); PGF_1 at low concentrations increased the amplitude of the contractions although at concentrations >71 nM, the amplitude was decreased (Johnson *et al.*, 1971).

2. *Ovaries*

In female rats maintained on a fat-free diet, PGE_2 or methyl arachidonate was equally effective in preventing irregular estrous cycles and prolongation of gestation. However, neither prevented the delayed vaginal opening caused by essential fatty acid deficiency (Hafiez, 1974).

Rat ovarian homogenates catalyze the biosynthesis of prostaglandin-like compounds from labeled arachidonic acid. This synthesis is suppressed if the rat is injected with anti-LH serum, and the suppression is overcome by a simultaneous injection of LH or by addition of LH to the incubation medium (Chasalow and Pharriss, 1972). Biosynthesis of $PGF_{2\alpha}$ was observed in 4-hr incubation experiments using ovarian slices from rabbits and monkeys (Wilks *et al.*, 1972).

cAMP as well as LH increased the synthesis of both PGE and PGF in 5-hr incubations of rabbit Graafian follicles and the authors suggest that cAMP may be a mediator of this action of LH (Marsh *et al.*, 1974). A dose–response relation in mouse ovaries was observed with PGE_1 and PGE_2 as with LH in the stimulation of cAMP formation (Kuehl *et al.*, 1970).

A pure population of granulosa cells isolated from preovulatory follicles of estrous rabbits secreted prostaglandin F in tissue culture (Challis *et al.*, 1974). The rate of secretion was not influenced by LH-FSH, but dibutyryl cAMP caused a 50% inhibition and indomethacin caused a 100% inhibition. The authors suggest that granulosa cells could secrete the prostaglandin that accumulates in the follicle at ovulation and that prostaglandin secretion may be modified by cAMP.

Ovarian cholesterol esterification is also affected by prostaglandins. Prostaglandin E_1 added to incubations of mitochondrial and microsomal preparations from the interstitial and luteal portions of pregnant rabbit ovaries inhibited the incorporation of cholesterol, palmitic acid, and palmitoyl-CoA substrates into cholesteryl esters (Morin, 1973b). PGF_2 had a similar but smaller effect while polyphloretin phosphate was a more potent inhibitor. Since cAMP also decreases the rate of esterification of cholesterol in slices of rabbit ovarian interstitial tissue (Flint *et al.*, 1973) the suggestion is made that prostaglandin-induced inhibition of cholesterol esterification is mediated by an increased level of cAMP.

B. *Glycolipids*

Glycogalactolipids have been identified in and isolated from male reproductive tissue. The major glycolipid of mammalian testis is 1-*O*-alkyl-2-*O*-acyl-3-β(3′-sulfogalactosyl)-glycerol and a minor component is the desulfated form. Both alkyl and acyl moieties are primarily 16-carbon compounds (Kornblatt *et al.*, 1972; Kornblatt *et al.*, 1974; Ishizuka *et al.*, 1973). These glycolipids have been found in testes and spermatozoa of many mammalian species (Kornblatt *et al.*, 1973; Handa *et al.*, 1973) and in very low concentrations in immature and sterile animals and somewhat reduced in hypophysectomized rats. It was suggested that the sulfolipid may appear first in early spermatocytes and that once synthesized it is relatively stable during spermatogenesis (Kornblatt *et al.*, 1974). The Golgi apparatus fraction of the testis was shown to be enriched with a sulfotransferase that transfers sulfate from 3′-phosphoadenosine-5′-phosphosulfate to monoalkyl-monoacyl-glycerol-monogalactoside (Knapp *et al.*, 1973). *In vivo* labeling with [^{35}S]SO_4 and *in vitro* labeling using a particular fraction of boar testis solubilized with Triton X-100 to catalyze the transfer of the sulfate from the 3′-phosphoadenosine 5′-phosphosulfate to desulfo-seminolipid has also been reported (Handa *et al.*, 1974). The sulfolipid, biosynthesized with ^{35}S, was shown to be a substrate for pure human arylsulfatase A but was not hydrolyzed by arylsulfatase B (Fluharty *et al.*, 1974).

V. *Lipids and Lipid Metabolism in Spermatozoa*

Although this chapter is concerned with testicular lipids, a brief summary of lipid metabolism in spermatozoa is desirable. The major fatty acid composition of phosphatidylcholine and of acylglycerols of human seminal plasma, sperm tails, and sperm heads is given in Table 22 (Ahluwalia and Holman, 1969). Polyunsaturated fatty acids are abundant in the three fractions of human semen examined and in semen of bull, boar, and rabbit as well. The percentage of polyenoic acids in the sperm tails of each species was higher than that in the heads. Palmitic and stearic acids were the major saturated fatty acids found in choline, serine, and ethanolamine phosphatides of porcine spermatozoa and docosapentaenoic (38%) and docosahexaenoic (25%) the major unsaturated fatty acids (L. A. Johnson *et al.*, 1969).

Human sperm phospholipids consisted of about 29% phosphatidylcholine, 22% phosphatidylethanolamine, 21% sphingomyelin, 9% ethanolamine plasmalogen, 5% phosphatidylserine, 3% choline plasmalogen, 2% phosphatidylinositol, and 2% cardiolipin (Poulos and White, 1973). Of the total phospholipids of bovine sperm, 40%, 24% and 36% were contributed by the head, midpiece, and tail fractions, respectively (Clegg and Foote, 1973). Choline phosphatides (about half plasmalogen) was the major sperm phosphatide. About twice as much phosphatidylethanolamine was present in sperm heads and midpieces as plasmalogens, whereas they were about equal in sperm tails. The midpiece was relatively rich in phosphatidylethanolamine and cardiolipin and low in sphingomyelin. Small amounts of other phospholipids were also present.

In the rat the total lipids of sperm collected from the epididymis have a fatty acid composition similar to the testis as a whole (Davis *et al.*, 1966). Changes do occur in lipid composition of sperm as they travel from the testis through the genital tract. A decrease in phosphatidylcholine, phosphatidylethanolamine, ethanolamine plasmalogens, and in cholesteryl content of spermatozoa of ram was observed by Scott *et al.* (1967) in ejaculates compared to sperm collected directly from the testis. However, in the rat

TABLE 22
Fazty Acid Composition of Phosphatidyl Choline and Triacylglycerols of Human Seminal Plasma, Sperm Tails, and Sperm Heads[a]

Fatty acid	Seminal plasma		Sperm tails		Sperm heads	
	Phosphatidyl choline	Triacylglycerols	Phosphatidyl choline	Triacylglycerols	Phosphatidyl choline	Triacylglycerols
16:0	16.0	21.6	20.0	20.0	20.0	15.4
16:1	2.8	1.4	9.6	9.2	7.1	6.0
17:0	3.2	2.9	2.8	1.5	2.8	2.8
18:0	5.9	3.0	5.4	6.8	7.9	6.2
18:1	14.4	2.0	19.8	27.7	13.8	6.4
18:2	2.0	0.7	1.9	1.8	1.9	1.8
18:3(*n*-3)	3.6	3.4	2.8	2.8	3.9	3.0
20:3(*n*-6)	2.9	2.6	1.2	1.5	1.9	2.8
20:4(*n*-6)	8.7	9.6	4.0	3.2	5.7	6.0
20:5(*n*-3)	8.1	18.5	9.4	12.3	10.4	12.0
22:4(*n*-6)	9.3	6.0	1.4	Trace	2.4	5.2
22:5(*n*-6)	2.7	4.8	1.0	4.5	4.6	3.0
22:5(*n*-3)	7.4	3.6	10.6	0	10.6	2.8
22:6(*n*-3)	0.8	2.5	Trace	0	Trace	1.9
24:4(*n*-6)	4.8	0	Trace	0	Trace	0

[a]Data of Ahluwalia and Holman (1969).

the concentration of choline plasmalogens increased in sperm as they travelled through the epididymis (Scott *et al.*, 1963).

In some recent studies (Poulos *et al.*, 1975), ram testicular spermatozoa were found to have more than double the amount of phospholipid that was contained in ejaculated spermatozoa. There was a decrease in concentration of most phospholipid components (including choline plasmalogens) during transit through the genital tract. Phosphatidylserine, ethanolamine phospholipids, and cardiolipin suffered the greatest relative decreases. Of the phospholipid fatty acids palmitic acid changed the most in concentration during transit (a loss of about 500 μg for every 10^9 spermatozoa). A decrease in palmitic and linoleic acids of neutral lipids also occurred in ram spermatozoa with passage through the tract (Scott *et al.*, 1967). There was also loss of arachidonic acid that prompted a study of the content of prostaglandins in testicular and epididymal fluids. The concentration of $PGF_{2\alpha}$ in these fluids was considerably higher than in blood plasma. In bovine spermatozoa the most prominent finding in the analysis of the fatty acids was a large proportion of docosahexaenoic acid in the phospholipids (54% of total phospholipid fatty acids), and this was in the greatest amount (79%) in the choline phospholipids (Neill and Masters, 1972).

Incorporation of intratesticularly injected [1-^{14}C]palmitate into rams led to maximal labeling of sperm lipids on days 28 and 42 after injection (Neill and Masters, 1974b). The radioactivity was largely in the phospholipids, with the highest specific activity in phosphatidylinositol. About 10–12% of the ^{14}C in sperm lipids was present in monoenoic acids. About 19–26% of the phospholipid ^{14}C on day 28 was in plasmalogenic aldehydes, but only about 10% by day 42.

Bovine spermatozoa have active incorporation of fatty acids under *in vitro* conditions (Payne and Masters, 1970). Myristic acid was incorporated into diacylglycerols to a greater extent than other fatty acids. The greatest incorporation of fatty acids was into phosphatidylcholine, the major phosphatide of spermatozoa, and the fatty acid showing the greatest incorporation was linoleic acid. No differences were apparent in the oxidation of the various fatty acids to carbon dioxide.

In other similar experiments saturated acids, in general, were incorporated more actively than unsaturated ones. However, stearic acid was poorly incorporated into all lipids except phosphatidylinositol (Neill and Masters, 1972). Similar studies using ovine spermatozoa revealed only minor differences between these and the previous experiments with bovine spermatozoa (Neill and Masters, 1973).

The major source of substrate for endogenous respiration of spermatazoa appears to be the phospholipids; their released fatty acids are readily oxidized by the cell during breakdown of the phospholipids (Mann, 1964; Mills and Scott, 1969; Scott, 1973).

VI. Role of Lipids in Reproduction

Consideration of the aspects of lipid metabolism discussed in this chapter leads to the impression that lipids in general and certain lipids in particular have indispensable roles in reproduction. The general role that lipids play in the structure of membranes is well established, and this role should be no less important in the testes and ovaries than in any other organ or tissue. Similarly, whatever general physiological functions are served by the prostaglandins, they need to be served in cells of reproductive organs as well as in cells of other organs. In addition, specific roles for the prostaglandins in reproductive tissue seem to be definite possibilities.

Of considerable interest with regard to the role of specific lipids in male reproductive tissue are the group of polyenoic acids of 22-carbon chain length and the novel galactosulfolipid. The 22-carbon polyenes [22:5(*n*-6) in the rat; 22:4(*n*-6) in the rooster; and 22:6(*n*-3) in other species, including humans] accumulate in testicular tissue at the time of sexual maturation, and they are abundant in sperms. In addition, conditions or agents that cause destruction of the germinal cells also bring about severe decreases in the content of these polyenes. Similar remarks may be made for the galactosulfolipid recently isolated from testicular tissue and from sperm. This glycolipid appears to be unique to reproductive tissue, and its presence may indicate another possible specific role for lipids in reproductive tissue related to cell surface interaction between sperms and eggs.

VII. Concluding Remarks

Considerable progress has been made in recent years in elucidating the chemistry and metabolism of lipids in gonadal tissue. This has been due largely to our advances in lipid methodology. Future advances will continue to depend on additional progress in methodology, not only of lipids but also of general cell biology. Among the many techniques needed in this area are those of tissue and cell cultures, better methods for separation of cell types, better and faster methods of separation of various complex lipids, and specific chemical or other agents that will affect only certain cells.

Much of the chemistry and metabolism of lipids of gonadal tissue is similar or identical to that occurring in other organs and tissues. However, there also seem to be some aspects that may be unique to reproductive tissue, and there will be some pleasant experiences in store for those investigators who explore them.

ACKNOWLEDGMENT

This work has been supported in part by U.S.P.H.S. grants No. AM 06483, HD 06070, and HD 07694.

References

Aaes-Jørgensen, E., and Holman, R. T. 1958. Essential fatty acid deficiency. I. Content of polyenoic acids in testes and heart as an indicator of EFA status. *J. Nutr.* **65**:633–641.

Aftergood, L., and Alfin-Slater, R. B. 1971. Further studies of the effects of an anovulatory drug on lipid metabolism in the rat. *J. Lipid Res.* **12**:306–312.

Aftergood, L., Hernandez, H. J., and Alfin-Slater, R. B. 1968. Effect of large doses of the oral contraceptive Enovid on cholesterol metabolism in the rat. *J. Lipid Res.* **9**:447–452.

Ahern, D. G., and Downing, D. T. 1970. Inhibition of prostaglandin synthesis by eicosa-5,8,11,14 tetraynoic acid. *Biochim. Biophys. Acta* **210**:456–461.

Ahluwalia, B., and Holman, R. T. 1966. Fatty acid distribution in the bovine pre- and postpartum testis. *Lipids* **1**:197–201.

Ahluwalia, B., and Holman, R. T. 1969. Fatty acid composition of lipids of bull, boar, rabbit and human semen. *J. Reprod. Fertil.* **18**:431–437.

Ahluwalia, B., Pincus, G., and Holman, R. T. 1967. Essential fatty acid deficiency and its effects upon reproductive organs of male rabbits. *J. Nutr.* **92**:205–214.

Alfin-Slater, R. B., and Aftergood, L. 1971. Lipids and the pill. *Lipids* **6**:693–705.

Anggard, E., Larsson, C., and Samuelsson, B. 1971. The distribution of 15-hydroxy prostaglandin dehydrogenase and prostaglandin-Δ 13-reductase. *Acta Physiol. Scand.* **81**:396–404.

Armstrong, D. T., O'Brien, J., and Greep, R. O. 1964. Effects of luteinizing hormone on progestin biosynthesis in the luteinized rat ovary. *Endocrinology* **75**:488–500.

Ayala, S., Gaspar, G., Brenner, R. R., Peluffo, R. O., and Kunau, W. 1973. Fate of linoleic, arachidonic, and docosa-7,10,13,16-tetraenoic acids in rat testicles. *J. Lipid Res.* **14**:296–305.

Bartke, A., Musto, N., Caldwell, B. V., and Behrman, H. R. 1973. Effects of a cholesterol esterase inhibitor and of prostaglandin $F_{2\alpha}$ on testis cholesterol and on plasma testosterone in mice. *Prostaglandins* **3**:97–104.

Basinska, J., Sastry, P. S., and Stancer, H. C. 1973. Incorporation of orthophosphate $^{32}P_i$ into the phospholipids of slices of gonads, anterior pituitary and liver of calf. *Endocrinology* **92**:1786–1789.

Behrman, H. R., and Armstrong, D. T. 1969. Cholesterol esterase stimulation by luteinizing hormone in luteinized rat ovaries. *Endocrinology* **85**:474–480.

Behrman, H. R., Macdonald, G. J., and Greep, R. O. 1971. Regulation of ovarian cholesterol esters: evidence for the enzymatic sites of prostaglandin-induced loss of corpus luteum function. *Lipids* **6**:791–796.

Bernhard, K., Lindlar, F., Schwed, P., Vuilleumier, J. P., and Wagner, H. 1963. Fatty acid metabolism in vitamin E deficiency. *Z. Ernährungswiss* **4**:42–48.

Bieri, J. G., and Andrews, E. L. 1964. Fatty acids in rat testes as affected by vitamin E. *Biochem. Biophys. Res. Commun.* **17**:115–119.

Bieri, J. G., and Prival, E. L. 1965. Lipid composition of testes from various species. *Comp. Biochem. Physiol.* **15**:275–282.

Bieri, J. G., and Prival, E. L. 1966. Effect of deficiencies of α-tocopherol, retinol and zinc on the lipid composition of rat testes. *J. Nutr.* **89**:55–61.

Bieri, J. G., Mason, K. E., and Prival, E. L. 1969. Essential fatty acid deficiency and the testis: lipid composition and the effect of preweaning diet. *J. Nutr.* **97**:163–172.

Blank, M. L., Wykle, R. L., and Snyder, F. 1973. The retention of arachidonic acid in ethanolamine plasmalogens of rat testes during essential fatty acid deficiency. *Biochim. Biophys. Acta* **316**:28–34.

Bridges, R. B., and Coniglio, J. G. 1970a. The biosynthesis of $\Delta^{9,12,15,18}$-tetracosatetraenoic and of $\Delta^{6,9,12,15,18}$-tetracosapentaenoic acids by rat testes. *J. Biol. Chem.* **245**:46–49.

Bridges, R. B., and Coniglio, J. G. 1970b. The metabolism of 4,7,10,13,16-5-^{14}C-docosapentaenoic acid in the testis of the rat. *Biochim. Biophys. Acta* **218**:29–35.

Bridges, R. B., and Coniglio, J. G. 1970c. The metabolism of linoleic and arachidonic acids in rat testis. *Lipids* **5**:628–635.

Burr, G. O., and Burr, M. M. 1930. On the nature and role of the fatty acids essential in nutrition. *J. Biol. Chem.* **86**:587–621.

Butler, W. R., Johnson, A. D., and Gomes, W. R. 1968. Effect of short term vitamin A deficiency on testicular lipids in the rat. *J. Reprod. Fertil.* **15**:157–159.

Carney, J. A., and Walker, B. L. 1971. Metabolism of 1-^{14}C linoleic acid in the vitamin E-deficient rat testis. *Nutr. Reports Int.* **4**:103–108.

Carney, J. A., and Walker, B. L. 1972. Ovarian lipids from normal and essential fatty acid-deficient rats during oestrus and dioestrus. *Comp. Biochem. Physiol.* **41B**:137–145.

Carpenter, M. P. 1971. The lipid composition of maturing rat testis; the effect of α-tocopherol. *Biochim. Biophys. Acta* **231**:52–79.

Carpenter, M. P. 1974. Prostaglandins of rat testis. *Lipids* **9**:397–406.

Carpenter, M. P., and Manning, L. M. 1975. Testis microsomal prostaglandin synthesis. *Fed. Proc.* **34**:672.

Carpenter, M. P., Manning, L., and Wiseman, B. 1971. Prostaglandin synthesis in rat testis. *Fed. Proc.* **30**:1081.

Challis, J. R. G., Erickson, G. F., and Ryan, K. J. 1974. Prostaglandin F production in vitro by granulosa cells from rabbit pre-ovulatory follicles. *Prostaglandins* **7**:183–193.

Chasalow, F. I., and Pharriss, B. B. 1972. Luteinizing hormone stimulation of ovarian prostaglandin biosynthesis. *Prostaglandins* **1**:107–117.

Christie, W. W., and Moore, J. H. 1972. The lipid components of the plasma, liver and ovarian follicles in the domestic chicken (gallus gallus). *Comp. Biochem. Physiol* **41B**:287–295.

Claesson, L., and Hillarp, N. 1947. The formation mechanism of oestrogenic hormones. II. The presence of the oestrogen-precursor in the ovaries of rats and guinea pigs. *Acta Physiol. Scand.* **14**:102–119.

Clark, J. H., and Zarrow, M. X. 1967. A circadian rhythm in ovarian cholesterol. *Acta Endocrinol.* **56**:445–452.

Clegg, E. D., and Foote, R. H. 1973. Phospholipid composition of bovine sperm fractions, seminal plasma and cytoplasmic droplets. *J. Reprod. Fertil.* **34**:379–383.

Cole, B. T. 1956. Oxidation of unsaturated fatty acids in embryonic and adult tissues of golden hamsters. *Proc. Soc. Exp. Biol. Med.* **93**:290–294.

Coniglio, J. G., Zseltvay, R. R., Jr., and Whorton, R. A. 1971. Biosynthesis of stearic acid in rat testis. *Biochim. Biophys. Acta* **239**:374–375.

Coniglio, J. G., Bridges, R. B., Aquilar, H., and Zseltvay, R. R., Jr. 1972. The metabolism of 1-^{14}C-palmitic acid in the testis of the rat. *Lipids* **7**:368–371.

Coniglio, J. G., Bridges, R. B., and Ghosal, J. 1974a. Dietary influences on lipid composition and metabolism of testicular tissue. *La Riv. Ital. Sostanze Grasse* **LI**:312–314.

Coniglio, J. G., Grogan, W. M., Jr., and Rhamy, R. K. 1974b. Lipids of human testes removed at orchidectomy. *J. Reprod. Fertil.* **41**:67–73.

Coniglio, J. G., Grogan, W. M., Jr., Harris, D. G., and Fitzhugh, M. L. 1975a. Lipid and fatty acid composition of testes of quaking mice. *Lipids* **10**:109–112.

Coniglio, J. G., Grogan, W. M., Jr., and Rhamy, R. K. 1975b. Lipid and fatty acid composition of human testes removed at autopsy. *Biol. Reprod.* **12**:255–259.

Coniglio, J. G., Buch, D., and Grogan, W. M., Jr. 1976. Effect of eicosa-5,8,11,14-tetraynoic acid on fatty acid composition of selected organs in the rat. *Lipids* **11**:143–147.

Davis, J. T., and Coniglio, J. G. 1966. The biosynthesis of docosapentaenoic and other fatty acids by rat testes. *J. Biol. Chem.* **241**:610–612.

Davis, J. T., and Coniglio, J. G. 1967. The effect of cryptorchidism, cadmium and antispermatogenic drugs on fatty acid composition of rat testis. *J. Reprod. Fertil.* **14**:407–413.

Davis, J. T., Bridges, R. B., and Coniglio, J. G. 1966. Changes in lipid composition of the maturing rat testis. *Biochem. J.* **98**:342–346.

Ellis, L. C., Johnson, J. M., and Hargrove, J. L. 1972. Cellular aspects of prostaglandin synthesis and testicular function, pp. 385–398. *In* P. W. Ramwell and B. B. Pharriss (eds.). ALZA Conference on Prostaglandins in Cellular Biology. Plenum Press, New York.

Evans, H. M., Lepkovsky, S., and Murphy, E. A. 1934. Vital need of the body for certain unsaturated fatty acids. VI. Male sterility on fat-free diets. *J. Biol. Chem.* **106**: 445–449.

Evans, O. B., Jr., Zseltvay, R., Whorton, R., and Coniglio, J. G. 1971. Fatty acid synthesis in rat testes injected intratesticularly or incubated with 1-^{14}C acetate. *Lipids* **6**:706–711.

Fleeger, J. L., Bishop, J. P., Gomes, W. R., and VanDemark, N. L. 1968. Testicular lipids. I. Effect of unilateral cryptorchidism on lipid classes. *J. Reprod. Fertil.* **15**:1–7.

Flint, A. P. F., Grinwich, D. L., and Armstrong, D. T. 1973. Control of ovarian cholesterol ester biosynthesis. *Biochem. J.* **132**:313–321.

Fluharty, A. L., Stevens, R. L., Miller, R. T., and Kihara, H. 1974. Sulfoglycerogalactolipid from rat testis: A substrate for pure human arylsulfatase A. *Biochem. Biophys. Res. Commun.* **61**:348–354.

Gambal, D. 1967. Antiognadotropic hormones and lipogenesis in the testis and seminal vesicles of the rat. *Arch. Biochem. Biophys.* **118**:709–715.

Gambal, D., and Ackerman, R. J. 1967. Hormonal control of rat testicular phospholipids. *Endocrinology* **80**:231–239.

Gondos, B., and Hobel, C. J. 1973. Germ cell degeneration and phagocytosis in the human foetal ovary, pp. 77–83. *In* H. Peters (ed.). The Development and Maturation of the Ovary and its Functions. Excerpta Medica, Amsterdam.

Goswami, A., and Williams, W. L. 1967. Effect of hypophysectomy and replacement therapy on fatty acid metabolism in the rat testis. *Biochem. J.* **105**:537–543.

Goswami, A., Skipper, J. K., and Williams, W. L. 1968. Stimulation of fatty acid synthesis *in vitro* by gonadotrophin-induced testicular ribonucleic acid. *Biochem. J.* **108**:147–152.

Grogan, W. M., Jr., Coniglio, J. G., and Rhamy, R. K. 1973. Identification of some polyenoic acids isolated from human testicular tissue. *Lipids* **8**:480–482.

Haeffner, E. W., and Privett, O. S. 1971. Studies on the in vitro incorporation of 2-^{14}C-acetate and 1,5-^{14}C-citrate into lipids using rat testis subcellular preparations. *Fed. Proc.* **30**:520.

Hafiez, A. A. 1974. Prostaglandin E_2 prevents impairment of fertility in rats fed a diet deficient in essential fatty acids. *J. Reprod. Fertil.* **38**:273–286.

Hall, D. A., and Burdett, P. E. 1975. Age changes in the metabolism of essential fatty acids. *Biochem. Soc. Trans.* **3**:42–46.

Hall, P. F., Nishizawa, E. E., and Eik-Nes, K. B. 1963. Synthesis of fatty acids by testicular tissue in vitro. *Can. J. Biochem. Physiol.* **41**:1267–1274.

Handa, S., Ishizuka, I., Suzuki, M., Ueta, N., Yamato, K., and Yamakawa, T. 1973. Metabolism of seminolipid in mammalian testis, p. 396. *In* IXth International Congress of Biochemistry (Abstract).

Handa, S., Yamato, K., Ishizuka, I., Suzuki, A., and Yamakawa, T. 1974. Biosynthesis of seminolipid: sulfation in vivo and in vitro. *J. Biochem. (Tokyo)* **75**:77–83.

Hargrove, J. L., Johnson, J. M., and Ellis, L. C. 1971. Prostaglandin E_1 induced inhibition of rabbit testicular contractions in vitro. *Proc. Soc. Exp. Biol. Med.* **136**:958–961.

Herbst, A. L. 1967. Response of rat ovarian cholesterol to gonadotropins and anterior pituitary hormones. *Endocrinology* **81**:54–60.

Holman, R. T., and Greenberg, S. I. 1953. Highly unsaturated fatty acids. I. A survey of possible animal sources. *J. Am. Oil Chem. Soc.* **30**:600–601.

Holman, R. T., and Hofstetter, H. H. 1965. The fatty acid composition of the lipids from bovine and porcine reproductive tissues. *J. Am. Oil Chem. Soc.* **42**:540–544.

Hølmer, G., and Aaes-Jørgensen, E. 1969. Essential fatty acid deficient rats: III. Distribution of lipid classes in rat testes after feeding partially hydrogenated oils. *Lipids* **4**:515–521.

Horn, E. H. 1955. Nutritional and hormonal influences upon reproductive maturation, organ weights and histochemistry of the immature male rat. *Endocrinology* **57**:399–408.

Ishizuka, I., Suzuki, M., and Yamakawa, T. 1973. Isolation and characterizations of a novel sulfoglycolipid, "seminolipid", from boar testis and spermatozoa. *J. Biochem. (Tokyo)* **73**:77–87.

Johnson, A. D. 1970. Testicular lipids, pp. 193–258. *In* A. D. Johnson, W. R. Gomes, and N. L. VanDemark (eds.) The Testis, V. II. Academic Press, New York.

Johnson, A. D., VanDemark, N. L., Gomes, W. R., Butler, W. R., and Hodgen, G. D. 1967. Effect of antispermatogenic and hormone treatments on testicular cholesterol. *Fed. Proc.* **26**:645.

Johnson, J. M., Hargrove, J. L., and Ellis, L. C. 1971. Prostaglandin $F_{1\alpha}$ induced stimulation of rabbit testicular contraction in vitro. *Proc. Soc. Exp. Biol. Med.* **138**:378–381.

Johnson, L. A., Gerrits, R. J., and Young, E. P. 1969. The fatty acid composition of porcine spermatozoa phospholipids. *Biol. Reprod.* **1**:330–334.

Kar, A. B., and Roy, S. N. 1955. The effect of testosterone propionate on total cholesterol content of the testis in young rats. *Acta Endocrinol.* **18**:267–272.

Kar, A. B., Jehan, Q., Kamboj, V. P., Chowdhury, S. R., and Chowdhury, A. R. 1968. Effect of busulphan on biochemical composition of rat seminiferous tubules. *Indian J. Exp. Biol.* **6**:9–12.

Keenan, T. W., Nyquist, S. E., and Mollenhauer, H. H. 1972. Lipid composition of subcellular fractions from rat testis. *Biochim. Biophys. Acta* **270**:433–443.

Kirschman, J. C., and Coniglio, J. G. 1961. Polyunsaturated fatty acids in tissues of growing male and female rats. *Arch. Biochem. Biophys.* **93**:297–301.

Knapp, A., Kornblatt, M. J., Schachter, H., and Murray, R. K. 1973. Studies on the biosynthesis of testicular sulfoglycerogalactolipid: demonstration of a golgi-associated sulfotransferase activity. *Biochem. Biophys. Res. Commun.* **55**:179–186.

Komatsu, Y., and Takahashi, Y. 1971. Studies on lipid and fatty acid composition in the human testis. *Acta Urol. Jap.* **17**:705–721.

Kornblatt, M. J., Schachter, H., and Murray, R. K. 1972. Partial characterization of a novel glycerogalactolipid from rat testis. *Biochem. Biophys. Res. Commun.* **48**:1489–1494.

Kornblatt, M. J., Schachter, H., and Murray, R. K. 1973. The sulfated glycerogalactolipid of mammalian testis: Its occurrence in germinal cells, p. 396. *In* IXth International Congress of Biochemistry (Abstracts).

Kornblatt, M. J., Knapp, A., Levine, M., Schachter, H., and Murray, R. K. 1974. Studies on the structure and formation during spermatogenesis of the sulfoglycerogalactolipid of rat testis. *Can. J. Biochem.* **52**:689–697.

Krause, R. F., and Beamer, K. C. 1974. Lipid content and phospholipid metabolism of subcellular fractions from testes of control and retinol-deficient rats. *J. Nutr.* **104**:629–637.

Kuehl, F. A., Jr., Humes, J. L., Tarnoff, J., Cirillo, V. J., and Ham, E. A. 1970. Prostaglandin receptor site: Evidence for an essential role in the action of luteinizing hormone. *Science* **169**:883–886.

Kunau, W. H., and Couzens, B. 1971. Studies on the partial degradation of polyunsaturated fatty acids in subcellular fractions of rat liver. *Z. Physiol. Chem.* **352**:1297–1305.

Lynch, K. M., Jr., and Scott, W. W. 1951. Lipid distribution in the Sertoli cell and Leydig cell of the rat testis as related to experimental alterations of the pituitary-gonad system. *Endocrinology* **49**:8–14.

Major, P. W., Armstrong, D. T., and Greep, R. O. 1967. Effects of luteinizing hormone *in vivo* and *in vitro* on cholesterol conversion to progestins in rat corpus luteum tissue. *Endocrinology* **81**:19–28.

Mann, T. 1964. The Biochemistry of Semen and of the Male Reproductive Tract. Methuen, London, p. 222.

Marsh, J. M., Yang, N. S. T., and LeMaire, W. J. 1974. Prostaglandin synthesis in rabbit Graafian follicles in vitro. Effect of luteinizing hormone and cyclic AMP. *Prostaglandins* **7**:269–283.

McDowell, L. R., Froseth, J. A., Kroening, G. H., and Haller, W. A. 1974. Effects of dietary vitamin E and oxidized cottonseed oil on SGOT, erythrocyte hemolysis, testicular fatty acids and testicular selenium in swine fed peas (*Pisum sativum*). *Nutr. Rep. Int.* **9**:359–369.

McMullin, G. F., Smith, S. C., and Wright, P. A. 1968. Tissue fatty acid composition in four diverse vertebrate species. *Comp. Biochem. Physiol.* **26**:211–221.

Michael, C. M. 1973. Prostaglandins in swine testes. *Lipids* **8**:92–93.

Mills, S. C., and Scott, T. W. 1969. Metabolism of fatty acids by testicular and ejaculated ram spermatozoa. *J. Reprod. Fertil.* **18**:367–369.

Morin, R. J. 1967. In vitro incorporation of acetate-1-^{14}C into sphingomyelin, phosphatidyl choline and phosphatidyl ethanolamine of rabbit testes. *Proc. Soc. Exp. Biol. Med.* **126**:229–232.

Morin, R. J. 1968. Ovarian phospholipid composition and incorporation of 1-^{14}C-acetate into the phospholipid fatty acids of ovaries from non-pregnant and pregnant rabbits. *J. Reprod. Fertil.* **17**:111–117.

Morin, R. J. 1969. Incorporation of formate-^{14}C, acetate-1-^{14}C, choline-1,2-^{14}C and ethanolamine-1,2-^{14}C into rabbit ovarian phospholipids, Part II. *Life Sci.* **8**:613–616.

Morin, R. J. 1971. Cholesterol esterification and cholesteryl ester hydrolysis by rabbit and human ovaries. *Lipids* **6**:815–819.

Morin, R. J. 1973a. Cholesteryl ester synthesis and hydrolysis by subcellular fractions of corpora lutea and interstitial tissue from ovaries of pregnant rabbits. *Biochim. Biophys. Acta* **296**:203–208.

Morin, R. J. 1973b. Effects of prostaglandins E_1 and $E_{2\alpha}$ and of polyphloretin phosphate on cholesterol esterification by rabbit ovarian subcellular fractions. *Res. Commun. Chem. Pathol. Pharmacol.* **6**:195–206.

Morin, R. J., Eras, J., and Martin, S. C. 1969. Hydrolysis of cholesteryl oleate in vitro by ovarian homogenates from non-pregnant and pregnant rabbits, Part II. *Life Sci.* **8**:995–999.

Nakamura, M., and Privett, O. S. 1969a. Metabolism of lipids in rat testes: Interconversions and incorporation of linoleic acid into lipid classes. *Lipids* **4**:41–49.

Nakamura, M., and Privett, O. S. 1969b. Metabolism of glyceryl 1-^{14}C-trilinoleate in rat testis. *Lipids* **4**:93–98.

Nakamura, M., and Privett, O. S. 1973. Studies on the metabolism of linoleic-1-^{14}C acid in testes of hypophysectomized rats. *Lipids* **8**:224–231.

Nakamura, M., Jensen, B., and Privett, O. S. 1968. Effect of hypophysectomy on the fatty acids and lipid classes of rat testes. *Endocrinology* **82**:137–142.

Nakano, J., Montague, B., and Darrow, B. 1971. Metabolism of prostaglandin E_1 in human plasma, uterus, and placenta, in swine ovary and in rat testicle. *Biochem. Pharmacol.* **20**:2512–2514.

Neill, A. R., and Masters, C. J. 1972. Metabolism of fatty acids by bovine spermatozoa. *Biochem. J.* **127**:375–385.

Neill, A. R., and Masters, C. J. 1973. Metabolism of fatty acids by bovine spermatozoa. *J. Reprod. Fertil.* **34**:279–287.

Neill, A. R., and Masters, C. J. 1974a. Fatty acid incorporations and lipid compositions of mature and immature bovine testes. *Int. J. Biochem.* **5**:137–156.

Neill, A. R., and Masters, C. J. 1974b. The distribution of ^{14}C-label in the lipids of ram semen following the intratesticular injection of 1-^{14}C-palmitic acid. *J. Reprod. Fertil.* **38**:311–323.

Neill, A. R., and Masters, C. J. 1975. Observations on the plasmalogen content and aldehyde composition of the phosphoglycerides of ovine and porcine testes. *Comp. Biochem. Physiol.* **51B**:99–101.

Nugara, D., and Edwards, H. M., Jr. 1970. Changes in fatty acid composition of cockerel testes due to age and fat deficiency. *J. Nutr.* **100**:156–160.

Oshima, M., and Carpenter, M. P. 1968. The lipid composition of the prepubertal and adult rat testis. *Biochim. Biophys. Acta* **152**:479–497.

Payne, E., and Masters, C. J. 1970. Fatty acid metabolism in bovine spermatozoa. *Int. J. Biochem.* **1**:409–421.

Peluffo, R. O., Ayala, S., and Brenner, R. R. 1970. Metabolism of fatty acids of the linoleic acid series in testicles of diabetic rats. *Am. J. Physiol.* **218**:669–673.

Poulos, A., and White, I. G. 1973. The phospholipid composition of human spermatozoa and seminal plasma. *J. Reprod. Fertil.* **35**:265–272.

Poulos, A., Brown-Woodman, P. D. C., White, I. G., and Cox, R. I. 1975. Changes in phospholipids of ram spermatozoa during migration through the epididymis and possible origin of prostaglandin $F_{2\alpha}$ in testicular and epididymal fluid. *Biochim. Biophys. Acta* **388**:12–18.

Schlenk, H., Sand, D. M., and Gellerman, J. L. 1969. Retroconversion of docosahexaenoic acid in the rat. *Biochim. Biophys. Acta* **187**:201–207.

Scott, T. W. 1973. Lipid metabolism of spermatozoa. *J. Reprod. Fertil. Suppl.* **18**:65–76.

Scott, T. W., and Setchell, B. P. 1968. Lipid metabolism in the testis of the ram. *Biochem. J.* **107**:273–278.

Scott, T. W., Dawson, R. M. C., and Rowlands, I. W. 1963. Phospholipid interrelationships between epididymis and spermatozoa. *Biochem. J.* **87**:507–512.

Scott, T. W., Voglmayr, J. K., and Setchell, B. P. 1967. Lipid composition and metabolism in testicular and ejaculated ram spermatozoa. *Biochem. J.* **102**:456–461.

Sharma, D. P., and Venkitasubramanian, T. A. 1973. Ovarian lipids in guinea pigs. *J. Reprod. Fertil.* **35**:191–193.

Sprecher, H., and Duffy, M. P. 1974. The biosynthesis and turnover of different molecular species of rat testicular choline phosphoglycerides and triacylglycerols following intratesticular injection with 1(3)-^{14}C-glycerol. *Biochim. Biophys. Acta* **380**:21–30.

Stegner, H.-E. 1973. Electron microscopic studies on the development of the ovarian interstitial cell system in the foetal guinea pig, pp. 84–94. *In* H. Peters (ed.) The Development and Maturation of the Ovary and Its Functions. Excerpta Medica. Amsterdam.

Stoffel, W., Ecker, W., Assad, H., and Sprecher, H. 1970. Enzymatic studies on the mechanism of the retroconversion of C_{22}-polyenoic acids to their C_{20}-homologues. *Z. Physiol. Chem.* **351**:1545–1554.

Tinsley, I. J., Arscott, G. H., and Lowry, R. R. 1971. Fertility and testicular fatty acid composition in the chicken as influenced by vitamin E and ethoxyquin. *Lipids* **6**:657–660.

Turchetto, E., Martinelli, M., and Weiss, H. 1969. Rat gonad lipids at different ages, Part II. *Life Sci.* **8**:271–277.

Verdino, B., Blank, M. L., Privett, O. S., and Lundberg, W. O. 1964. Metabolism of 4,7,10,13,16-docosapentaenoic acid in the essential fatty acid deficient rat. *J. Nutr.* **83**:234–238.

Whorton, A. R. and Coniglio, J. G. 1974. Fatty acid synthesis in testes from fat-deficient rats. *Fed. Proc.* **33**:685.

Whorton, R., and Coniglio, J. 1975a. Metabolism of ^{14}C-arachidonate in testes of vitamin E-deficient rats. *Fed. Proc.* **34**:938.

Whorton, A. R. and Coniglio, J. G. 1975b. Studies *in vitro* of lipogenesis in rat testicular tissue. *Lipids* **10**:783–789.

Whorton, A. R., Antalis, T., and Coniglio, J. G. 1976. The metabolism of [1-^{14}C]stearic acid in rat testicular tissue. *Biochim. Biophys. Acta* **424**:66–72.

Wilks, J. W., Forbes, K. K., and Norland, J. F. 1972. Synthesis of prostaglandin $F_{2\alpha}$ by the ovary and uterus. *J. Reprod. Med.* **9**:271–276.

Witting, L. A., Likhite, V. N., and Horwitt, M. K. 1967. The effect of antioxidant deficiency on tissue lipid composition in the rat. III. Testes. *Lipids* **2**:103–108.

Wykle, R. L., Blank, M. L., and Snyder, F. 1973. The enzymic incorporation of arachidonic acid into ether-containing choline and ethanolamine phosphoglycerides by acylation-deacylation reactions. *Biochim. Biophys. Acta* **326**:26–33.

Yokoe, Y., Means, A. R., and Hall, P. F. 1969. The influence of follicle-stimulating hormone on testicular phospholipids. *Biochim. Biophys. Acta* **187**:278–280.

Mammary Glands

R. R. DILS

I. Introduction

There have recently been three excellent reviews of lipid metabolism in the mammary gland. Bauman and Davis (1974) have reviewed the literature (about 230 references) to 1971–1972 concerned with the biosynthesis of fatty acids, triacylglycerols, phospholipids, and cholesterol in the gland. The composition and biosynthesis of milk triacylglycerols has also been reviewed by Smith and Abraham (1975), who cover about 120 papers up to 1974. A somewhat broader area has been covered by Patton and Jensen (1975), who have reviewed nearly 400 papers to 1973 which relate lipid metabolism in the gland to (a) general body metabolism, (b) the structure and function of mammary epithelium, (c) biochemical pathways involved in milk fat synthesis, and (d) the composition of milk lipids.

The purpose of this chapter is solely to update these reviews. The coverage is not as exhaustive, and it inevitably reflects the interests of the author.

II. Fatty Acid Synthesis and Oxidation in Lactating Mammary Glands

A. Source of Carbon Atoms for Milk Fat Synthesis

Bickerstaffe *et al.* (1974) have used specialized techniques developed with goat udder to study the uptake of milk constituents and their transfer into the milk of the lactating dairy cow. The data show substantial arterio-

R. R. DILS • Department of Physiology and Biochemistry, University of Reading, Whiteknights, Reading RG6 2AJ, United Kingdom.

venous differences for glucose, acetate, and β-hydroxybutyrate, as well as for triacylglycerols; these differences were similar between breeds and not related to milk yields.

The rates of synthesis of fatty acids from acetate and from glucose by slices of lactating mammary gland from a number of species have been compared (Strong and Dils, 1972a). The rate of acetate incorporation increases in the order rat < sow < rabbit < guinea pig < cow, whereas the rate of glucose incorporation shows the reverse order. The results indicate that mammary slices from lactating guinea pig may have a capacity to utilize acetate alone for fatty acid synthesis, which approaches that of slices of ruminant mammary gland. Experiments with dispersed cells from lactating-bovine mammary glands have shown that butyrate and hexanoate can act as primers for fatty acid synthesis *de novo* (Kinsella, 1972a,b).

B. *Purification of Enzymes Involved in Fatty Acid Synthesis de Novo*

Acetyl-CoA carboxylase (EC 6.4.1.2) has been purified from lactating-rat (Miller and Levy, 1974) and lactating-rabbit (Manning *et al.*, 1976) mammary gland. Fatty acid synthase has now been isolated from lactating-bovine (Knudsen, 1972; Kinsella *et al.*, 1975) and -rat mammary glands (Smith and Abraham, 1974). A comparison of the properties of synthases isolated from the glands of a number of species show few biochemical differences (Dils and Carey, 1974). However, Smith (1973) has used immunological techniques to show that mammary and liver synthases from within the same species (but not between species) are similar, if not identical, proteins. Glucose-6-phosphate dehydrogenase (EC 1.1.1.49) has been isolated from lactating-rat mammary gland (Nevaldine *et al.*, 1974) and the molecular weight of the dimer, monomer, and subunits established. 6-Phosphogluconate dehydrogenase (EC 1.1.1.44) has been purified from lactating-rabbit mammary glands using affinity chromatography. It was homogeneous bv the criterion of SDS-gel electrophoresis (Betts and Mayer, 1975).

C. *Fatty Acid Chain Termination*

The mammary gland is unique among animal tissues in being able to terminate fatty acid chain elongation for the synthesis of short (C_4–C_6) and medium-chain (C_8–C_{12}) fatty acids. This ability varies between species; for example, lactating-rabbit-mammary glands synthesize exclusively $C_{8:0}$ and

$C_{10:0}$ acids, whereas the guinea pig tissue cannot synthesize acids shorter than $C_{16:0}$ (Strong and Dils, 1972b). Lactating-rabbit mammary gland contains acyl-thioester hydrolase(s), which are active towards medium- as well as long-chain acyl-CoA esters (Knudsen and Dils, 1975; Knudsen *et al.*, 1975). It is suggested that this enzyme(s) could cleave medium-chain acyl groups from the acyl carrier protein of fatty acid synthase. Smith and Abraham (1971) had also found evidence for this enzyme(s) in the cytosol of lactating-rat mammary glands. The cytosol had thioesterase activity towards medium-chain ($C_{10:0}$-$C_{14:0}$) acyl-CoA esters that was sufficient to account for the synthesis of these fatty acids in this fraction. Chain termination of the growing acyl group on the synthase does not appear to be directly related to the chain-length specificity of the enzymes involved in the esterification of the synthesized fatty acids (Carey and Dils, 1973; Tanioka *et al.*, 1974; Breach and Dils, 1975). There is an interesting recent report of a factor in rat liver cytosol that terminates chain lengthening at C_9–C_{11} acids (Divakaran and Kumar, 1975).

D. *Fatty Acid Desaturation*

Microsomal fractions of lactating-goat, -sow (Bickerstaffe and Annison, 1971), -cow (McDonald and Kinsella, 1973), and -mouse (Rao and Abraham, 1974) mammary glands actively desaturate long-chain fatty acids to the corresponding monoenoic acids. This may provide an important source of, for example, oleic acid for milk fat synthesis. Though the lactating-guinea-pig mammary gland has an active desaturase system, this is entirely absent in lactating-rabbit mammary glands (Strong and Dils, 1972a).

E. *Fatty Acid Oxidation*

When lactating mammary glands of sheep and goats are perfused with [1-^{14}C]myristic acid ($C_{14:0}$), part of the activity is recovered as $^{14}CO_2$ and a small proportion is converted to volatile fatty acids (Massart-Leën *et al.*, 1974). Medium-chain ($C_{10:0}$ and $C_{12:0}$) fatty acids are also oxidized by lactating-goat mammary glands *in vivo*. The products contribute 4-carbon units (rather than exclusively 2-carbon units as expected from β-oxidation) to the short-chain acids of milk lipids (Swenson and Dimick, 1974). By contrast, Kinsella (1972b) has reported that exogenous hexanoic acid ($C_{6:0}$) is initially catabolized via β-oxidation to acetate by isolated cells or by slices from lactating-bovine mammary glands. The acetate was subsequently reincorporated into fatty acids synthesized *de novo*.

III. Triacylglycerol and Phospholipid Synthesis in Lactating Mammary Glands

A. Lipoprotein Lipase and the Uptake of Lipid

Scow and his colleagues (1973) have reviewed the role of lipoprotein lipase in the uptake of blood triacylglycerols during pregnancy and lactation, and the hormonal control of this enzyme during lactation. The activity in rabbit (Falconer and Fiddler, 1970) and rat (Zander *et al.,* 1974) mammary gland is highly sensitive to control by prolactin.

McCarthy and Coccodrilli (1975) have presented evidence that lipoprotein lipase is involved in the formation of 2-acylglycerols for milk fat synthesis. However, the stereochemical course of the hydrolysis of triacylglycerols suggests that, though the enzyme attacks preferentially at position *sn*-1, this is followed by hydrolysis at positions *sn*-2 and *sn*-3 (Morley and Kuksis, 1972).

The contributions of the very-low-density and low-density serum lipoprotein fractions of lactating cows to milk triacylglycerol synthesis have been assessed by Raphael *et al.* (1973) and by Glascock and Welch (1974).

B. sn-Glycerol 3-P Pathway of Glyceride Synthesis

Within 10 minutes of injecting [^{32}P]P_i into lactating rats, the pool of phosphatidate is intensely labeled in the mammary tissue (Patton, 1975). This provides further evidence for the *sn*-glycerol 3-P pathway in the synthesis of milk triacylglycerols *in vivo.* Gross and Kinsella (1974) have characterized the palmitoyl-CoA : *sn*-glycerol 3-P acyltransferase (EC 2.3.1.15) activity (see the introductory chapter) in microsomes of lactating-bovine mammary glands; the major product was phosphatidic acid. This suggests that [in addition to phosphatidate phosphatase (EC 3.1.3.4)] this could be a key enzyme in the synthesis of milk triacylglycerols. The chain-length specificity of this acyltransferase has been investigated using subcellular fractions of lactating-rat (Tanioka *et al.,* 1974) and -rabbit (Breach and Dils, 1975) mammary glands. Overall, there was a preference for long-chain acyl groups, which reflects the specificities of glycerolphosphate acyltransferase and lysophosphatidate acyltransferase for the *sn*-1 and *sn*-2 positions respectively of glycerol.

A very active acyl-CoA: 1-acyl-*sn*-glycerol 3-phosphorylcholine acyltransferase (EC 2.3.1.1) is present in microsomal fractions of lactating-bovine mammary glands (Kinsella and Infante, 1974). This enzyme may be involved in phosphatidylcholine turnover, which in turn may be an integral part of the assembly of milk fat droplets.

C. Monoacyglycerol Pathway of Glyceride Synthesis

Stereospecific analysis of high-melting-point triacylglycerols of bovine milk fat supports the view that the monoacylglycerol pathway may be important in the biosynthesis of these lipids (Barbano and Sherbon, 1975). Additional evidence comes from the work of McCarthy and Coccodrilli (1975), who have shown that lipoprotein lipase acts on blood triacylglycerols to yield 2-monoacylglycerols, which enter the gland and are utilized for the synthesis of milk fat.

D. Esterification in Milk

Kinsella (1972c) and Christie (1974) have shown that freshly secreted bovine and goat milk contain all the enzymes required to synthesize triacylglycerols from glycerol and fatty acids. Milk may therefore be a useful alternative to mammary biopsy material for studies on the synthesis of milk fat (Christie, 1974).

E. Fatty Acid Composition of Milk

Jenness (1974) has written a short review of the biosynthesis and composition of milk fat in which he points out that the melting point of milk fat is lowered by a number of factors. The pattern of fatty acids in milk fat is a distinctive species characteristic. The production of a milk fat that is liquid at body temperature must therefore entail a delicately controlled balance of fatty acids available for esterification. For each species, the balance must be maintained between (1) the fatty acids taken up from the circulation (which will reflect dietary intake), (2) the synthesis of medium-chain acids by chain termination in the gland, (3) the extent of desaturation and elongation of fatty acids in the gland, and (4) the asymmetric positioning of fatty acids on the glycerol backbone.

IV. Hormonal Control of Milk Fat Synthesis

A. Adaptive Changes in Lipogenic Enzymes during the Lactation Cycle

There has been an extensive investigation of adaptive changes in enzymes related to lipogenesis in rat mammary glands from pregnancy, through lactation to weaning (Gumma *et al.*, 1973). The activities of a "satellite" system of lipogenic enzymes [such as glucose 6-phosphate dehydrogenase, citrate synthase (EC 4.1.3.7), ATP citrate lyase (EC

4.1.3.8), acetyl-CoA carboxylase and fatty acid synthase] increase rapidly with the onset and progression of lactation compared with a number of tricarboxylic acid cycle enzymes. The results are compared with the activities of enzymes of lipogenesis in lactating-sheep mammary glands, and the consequences of the lack of incorporation of glucose into fatty acids in this latter tissue are discussed.

At the termination of lactation in sheep, the rate of fatty acid synthesis and the activity of acetyl-CoA carboxylase decrease by 99% in mammary tissue (Bauman *et al.*, 1974). Changes in the activities of lipogenic enzymes in rat mammary glands on weaning have also been described (Gumma *et al.*, 1973; Carlsson *et al.*, 1973).

Between days 18 and 21 of pregnancy in the rabbit, the gland undergoes intense morphological differentiation (Bousquet *et al.*, 1969). During this period there is increased lipogenesis due to the ability to synthesize triacylglycerols containing $C_{8:0}$ and $C_{10:0}$ fatty acids characteristic of the lactating tissue (Strong and Dils, 1972b; Mellenberger and Bauman, 1974). The enzyme system involved in chain termination (see section II.C) must become operative at this stage. The kinetic characteristics of acyltransferases involved in fatty acid esterification also change at this time (Caffrey *et al.*, 1975). This differentiation correlates with a fall in plasma progesterone at day 18 of pregnancy and a very abrupt rise in plasma cortisol (Baldwin and Stabenfeldt, 1974). Though plasma prolactin concentrations are not available, Assairi (1974) has shown that progesterone inhibits the lactogenic action of prolactin at this critical period of differentiation in the pseudopregnant rabbit mammary gland.

At parturition, a second lipogenic stimulus occurs in the gland, though the pattern of fatty acids synthesized is unaltered (Strong and Dils, 1972b; Mellenberger and Bauman, 1974). The activities of enzymes involved in milk fat synthesis increase, and there is a close temporal correlation during pregnancy and lactation between the rate of fatty acid synthesis and the activities of ATP : citrate lyase and acetyl-CoA carboxylase in the gland (Mellenberger and Bauman, 1974). The malate-transhydrogenation cycle was found to play little part in the generation of reducing equivalents.

There is also a highly significant temporal correlation between the activities of acetyl-CoA synthetase (EC 6.2.1.1) and acetyl-CoA carboxylase and the lipogenic capacity of cow mammary glands between day 30 prepartum and day 7 postpartum. The increased triacylglycerol synthesis is related to the increased synthesis of short- and medium-chain fatty acids characteristic of the lactating gland (Mellenberger *et al.*, 1973; Kinsella, 1975). Shirley *et al.* (1973) have correlated changes in the activities of bovine mammary enzymes involved in the synthesis of milk fat with serum and mammary tissue concentrations of hormones. Cortisol and progesterone in mammary tissue decreased when lactation was induced, whereas corticosterone remained constant. For a review of serum hormone concentrations in ruminants during mammary growth, lactogenesis, and lactation, see Convey (1974).

The activity of pyruvate dehydrogenase (EC 1.2.4.1) in rat mammary glands shows many of the regulatory properties displayed by this enzyme in other tissues (Coore and Field, 1974). Its total activity (i.e., after dephosphorylation *in vitro*) in the gland and the fraction in the active form increase from pregnancy to midlactation and remain elevated until the end of lactation. They then fall steeply within 3 days of weaning. These authors have presented preliminary evidence that prolactin plays a major role in regulating the synthesis and phosphorylation of the mammary enzyme at midlactation *in vivo* (Field and Coore, 1975).

A study of rates of enzyme synthesis (rather than merely of enzyme activity) has shown that cortisol regulates the rates of synthesis of a number of enzymes involved in the synthesis of milk fat in lactating-rat mammary gland *in vivo* (Korsrud and Baldwin, 1972a,b).

B. *The Role of Cyclic Nucleotides*

There is a coordinate change in rat mammary glands between the activities of adenylate cyclase (EC 4.6.1.1) and cAMP phosphodiesterase (EC 3.1.4.17) and the tissue concentrations of cAMP and cGMP at different stages of the lactation cycle (Sapag-Hagar and Greenbaum, 1974a,b). The concentration of cAMP rises to a peak at the end of pregnancy and then falls after parturition to reach a low level on the sixteenth day of lactation; the reverse pattern holds for cGMP and for cAMP phosphodiesterase activity. The results suggest that the growth and development of the gland is related to the dual system of control by tissue concentrations of cAMP and cGMP. The initiation and the scale of lactation, and the changes seen at involution, may be related to the removal of cAMP.

Mammary explants from midpregnant rats have been cultured (see also section IV.C.3) with insulin, corticosterone, and prolactin in the presence and absence of dibutyryl-cAMP (Sapag-Hagar *et al.*, 1974). The nucleotide inhibited the increase in enzyme activities normally associated with lactogenesis. It had a pronounced inhibitory effect on fatty acid synthase activity and on fatty acid synthesis from acetate. Speake *et al.* (1976) have extended this work. When mammary explants from 16-day pregnant rabbits are cultured with insulin, cortisol, and prolactin, the addition of dibutyryl cAMP plus theophylline delays the accumulation of fatty acid synthase. This is achieved by a decreased rate of synthesis of the enzymically active enzyme (which might be due to the synthesis of inactive precursors of the complex) together with an increased rate of degradation of the enzyme. These results give some insight into the mechanism by which the cyclic nucleotides may operate.

There appears to be no direct evidence as yet that prolactin exerts its effect on the mammary gland receptors via adenylate cyclase. Falconer and Rowe (1975) have suggested that the action of prolactin may be derived from its earlier (i.e., in evolution) osmoregulatory control of sodium transport.

C. *Use of Cell and Organ Cultures for Studies on the Hormonal Control of Milk Fat Synthesis*

1. *Isolated Cells*

Suspensions of adipose-free parenchymal cells from lactating-mouse mammary glands are superior to slices in their metabolic activities. The lipids synthesized by the cells are predominantly triacylglycerols, which contain medium-chain fatty acids characteristic of the lactating gland (Abraham *et al.*, 1972). Similarly, dispersed mammary cells from lactating rats synthesize tissue-specific fatty acids esterified as triacylglycerols (Kinsella, 1974), esterify exogenous long-chain fatty acids into triacylglycerols (Kinsella, 1973), and metabolize butyrate as does the gland *in vivo* (Kinsella, 1972a).

Isolated rat mammary secretory cells have been used to study the effects of insulin on fatty acid synthesis from pyruvate or from glucose (Yang and Baldwin, 1975). There was an increased rate of synthesis in cells with a low redox state.

The ability of bovine mammary cells to synthesize fatty acids diminishes rapidly after six days in culture, as does the proportion of short- and medium-chain acids formed (Kinsella, 1972d). This "dedifferentiation" obviously limits the usefulness of isolated cell preparations for studying some aspects of the hormonal control of milk fat synthesis. On the other hand, it could well be an excellent model with which to mimic changes in milk fat synthesis at involution. The maintenance of monolayers of mammary epithelial cells in culture does not appear to have been reported yet.

2. *Intact Acini*

Katz *et al.* (1974) have published an interesting technique for the isolation of intact acini from the mammary tissue of lactating rat. The acini were metabolically active and offer a distinctive preparation with which to study hormonal responsiveness. A detailed examination has been made of rates of lipogenesis from pyruvate, lactate, acetate, and glucose.

3. *Results Obtained Using Mammary Explants*

Explants of lobulo-alveolar mammary tissue can be maintained in organ culture for several days. This technique [see Forsyth (1971) for review] has been used to study the hormonal control of differentiation in the gland during pregnancy. Differentiation can be monitored by measuring the biochemical changes that lead to the unique synthesis of milk fat. These can then be related to morphological changes, which have been briefly reviewed by Heald (1974).

Mammary explants from 10- to 14-day pregnant mice and rats respond to culture with insulin, cortisol or corticosterone, and prolactin by increasing the rate of synthesis of fatty acids that are characteristic of the lactating tissue (Wang *et al.,* 1972; Hallowes *et al.,* 1973; Cameron and Rivera, 1975). The activities of glucose 6-phosphate dehydrogenase and 6-phosphogluconate dehydrogenase increase twofold; actinomycin D and cycloheximide inhibit these increases when they are present from the start of culture (Rivera and Cummins, 1971).

Mammary explants from 16-day pregnant and 11-day pseudopregnant rabbits also differentiate when cultured with these three hormones. There is a 30- to 40-fold increase in the rate of fatty acid synthesis after 2 days in culture to values observed in early lactation. This is accompanied by chain termination leading to the synthesis of characteristic $C_{8:0}$ and $C_{10:0}$ acids esterified as triacylglycerols (Forsyth *et al.,* 1972; Strong *et al.,* 1972; Speake *et al.,* 1975). These changes are not observed if prolactin is omitted from the hormone combination. The low progesterone and high cortisol concentrations *in vivo* which signal the onset of differentiation may be being mimicked in culture. Removal of hormones from the medium after explants have been cultured with insulin, cortisol, and prolactin for 1 to 3 days causes a decrease, within 5–10 hr, to very low values in the rate of fatty acid synthesis and in the proportion of fatty acids formed in milk. This may be useful as a model system to study milk fat synthesis during involution.

Organ culture experiments with goat mammary glands (Lynch and Dils, 1976) show that the maximum response to insulin, cortisol, and prolactin in terms of increased fatty acid synthesis occurs in early pregnancy. The tissue-specific pattern of fatty acids is less easy to detect in this case.

4. *Hormonal Control of Turnover of Fatty Acid Synthase*

Organ cultures have recently been used to study the rates of synthesis, accumulation, and degradation of fatty acid synthetase during differentiation in rabbit mammary gland (Speake *et al.,* 1975, 1976). When mammary explants from 16-day pregnant rabbits are cultured with insulin, cortisol, and prolactin, the increased synthesis of milk fatty acids after 40 hr is paralleled by a fivefold increase in the rate of synthesis of fatty acid synthase and a 40-fold increase in the amount of the enzyme in the explants. The enzyme is rapidly degraded during this period unless prolactin is present, whereas explants cultured with prolactin alone appear to synthesize enzymically inactive precursors of fatty acid synthase. When explants are cultured with hormone combinations that include prolactin, there is a striking decrease, or even cessation, in the degradation of synthase during the period (day 2 in culture) of maximum accumulation of the enzyme. Continued culture results in the synthase being degraded again.

This behavior of synthase in response to hormones *in vitro* may represent that of other enzymes involved in the synthesis of cell-specific products *in vivo*. These enzymes must accumulate rapidly in the tissue during differentiation so that milk components such as milk fat can be synthesized as efficiently as possible when the tissue is hormonally stimulated. When sufficient enzyme has accumulated, degradation recommences.

References

Abraham, S., Kerkof, P. R., and Smith, S. 1972. Characteristics of cells dissociated from mouse mammary glands II. Metabolic and enzymatic activities of parenchymal cells from lactating glands. *Biochim. Biophys. Acta* **261**:205–208.

Assairi, L., Delouis, C., Gaye, P., Houdebine, L. M., Oliver-Bousquet, M., and Denamur, R. 1974. Inhibition by progesterone of the lactogenic effect of prolactin in the pseudopregnant rabbit. *Biochem. J.* **144**:245–252.

Baldwin, D. M., and Stabenfeldt, G. H. 1974. Plasma levels of progesterone, cortisol and corticosterone in the pregnant rabbit. *Biol. Reprod.* **10**:495–501.

Barbano, D. M., and Sherbon, J. W. 1975. Sterospecific analysis of high melting triglycerides of bovine milk fat and their biosynthetic origin. *J. Dairy Sci.* **58**:1–8.

Bauman, D. E., and Davis, C. L. 1974. Biosynthesis of milk fat, pp. 31–75. *In* B. L. Larson and V. R. Smith (eds.). Lactation: A Comprehensive Treatise, Vol. 2. Academic Press, New York and London.

Bauman, D. E., Mellenberger, R. W., and Ingle, D. L. 1974. Metabolic adaptations in fatty acid and lactose biosynthesis by sheep mammary tissue during cessation of lactation. *J. Dairy Sci.* **57**:719–723.

Betts, S. A., and Mayer, R. J. 1975. Purification and properties of 6-phosphogluconate dehydrogenase from rabbit mammary gland. *Biochem. J.* **151**:263–270.

Bickerstaffe, R., and Annison, E. F. 1971. The desaturase activity of goat and sow mammary gland. *Comp. Biochem. Physiol.* **35**:653–665.

Bickerstaffe, R., Annison, E. F., and Linzell, J. L. 1974. The metabolism of glucose, acetate, lipids and amino acids in lactating dairy cows. *J. Agric. Sci., Camb.* **82**:71–85.

Bousquet, M., Fléchon, J. E., and Denamur, R. 1969. Aspects ultrastructuraux de la glande mammaire de lapine pendant la lactogénèse. *Z. Zellforsch. Mikrosk. Anat.* **96**:418–436.

Breach, R. A., and Dils, R. 1975. Fatty acyl specificity of glycerol 3-phosphate esterification in lactating rabbit mammary gland. *Int. J. Biochem.* **6**:329–340.
Caffrey, M., Infante, J. P., and Kinsella, J. E. 1975. Isoenzymes of an acyl transferase from rabbit mammary gland: evidence from biphasic substrate saturation kinetics. *FEBS Lett.* **52**:116–120.
Cameron, J. A., and Rivera, E. M. 1975. Hormone-stimulated lipid synthesis in mammary culture. *Am. Zool.* **15**:285–293.
Carey, E. M., and Dils, R. 1973. Regulation of the chain-length of fatty acids synthesized by cell-free preparations of lactating rabbit, rat and guinea-pig mammary gland. *Comp. Biochem. Physiol.* **44B**:989–1000.
Carlsson, E. I., Karlsson, B. W., and Waldemarson, K. H. C. 1973. Dehydrogenases and nucleic acids in rat mammary glands during involution initiated at various stages of lactation. *Comp. Biochem. Physiol.* **44B**:93–108.
Christie, W. W. 1974. Biosynthesis of triglycerides in freshly secreted milk from goats. *Lipids* **9**:876–882.
Convey, E. M. 1974. Serum hormone concentrations in ruminants during mammary growth, lactogenesis and lactation: A review. *J. Dairy Sci.* **57**:905–917.
Coore, H. C., and Field, B. 1974. Properties of pyruvate dehydrogenase of rat mammary tissue and its changes during pregnancy, lactation and weaning. *Biochem. J.* **142**:87–95.
Dils, R., and Carey, E. M. 1974. Fatty acid synthetase from rabbit mammary gland, pp. 74–83. *In* J. M. Lowenstein (ed.). Methods in Enzymology. Vol. XXXV, part B. Academic Press, New York and London.
Divakaran, P., and Kumar, S. 1975. Formation of acids shorter than palmitic by rat liver cytosol. *Biochem. Biophys. Res. Commun.* **66**:1042–1047.
Falconer, I. R., and Fiddler, T. J. 1970. Effects of intraductal administration of prolactin, actinomycin D and cycloheximide on lipoprotein lipase activity in the mammary glands of pseudopregnant rabbits. *Biochim. Biophys. Acta* **218**:508–514.
Falconer, I. R., and Rowe, J. M. 1975. Possible mechanism for action of prolactin on mammary cell sodium transport. *Nature* **256**:327–328.
Field, B., and Coore, H. G. 1975. Effects of prolactin withdrawal on activity of pyruvate dehydrogenase of rat mammary gland. *Biochem. Soc. Trans.* **3**:258–260.
Forsyth, I. A. 1971. Organ culture techniques and the study of hormone effects on the mammary gland. *J. Dairy Sci.* **3**:419–444.
Forsyth, I. A., Strong, C. R., and Dils, R. 1972. Interactions of insulin, corticosterone and prolactin in promoting milk-fat synthesis by mammary explants from pregnant rabbits. *Biochem. J.* **129**:929–935.
Glascock, R. F., and Welch, V. A. 1974. Contribution of the fatty acids of three low density serum lipoproteins to bovine milk fat. *J. Dairy Sci.* **57**:1364–1370.
Gross, M. J., and Kinsella, J. E. 1974. Properties of palmitoyl-CoA: *L*-α-glycerophosphate acyl transferase from bovine mammary microsomes. *Lipids* **9**:905–912.
Gumma, K. A., Greenbaum, A. L., and McLean, P. 1973. Adaptive changes in satellite systems related to lipogenesis in rat and sheep mammary gland and in adipose tissue. *Eur. J. Biochem.* **34**:188–198.
Hallowes, R. C., Wang, D. Y., Lewis, D. J., Strong, C. R., and Dils, R. 1973. The stimulation by prolactin and growth hormone of fatty acid synthesis in explants from rat mammary glands. *J. Endocrinol.* **57**:265–276.
Heald, W. C. 1974. Hormonal effects on mammary cytology. *J. Dairy Sci.* **57**:917–925.
Jenness, R. 1974. Biosynthesis and composition of milk. *J. Invest. Dermatol.* **63**:109–118.
Katz, J., Wals, P. A., and Van de Velde, R. L. 1974. Lipogenesis by acini from mammary glands of lactating rats. *J. Biol. Chem.* **249**:7348–7357.
Kinsella, J. E. 1972a. Lipid biosynthesis in bovine mammary cells from [1-^{14}C]butyrate. *Int. J. Biochem* **3**:637–641.

Kinsella, J. E. 1972b. Utilization of 1-^{14}C-hexanoic acid for fatty acid and lipid synthesis by bovine mammary tissue. *J. Dairy Sci.* **55**:1181–1184.

Kinsella, J. E. 1972c. Glycerolipid synthesis in milk: Evidence of glycerol kinase and other biosynthetic enzymes. *Int. J. Biochem.* **3**:89–92.

Kinsella, J. E. 1972d. Lipogenesis in bovine mammary cells: Progressive changes with age in culture. *Biochim. Biophys. Acta* **270**:296–300.

Kinsella, J. E. 1973. Lipid composition and fatty acid metabolism by mammary cells of rat. *Int. J. Biochem.* **4**:549–556.

Kinsella, J. E. 1974. Biosynthesis of fatty acids in rat mammary cells. *Int. J. Biochem.* **5**:417–421.

Kinsella, J. E., 1975. Coincident synthesis of fatty acids and secretory triglycerides in bovine mammary tissue. *Int. J. Biochem.* **6**:65–67.

Kinsella, J. E., and Infante, J. P. 1974. Acyl-CoA acyl-*sn*-glycerol-3-phosphorylcholine acyltransferase of bovine mammary tissue. *Lipids* **9**:748–751.

Kinsella, J. E., Bruns, D., and Infante, J. P. 1975. Fatty acid synthetase of bovine mammary: Properties and products. *Lipids* **10**:227–237.

Knudsen, J. 1972. Fatty acid synthetase from cow mammary gland tissue cells. *Biochim. Biophys. Acta* **280**:408–414.

Knudsen, J., and Dils, R. 1975. Partial purification from rabbit mammary gland of a factor which controls the chain length of fatty acids synthesized. *Biochem. Biophys. Res. Commun.* **63**:780–785.

Knudsen, J., Clark, S., and Dils, R., 1975. Acyl-CoA hydrolase(s) in rabbit mammary gland which control the chain length of fatty acids synthesized. *Biochem. Biophys. Res. Commun.* **65**:921–926.

Korsrud, G. O., and Baldwin, R. L. 1972a. Effects of adrenalectomy, adrenalectomy-oviarectomy, and cortisol and estrogen therapies upon enzyme activities in lactating rat mammary glands. *Can. J. Biochem.* **50**:366–376.

Korsrud, G. O., and Baldwin, R. L. 1972b. Hormonal regulation of rat mammary gland enzyme activities and metabolite patterns. *Can. J. Biochem.* **50**:377–385.

Lynch E. J., and Dils, R. 1976. Differentiation of mammary gland during pregnancy: The response to hormones of rabbit and goat mammary gland explants in organ culture. *J. Endocrinol.* **68**:32P.

Manning, R., Dils, R., and Mayer, R. J. 1976. Purification and some properties of acetyl-CoA carboxylase from lactating rabbit mammary gland. *Biochem. J.* **153**:463–468.

Massart-Leën, A. M., Roets, E., Verbeke, R., and Peeters, G. 1974. Incorporation de l'acide myristique-1-^{14}C dans les constituants du lait par la glande mammaire isolée. *Ann. Biol. Anim. Biochim. Biophys.* **14**:459–469.

McCarthy, R. D., and Coccodrilli, G. D. 1975. Structure and synthesis of milk fat XI. Effects of heparin on paths of incorporation of glucose and palmitic acid into milk fat. *J. Dairy Sci.* **58**:164–168.

McDonald, T. M., and Kinsella, J. E. 1973. Stearyl-CoA desaturase of bovine mammary microsomes. *Arch. Biochem. Biophys.* **156**:223–231.

Mellenberger, R. W., and Bauman, D. E. 1974. Metabolic adaptations during lactogenesis. Fatty acid synthesis rabbit mammary gland tissue during pregnancy and lactation. *Biochem. J.* **138**:373–379.

Mellenberger, R. W., Bauman, D. E., and Nelson, D. R. 1973. Metabolic adaptations during lactogenesis. Fatty acid and lactose synthesis in cow mammary tissue. *Biochem. J.* **136**:741–748.

Miller, A. L., and Levy, H. R. 1974. Acetyl-CoA carboxylase from rat mammary gland, pp. 11–17. *In* J. M. Lowenstein (ed.). Methods in Enzymology, Vol. XXXV, part B. Academic Press, New York and London.

Morley, N., and Kuksis, A. 1972. Positional specificity of lipoprotein lipase. *J. Biol. Chem.* **247**:6389–6393.

Nevaldine, B. H., Hyde, C. M., and Levy, H. R. 1974. Mammary glucose 6-phosphate dehydrogenase. *Arch. Biochem. Biophys.* **165**:398–406.

Patton, S. 1975. Detection of rapidly labelled phosphatidic acid in lactating mammary gland of the intact rat. *J. Dairy Sci.* **58**: 560–563.

Patton, S., and Jensen, R. 1975. Lipid metabolism and membrane functions of the mammary gland, pp. 163–277. *In* R. T. Holman (ed.). Progress in the Chemistry of Fats and other Lipids, Vol. XIV, Part 4. Pergamon Press, Oxford.

Rao, G. A., and Abraham, S. 1974. Fatty acid desaturation by mammary gland microsomes from lactating mice. *Lipids* **4**:269–271.

Raphael, B. C., Dimick, P. S., and Puppione, D. L. 1973. Lipid characterization of bovine serum lipoproteins throughout gestation and lactation. *J. Dairy Sci.* **56**:1025–1032.

Rivera, E. M., and Cummins, E. P. 1971. Hormonal induction of dehydrogenase enzymes in mammary gland *in vitro*. *Gen. Comp. Endocrinol.* **17**:319–326.

Sapag-Hagar, M., and Greenbaum, A. L. 1974a. Adenosine 3′:5′-monophosphate and hormone interrelationships in the mammary gland of the rat during pregnancy and lactation. *Eur. J. Biochem.* **47**:303–312.

Sapag-Hagar, M., and Greenbaum, A. L. 1974b. The role of cyclic nucleotides in the development and function of rat mammary tissue. *FEBS Lett.* **46**:180–183.

Sapag-Hagar, M., Greenbaum, A. L., Lewis, D. J., and Hallowes, R. C. 1974. The effects of di-butyryl cAMP on enzymatic and metabolic changes in explants of rat mammary gland. *Biochem. Biophys. Res. Commun.* **59**:261–268.

Scow, R. O., Mendelson, C. R., Zinder, O., Hannosh, M., and Blanchette-Mackie, E. J. 1973. Role of lipoprotein lipase in the delivery of dietary fatty acids to lactating mammary tissue, pp. 91–114. *In* C. Galli, G. Jacini, and A. Pecite (eds.). Dietary Lipids and Postnatal Development. Raven Press, New York.

Shirley, J. E., Emery, R. S., Convey, E. M., and Oxender, W. D. 1973. Enzymic changes in bovine adipose and mammary tissue, serum and mammary tissue hormonal changes with initiation of lactation. *J. Dairy Sci.* **56**:569–574.

Smith, S. 1973. Studies on the immunological cross-reactivity and physical properties of fatty acid synthetases. *Arch. Biochem. Biophys.* **156**:751–758.

Smith, S., and Abraham, S. 1971. Fatty acid synthetase from lactating rat mammary gland. Studies on the termination sequence. *J. Biol. Chem.* **246**:2537–2542.

Smith, S., and Abraham, S. 1974. Fatty acid synthetase from lactating rat mammary gland, pp. 65–74. *In* J. M. Lowenstein (ed.). Methods in Enzymology. Vol. XXXV, Part B. Academic Press, New York and London.

Smith, S., and Abraham, S. 1975. The composition and biosynthesis of milk fat, pp. 195–239. *In* R. Paoletti and D. Kritchevsky (eds.). Advances in Lipid Research, Vol. 13. Academic Press, New York and London.

Speake, B., Dils, R., and Mayer, R. J. 1975. Regulation of enzyme turnover during tissue differentiation. Studies on the effects of hormones on the turnover of fatty acid synthetase in rabbit mammary gland in organ culture. *Biochem. J.* **148**:309–320.

Speake, B., Dils, R., and Mayer, R. J. 1976. Regulation of enzyme turnover during tissue differentiation. Interactions of insulin, prolactin and cortisol in controlling the turnover of fatty acid synthetase in rabbit mammary gland in organ culture. *Biochem. J.* **154**:359–370.

Strong, C. R., and Dils, R. 1972a. Fatty acids synthesized by mammary gland slices from lactating guinea pig and rabbit. *Comp. Biochem. Physiol.* **43B**:643–652.

Strong, C. R., and Dils, R. 1972b. Fatty acid biosynthesis in rabbit mammary gland during pregnancy and early lactation. *Biochem. J.* **128**:1303–1309.

Strong, C. R., Forsyth, I., and Dils, R. 1972. The effects of hormones on milk-fat synthesis in mammary explants from pseudopregnant rabbits. *Biochem. J.* **128**:509–519.

Swenson, P. E., and Dimick, P. S. 1974. Oxidation of medium chain fatty acids by the ruminant mammary gland. *J. Dairy Sci.* **57**:290–295.

Tanioka, H., Lin, C. Y., Smith, S., and Abraham, S. 1974. Acyl specificity in glyceride synthesis by lactating rat mammary gland. *Lipids* **9**:229–234.

Wang, D. Y., Hallowes, R. C., Bealing, J., Strong, C. R., and Dils, R. 1972. The effect of prolactin and growth hormone on fatty acid synthesis by pregnant mouse mammary gland in organ culture. *J. Endocrinol.* **53**:311–322.

Yang, Y. T., and Baldwin, R. L. 1975. Effects of insulin upon fatty acid synthesis from pyruvate, lactate, and glucose in rat mammary cells. *J. Dairy Sci.* **58**:337–343.

Zander, O., Hannosh, M., Fleck, T. R. C., and Scow, R. O. 1974. Effect of prolactin on lipoprotein lipase in mammary gland and adipose tissue of rats. *Am. J. Physiol.* **226**:744–748.

The Eye

R. M. Broekhuyse and F. J. M. Daemen

I. Introduction

Currently, the study of ocular lipids has become of great value for basic science as well as for clinical pathology. Complicated processes of lipid metabolism, lipid storage, and recycling are involved in the mechanism of light perception, which starts when photons reach the retina. Mechanisms of differentiation and aging can be studied in the lens, where fiber cells are formed and stored during the lifespan. In the cornea and sclera, lipid deposition occurs resembling that seen in the aging aorta. In research on specialized membranes, retinal rod outer segments and lens fiber membranes are used as model systems of high purity. The iris can be mentioned as one of the tissues in which prostaglandins were discovered in the first stages of their still young history. The eye is a favorite subject for hormonal research because not only do these substances have profound effects on ocular physiology and biochemistry, but such effects are often easily observable; for instance, by measurement of ocular pressure, aqueous flow, by slit lamp observation (anterior segment), and by funduscopy (posterior segment, retina). For these reasons we have collected together the most striking results of the research on ocular lipids.

II. Ocular Anatomy

Ocular anatomy is illustrated in Fig. 1. The bulbus is formed by the sclera and cornea. In the posterior segment, the cornea is the first transpar-

R. M. Broekhuyse • Department of Ophthalmology, University of Nijmegen, Nijmegen, The Netherlands. F. J. M. Daemen (coauthor for the section on the retina) • Department of Biochemistry, University of Nijmegen, Nijmegen, The Netherlands.

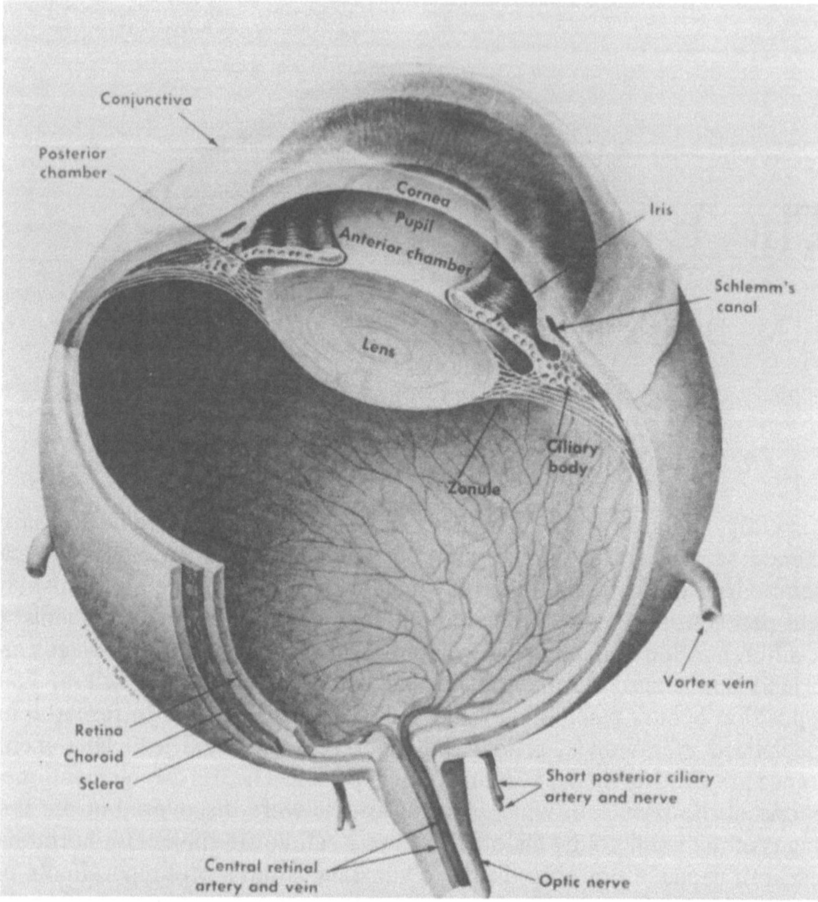

FIGURE 1. The human eye (from Newell, F. W., and Terry, E. J. 1974. Ophthalmology, 3rd ed. The C. V. Mosby Co., St. Louis, Missouri, 491 pp.).

ent tissue through which the light enters the eye, subsequently passing through the aqueous humor, the lens, and the vitreous body, and finally reaching the retina. The vitreous body is surrounded for about two thirds by the retina. The choroid is situated between the retina and sclera and shades off into the ciliary body and iris at the front of the eye. At the back, the retina is connected to the brain via the optic nerve. Further details about the morphology of the ocular tissues are given in the relevant sections of this chapter.

III. Lens

A. Morphology and General Metabolism

Figure 2 shows a cross section of the lens together with the zonular fibers by which it is positioned behind the iris. The lens is surrounded by a capsule and grows within it. There is a single layer of nucleated epithelial cells at the front surface. In the equatorial region the cells differentiate and elongate into fibers. When they are full grown, they lose their organelles. New fibers are layered on top of previous fibers. Thus, the oldest cells are situated in the lens nucleus and fibers containing subcellular organelles are largely confined to the equator.

The lens is avascular and normally completely clear. Its sole nutritional supply comes from the aqueous humor, which flows slowly along the equatorial and front surfaces. The main nutrient is glucose, which is mainly converted into lactic acid, although the epithelial cells also have an operational citric acid cycle. These and other pathways by which glucose is converted and/or by which ATP is generated have been reviewed by Van Heyningen (1969) and Kuck (1970a,b). Metabolism and transport processes in the lens are largely confined to the epithelium. In the equator, where growth of fiber cells is the dominating process, the biosynthetic apparatus is

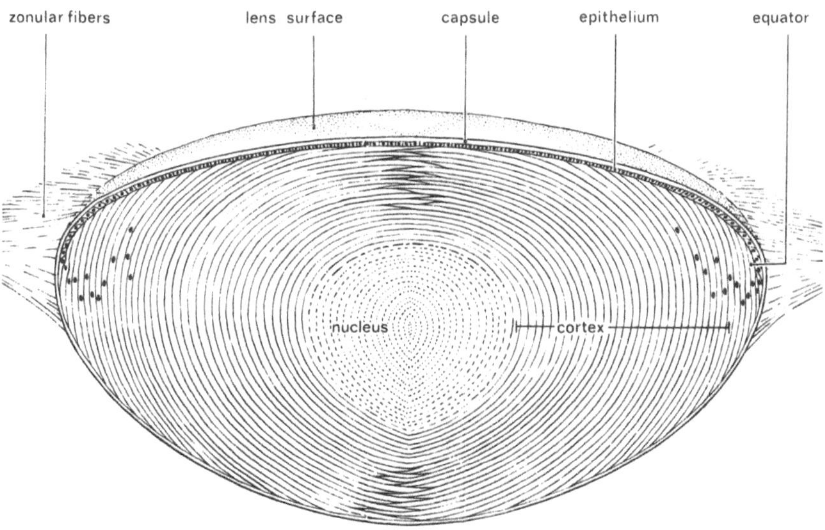

FIGURE 2. Cross section of lens.

very active. At this site the constituents of the cell membranes and the cytoplasm are formed. The cytoplasm of the fibers acquires a high protein concentration, α, β, and γ crystallins being the main components. During the aging process α and β crystallins tend to form high-molecular-weight aggregates, which have been found in lower animal (Liem-The *et al.*, 1975) as well as in human lens extracts (Jedziniak *et al.*, 1975). In view of findings that the water-insoluble fraction of the lens fiber cells can be split into a urea-soluble fraction containing crystallins and a urea-insoluble fraction containing mainly plasma membranes, it may be assumed that an association between these lens membranes and crystallins exists (Dische *et al.*, 1967; Lasser and Balazs, 1972; Bloemendal *et al.*, 1972; Broekhuyse and Kuhlmann, 1974).

Transport of substances seems to be limited to small molecules. The lens has an intracellular cation composition and the epithelium is the site of active transport of cations, sugars, amino acids, ascorbic acid (van Heyningen, 1969), and myo-inositol (Broekhuyse, 1968a). The capsule, which is a type of basement membrane, functions as a molecular sieve and obstructs the entrance of proteins and other large molecules. Surprisingly, horseradish peroxidase has been shown to penetrate the lens (Gorthy *et al.*, 1971), thereby passing the fine meshwork of the capsule.

B. *Lipid Composition*

For the study of lipid composition of aging lens and lens parts, the bovine lens is a useful model, first, because its morphology is very similar to that of human and other mammalian lenses and, second, because it is readily available and has large dimensions. The predominant lipids in mammalian lens are cholesterol and phospholipids, while in human lens glycolipids are also present in considerable amounts. The concentration of cholesterol in lens increases with age (Broekhuyse, 1973; Table 1). This can also be detected by analyzing lens parts going from peripheral layers (cortex) toward the central part (nucleus) as is shown in Table 2. The latter values are based on the wet weight of the parts concerned. However, the

TABLE 1

Cholesterol Concentrations in the Aging Human Lens

Age (years)	Number of lenses	Cholesterol concentration (mg/g wet wt)	Lens weight[a] (g)	Cholesterol content (mg/lens)
20–30	4	3.69 ± 0.61	0.176	0.65
60–70	6	5.70 ± 0.65	0.236	1.35

[a]From Broekhuyse (1973).

TABLE 2
The Concentration of Cholesterol in the Cortex and Nucleus of the Lens[a]

	Human lens		Calf lens	
	Cortex	Nucleus	Cortex	Nucleus
Cholesterol (mg/g wet wt)	5.66 ± 0.61	5.93 ± 0.69	0.67 ± 0.07	0.92 ± 0.05
Number of lenses	6		5	
Weight ratio cortex:nucleus	2.03:1		1.58:1	
Mean lens weight (g)	0.236		1.40	
Mean lens age (years)	65		0.25	

[a] From Broekhuyse (1973).

water content decreases with the age of fiber cells and amounts to 70% for cortex and 58% for nucleus in this material. Hence, it can be calculated that the cholesterol concentration is 2.2 mg/g dry wt in both lens parts. Assuming that this dry material is almost entirely protein and that in the fibers no protein is synthesized or lost, one can conclude that the cholesterol concentration is constant. These figures do not allow us to draw the conclusion that the cholesterol concentration in older lenses (on dry weight basis) will be the same. The possibility exists that an age-dependent cholesterol synthesis in the equator causes an increase in cholesterol in the aging lens. In fact this is the case with another very stable lipid in lens, namely sphingomyelin.

From Table 3 it appears that the relative sphingomyelin concentration

TABLE 3
The Proportional Phospholipid Composition of Parts of the Bovine Lens as a Function of Age[a]

	0.25-year-old calf lens							6-year-old bovine lens:
			Cortex			Epithelial cells with capsule[b]		
Phospholipid	Total lens	Nucleus	Inner layer	Outer layer	Equator	Center	Periphery	Total lens
Sphingomyelin	18.2	23.4	21.9	15.8	12.9	8.4	9.0	28.2
Phosphatidylcholine	31.8	28.9	26.4	32.7	39.1	45.0	45.4	20.7
Phosphatidylethanolamine	33.2	30.2	33.0	36.0	34.7	31.5	31.9	34.4
Phosphatidylserine	12.3	10.4	12.7	11.3	8.2	7.6	7.3	12.5
Phosphatidylinositol	1.9	0.8	0.8	2.2	4.3	6.0	5.6	3.1
Phosphatidic acid	1.0	1.6	1.0	0.4	0.7	0.6	0.5	0.8
Lysophosphatidylcholine	0.5	0.8	—	—	—	—	—	—
Others[c]	1.0	—	2.5	1.4	0.3	0.8	0.5	0.5
Total phospholipid (mg/g)	1.80	1.02	1.51	2.16	2.77	3.71	3.75	1.19
Weight % of total lens	100	35	22.5	42.5	4.0	1.12	0.56	100

[a] From Broekhuyse (1973).
[b] Weight ratio capsule:epithelial cells = 3:1.
[c] Includes 0–1 % lysophosphatidylethanolamine, especially in the older material.

increases from epithelium towards the nucleus and is higher in 6-year-old bovine lens than in calf lens (Broekhuyse, 1973). Calculated on a dry weight basis, the sphingomyelin concentration is higher in cow lens and its various parts than in calf lens and its corresponding parts (Roelfzema *et al.*, 1976; Table 4). This can be explained by assuming that during the formation of new fibers in the equator, the *de novo* synthesis of sphingomyelin increases with age. The concentrations of the other phospholipids (the phosphoglycerides) decrease (Tables 3 and 4) during aging of the lens fibers owing to continuous breakdown by phospholipases (see section III.C), which is apparent from Table 3. Human lens attains a much higher age and displays this breakdown very clearly (Broekhuyse, 1969a, 1973). Unusually high lysophospholipid concentrations are present in 70-year-old lens, especially in the nucleus, where phosphatidylcholine and phosphatidylethanolamine are absent or very low and where sphingomyelin is the dominating phospholipid (72%). Because all of the lipid is concentrated in the plasma membranes of the fibers, one could speculate that the high lyso lipid concentrations have a lytic action on these membranes. This hypothesis is the more tempting because some electron microscopists have found very irregular membrane structures in aged human lenses (Hogan *et al.*, 1971; Yamada and Shikano, 1973; Philipson, 1973). Figure 3 shows irregular fiber membranes in a normal human lens that was fixed immediately after obtaining the fresh lens. Degenerative processes seem to occur in cataractous lens (section III.D). The lytic action of lysophospholipids can, however, probably be neutralized by cholesterol (Rand *et al.*, 1975). This sterol might act as a mobile molecule available to occupy free volumes created in the membrane, thus stabilizing the lamellar structure. This is based on the assumption that the lysolecithin–cholesterol complex exists within the plane of the diacylphospholipid bilayer. This assumption is particularly important in the case of the lens because unusually high molar cholesterol–phospholipid ratios (> 2:1) are attained at greater ages (Broekhuyse, 1973). It appears that research on artificial and biomembranes has not yet progressed enough to be able to explain the structures revealed.

Investigating the molecular structure of sphingomyelin in bovine lens, Broekhuyse *et al.* (1974) found that sphingosine (D-*erythro*-1,3-dihydroxy-2-amino-*trans*-4-octadecene) is the only long-chain base. The fatty acid spectrum resembles that of sphingomyelin from other tissues in that only saturated and monounsaturated fatty acids are present, in particular 16- and 24-carbon-atom chain lengths. During differentiation from epithelial cell to cortex fiber the 24:1 (nervonic acid) content of sphingomyelin increases greatly, which is in the main compensated by a decrease in C_{16} and C_{18} fatty acids. This causes a marked increase in the mean chain length and monounsaturation of the paraffin chains. During the aging of the fibers the molecular composition of sphingomyelin is hardly changed, which together with its

TABLE 4

Sphingomyelin Concentration of the Bovine Lens[a]

Tissue	Sphingomyelin (μg per g tissue)			Sphingomyelin[b] (mg per g dry weight)			Relative sphingomyelin concentration		
	Fetus	Calf	Cow	Fetus	Calf	Cow	Fetus	Calf	Cow
Total Lens	157 ± 33	328 ± 1	470 ± 60	0.62 ± 0.12	1.02 ± 0.01	1.35 ± 0.17	8.2 ± 1.1	18.2 ± 1.1	27.0 ± 3.3
Epithelium		1360 ± 40	410 ± 20		—	—		8.9 ± 0.3	7.7 ± 0.4
Equator (Eq)		261 ± 12	340 ± 25		1.32 ± 0.06	1.56 ± 0.12		12.4 ± 0.2	16.6 ± 0.9
Cortex (Co)		370 ± 9	465 ± 56		1.15 ± 0.07	1.38 ± 0.16		20.3 ± 0.6	25.4 ± 2.0
Nucleus (Nu)		324 ± 28	388 ± 46		0.73 ± 0.05	0.81 ± 0.10		23.7 ± 3.0	37.7 ± 2.6

[a]The sphingomyelin concentration was determined in duplicate in pools of 5–10 lenses or lens parts. Results are given with the SD ($N = 3$ for calf and cow lens). For the fetal lens results of 6 determinations on single lenses are given. The relative sphingomyelin concentration is given as percent of total phospholipids.

[b]The Student test was applied for two samples and for paired comparison: Calf lens: Eq–Co: $P = 0.02$; Co–Nu: $P = 0.01$. Cow lens: Eq–Co: $P = 0.04$; Co–Nu: $P = 0.02$. Calf lens–cow lens: Eq–Eq: $P = 0.04$; Co–Co: $P = 0.08$; Nu–Nu: $P = 0.40$. Fetus: Calf: Cow (whole lens) $P = 0.001$; $P = 0.01$.

constant concentration points to a great metabolic stability. Molecules of this type together with cholesterol yield a very rigid membrane (Chapman, 1973), as with myelin. Both tissues probably need this rigidity and metabolic stability in view of their function: myelin as an insulator, and the lens as an optical system that depends on the structural integrity of the fibers during the whole life span.

Very little is known about age-dependent changes in composition or concentration of other lens lipids. Windeler and Feldman (1970) analyzed lens gangliosides and found a very low content of a sialo-gluco-galacto-ceramide in bovine lens. In human lens the ganglioside concentration was 83 times higher and also included a component containing sialic acid, glucose, galactose, galactosamine, and ceramide in the molar ratio 1:1:2:1:1. The fatty acid spectrum of the ganglioside fraction contained 24:1, 24:0, and 16:0 as major components, with 24:1 reaching 55.6% in human lens ganglioside. From several reports it appears that triacylglycerols and neutral glycolipids are minor lipid components in the lens, cholesterol esters being practically absent (Culp *et al.*, 1968; Feldman, 1968; Broekhuyse and Kuhlmann, 1974). Species differences have been detected but seem to be increased by age differences (Broekhuyse, 1970). However, 15-year-old monkey lens most resembled young human lens with respect to its phospholipid composition.

Esterified fatty acid composition has been studied by several groups. Their conclusions are similar with regard to the high content of 16:0 and 18:1 in total lipids (Bartley *et al.*, 1962; Culp *et al.*, 1970a). The latter group reported the fatty acid composition of total lipids from epithelium, cortex, and nucleus of rabbit lens. The differences in the fatty acid spectra were found to be of little significance. In epithelium, small amounts of 18:2, 20:2, and 20:4 were detected; 20:4 was also reported to be present by Bartley *et al.* (1962). Analyses of individual phospholipids of total lens by Feldman *et al.* (1964), Anderson *et al.* (1969), and Bernhard *et al.* (1969) confirmed the high content of 16:0 and 18:1, especially in phosphatidylcholine. Anderson *et al.* (1969), however, did not confirm the presence of polyunsaturated fatty acids. Each phospholipid class appeared to have its own characteristic composition. The same lipid classes from the mature rabbit and bovine lens are similar, but not identical, in fatty acid concentrations. The phosphatidylethanolamines contain 29 and 35 mole % of ethanolamine plasmalogens, the 16:0 and 18:1 aldehyde chains being the most abundant.

C. *Lipid Metabolism during Growth*

An active lipid metabolism can theoretically be expected in lens epithelium and equator: their cells have an active general metabolism, and the functional integrity of the epithelial membranes has to be maintained.

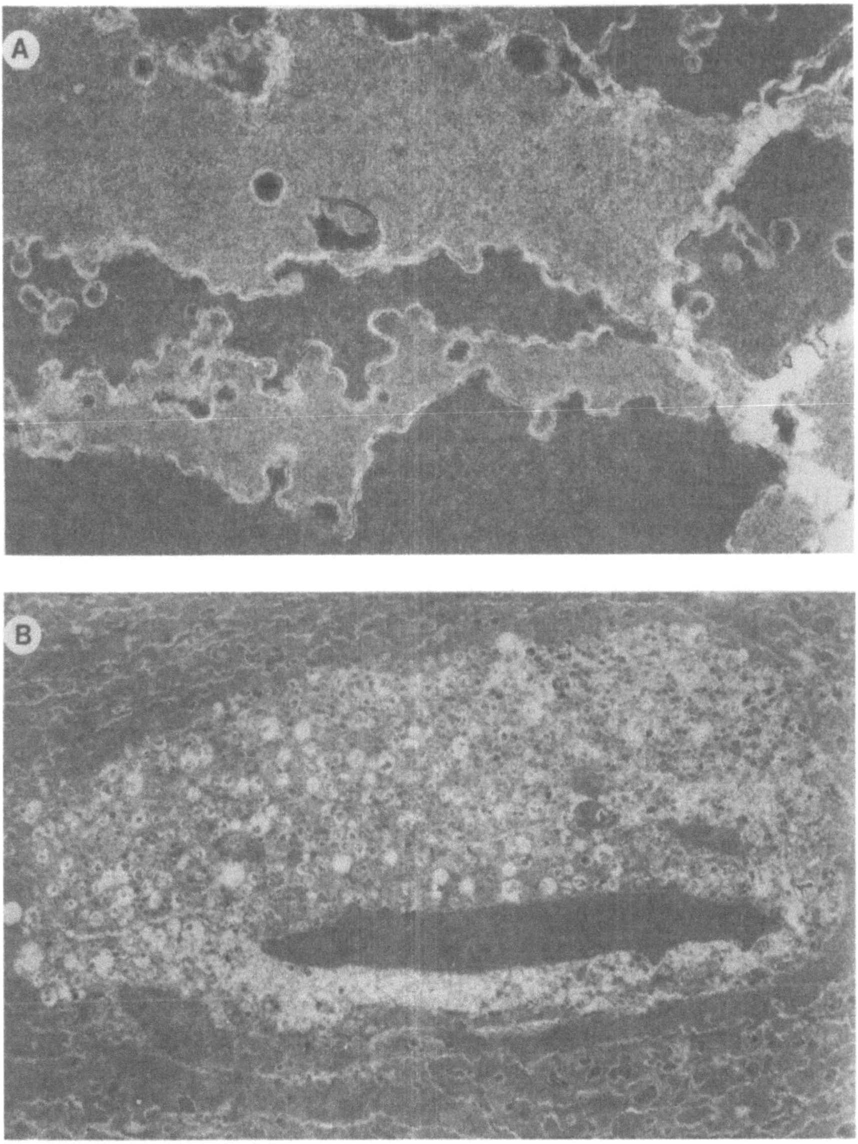

FIGURE 3. Electron micrographs of human lens. (A) Nucleus of normal lens (age of patient—51 years). Magnification 14,700×. (B) Nucleus in senile nuclear cataract (age of patient—65 years). Degenerative changes consisting of the formation of two types of vesicles from the plasma membrane: an electron-dense small type with rather uniform diameter (about 150 nm) and a large type (diameter 200–500 nm), which seems to be derived from the small vesicles by swelling and protein aggregation. Both types are large enough to scatter light and to cause opacification (fixation:OsO_4 in collidine buffer) (Broekhuyse and Tap, unpublished results).

Active lipid metabolism in lens equator is to be expected because at this site continuous growth occurs. The plasma membrane enlarges its surface area by a factor of 1500 during differentiation from epithelial cell to full-grown fiber (Broekhuyse, 1973). Incorporation studies with [^{32}P]P_i- and [^{3}H]*myo*-inositol in calf and human lens have shown that these phospholipid precursors are indeed rapidly incorporated in these regions (Broekhuyse and Veerkamp, 1968; Broekhuyse, 1968a). Phosphatidic acid and phosphatidylinositol show a high turnover *in vitro* (lens cultures) as well as *in vivo* (in rats). Plasmalogens, phosphatidylserine, and sphingomyelin exhibited very little or no ^{32}P incorporation, whereas phosphatidylcholine and phosphatidylethanolamine occupied an intermediate position. The central part of the lens incubated separately showed a very low incorporation rate. Intraperitoneal injections of prostaglandin PGE increase P_i incorporation into all lens phospholipids (Ono *et al.*, 1972/73). Microsomal fractions from the equator exhibited a CTP and Mg^{2+} or Mn^{2+} stimulated incorporation of [^{3}H]*myo*-inositol into phosphatidylinositol (Broekhuyse, 1971). Moreover, CMP promoted the incorporation of *myo*-inositol in the absence of CTP and at a pH different from that of the latter. This pointed to the presence of an exchange reaction similar to that found by Paulus and Kennedy (1960) (cf. the introductory chapter, Vol. 1). By incubating homogenates of calf lens cortex with pure [^{32}P]phosphatidylinositol, the presence of the enzyme phosphatidylinositol inositol phosphohydrolase was demonstrated (Broekhuyse, 1969b, 1971).

A study with [^{14}C]acetate by Culp *et al.* (1970a) demonstrated that its activity was incorporated into phospholipids and neutral lipids, with a relatively high incorporation rate into phosphatidylcholine, phosphatidylethanolamine, and cholesterol. The possibility that amino acids can be precursors for lipids is clear for serine with regard to phosphatidylserine. Trayhurn and Van Heyningen (1973) found that of nine ^{14}C-labeled L-amino acids tested, only the label from alanine, leucine, and serine was incorporated to a certain extent into the lipid fraction.

Recent studies have provided information about the activities and localization of phospholipases in lens. Sphingomyelinase localization in epithelial and equatorial lysosomes was demonstrated by differential centrifugation. The activities of sphingomyelinase and marker enzymes, however, were scattered over the fractions to a greater degree than is found in other tissues (Roelfzema *et al.*, 1973, 1974). Lens sphingomyelinase hydrolyzes its substrate into ceramide and phosphorylcholine. It has a pH optimum at 4.6 and is activated by Triton X-100 like the corresponding enzyme in rat liver and other tissues (introductory chapter). It displays no specificity towards sphingomyelins isolated from epithelium or from the rest of the lens, in spite of their differences in fatty acid composition (see section III.B).

Cytochemical tests for acid phosphatase activity on isolated lysosomal fractions from lens epithelium indicate that acid hydrolases are localized in lysosomes. In the fibers, where lysosomes could not be demonstrated, sphingomyelinase activity can be considered as a "rest" activity (Roelfzema, Broekhuyse and Veerkamp, unpublished; Roelfzema, 1975). The same authors also found that the synthesis of sphingomyelin is restricted to epithelium and equatorial zone. The activities of three synthetic pathways could be demonstrated: ceramide plus CDP-choline, sphingosylphosphorylcholine plus stearoyl-CoA, and ceramide plus phosphatidylcholine. The rate of synthesis measured with *N*-acetyl-DL-*threo-trans*-sphingosine as substrate and CDP-choline as phosphorylcholine donor (cf. introductory chapter), appeared to be sufficient for the formation of the sphingomyelin required for the enlargement of the plasma membrane during the constitution of new fibers. The other synthetic routes are probably of minor importance. The results are in agreement with those of Broekhuyse and Veerkamp (1968), showing that the incorporation of [^{32}P]phosphate into sphingomyelin was very low *in vivo* as well as *in vitro*. The metabolic stability of the fiber sphingomyelins was also established by studies of their concentration and their fatty acid and base composition during aging (see section III.B).

From the data assembled in section III.B the conclusion can be drawn that the phosphoglycerides, once synthesized during equatorial growth, are gradually degraded during aging of the fibers. In animal as well as in human lens the phosphoglyceride concentration decreases with age, while lysophospholipids increase to a certain extent. These lysolipids are apparently intermediates in the degradation process. Their increase only partially compensates for the decrease in the diacyl phosphoglycerides, indicating that they are in turn hydrolyzed. In aged human lens, nuclear phosphatidylcholine has practically disappeared (Table 5). Although it could be conceivable that phosphatidylethanolamine was not recovered in the lipid extracts as a result of autoxidation or other changes, this explanation of its disappearance is unlikely, firstly, because polyunsaturated fatty acids are absent or very low in concentration in lens (section III.B) and, secondly, because as Broekhuyse (1969a) showed, lysophosphatidylethanolamine was formed neither during postmortem storage nor during lipid isolation. The presence of phospholipase activities was recently demonstrated in the various lens layers (Broekhuyse and Roelfzema, unpublished results) with [^{14}C-methyl]phosphatidylcholine as substrate. This lipid could be fully deacylated, indicating the presence of phospholipase as well as lysophospholipase activity. Both activities decreased from epithelium towards lens nucleus.

From the data described it appears that lipid metabolism in lens epithelium and equator has many features in common with that in brain and liver (cf. introductory chapter). With regard to sphingomyelin metabolism this finding is supported by the storage of this lipid in lens epithelium in Niemann-Pick's disease (Robb and Kuwabara, 1973). Clear differences exist with regard to the stability of the fiber cells. In contrast with liver cells, the lens fibers are retained during the life span, resulting in a typical pattern of age-dependent lipid composition (section III.B) due to very low or absent lipid synthesis and slow hydrolysis of phosphoglycerides in the fibers.

D. *Lipids in Cataract*

Previously, several authors have described an increase in the concentration of cholesterol in the cataractous lens; cholesterol crystals have occasionally been detected by biomicroscopy. However, recent analyses with improved techniques have demonstrated that neither the cholesterol concentration (Feldman, 1968) nor the phospholipid composition of the human lens is affected in immature cataract (Broekhuyse, 1969a,b), although both authors found indications of a change in the structure of the membranes. The latter finding has recently been confirmed by electron-microscopic examinations (Y. Kobayashi, as shown by Yamada and Shikano, 1973; Philipson, 1973). By means of extraction of normal and cataractous lenses first with a solvent of low polarity and then with one of high polarity, it appeared that the amount of weakly bound cholesterol and phospholipids in human senile cataract was increased. In the plasma membranes of senile and of experimental galactose cataract, the amount of proteolipid-protein extracted by the polar solvent was less than normal lenses (Broekhuyse, 1969a). This was interpreted as a sign of a disintegration of these membranes. Lipoprotein complexes could be split off and become extractable by the nonpolar solvent. Hence the residual membranes contained less proteolipid complexes, which resulted in a lowered amount of proteolipid protein in the polar solvent. These alterations could be the result of tension in the membranes caused by swelling of the cells or by the presence of water in clefts. The fluid in the vacuoles could also initiate disintegration owing to its abnormal composition (chaotropic agents). In both instances the disintegration would be secondary to the formation of the cataract.

In galactose, diabetic and senile cortical cataracts, alterations in the incorporation pattern of [^{32}P]orthophosphate into the phospholipids have been demonstrated. Lens swelling inhibits the formation of phosphatidylinositol from phosphatidic acid (Broekhuyse, 1969a). Hypotonic media, toxic concentrations of galactose and glucose, and senility lead to similar results. The effect is probably caused by the lower CTP concentration resulting from water uptake by the lens. It is known that the ATP concentration in cataractous lenses is lowered, which leads to a decreased synthesis of CTP. It is still not clear in what way this part of phospholipid metabolism is involved in transport or other processes and hence it is not possible to evaluate the consequences of the effect. The active transport of *myo*-inositol in the lens, which is inhibited by ouabain, and in sugar cataracts, leads to *myo*-inositol loss from the lens (Broekhuyse, 1968a). At the same time *myo*-inositol accumulates in the epithelial cells in the initial stages of an induced experimental cataract (Broekhuyse, unpublished results). Up to now it has not been possible to demonstrate that the changed inositol concentrations influence lenticular lipid metabolism.

Studies of the lens in alloxan diabetic rabbits revealed practically no differences between the various phospholipid concentrations of normal and affected lenses, although the choline phosphoglyceride concentrations were lowered in aqueous humor and the phospholipid concentrations in plasma were increased (Varma and Reddy, 1972). Changes in fatty acid composition during galactose cataract formation have been reported (Hatcher and Andrews, 1970). Growth retardation caused by the cataractous condition might influence lipid composition as well.

In rats fed triparanol, a drug which depresses blood cholesterol levels, it was found that desmosterol increased from 1% to 10% of the sterols during cataract development, indicating an undesirable side effect of this drug (Mizuno *et al.,* 1974). In view of its detergent-like structure (1-*p*-tolyl, 1-*p*[β-diethylaminoethoxyl]-phenyl, 2-*p*-cholorophenyl-ethanol), it could disturb the lens membrane structure. Experiments with various surface active agents were performed by Sanders *et al.* (1974). Intravitreously administered anionic, cationic, and neutral surface active agents (including lysolecithin) all appeared to be capable of producing cataracts. These opacities can probably be attributed to membrane disorganization as a result of dissolution of membrane lipids and proteins leading to increased light scattering. Hence, although no alterations have been found in the lipid composition of certain types of cataract, it is likely that in opaque lens regions, membrane lipid and protein dislocations occur, which are sometimes observable by electron microscopy. Specialized techniques of lipid and membrane examination on the micro or submicro scale will be necessary to detect these dislocations or membrane degenerations. An example of such a detection for the lens is a very peculiar phenomenon that has been found in Niemann-Pick's disease (type A). Electron microscopy revealed membranous cytoplasmic bodies in lens epithelium and other ocular tissues, apparently due to sphingomyelin storage (Robb and Kuwabara, 1973).

IV. Retina and Optic Nerve

A. Morphology and General Metabolism

The retina is a thin layer of tissue (50–100 μm) lining the posterior two-thirds of the vertebrate eye cup. It is the actual site of light conversion into nervous excitation, which ultimately leads to the subjective experience of "vision" in the visual cortex of the brain. Going from inner to outer layers, in the direction of entering light, the retina contains several layers of interneurons and their fibers, then the photoreceptor cells, and finally the pigment epithelium (Fig. 4). The impulses generated by the receptor cells are provisionally processed in four types of interneurons: the bipolar and horizontal cells, both in direct contact with the photoreceptors, and the amacrine and ganglion cells (Dowling, 1970). The axons of the ganglion cells run along the inner margin of the retina to form the optic nerve, which leaves the retina entering into the brain. All these cells are held together by a framework of neuroglia cells, the Müller cells, which span the full thickness of the retina. The pigment epithelium is a brown or black layer that, among other functions, serves as a "sink" for light not absorbed by the photoreceptors. In normal isolation procedures, in which the retina can be collected as a coherent tunic of tissue from the opened eye cup after removal of the vitreous, most of the pigment epithelium remains attached to a nonretinal tissue, the choroid. This layer contains the major vascular system that supplies the retina, without actually invading it. A second system, similarly branching off from the ophthalmic artery, occurs in many vertebrates. It penetrates the photoreceptor cell layer at the blind spot to supply the inner retinal layers. Many excellent reviews on the anatomy of the retina exist (Polyak, 1957; Dowling, 1970; Dubin, 1974).

Two types of photoreceptor cells are found in vertebrates: rod cells and cone cells. They are both elongated cells, in which morphologically a synaptic ending, an inner segment, and an outer segment (the latter two connected by a nonmobile cilium) can be distinguished. The inner segment contains the normal organelles (mitochondria, endoplasmic reticulum, Golgi apparatus). The outer segment is built of a dense system of parallel membranes, containing visual pigment and in case of the rods enclosed by a plasma membrane (Fig. 5).

Though only studied in a limited number of species, by now most major pathways of metabolism have been shown to occur in the retina. Extensive attention has been paid to its carbohydrate metabolism, but a really coherent picture is still lacking (Graymore, 1969; Lolley and Schmidt, 1974). The need of the retina for glucose and O_2 is extremely high, reflecting the fact that it has the highest rate of respiration of any mammalian tissue, and a high rate of glycolysis as well. Even under conditions of

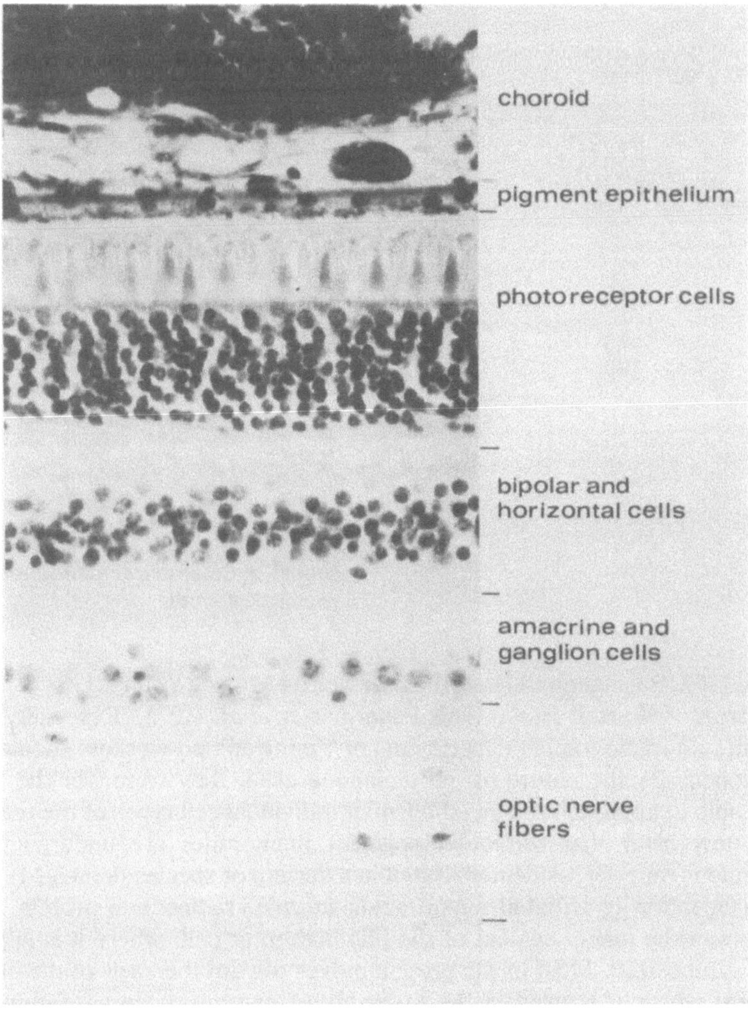

FIGURE 4. Cross section of human retina, fixed with Susa solution, stained by Heidenhain's Azan method. Magnification 1300 ×. Provided by C. Jerusalem.

adequate oxygenation, considerable lactic acid production occurs. The pentose phosphate pathway is operative, while very active CO_2-fixation has been observed probably by the way of malic enzyme. The retinal stores of glycogen are small, so that a continuous supply of metabolites seems important.

Much less is known about the protein metabolism of the retina, but protein synthesis has been shown to occur *in vitro* (Matuk, 1972; O'Brien

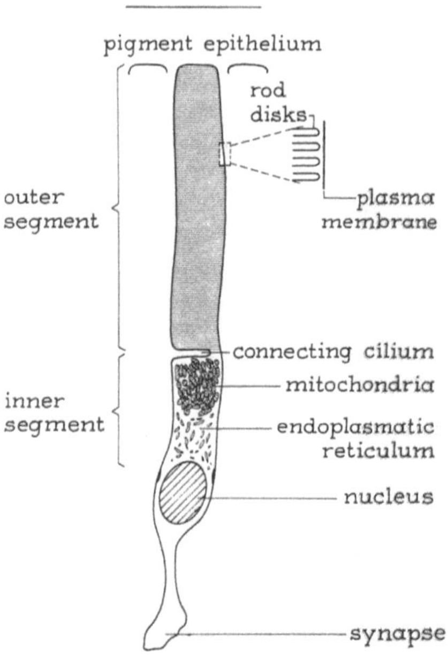

FIGURE 5. Schematic diagram of the rod photoreceptor cell.

et al., 1972; Basinger and Hall, 1973) as well as *in vivo* (Young, 1967; Young and Droz, 1968; Hall *et al.*, 1969; Papermaster *et al.*, 1975). Especially the elegant, autoradiographic experiments of Young and co-workers should be mentioned. By the nature of the technique used, they were, for the first time, able to approach the metabolism of individual cell types of the retina. It is now clear that rod outer segment membranes are undergoing a continuous renewal, as demonstrated in a variety of species (Young, 1974). Upon injection of tritiated amino acids *in vivo,* radioactive protein first appears in the inner segment of the photoreceptor cell, where it is apparently synthesized. Most of the protein moves toward the basis of the outer segment where it is used in the assembly of new disks. In a continuous process new rod disks are formed, thereby displacing older disks towards the apical end of the outer segment, where groups of 5–30 disks are detached, and scavenged and digested by pigment epithelial cells. Similar experiments with phospholipid precursors will be dealt with in the next section.

B. *Lipid Metabolism*

The overall lipid composition of the retina is rather extensively studied. Data on the phospholipid composition of calf retina (Broekhuyse, 1968b) and of pig, sheep, human, dog, cow, and rabbit (Anderson *et al.*,

1970a; Anderson, 1970) agree fairly well and are summarized in the first two columns of Table 5, for the quantitatively most important classes. This phospholipid pattern is common for nervous tissue, phosphatidylcholine, phosphatidylethanolamine, and phosphatidylserine being most abundant, in this order. While between 10% and 30% of ethanolamine-containing phospholipids is in the form of plasmalogens, hardly any phosphatidylcholine plasmalogen is found. The overall concentration of phospholipids in the retina may be about 15 mg/g of wet tissue.

Anderson and co-workers (1970a) have also carried out detailed fatty acid analyses of the individual classes of phospholipids (Table 5), which agree fairly well with an earlier overall analysis (Bartley *et al.*, 1962). The differences in fatty acid composition between the different phospholipids is conspicuous. Of other lipids, all occurring in minor amounts, data are available on gangliosides (Holm, 1972; Holm *et al.*, 1972; Holm and Månsson, 1974), cholesterol (Hendriks *et al.*, 1976), diacylglycerols, and triacylglycerols, and free fatty acids (Aveldano and Bazán, 1973; 1974a,b) in various species.

The only subcellular fraction of retina the phospholipid and fatty acid composition of which is extensively studied are the rod outer segments (for a review see Daemen, 1973). They have a very high lipid content and contain, as compared to the retina, less sphingomyelin, phosphatidylinositol, and cholesterol, while the percent of the highly unsaturated docosahexaenoic acid is higher, especially in the major phospholipids.

Various aspects of the phospholipid metabolism of the retina have been studied by Swartz and Mitchell (1970, 1973, 1974). First the presence of synthesis *de novo* via the diacylglycerol pathway was shown to occur in rat retina by incubation of retinal homogenates with radioactive phosphorylcholine and CDP-choline (Swartz and Mitchell, 1970). Subsequent work demonstrated phospholipase A_1 and A_2 activities in various fractions of bovine retina and pigment epithelium (Swartz and Mitchell, 1973). Most recently, acyl transfer reactions in the same preparations were found to be dependent on ATP and CoA, with varying degrees of substrate preference

TABLE 5
Overall Phospholipid and Fatty Acid Composition of Mammalian Retina

	% of phospholipid	Fatty acid composition				
		16:0	18:0	18:1	20:4	22:6
Phosphatidylcholine	41	41	18	18	5	11
Phosphatidylethanolamine	34	10	32	8	8	29
Phosphatidylserine	10	6	34	15	3	24
Phosphatidylinositol	6	7	36	7	42	3
Sphingomyelin	4	25	42	11	<1	<1

and positional specificity of the transferred fatty acids for both lysophosphatidylcholine and lysophosphatidylethanolamine (Swartz and Mitchell, 1974). These data, though far from being complete, seem to suggest that the normal routes of phospholipid metabolism are operative in the retina. As to subcellular fractions isolated from retina homogenates, microsomal, and to a lesser extent, mitochondrial fractions are most active, but no consistent enzymatic activity related to phospholipid metabolism has so far been observed in rod outer segments.

More detailed information on the phospholipid metabolism of frog photoreceptor and pigment epithelium cells comes from autoradiographic studies of Bibb and Young (1974a,b) after injection *in vivo* of the tritiated phospholipid precursors, glycerol (1974a) and fatty acids (1974b). Glycerol is rapidly incorporated, predominantly into phospholipids, in the inner segment of both rods and cones, indicating that this is the site of phospholipid synthesis in visual cells. The visual cell synaptic bodies (and cone oil droplets) become gradually labeled. Between 1 and 4 hr after injection, labeled molecules begin to appear throughout the outer segments, presumably by replacing "older" constituents. In rods some accumulation of labeled material occurs at the base of the outer segment (probably lipid *and* protein). This labeled band is displaced along the rod outer segment as described in the previous paragraph. The experiments suggest that membrane renewal by molecular replacement is more rapid for lipids than for proteins.

Replacement of intact phospholipid molecules in existing outer segment membranes is also demonstrated with labeled palmitic acid as precursor, but in addition acyl exchange seems to occur (Bibb and Young, 1974b). Remarkably, palmitic acid is initially concentrated in the oil droplets of the pigment epithelium. Gradually the concentration of radioactivity in the outer segments increases, apparently as a result of the addition of new phospholipids, possibly augmented by the transfer from pigment epithelium of retinyl palmitate (see next section). Accumulation of radioactivity is short-lived, suggesting a rapid exchange of fatty acid chains of the phospholipids. Some experiments with stearic and arachidonic acids suggest the same pattern. Indications for an exchange of fatty acids between lysophosphatidylcholine molecules have been reported (Swartz and Mitchell, 1974). Clearly, our knowledge of the lipid metabolism of the retina is fragmentary and much more research is needed to ascertain the vague, general patterns presently visible.

C. *Lipids and the Visual System*

In the visual process, light captured by the visual pigments in the outer segments of vertebrate photoreceptor cells leads to photolysis of these pigments (Wald, 1968). The conformational changes accompanying photolysis generate a response, possibly the release of Ca-ions (Hagins, 1972), which in turn increases the resistance for sodium ions of the plasma membrane of the outer segments (Tomita, 1970). This causes a light-dependent reduction in an intensive current of sodium ions, which in darkness continuously flows from the inner segment along the photoreceptor cell to and into the outer segment and from there diffuses back to the inner segment (Hagins *et al.*, 1970). Probably the dark current is maintained by a highly active Na-K activated ATPase system, located on the plasma membrane of the inner segment. Thus the result of illumination, both in rods and cones, is hyperpolarization of the photoreceptor cell, observable as a graded potential, which passively spreads to the synaps, where bipolar and horizontal cells are stimulated (Dowling, 1970; Dowling and Ripps, 1973). A relatively complete and up-to-date treatise on the visual process can be found in *Handbook of Sensory Physiology,* Vols. VII/1 and VII/2 (edited by H. J. A. Dartnall and M. G. F. Fuortes, respectively, Springer-Verlag, Berlin, 1972).

In various respects, the involvement of lipids in the visual process seems to exceed the usuai role of lipids in a normal cell as important components of plasma and intracellular membranes. These include the very high phospholipid content of outer segments, the lipid prosthetic group of visual pigments, 11-*cis*-retinaldehyde, and the extremely high content of polyunsaturated fatty acid chains in outer segments.

The very regular alignment in the outer segments of thousands of membranes, parallel to one another (Fig. 5) with a disk repeat distance of 300 ± 10 Å (Clark and Branton, 1968; Worthington, 1971; Korenbrot *et al.*, 1973) leads to an extremely high phospholipid concentration of 170–200 mM in intact rod outer segments. It means that about 15% of the wet weight of rod outer segment is lipid. This arrangement seems to provide the fairly stable hydrophobic matrix for the highly organized and concentrated visual pigment system that permits the extremely sensitive and efficient transduction of light. From an experimental point of view it makes possible the application of a broad variety of especially spectrophotometric techniques with a sensitivity and/or accuracy dramatically beyond that attained with less oriented membranes. In this way very reliable data have become available about orientation and rotational and lateral movements of rhodopsin molecules in disk membranes (Brown, 1972; Cone, 1972; Poo and Cone, 1974; Liebman and Entine, 1974). Since a previous discussion of the interaction between visual pigments and membrane phospholipids (Daemen, 1973), some interesting papers have appeared related to this topic (Hong and Hubbell, 1973; Shichi and Somers, 1974; Saari, 1974). However, in spite of the very strong physicochemical interaction between these major components of the photoreceptor membrane, there is little evidence for a more specific role of any phospholipid in the visual process.

Rhodopsin itself is a membrane protein, which carries the lipid 11-*cis*-

retinaldehyde (vitamin A aldehyde) as chromophoric group (Wald, 1968). Upon photolysis, following illumination, 11-*cis*-retinaldehyde, probably buried within the native pigment molecule, is isomerized to the all-*trans*-configuration, which involves a very significant change in the shape of the polyene from a bent to an extended form. This leads to conformational changes in the apoprotein part of rhodopsin (opsin), which, in an as yet poorly understood way, have to do with the later events in visual excitation. At the same time, all-*trans*-retinaldehyde becomes exposed and may enter a pathway, known as the visual cycle (Wald, 1968). All-*trans*-retinaldehyde from photolyzed rhodopsin is reduced to all-*trans*-retinol, most of which leaves the outer segment and is esterified with long-chain fatty acids to retinyl esters in the pigment epithelium (Dowling, 1960; Zimmerman, 1974). During regeneration a reverse process of hydrolysis and oxidation to the aldehyde occurs (Fig. 6). The reisomerization to the 11-*cis*-configuration takes place in the outer segment, probably from all-*trans*-retinaldehyde (Daemen *et al.*, 1974; Lion *et al.*, 1975), whereupon spontaneous condensation of opsin and 11-*cis*-retinaldehyde leads to regeneration of visual pigment.

A most striking feature of the lipids of rod outer segment membranes is their high content of long-chain, highly unsaturated acyl residues. Up to 40% of the fatty acids in these membranes is docosahexaenoic acid (C22:6),

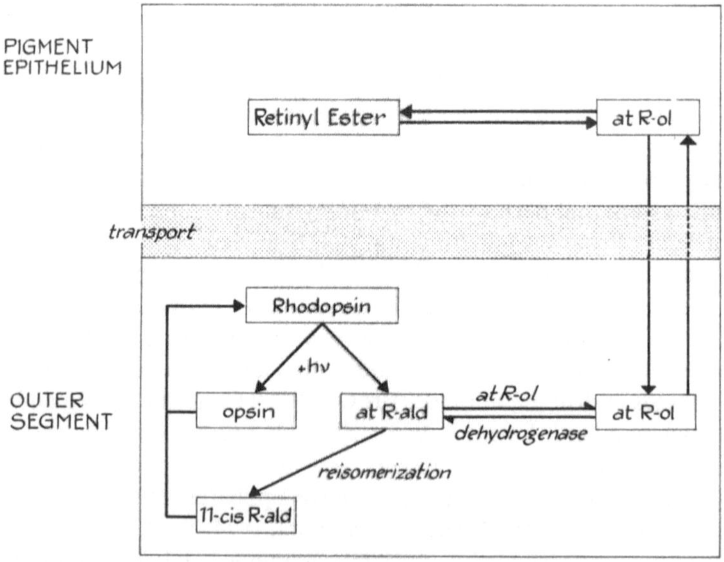

FIGURE 6. Simplified scheme of the visual cycle according to Lion *et al.* (1975). Abbreviations: at R-ol: all-*trans*-retinol; at R-ald: all-*trans*-retinaldehyde.

a unique situation among natural membranes, investigated so far (Daemen, 1973). This high unsaturation nicely correlates with a low membrane viscosity, expected from calculations on the rotational and lateral movements of rhodopsin (Cone, 1972; Liebman and Entine, 1974). However, the functional implications of this unsaturation are not yet clear. Attempts to reduce the high unsaturation of the retina and of rod outer segments by means of a diet low in essential fatty acids have met with little success (Futterman *et al.*, 1971; Anderson and Maude, 1972; Anderson *et al.*, 1974), but it is not clear whether an adequate distinction has been made between deficiency in the linoleic and the α-linolenic acid families of polyunsaturated fatty acids.

It is well known that polyunsaturated fatty acids are very sensitive to oxidation. The same is true for retinol, the photolysis product of rhodopsin *in vivo*. On the other hand, one has to expect a relatively high oxygen tension in photoreceptor cells in view of the high respiration of numerous and large mitochondria in the inner segment. Their supply occurs by diffusion from the choroidal circulation and oxygen is known to diffuse easily through membranes. In addition, the presence of light and high concentrations of (visual) pigment may create favorable conditions for so called photodynamic action (Spikes, 1968). This light-induced, mostly oxygen-mediated radical formation, leads to a broad variety of deleterious reactions in biological systems. Some amino acid and thymine residues, but also unsaturated lipids, are the substrates for this type of photo-oxidation. In fact, prolonged illumination with visible light at normal light levels damages the retina, as measured by a variety of parameters (Noell *et al.*, 1966; Kuwabara, 1970; Noell and Albrecht, 1971), while Russian workers have detected light-induced free radial oxidation in photoreceptor membranes (Kagan *et al.*, 1973; Novikov *et al.*, 1974). These considerations lead to a number of interesting questions. Do the rod outer segments contain an active antioxidant system? Could the large excess of lipid double bonds play a role in the protection of visual pigment? Are fatty acid repair mechanisms operative in the outer segment? Answers to these and other related questions might provide clues with regard to the importance of the highly unsaturated nature of vertebrate photoreceptor membranes.

D. Lipids of Optic Nerve

Myelin was isolated from bovine optic nerve by MacBrinn and O'Brien (1969). It contained 76.3% lipid, mainly cholesterol (18.4%), phosphatidylethanolamine (12.5%), and cerebroside (16.7%) (percentages expressed on a dry weight basis); 34.6% of the phosphatidylethanolamine was plasmalogen. Phosphatidylserine, sphingomyelin, and cerebroside sulfate were present in smaller proportions. The fatty acids of the glycerophospholipids were chiefly 16:0, 18:0, and 18:1, with small proportions of 20- and 22-carbon polyunsaturates. The sphingolipids contained predominantly saturated and monounsaturated fatty acids of chain lengths of 20–26 carbon atoms. Optic nerve myelin and white matter myelin resembled one another closely in overall lipid composition and in the fatty acid compositions of their constituent lipids. The authors concluded further that optic nerve myelin and white matter myelin are chemically similar membranes, but both of these differ in their lipid composition from spinal root myelin.

The gangliosides of bovine optic nerve were studied by Holm and Mansson (1974). The ganglioside concentration expressed as lipid-sialic acid amounted to approximately 500 nmoles/g fresh wt. The hexosamine was exclusively galactosamine. About 90% of the total lipid-sialic acid was bound to four gangliosides, G_{M1}, G_{D1a}, G_{D1b}, and G_{T1}. The main long-chain bases in all components were d18:1 and d20:1, while 18:1 and 20:0 were the major fatty acids. The fatty acid and long-chain base pattern are different from those of the corresponding retina gangliosides. The ganglioside composition of optic nerve is different from that of retina in that no G_{D3} is present, while in retina this component comprises approximately 50% of the gangliosides. The authors conclude that in the retina G_{D3} is localized in cells other than the retina ganglion cells, which confirms what was found in their earlier studies. In contrast to previous suggestions, it appears that G_{D3} is not transported through the optic pathway. This is also valid for the other gangliosides in view of the mentioned differences in fatty acid and long-chain base pattern of the common retina and optic nerve gangliosides. They are apparently synthesized *in loco*. It has been demonstrated that intraocular injection of [^{3}H]acetate or [1-^{14}C]glucose leads to the rapid labeling of the gangliosides of all parts of the optic pathways (retina, optic nerve, optic tract, and lateral geniculate body) (Holm, 1972). A study of interest about these gangliosides has been made by Schmidt (1965).

E. Lipids Involved in Pathological Conditions

The morphology and physiology of the retina and optic nerve are very much like that of the brain, because they are derivatives of the forebrain. This lipid-rich tissue frequently demonstrates accumulations of lipoidal material that has been liberated from dead cells. The lipoidal accumulations characteristic of the cerebral lipidoses are frequently observed in the retina. The following retinal changes directly related to disturbed lipid metabolism have been observed in humans: sphingomyelin storage in Niemann-Pick's disease (types A and D); G_{M2} ganglioside storage inside and outside the distended ganglionic cells, demyelination, and membranous lamellar vacuoles (glycolipid) in Tay-Sachs disease; glycolipid in the macula in Farber's lipogranulomatosis; swollen ganglion cells, probably neutral lipid, in Wolman's disease; lipofucsin or lipid storage in the ganglion cells in Batten's disease; macular atrophy in a-β-lipoproteinemia; storage of intracellular metachromatic material in metachromatic leucodystrophy; optic atrophy (demyelination, stored galactocerebroside) in globoid cell leucodystrophy. Most of the data have been collected together by Goux and Kallay (1971), Warburg (1972), and Kenyon (1974). The latter author has described many ultrastructural investigations in metabolic disorders in detail, together with a review of the deficient enzymes. It appears that retinal storage of a variety of glycolipids, rather than one particular substance, leads to clinically apparent maculapathy. Ultrastructural studies reveal membranous lamellar vacuoles in the ganglion cells in G_{M1} gangliosidosis (keratan sulfate plus G_{M1} ganglioside) and lipid deposits in the vessel walls in the Refsum syndrome. In diabetic retinopathy lipid infiltrations are ophthalmoscopically observed in areas with increased capillary permeability. Lipemia retinalis occurs in individuals with high blood triacylglycerol levels.

The effects of vitamin A deficiency on the visual system are well known (McLaren, 1970) and its biochemical basis is thoroughly investigated (Dowling and Wald, 1958; 1960). More recently the effects of vitamin A deficiency appeared to be critically dependent on the levels of illumination to which the animals were daily exposed (Noell *et al.*, 1971).

V. Cornea and Sclera

A. Morphology and General Metabolism

The cornea forms the anterior part of the ocular globe (Fig. 1). Its stroma constitutes 90% of its structure and is composed almost entirely of extracellular materials such as collagen and proteoglycans. The stroma also contains a scattering of flat keratocytes. The epithelium at the anterior surface is multilayered, while the endothelium at the posterior surface is single-layered. Both cell layers are separated from the stroma by a basement membrane. Functionally the cornea constitutes the major refractive element of the eye.

The sclera forms the rest of the outer coat of the eye. It consists of white, collagenous material to which eye muscles are attached and which is penetrated by vessels and nerves. The transition zone between cornea and sclera is the limbus.

The cornea exchanges metabolites across the limbus with the blood capillaries of the sclera, across the endothelium with the aqueous humor in the anterior chamber, and across the epithelium with the tear film. The cornea has an active control of its thickness, which is inhibited by ouabain, and in which the endothelium plays an important role. This aspect of the subject, together with corneal and scleral metabolism and structure, has been reviewed by Maurice (1969) and by Maurice and Riley (1970). Glucose and oxygen are the basic metabolites of the cornea. Glucose is incompletely oxidized and it is partly (84% in the rabbit) converted into lactic acid. In view of the high activity of the epithelium, with its replacement time of a few days, it must have a high metabolic capacity. However, little is known about the proportionality between the metabolic activities of epithelial, stromal, and endothelial cells under various conditions. Reim *et al.* (1971) concluded from their extensive enzyme studies that the synthetic reactions are well developed in epithelium, while they play a minor role in endothelium. This confirms the supposition that epithelium has an important protecting function, while the energy of the endothelium is mainly used for transport processes.

The stroma contains keratocytes, which are in a rather inactive state; however, synthetic activities have been detected. The synthesis of the various components of the stromal matrix has been studied by Robert and Robert (1974, review). They found a relatively high incorporation of precursors into the fraction of proteins containing the soluble precursors of the structural glycoproteins, which are associated with collagen, tropocollagen, and the proteoglycan precursors. The insoluble matrix proteins exhibited a low turnover. From their studies and from those of other authors (Van der Eerden and Broekhuyse, 1974; Van der Eerden, 1975), it appears that the structural glycoproteins are of great immunological significance, and much of these data are probably valid for scleral matrix as well. A review dealing with the proteins of the cornea was recently presented by Barber (1974).

B. *Lipids*

Data regarding lipid metabolism in the cornea are scarce. Characterizing the phospholipids in the various bovine ocular tissues, Broekhuyse (1968) found that cornea, as well as the other tissues of the eye, incorporate [^{32}P]P_i into its phospholipids. High incorporation rates were found for phosphatidic acid and phosphatidylinositol, moderate incorporation occurred into phosphatidylcholine and phosphatidylethanolamine, while the other phospholipids incorporated little ^{32}P. Andrews (1966) demonstrated that calf cornea probably has pathways for lipid metabolism similar to those found in liver. [1-^{14}C]Acetate and activity from [U-^{14}C]glucose were incorporated into fatty acids and cholesterol *in vitro*. A similar study was performed by Culp *et al.* (1970c) *in vivo;* [1-^{14}C]acetate was introduced into the anterior chambers of rabbit eyes. The activity was rapidly incorporated into the fatty acids and to a lesser extent into cholesterol. Esterified fatty acids of neutral lipids showed a higher incorporation rate than those of phospholipids. Among the phospholipids, phosphatidylcholine was most rapidly labeled. Free fatty acids of bovine cornea was analyzed by Smidt-Martens and Hohorst (1972); the concentration in whole cornea amounted to 226 ± 15 μmoles/g fresh wt and in epithelium 275 ± 12 μmoles/g fresh wt, palmitic, stearic, and oleic acids being the major constituents of this lipid fraction. The presence of linoleic acid was not reported. This acid is, however, a major constituent of neutral and phospholipids in epithelium, stroma, and endothelium, as was found by Thiele and Denden (1967). They also analyzed the phospholipid composition of these bovine corneal layers. Their values deviate from those reported in Table 6 (see below) and from those of Broekhuyse (1968b), the phosphatidylcholine value being much lower while the phosphatidylserine and sphingomyelin are higher. The age of the material used by Thiele and Denden (1967) is unknown. The same authors found a seasonal influence on total lipids and plasmalogen content. In stroma and endothelium a high percentage of cholesterol esters was found, while cholesterol in the epithelium was mainly in the free form. This finding is not in agreement with the neutral lipid pattern shown by Smidt-Martens and Hohorst (1972) in which the cholesterol ester component dominates. The phospholipid compositions of calf cornea and sclera have been reported to be similar (Broekhuyse, 1968b; Table 7). More details are available about the lipid composition of human cornea and sclera, and these will be discussed in section C.

C. *Lipids during Aging and in Pathological Conditions*

The age-dependent increase in the lipid content of the sclera has been known for a long time. Krekeler (1923) reported the relatively high lipid content of senile sclera and Löwenstein (1940) supposed this increase to be a process that would develop parallel to the formation of the arcus senilis in the cornea. Cogan and Kuwabara (1959) noticed that sudanophilia of the sclera consisted of red dots between the laminae while the cells appeared normal. Andrews (1962) obtained indications that in arcus senilis cholesterol esters accumulated in the cornea and sclera. Du Chesne (1967) found lipid deposits in the inner layers of the sclera as early as 40 years of age, which increased progressively, especially in arcus senilis; arteriosclerosis was not always present. Broekhuyse and Kuhlmann (1972) showed that in senile human sclera the sphingomyelin and cholesterol ester concentrations were considerably increased. They noticed that in the aging process the sclera showed striking similarities with aorta and cornea. The major phospholipids found in human sclera are sphingomyelin and the choline and ethanolamine glycerophospholipids, together amounting to about 90% of

TABLE 6

The Lipid Composition of Normal and Arcus Containing Cornea[a]

	Clear human corneas				Human corneas with arcus				Bovine corneas	
	21 years old	34 years old	70 years old		69 years old		73 years old		0.3 years old	6 years old
			c.p.[b]	p.p.[c]	c.p.	arcus[c]	c.p.	arcus		
Sphingomyelin	15	18	33	39	32	42	32	40	14	19
Phosphatidylcholine	44	41	32	36	34	29	33	30	61	53
Phosphatidylethanolamine	26	29	18	12	21	17	20	14	16	18
Phosphatidylinositol	6	5	8	7	2	3	5	4	2	2
Phosphatidylserine	7	6	5	5	6	6	6	6	7	7
Lysophosphatidylcholine	1	—[d]	—	—	2	—	3	4	—	—
Lysophosphatidylethanolamine	—	1	3	2	1	2	2	1	—	—
Phosphatidic acid	—	—	1	—	2	1	—	1	—	—
Total phospholipid (mg/g wet wt)	1.4	1.1	0.8	1.3	1.7	3.2	1.1	3.4	1.4	1.8
Cholesterol (mg/g wet wt)	0.2	0.3	0.3	0.4	1.0	1.5	1.2	1.6	0.3	0.3
Cholesterol esters (mg/g wet wt)	0.5	1.0	0.3	1.4	1.8	2.2	2.0	2.5	0.4	0.4
Number of corneas per extract	2	2	5		6		5		22	13

[a] Phospholipid composition is presented in weight percents phosphorus of total lipid phosphorus recovered. Analyses were performed as described previously (Broekhuyse, 1969b, 1972), with duplicate determinations on each extract; one extract per column of results. Corneal epithelia were removed before use. Triacylglycerol and free fatty acid fractions amounted to 0.1–0.3 mg/g each, and squalene to 0.03–0.08 mg/g in all cornea tissues.

[b] c.p. = central part.

[c] p.p. = peripheral (annular) part; weight ratio c.p.:p.p. = 1:1.7.

[d] — = not detected

TABLE 7
The Phospholipids of Calf Eye[a]

Phospholipid	Retina	Iris	Choroid	Lens	Cornea	Sclera	Vitreous body
Lysophosphatidylcholine	0.1	0.6	0.8	0.5	1.7	0.2	0.5
Sphingomyelin	5.8	14.3	17.4	18.2	16.3	18.2	30.5
Phosphatidylcholine	40.1	31.6	35.6	31.8	35.2	39.4	36.0
Phosphatidylinositol	6.4	6.8	5.4	1.9	5.8	4.1	3.6
Phosphatidylserine[b]	10.0	9.8	9.9	12.3	12.2	10.3	3.7
Phosphatidylethanolamine	33.8	31.8	27.6	33.2	25.0	24.8	15.3
Diphosphatidylglycerol	3.0	2.7	1.7	0.3	2.2	1.5	1.6
Phosphatidic acid	0.3	0.2	0.7	1.0	0.4	0.5	0.4
Phosphatidylglycerol	0.2	0.3	0.5	0.2	0.3	0.1	0.0
Unidentified	0.2	0.4	0.3	0.5	0.9	0.4	3.1
μg lipid-P per g wet wt	627	310	209	72	70	48	0.1

[a] The values, obtained by quantitative two-dimensional thin-layer chromatography in duplicate, are expressed in percent of total lipid phosphorus. Vinyl-ether phospholipids are included in the corresponding diacyl compounds. Age of animals: 0–3 months (Broekhuyse, 1968b).
[b] Phosphatidylserine on the plate contained an unidentified phospholipid.

total phospholipids (Broekhuyse, 1972). In the aging sclera, a gradual continuous shift occurs in the proportional distribution of these lipids (Fig. 7). The main alteration is found in the percentage of sphingomyelin, which increases from 20% to 50% (a factor of 2.5) between the ages of 3 and 68 years. The percentages of the other phospholipids decrease linearly over the same period. At an early age (between 3 and 17 years) no lysophospholipids can be found. However, from 24 years of age onwards varying amounts of lysophosphatidylcholine (3.9–19.5%) and lysophosphatidylethanolamine (4.0–9.9%) occur in human sclera. The senile sclera contains the highest percentages of these lipids.

A similar pattern of alterations has been found for the phospholipid concentrations (Fig. 8). The sphingomyelin concentration increases from 0.45 to 1.30 μg/mg wet wt. between 3 and 68 years of age. The threefold increase is almost entirely compensated by decreasing concentrations of the other phospholipids. Hence, the total phospholipid content undergoes only a small increase during life (2.25–2.60 μg/mg wet wt). The concentrations of free fatty acids, cholesterol, and triacylglycerols increase (0.30–1.17, 0.81–1.49, and 0.60–1.65 μg/mg wet wt, respectively). Mono- and diacylglycerols were present in trace amounts. The cholesterol esters show the most prominent change in concentration, 0.16–3.70 μg/mg wet wt. Until about 33 years of age the change is mininal, but thereafter it becomes very rapid. This provides the major contribution to the increase in total lipids of the human sclera during life: 4.1–10.6 μg/mg wet wt or from 0.41% to 1.06% (w/w).

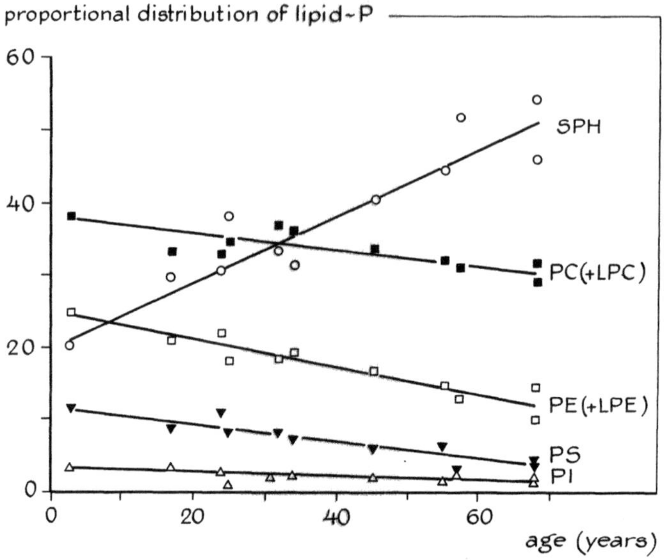

FIGURE 7. Age-dependent changes in the proportional distribution of the phospholipids of human sclera. SPH: sphingomyelin; (L)PC: (lyso)phosphatidylcholine; (L)PE: (lyso)phosphatidylethanolamine; PS; phosphatidylserine; PI: phosphatidylinositol.

Gas chromatographic analyses revealed that in the phospholipids only slight changes in the fatty acid pattern occur during the aging process. However, in phosphatidylcholine, and especially in the cholesterol esters, the linoleate content increases (8–12% and 10–40%, respectively). Cholesterol palmitate and stearate decrease (21–11% and 21–1%, respectively).

Analyses of rabbit and bovine scleras show that over the relatively short period of life that has been investigated the changes in phospholipid pattern resemble those in the human sclera. The percentage of sphingomyelin increases during the first years of life in bovine sclera from 18.0% to 33.7% and in rabbit sclera from 18.0% to 29.9%. Lysophosphatidylcholine and lysophosphatidylethanolamine were present in small amounts (< 2%) in contrast to the values found for human sclera above 25 years of age. The percentages of the phosphoglycerides decrease, and the concentrations show a similar pattern of change. The total phospholipid concentration decreases. The concentrations of the neutral lipids and total phospholipids are of the same order of magnitude as in the infantile human sclera. Changes with age are minimal or nonexistent (Broekhuyse, 1972).

Most studies on corneal lipids are of recent date. Feldman (1967) analyzed human cornea and compared its lipid composition with that of brain (Table 8). Neutral and sphingolipids (sphingomyelin and gangliosides) together comprise 74% of the total lipid, while the phosphoglycerides are present in low amounts, in contrast to brain. Data from Broekhuyse (unpublished results; Table 6) confirm the relatively high sphingomyelin content of human cornea. The high value for neutral lipids is mainly due to high cholesterol ester concentrations, as was also found in human sclera (see above). From Table 6 it appears that the central part and the annular outer part of the human cornea have different compositions in normal as well as in the arcus-containing cornea. Sphingomyelin is increased in the outer part of both cornea types, but especially in the arcus, which constitutes the outer annular part of the pathological cornea. In both parts of this cornea the total phospholipid concentration is markedly increased as well. This is in agreement with the findings for sclera, in which sphingomyelin was also the major phospholipid in individuals above 40 years of age (Fig. 7). Comparing the various data for the corneal parts and the sclera it looks as if the cornea centrum has a "younger" phospholipid composition than the peripheral part. The total phospholipid and cholesterol ester concentrations in human cornea are age dependent as is the case in the sclera (Table 6). The figures show that the arcus has a higher cholesterol and cholesterol

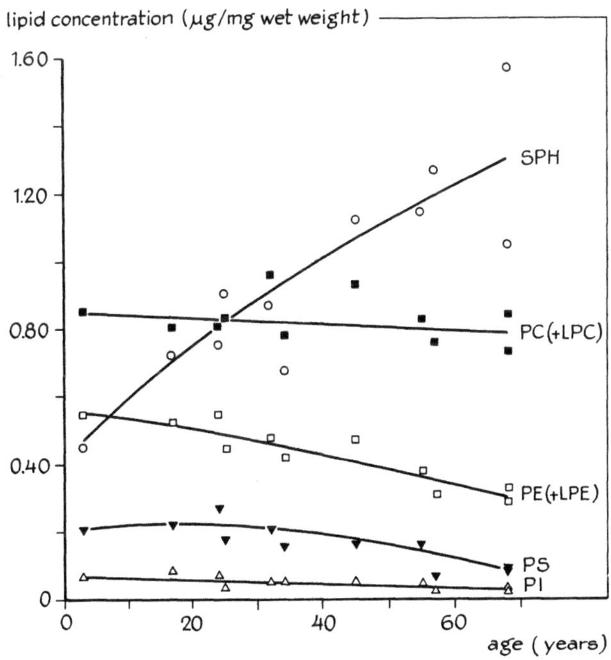

FIGURE 8. Age-dependent changes in the phospholipid concentrations of human sclera. For abbreviations, see Fig. 7.

TABLE 8
Comparison of the Composition of Human Corneal and Brain Lipids[a]

Lipid	Total lipid (%)	
	Cornea	Brain
Neutral lipids	34.1	22.1
Gangliosides	16.1	10.6
Sphingomyelin	24.3	5.4
Phosphatidylcholine	2.2	12.0
Phosphatidylethanolamine	0.7	11.2
Phosphatidylserine	5.0	4.0
Acidic lipids	13.0	1.6
Unidentified compounds	4.6	33.1[b]
Total	100.0	100.0

[a] Analysis made on an extract from 15 corneas removed at autopsy and found to be free of gross evidence of pathology. Wet weights of tissue are unreliable. The average dry weight was 21.76 mg, with an average lipid content of 1.08 mg/cornea (Feldman, 1967).
[b] Includes cerebrosides, 15.8%; sulfatides, 6.2%.

ester concentration than the central part and the clear peripheral part as well. The arcus represents a more "aged" cornea part in view of its lipid composition and thus in this respect most resembles the sclera of the same age.

Although it has been suggested that cholesterol ester accumulation is the cause of the opacity of the arcus, it seems more reasonable to suppose that high phospholipid or total lipid concentrations are involved in the opacification. The supposition that the extracted lipoidal material originates mainly from the epithelial cell membranes (Feldman, 1967) is not confirmed by the results presented in Table 6, because all analyses were done on epithelium-free corneas. In fact, histological staining (Cogan and Kuwabara, 1959) points to lipid accumulation in stroma and Descemet's membrane. Whether gangliosides are involved in the arcus is unknown. Feldman (1967) reported them to be present in normal cornea in a higher proportion of total lipids (16%) than in brain (11%).

The pattern of age-dependent changes of the various lipid concentrations in the human sclera shows a striking similarity with that of aorta and, as far as can be seen from recent reports, also with some changes in certain other connective tissues. The studies of Böttcher and Woodford (1962) and of Rouser and Solomon (1969) established that sphingomyelin increased steadily with age in normal aorta. The latter authors also analyzed grossly sclerotic aortas; the level of sphingomyelin did not appear to be greatly elevated. They obtained a similar pattern of straight lines for the phospholipid percentages of normal aorta as found in the sclera. Moreover, the proportional distribution of the various phospholipids at corresponding ages showed great resemblances to those in the sclera. In a study of aorta lipids Smith *et al.* (1967) reported age-dependent changes in the concentrations of cholesterol esters, phospholipids, triacylglycerols, and free cholesterol, which are also similar to our results. The marked increase in cholesterol linoleate is another point of resemblance. Until recently the findings for aorta seemed rather unique, but analyses by Crouse *et al.* (1972) of various connective tissues have revealed that dense connective tissues appear to act as a trap for cholesterol in the esterified form. A five- to tenfold increase in cholesterol esters was found over the whole life span, while free cholesterol increased more slowly, as in aorta and sclera. Relatively high concentrations of cholesterol esters are present in the skin as well, although no influence of aging could be demonstrated (Hsia, 1971). A hitherto unknown component of sclera is squalene, which was identified by gas chromatographic analysis of the hydrocarbon fraction (Broekhuyse, 1972). Although squalene was not quantified, gas chromatographic analyses suggested it was present at a concentration about equal to that of cholesterol. Hence, it seems to be one of the major neutral lipids in sclera, as it is in the surface lipids of the skin (Hsia, 1971) and to a lesser extent in the aorta (Stefanovitch and Kajiyama, 1970).

The lipid accumulation in aging sclera and cornea appears to be comparable, both in its chemical and morphological aspects, with that in the intima of the aorta. The decreasing number of cells in the senile sclera seems to exclude intracellular synthesis of large amounts of cholesterol esters. These components are found perifibrously and may have originated from the serum. The increasing content of cholesterol linoleate, the major serum cholesterol ester, makes this assumption very probable. This situation in sclera supports the corresponding theory for aorta, which is more difficult to prove because a possibility exists for local synthesis. However, extracellular intima lipid contains cholesterol linoleate and intracellular lipid contains cholesterol oleate as the dominant sterol ester, and various other data also point to serum β-lipoprotein as a donor of the cholesterol linoleate (Smith, 1968). Whether this lipid deposit in the connective tissues must be characterized as "normal" cannot yet be established with certainty. Owing to the high correlation between the severity of atherosclerosis and age, it is difficult to differentiate between lipid that accumulates specifically as a result of the disease process and lipid that accumulates as a result of "normal" aging.

The origin of the high sphingomyelin concentrations is still obscure. Portmann and Alexander (1970) found, in atherosclerotic aortas of squirrel monkeys, not only increased synthesis of sphingomyelin but also increased uptake from plasma. Sclera and cornea stroma, however, contain a very limited number of cells and consequently sphingomyelin might originate from the plasma. The scleral cells have a normal appearance in senility and intracellular lipid deposits are infrequently found. Most elderly people have a more or less pronounced arcus senilis in the cornea, which is comparable in composition to the sclera. The sudanophilia of the corneal stroma resembles that of the sclera. This finding caused several investigators to suppose that the arcus of the cornea is part of a paralimbal lipid infiltration of the corneo-scleral coat (Cogan and Kuwabara, 1959). Andrews (1962) obtained indications that the content of cholesterol esters in the sclera and the cornea might represent actual increase in arcus senilis. The normal values, however, were obtained from tissues from children and, as appears from my study (Broekhuyse, 1972), these tissues have much lower cholesterol ester concentrations. The severity of the arcus is closely related to the serum cholesterol or β-lipoprotein concentration. For that reason a pronounced arcus is always found in cases of hypercholesterolemia. In "normal" persons, however, the arcus consists usually of a slight opacification. The high cholesterol ester concentrations in sclera and cornea will probably be absent in certain populations of nonindustrialized countries, e.g., the Negroes of Uganda, since they have very low arterial concentrations of these lipids (Scott *et al.*, 1966). There is an extremely low incidence of coronary artery disease in Africans and yet corneal arcus in quite common among them (Davidson and Kolbe, 1965). In these subjects there was no association between arcus and serum cholesterol, as contrasted to populations in industrialized countries. A review regarding the incidence of arcus in African and other populations and its relation to systemic disease has recently appeared (Lemp, 1973). The conclusion based on these reports is largely in agreement with our previous supposition based on the lipid studies of aging cornea and sclera described above. The corneal arcus occurring in the elderly is probably a reflection of a "normal" aging process.

The problem was studied in cholesterol-fed rabbits by Rodger (1971). The corneas of these rabbits showed lipid deposits infiltrating the keratocytes. In the deeper stromal layers lipid droplets tended to become solid crystals. These crystals consisted of cholesterol or cholesterol esters. Some of the droplets consisted of cholesterol as well as phospholipids, but they were mostly cholesterol esters. Triacylglycerols were increased around the blood vessels in the periphery.

Other pathological conditions also involve corneal lipids: lipid deposits occur in corneal epithelium, stroma, and in sclera in Fabry's disease; metachromatic droplets (or granules) in corneal tissues in G_{M1} gangliosidosis; cholesterol containing birefringent crystals in corneal stroma in hereditary crystalline corneal degeneration; corneal stromal opacities with arcus in plasma-lecithin : cholesterol acyltransferase deficiency. The literature concerning these symptoms has been collected by Warburg

(1972). Primary lipoidal degeneration of the cornea was recently investigated by Fine *et al.* (1974). Bron and Tripathi (1974) reviewed the occurrence of arcus lipoides in hyperlipoproteinemia, type I and II and in hypolipoproteinaemia (Tangier disease) together with other corneal changes in inherited metabolic disorders. Recently, sphingomyelin storage in cornea stroma and endothelium has been found in Niemann-Pick disease (type A) (Robb and Kuwabara, 1973).

VI. Iris, Ciliary Body, and Choroid

A. Morphology and General Metabolism

Iris, ciliary body, and choroid form the uvea (Fig. 1). The iris is located behind the cornea and regulates the amount of light entering the eye. The ciliary body connects the iris with the choroid, which is situated between the sclera and the retina. The choroid supplies blood to the outer layer of the retina and has a rich nerve supply. The loose stroma fills the spaces between the vessels. The ciliary body consists of stroma tissue, which is surrounded by the pigmented epithelium and the unpigmented epithelium (outer layer). Its pars plicata is situated near the iris and its ciliary processes (prominent folds) are very close to the lens equator, which is attached to them by very thin zonular fibers. The ciliary body secretes aqueous humor and contains the ciliary muscle that governs accommodation. Detailed descriptions of the morphology of iris, ciliary body, and choroid are given by Hogan *et al.* (1971).

Little is known concerning the general metabolism of these tissues. Most investigators have concentrated on the ciliary epithelium and the aqueous humor secretion. The citrate cycle plays a major role in the energy supply of the ciliary epithelium. Active transport and oxygen uptake are both affected by metabolic inhibitors, and probably sodium transport is chiefly energized by ATP derived from citrate cycle oxidations. Histochemical studies lead to the conclusion that these activities are predominantly associated with the nonpigmented epithelial cells (Cole, 1970; Barber, 1974).

B. Lipids

The phospholipid composition of the various bovine ocular tissues including iris is shown in Table 7. Iris and retina have the highest phospholipid concentrations, while the phospholipid compositions of the various structures are not very different from those of most other tissues. Ethanolamine plasmalogens are present in considerable quantities (Table 9); the phospholipid values are in good agreement with those of Anderson *et al.*

TABLE 9
Plasmalogens of Calf Eye[a]

Compound	Retina	Iris	Choroid	Lens	Cornea	Sclera	Vitreous body
Ethanolamine plasmalogen	32.6	52.4	49.9	50.2	43.4	67.1	31.4
Choline plasmalogen	1.2	2.5	2.1	1.1	2.3	3.0	1.5

[a] From Broekhuyse (1968b).

(1970a). These authors found the major fatty acids of the various iris phospholipids to be 16:0, 18:0, and 18:1, and 20:4 was especially important in the ethanolamine, serine, and inositol phosphoglycerides. Whereas 22:6 was not detected, other polyunsaturated fatty acids were present in small amounts. Most of the fatty acid data are in agreement with those of Culp *et al.* (1970b). They found, however, high 20:4 values in iris phosphatidylethanolamine (26%) and phosphatidylcholine (14%). In addition, cerebrosides, gangliosides, and the various neutral lipids were also analyzed. Striking features were the low 24:0 and 24:1 contents (0–4.8%) of the sphingolipids and the high 22:0 content (20%) of the cerebrosides. Human iris total lipids contained similar types and amounts of fatty acids.

Lipid metabolism in iris has been demonstrated in various ways. The phospholipids incorporate [^{32}P]P_i (Broekhuyse, 1968b), while activity from [1-^{14}C]acetate is incorporated by the neutral lipids, as well as by the phospholipids (Culp *et al.*, 1970c).

Since the first characterization by Ambache (1957) of "irin," a prostaglandin preparation from iris, much research has been carried out on the significance of prostaglandins in ocular physiology and biochemistry. Prostaglandins (PG) are synthesized in the iris from linolenic acid (Van Dorp *et al.*, 1967) and secreted into the aqueous humor (PGE_2). PGE_1, E_2, F_2, and A_1 injected directly into the rabbit anterior chamber produce miosis and a sustained rise in intraocular pressure together with vasodilatation and increased permeability of the blood–aqueous barrier (Beitch and Eakins, 1969; Karim and Hillier, 1972). An elevation of the intraocular pressure in rabbits was also effected by arachidonic acid, the precursor of PGE_2 (Podos *et al.*, 1973).

VII. Vitreous Body, Aqueous Humor, and Tear Fluid

A. Vitreous Body

The vitreous body is a transparent gel that fills the space bordered by retina, ciliary body, and lens (Fig. 1). Its water content is approximately

99%, and it contains 0.1% colloid materials, mainly hyaluronic acid dispersed in an insoluble network of collagen fibers. The gel contains a small number of cells, which are confined to the peripheral layer. The vitreous body can be separated into its liquid and solid phases by centrifugation; the liquid one is called the vitreous humor and the solid phase is the residual protein. Cations seem to enter the gel from the ciliary body, whereas glucose enters principally from the retina. Small amounts of albumin, globulin, and soluble glycoproteins are present (Berman and Voaden, 1970).

The phospholipid composition of the vitreous body, separated from its membrane-bound hyalocytes by filtration, was analyzed by Broekhuyse (1968b). Table 7 shows that the phospholipid concentration is extremely low. Sphingomyelin, phosphatidylcholine, and phosphatidylethanolamine are the major phospholipid constituents. The latter component consisted of approximately one-third plasmalogen (Table 9). Cholesterol is found in the vitreous body in a concentration of 0.5 mg/100 ml, which is more than 200 times less than in blood serum. In cases of asteroid hyalopathy, asteroid bodies, which have been thought to consist of cholesterol, are deposited in the vitreous body. Cholesterol feeding of rabbits fails to produce asteroid bodies (Lamba and Shukla, 1971). Theories regarding the penetration of the vitreous body by cholesterol esters followed by breakdown to cholesterol, the cataractous lens as a source of cholesterol, and the precipitation of calcium–lipid complexes during hyaluronic acid loss (Jaffe, 1972), have yet to be carefully examined. Recently, the asteroid bodies in the vitreous body of a 79-year-old man have been shown to consist of calcium oxalate monohydrate and calcium hydroxyphosphate (hydroxypatite) by electron diffraction studies (March and Shoch, 1975). Increased cholesterol concentrations have been found in cholesteroisis bulb by Andrews *et al.* (1973). They analyzed the crystalline and fluid fractions of both the aqueous and vitreous humor from an eye of a patient with this disease. The crystals appeared to be pure cholesterol, while abnormally high cholesterol ester concentrations were found in the fluid. They suggested that the vitreous body may act as a "cholesterol sink."

B. *Aqueous Humor*

The aqueous humor is secreted by the ciliary body and flows from the posterior chamber through the pupil into the anterior chamber (Fig. 1). It leaves this chamber mainly via the Schlemm's canal, although other pathways have also been detected (Cole, 1970). All tissues bordering the aqueous humor exchange metabolites with it. For lens and cornea, the aqueous humor serves as a nutrient and supporting medium. Aqueous humor can be considered, to a certain extent, as an ultrafiltrate of plasma. The bicarbonate concentration, however, depends on the relative dimensions of the lens. An eye from a small animal usually contains a large lens and produces a relatively large amount of lactate, which is buffered by the bicarbonate of the aqueous humor. The protein concentration is very low, 5–16 mg/100 ml in man, 50 mg/100 ml in the rabbit, compared with 6000–7000 mg/ml in the plasma. The various globulin fractions and albumin have been analyzed with varying success. Prealbumin and α-, β-, and γ-globulin were present (Davson, 1969), although the blood–aqueous barrier inhibits the penetration of large molecules. Small lipid soluble molecules (ethanol) rapidly pass the ciliary body and are not filtered back during the formation of the primary secretion. Actively transported substances like ascorbate and *myo*-inositol are present in high concentrations in aqueous humor as well as in lens (Cole, 1970; Broekhuyse, 1968a). It is not surprising that a large molecule like β-lipoprotein (mol wt 3,200,000) cannot pass the blood–aqueous barrier. In connection with this finding it was demonstrated that α-lipoprotein was the only lipoprotein in human aqueous humor (Schmut and Zirm, 1974). The same authors also found that aqueous humor was deficient in IgM and α_2-macroglobulin.

The phospholipids of rabbit aqueous humor were analyzed by Varma and Reddy (1972). Total phospholipid amounted to 2.55 ± 1.3 mg/100 ml, approximately 61% being phosphatidylcholine; the rest consisted of lysophosphatidylcholine, sphingomyelin, and an unidentified phosphorus-containing lipid. The proportion of the phospholipids resembles that of serum and was drastically changed in diabetic conditions. The total phospholipid value found by the authors for rabbit is very high in comparison to results obtained by Broekhuyse and Kuhlmann (unpublished results). After collection of 100 ml aqueous humor from calf eyes and removal of traces of cellular material by centrifugation and ultrafiltration (through 0.45-μm filters) they obtained a total phospholipid value of 0.13 mg/100 ml. Other lipid components of aqueous humor are cholesterol and prostaglandins. The latter substances are synthesized by the iris (see section VI.B) and an active process for the removal of PGE_1 is located posterior to the iris (Bito and Salvador, 1972). The concentration of cholesterol in normal human aqueous humor was 10 μg/ml, while cholesterol esters were not detected. In cholesterosis bulbi, however, 4 μg/ml cholesterol ester was found and only 0.6 μg/ml cholesterol (Andrews *et al.*, 1973). The authors suggest that in this disease the mechanism for maintaining cholesterol in solution is impaired in the eye (see section VII.A).

C. *Tear Fluid*

The cornea is protected by a thin (7 μm) tear film, which has a complex structure. It is prevented from evaporating by a superficial oily layer, while at its base it has an intimate contact with the corneal epithelial surface via the microvilli of the plasma membranes (Brauninger *et al.*, 1972). It is probable that the superficial oily layer consists of mixed waxes and cholesterol esters as these lipids have been isolated from human Meibomian secretion (Andrews, 1970), but cholesterol has also been shown to be present (Van Haeringen and Glasius, 1975). A considerable amount of branched-chain fatty acids and branched-chain aliphatic alcohols were present in this product. Although it is possible that the lipids are bound to protein, the need for a closed, hydrophobic nonvolatile superficial layer leads one to suppose that they form a lipid film. The absence of protein or mucin in this film has not been proved. The experiments of Brauninger *et al.* (1972), however, show that the presence of hydrophilic substances, at least at the interface air/tear film, is improbable. Recent experiments of Holly (1973) have shown that upon spreading of mucin–lipid mixtures, spreading of the lipid is the primary event, which is immediately followed by the spreading of the mucin originating from the conjunctival goblet cells. The rapid lipid spreading would restrict protein–lipid interaction. The superficial lipid layer is resilient and the mucin–lipid interactions lower the surface tension of the tear film and stabilize it. Lipid "contamination" of the mucin layer coating the corneal epithelium gradually destabilizes the tear film. This causes the formation of dry spots, which irritate the nerve endings in the epithelium and trigger the next blink (Holly, 1973).

References

Ambache, N. 1957. Properties of irin, a physiological constituent of the rabbit's iris. *J. Physiol. (London)* **135**:114–132.

Anderson, R. E. 1970. Lipids of ocular tissues IV. A comparison of the phospholipids from the retina of six mammalian species. *Exp. Eye Res.* **10**:339–344.

Anderson, R. E., and Maude, M. B. 1972. Lipids of ocular tissues. VIII. The effects of essential fatty acid deficiency on the phospholipids of the photoreceptor membranes of rat retina. *Arch Biochem. Biophys.* **151**:270–276.

Anderson, R. E., Maude, M. B., and Feldman, G. L. 1969. Lipids of ocular tissues I. The phospholipids of mature rabbit and bovine lens. *Biochim. Biophys. Acta* **187**:345–353.

Anderson, R. E., Feldman, L. S., and Feldman, G. L. 1970a. Lipids of ocular tissues II. The phospholipids of mature bovine and rabbit whole retina. *Biochim. Biophys. Acta* **202**:367–373.

Anderson, R. E., Maude, M. B., and Feldman, G. L. 1970b. Lipids of ocular tissues III. The phospholipids of mature bovine iris. *Exp. Eye Res.* **9**:281–284.

Anderson, R. E., Benolken, R. M., and Dutley, P. A., Londis, D. J., and Wheeler, T. G. 1974. Polyunsaturated fatty acids of photoreceptor membranes. *Exp. Eye Res.* **18**:205–213.

Andrews, J. S. 1962. The lipids of arcus senilis. *Arch. Ophthalmol.* **68**:264–266.

Andrews, J. S. 1966. Corneal lipids I. Sterol and fatty acid synthesis in the intact calf cornea. *Invest. Ophthalmol.* **5**:367–370.

Andrews, J. S. 1970. Human tear film lipids. I. Composition of the principal non-polar component. *Exp. Eye Res.* **10**:223–227.

Andrews, J. S., Lynn, C., Scobey, J. W., and Elliot, J. H. 1973. Cholesterosis bulbi—a case report with modern chemical identification of the ubiquitous crystals. *Br. J. Ophthalmol.* **57**:838–844.

Aveldano, M. I., and Bazan, N. G. 1973. Fatty acid composition and level of diacylglycerols and phosphodiglycerides in brain and retina. *Biochim. Biophys. Acta* **296**:1–9.

Aveldano, M. I., and Bazan, N. G. 1974a. Free fatty acids, diacyl- and triacylglycerols and total phospholipids in vertebrate retina: Comparison with brain, choroid and plasma. *J. Neurochem.* **23**:1127–1135.

Aveldano, M. I., and Bazan, N. G. 1974b. Displacement into incubation medium by albumin of highly unsaturated free fatty acids arising from membrane lipids. *FEBS Lett.* **40**:53–56. **40**:53–56.

Barber, G. W. 1974. Physiological chemistry of the eye. *Arch. Ophthalmol.* **91**:141–159.

Bartley, W., van Heyningen, R., Notton, B. M., and Renshaw, A. 1962. Fatty acid composition of lipids present in different parts of the ox eye. *Biochem. J.* **85**:332–335.

Basinger, S. F., and Hall, M. O. 1973. Rhodopsin biosynthesis *in vitro*. *Biochemistry* **12**:1996–2003.

Beitch, B. R., and Eakins, K. E. 1969. The effects of prostaglandins on the intraocular pressure of the rabbit. *Br. J. Pharmacol.* **37**:158–167.

Berman, E. R., and Voaden, M. 1970. The vitreous body, pp. 373–463. *In* C. N. Graymore (ed.). Biochemistry of the Eye. Academic Press, London.

Bernhard, K., Lesch, P., and Graub, H.-R. 1969. Beitrag zur Charakterisierung der Lipide aus den Linsen einiger Säugetiere. *Helv. Chim. Acta* **52**:801–806.

Bibb, C., and Young' R. W. 1974a. Renewal of glycerol in the visual cells and pigmentepithelium of the frog retina. *J. Cell Biol.* **62**:378–389.

Bibb, C., and Young, R. W. 1974b. Renewal of fatty acids in the membranes of visual cell outer segments. *J. Cell Biol.* **61**:327–343.

Bito, L. Z., and Salvador, E. V. 1972. Intraocular fluid dynamics III. The site and mechanism of prostaglandin transfer across the blood intraocular fluid barriers. *Exp. Eye Res.* **14**:233–241.

Bloemendal, H., Zweers, A., Vermorken, F., Dunia, I., and Benedetti, E. L. 1972. The plasma membranes of eye lens fibres. Biochemical and structural characterization. *Cell Differ.* **1**:91–106.

Böttcher, C. J. F., and Woodford, S. P. 1962. Chemical changes in the arterial wall associated with atherosclerosis. *Fed. Proc.* Suppl. 11, **21**:15–21.

Brauninger, G. E., Shah, D. D., and Kaufman, H. E. 1972. Direct physical demonstration of oily layer on tear film surface. *Am. J. Ophthalmol.* **73**:132–134.

Broekhuyse, R. M. 1968a. Changes in myo-inositol permeability in the lens due to cataractous conditions. *Biochim. Biophys. Acta* **163**:269–272.

Broekhuyse, R. M. 1968b. Phospholipids in tissues of the eye. I. Isolation, characterization and quantitative analysis by two-dimensional thin-layer chromatography of diacyl and vinyl-ether phospholipids. *Biochim. Biophys. Acta* **152**:307–315.

Broekhuyse, R. M. 1969a. Phospholipids in tissues of the eye III. Composition and metabolism of phospholipids in human lens in relation to age and cataract formation. *Biochim. Biophys. Acta* **187**:354–365.

Broekhuyse, R. M. 1969b. The phospholipids of the lens. Thesis, Nijmegen, 97 pp.

Broekhuyse, R. M. 1970. Lipids in tissues of the eye. IV. Influence of age and species differences on the phospholipid composition of the lens. *Biochim. Biophys. Acta* **218**:546–548.

Broekhuyse, R. M. 1971. Lipids in tissues of the eye V. Phospholipid metabolism in normal and cataractous lens. *Biochim. Biophys. Acta* **231**:360–369.

Broekhuyse, R. M. 1972. Lipids in tissues of the eye VII. Changes in concentration and composition of sphingomyelins, cholesterol esters and other lipids in aging sclera. *Biochim. Biophys. Acta* **280**:637–645.

Broekhuyse, R. M. 1973. Membrane lipids and proteins in ageing lens and cataract, pp. 135–149. *In* The Human Lens—In Relation to Cataract. Ciba Foundation Symposium 19. Elsevier, Amsterdam.

Broekhuyse, R. M., and Kuhlmann, E. D. 1972. Lipids in tissues of the eye VI. Sphingomyelins and cholesterol esters in human sclera. *Exp. Eye Res.* **14**:111–113.

Broekhuyse, R. M., and Kuhlmann, E. D. 1974. Lens membranes 1. Composition of urea-treated plasma membranes from calf lens. *Exp. Eye Res.* **19**:297–302.

Broekhuyse, R. M., and Veerkamp, J. H. 1968. Phospholipids in the tissues of the eye II. Composition and incorporation of $^{32}P_i$ of phospholipids of normal rat and calf lens. *Biochim. Biophys. Acta* **152**:316–324.

Broekhuyse, R. M., Roelfzema, H., Breimer, M. E., and Karlsson, K.-A. 1974. Lipids in tissues of the eye X. Molecular species of sphingomyelin from different parts of calf lens in relation to differentiation and aging. *Exp. Eye Res.* **19**:477–484.

Bron, A. J., and Tripathi, R. C. 1974. Corneal disorders, pp. 281–323. *In* M. G. Goldberg (ed.). Genetic and Metabolic Eye Diseases. Little, Brown and Company, Boston.

Brown, P. K. 1972. Rhodopsin rotates in the visual receptor membrane. *Nature (London) New Biol.* **236**:35–38.

Chapman, D. 1973. Studies of lipid-cholesterol interaction, pp. 118–144. *In* D. Chapman and D. F. H. Wallach (eds.). Biological Membranes, Vol. 2. Academic Press, New York.

Clark, A. W., and Branton, D. 1968. Fracture faces of frozen outer segments from the guinea pig retina. *Z. Zellforsch.* Mikroskop. Anat. **91**:586–603.

Cogan, D. G., and Kuwabara, T. 1959. Arcus senilis. *Arch. Ophthalmol.* **61**:553–560.

Cole, D. F. 1970. Aqueous and ciliary body, pp. 105–172. *In* C. N. Graymore (ed.). Biochemistry of the Eye. Academic Press, London.

Cone, R. A. 1972. Rotational diffusion of rhodopsin in the visual receptor membrane. *Nature (London) New Biol.* **236**:39–43.

Crouse, J. R., Grundy, S. M., and Ahrens, Jr., E. H. 1972. Cholesterol distribution in the bulk tissues of man: Variation with age. *J. Clin. Invest.* **51**:1292–1296.

Culp, T. W., Creger, C. R., Swanson, A. A., Couch, J. R., and Harlow, R. D. 1968. Identification of triglycerides in the bovine lens by chromatographic analysis. *Exp. Eye Res.* **7**:134–141.

Culp, T. W., Hall, F. F., Jeter, J., and Ratliff, C. R. 1970a. Lens lipids: Biosynthesis and histological distribution in rabbit lens. *Ophthal. Res.* **1**:313–320.

Culp, T. W., Tucker, P. W., Ratliff, C. R., and Hall, F. F. 1970b. Chromatographic analysis of ocular lipids. I. Bovine and human iris tissue. *Biochim. Biophys. Acta* **218**:259–268.

Culp, T. W., Cunningham, R. D., Tucker, P. W., Jeter, J., and Determan, Jr., L. H. 1970c. In vivo synthesis of lipids in rabbit iris, cornea and lens tissues. *Exp. Eye Res.* **9**:98–105.

Daemen, F. J. M. 1973. Vertebrate red outer segment membranes. *Biochim. Biophys. Acta* **300**:255–288.

Daemen, F. J. M., Rotmans, J. P., and Bonting, S. L. 1974. On the rhodopsin cycle. *Exp. Eye Res.* **18**:97–103.

Davidson, J. C., and Kolbe, R. J. 1965. Arcus senilis and ischaemic heart disease. *Lancet* **1**:707.

Davson, H. 1969. The intraocular fluids, pp. 67–187. *In* H. Davson (ed.). The Eye, Vol. 1. Academic Press, New York.

Dische, Z., Hairstone, M. A., and Zelmenis, G. 1967. Glyco- and glycolipoproteins in cell surface from bovine lens fibers, pp. 123–129. *In* H. Peeters (ed.). Protides of the Biological Fluids, Vol. 15. Elsevier, Amsterdam.

Dowling, J. E. 1960. Chemistry of visual adaptation in the rat. *Nature* **188**:114–118.

Dowling, J. E. 1970. Organization of vertebrate retinas. *Invest. Ophthalmol.* **9**:655–680.
Dowling, J. E., and Ripps, H. 1973. Effect of magnesium on horizontal cell activity in the skate retina. *Nature* **242**:101–103.
Dowling, J. E., and Wald, G. 1958. Vitamin A deficiency and night blindness. *Proc. Nat. Acad. Sci. (USA)* **44**:648–661.
Dowling, J. E., and Wald, G. 1960. The biological function of vitamin A acid. *Proc. Nat. Acad. Sci (USA)* **46**:587–608.
Dubin, M. W. 1974. Anatomy of the vertebrate retina, pp. 227–256. *In* H. Davson and L. T. Graham Jr. (eds.) The Eye, Vol. 6, Academic Press, New York.
Du Chesne, J. 1967. Fettgehalt und Morphologie der menschlichen Sclera in den verschiedenen Altersstufen. Thesis, Leipzig.
Feldman, G. L. 1967. Human ocular lipids: Their analysis and distribution. *Surv. Ophthal.* **12**:207–243.
Feldman, G. L. 1968. Lipids of the human lens, pp. 348–357. *In* M. U. Dardenne and J. Nordmann (eds.) Biochemistry of the Eye. Symposium Tutzing Castle 1966. S. Karger, Basel.
Feldman, G. L., Culp, T. W., Feldman, L. S., Grantham, C. K., and Jonsson, Jr., H. T. 1964. Phospholipids of the human, bovine, rabbit, and human lens. *Invest. Ophthalmol.* **3**:194–197.
Fine, B. S., Townsend, W. M., Zimmerman, L. E., and Lashkari, M. H. 1974. Primary lipoidal degeneration of the cornea. *Am. J. Ophthalmol.* **78**:12–23.
Futterman, S., Downer, E. L., and Hendrickson, A. 1971. Effect of essential fatty acid deficiency on the fatty acid composition, morphology and electroretinographic response of the retina. *Invest. Ophthalmol.* **10**:151–156.
Gorthy, W. C., Snavely, M. R., and Berrong, N. D. 1971. Some aspects of transport and digestion in the lens of the normal young adult rat. *Exp. Eye Res,* **12**:112–119.
Goux, J.-P., and Kallay, O. 1971. Les complications oculaires des erreurs congenitales de metabolisme. *Bull. Soc. Belge Ophtalmol.* **157**:95–278.
Graymore, C. N. 1969. General aspects of the metabolism of the retina, pp. 601–645. *In* C. N. Graymore (ed.) The Eye, Vol. 1. Academic Press, New York.
Hagins, W. A. 1972. The visual process: Excitatory mechanisms in the primary receptor cells. *Annu. Rev. Biophys. Bioeng.* **1**:131–158.
Hagins, W. A., Penn, R. D., and Yoshikami, S. 1970. Dark current and photocurrent in retinal rods. *Biophys. J.* **10**:380–412.
Hall, M. O., Bok, D., and Bacharach, A. D. E. 1969. Biosynthesis and assembly of the rod outer segment membrane system. Formation and fate of visual pigment in the frog retina. *J. Molec. Biol.* **45**:397–410.
Hatcher, H., and Andrews, J. S. 1970. Changes in lens fatty acid composition during galactose cataract formation. *Invest. Ophthalmol.* **9**:801–806.
Hendriks, Th., Klompmakers, A. A., Daemen, F. J. M., and Bonting, S. L. 1976. Movement of sodium ions through bilayers composed of retinal and rod outer segment lipids. Biochim. Biophys. Acta **433**:271–281.
Hogan, M. J., Alvarado, J. A., and Wedell, J. E. 1971. Histology of the Human Eye. Saunders Company, Philadelphia, 687 pp.
Holly, F. J. 1973. Formation and rupture of the tear film. *Exp. Eye Res.* **15**:515–525.
Holm, M. 1972. Gangliosides of the optic pathway: Biosynthesis and biodegradation studied in vivo. *J. Neurochem.* **19**:623–629.
Holm, M., and Mansson, J.-E. 1974. Gangliosides of bovine optic nerve. *FEBS Lett.* **45**:159–161.
Holm, M., Mansson, J. E., Venier, M. T., and Svennerholm, L. 1972. Gangliosides of human, bovine and rabbit retina, *Biochim. Biophys. Acta* **280**:356–364.
Hong, K., and Hubbell, W. L. 1973. Lipid requirements for rhodopsin regenerability. *Biochemistry* **12**:4517–4523.

Hsia, S. L. 1971. Potentials in exploring the biochemistry of human skin, pp. 1–38. *In* P. N. Campbell (ed.). Essays in Biochemistry, Vol. 7, Academic Press, New York.

Jaffe, N. S. 1972. The vitreous. *Arch. Ophthalmol.* **87**:599–611.

Jedziniak, J. A., Kinoshita, J. H., Yates, E. M., and Benedek, G. B. 1975. The concentration and localization of heavy molecular weight aggregates in aging normal and cataractous human lenses. *Exp. Eye Res.* **20**:367–369.

Kagan, V. E., Shedova, A. A., Novikov, K. N., and Kozlov, Yu. P. 1973. Light-induced free radical oxidation of membrane lipids in photoreceptors of frog retina. *Biochim. Biophys. Acta* **330**:76–79.

Karim, S. M. M., and Hillier, K. 1972. General introduction and some pharmacological actions of prostaglandins, pp. 1–46. *In* S. M. M. Karim (ed.). The Prostaglandins. Medical and Technical Publishing Company, Oxford.

Kenyon, K. R. 1974. Ocular ultrastructure of inherited metabolic disease, pp. 139–185. *In* M. F. Goldberg (ed.). Genetic and Metabolic Eye Diseases. Little, Brown and Company, Boston.

Korenbrot, J. I., Brown, D. T., and Cone, R. A. 1973. Membrane characteristics and osmotic behaviour of isolated rod outer segments. *J. Cell Biol.* **56**:389–398.

Krekeler, F. 1923. Die Struktur der Sclera in den verschiedenen Lebensaltern. *Arch. Augenheilk.* **93**:144–150.

Kuck, J. F. R. 1970a. Chemical constituents of the lens, pp. 183–260. *In* C. N. Graymore (ed.). Biochemistry of the Eye. Academic Press, London.

Kuck, J. F. R. 1970b. Metabolism of the lens, pp. 261–318. *In* C. N. Graymore (ed.). Biochemistry of the Eye. Academic Press, London.

Kuwabara, T. 1970. Retinal recovery from exposure to light. *Am. J. Ophthalmol.* **701**:187–198.

Lamba, P. A., and Shukla, K. N. 1971. Experimental asteroid hyalopathy. *Br. J. Ophthalmol.* **55**:279–283.

Lasser, A., and Balazs, E. A. 1972. Biochemical and fine structure studies on the water-insoluble components of the calf lens. *Exp. Eye Res.* **13**:292–308.

Lemp, M. A. 1973. Cornea and sclera. *Arch. Ophthalmol.* **90**:408–422.

Liebman, P. A., and Entine, G. 1974. Lateral diffusion of visual pigment in photoreceptor disc membranes. *Science* **185**:457–459.

Liem-The, K. N., Stols, A. L. H., and Hoenders, H. J. 1975. Further characterization of HM-crystallin in rabbit lens. *Exp. Eye Res.* **20**:307–316.

Lion, F., Rotmans, J. P., Daemen, F. J. M., and Bonting, S. L. 1975. Stereospecificity of ocular retinal dehydrogenases and the visual cycle. *Biochim. Biophys. Acta* **384**:283–292.

Lolley, R. N., and Schmidt, S. Y. 1974. Metabolism of the vertebrate retina, pp. 343–378. *In* H. Davson and L. T. Graham, Jr. (eds.). The Eye, Vol. 6. Academic Press, New York and London.

Löwenstein, A. 1940. Lipoid droplet in the episclera as a regular change with age. *Ophthalmologica* **100**:345–350.

MacBrinn, M. C., and O'Brien, J. S. 1969. Lipid composition of optic nerve myelin. *J. Neurochem.* **16**:7–12.

March, W. F. and Shoch, D, 1975. Electron diffraction study of asteroid bodies. *Invest. Ophthalmol.* **14**:399–401.

Matuk, Y. 1972. Studies on the incorporation of amino acids by a cell-free system obtained from beef retina. *Can. J. Biochem.* **50**:581–587.

Maurice, D. M. 1969. The cornea and sclera, pp. 489–585. *In* H. Davson (ed.). The Eye, Vol. 1, Academic Press, New York.

Maurice, D. M., and Riley, M. V. 1970. The cornea, pp. 1–95. *In* C. N. Graymore (ed.). Biochemistry of the Eye. Academic Press, London.

McLaren, D. S. 1970. Nutritional aspects of the eye, pp. 519–561. *In* C. N. Graymore (ed.). Biochemistry of the Eye. Academic Press, London.

Mizuno, G., Ellison, J. R., Chipault, J. R., and Harris, J. E. 1974. Lipids of the triparanol cataract in the rat. *Ophthal. Res.* **6**:206–215.

Noell, W. U., and Albrecht, R. 1971. Irreversible effects of visible light on the retina—role of vitamin A. *Science* **172**:76–80.

Noell, W. U., Walker, V. S., Kang, B. S., and Berman, S. 1966. Retinal damage by light in rats. *Invest. Ophthalmol.* **5**:450–473.

Noell, W. U., Delmelle, M. C., and Albrecht, R. 1971. Vitamin A deficiency effect on retina; Dependence on light. *Science* **172**:72–76.

Novikov, K. N., Shvedova, A. A., Kagan, V. Ye., Kozlov, Yu. P., and Ostrovskii, M. A. 1974. Study on the photo-induced changes in the photoreceptor membrane and rhodopsin by the method of inoculated copolymerization. *Biofizika* **19**:280–284.

O'Brien, P. J., Muellenberg, C. G., and Bungenberg de Jong, J. J. 1972. Incorporation of leucine into rhodopsin in isolated bovine retina. *Biochemistry* **11**:64–70.

Ono, S., Obara, Y., and Hatano, M. 1972/73. The effect of prostaglandin E_1 on the phospholipid metabolism in the lens. *Ophthal. Res.* **4**:281–283.

Papermaster, D. S., Converse, C. A., and Sin, J. 1975. Membrane biosynthesis in the frog retina: opsin transport in the photoreceptor cell. *Biochemistry* **14**:1343–1352.

Paulus, H., and Kennedy, E. P. 1960. The enzymatic synthesis of inositol monophosphatide. *J. Biol. Chem.* **235**:1303–1311.

Philipson, B. 1973. Changes in lens related to the reduction of transparency. *Exp. Eye Res.* **16**:29–39.

Podos, S. M., Becker, B., and Kass, M. A. 1973. Prostaglandin synthesis, inhibition, and intraocular pressure. *Invest. Ophthalmol.* **12**:426–433.

Polyak, S. 1957. The vertebrate visual system. The University of Chicago Press. 1220 pp.

Poo, M. M., and Cone, R. A. 1974. Lateral diffusion of rhodopsin in the photoreceptor membrane. *Nature* **247**:438–441.

Portman, O. W., and Alexander, M. 1970. Metabolism of sphingolipids by normal and atherosclerotic aorta of squirrel monkeys. *J. Lipid Res.* **11**:23–30.

Rand, R. P., Pangborn, W. A., Purdon, A. D., and Tinker, D. O. 1975. Lysolecithin and cholesterol interact stoichiometrically forming bimolecular lamellar structures in the presence of excess water, or lysolecithin or cholesterol. *Can. J. Biochem.* **53**:189–195.

Reim, M., Henninghausen, U., Hildebrandt, D., and Maier, R. 1971. Enzyme activities in cornea epithelium and endothelium of different species. *Ophthal. Res.* **2**:171–182.

Robb, R. M., and Kuwabara, T. 1973. The ocular pathology of Type A Niemann-Pick disease. *Invest. Ophthalmol.* **12**:366–377.

Robert, L., and Robert, B. 1974. Structural glycoproteins of connective tissue: Their role in morphogenesis and immunopathology, pp. 240–256. *In* R. Fricke and F. Hartmann (eds.). Connective Tissues. Biochemistry and Pathology. Springer-Verlag, Berlin.

Rodger, F. C. 1971. A study of the ultrastructure and cytochemistry of lipid accumulation and clearance in cholesterol-fed rabbit cornea. *Exp. Eye Res.* **12**:88–93.

Roelfzema, H. 1975. Sphingomyelinase in the lens. Thesis, Nijmegen, 110 pp.

Roelfzema, H., Broekhuyse, R. M., and Veerkamp, J. H. 1973. Lipids in tissues of the eye VIII. Sphingomyelinase of the lens. *Biochim. Biophys. Acta* **306**:329–339.

Roelfzema, H., Broekhuyse, R. M., and Veerkamp, J. H. 1974. Subcellular distribution of sphingomyelinase and other hydrolases in different parts of the calf lens. *Exp. Eye Res.* **18**:579–594.

Roelfzema, H., Broekhuyse, R. M., and Veerkamp, J. H. 1976. Phospholipid and sphingomyelin concentrations in bovine lenses in relation to differentiation and aging. *Exp. Eye Res.* **23**:409–415.

Rouser, G., and Solomon, R. D. 1969. Changes in phospholipid composition of human aorta with age. *Lipids* **4**:232–234.

Saari, J. C. 1974. The accessibility of bovine rhodopsin in photoreceptor membranes. *J. Cell Biol.* **63**:480–491.

Sanders, D., Cotlier, E., Wyhinny, G., and Millman, L. 1974. Cataracts induced by surface active agents. *Exp. Eye Res.* **19**:35–42.

Schmidt, J. G. H. 1965. Über Netzhaut-Ganglioside. III. Abspaltung von *N*-Acctyl- und *N*-Glykolyl-neuraminsaure aus Gangliosiden der Rindernetzhaut. *Albrecht von Graefes Arch. Ophthalmol.* **168**:70–79.

Schmut, O., and Zirm: M, 1974. Immunologische Bestimmung von Lipoproteinen im Kammerwasser. *Albrecht von Graefes Arch. Klin. Exp. Ophthalmol.* **191**:19–23,

Scott, R. F., Florentin, R. A., Daoud, A. S., Morrisson, E. S., Jones, R. M., and Hutt, M. S. R, 1966. Coronary arteries of children and young adults. *Exp. Molec. Pathol.* **5**:12–42.

Shichi, H., and Somers, R. L. 1974. Possible involvement of retinylidene phospholipid in photoisomerization of all-trans retinal to 11-cis retinal. *J. Biol. Chem.* **249**:6570–6577.

Smidt-Martens, F. W., and Hohorst, H. J. 1972. Free fatty acids in bovine cornea and corneal epithelium. *Ophthalmic Res.* **3**:126–128.

Smith, E. B. 1968. The origin and significance of the changes in the lipids of vascular tissue with age. *J. Atheroscler. Res.* **8**:197–199.

Smith, E. B., Evans, P. H., and Downham, M. D. 1967. Lipid in the aortic intima. The correlation of morphological and chemical characteristics. *J. Atheroscler. Res.* **7**:171–186.

Spikes, J. D. 1968. Photodynamic action. pp. 33–64. *In* A. C. Giese (ed.). Photophysiology. Academic Press, New York.

Stefanovich, V., and Kajiyama, G. 1970. Squalene and cholestanol in normal rabbit aorta. *Atherosclerosis* **11**:401–403.

Swartz, J. G., and Mitchell, J. E. 1970. Biosynthesis of retinal phospholipids: Incorporation of radio-activity from labeled phosphorylcholine and cytidine diphosphate choline. *J. Lipid Res.* **280**:356–364.

Swartz, J. G., and Mitchell, J. E. 1973. Phospholipase activity of retina and pigment epithelium. *Biochemistry* **12**:5273–5278.

Swartz, J. G., and Mitchell, J. E. 1974. Acyl transfer reactions of retina. *Biochemistry* **13**:5053–5059.

Thiele, O. W., and Denden, A. 1967. Über die Verteilung der Lipide in der Cornea des Rindes. *Z. Physiol. Chem.* **348**:1097–1101.

Tomita, T. 1970. Electrical activity of vertebrate photoreceptors. *Quart. Rev. Biophys.* **3**:179–222.

Trayhurn, P., and Van Heyningen, R. 1973. The metabolism of amino acids in the bovine lens. *Biochem. J.* **136**:67–75.

Van der Eerden, J. J. J. M. 1975. Ocular structural glycoproteins. Thesis, Nijmegen.

Van der Eeerden, J. J. J. M., and Broekhuyse, R. M. 1974. Ocular antigens V: Species specificity of structural glycoproteins isolated from ocular connective tissues. An immunofluorescence and immunochemical study. *Exp. Eye Res.* **19**:341–351.

Van Dorp, D. A., Jouvenaz: G, H., and Struijk, C. B. 1967. The biosynthesis of prostaglandin in pig iris. *Biochim. Biophys. Acta* **137**:396–399.

Van Haeringen, N. J., and Glasius, E. 1975. Cholesterol in human tear fluid. *Exp. Eye Res.* **20**:271–274.

Van Heyningen, R. 1969. The lens: Metabolism and cataract, pp. 381–488. *In* H. Davson (ed.). The Eye, Vol. 1. Academic Press, New York.

Varma, S. D., and Reddy, V. N. 1972. Phospholipid composition of aqueous humor, plasma and lens in normal and alloxan diabetic rabbits. *Exp. Eye Res.* **13**:120–125.

Wald, G. 1968. Molecular basis of visual excitation. *Science* **162**:230–239.
Warburg, M. 1972. Diagnosis of Metabolic Eye Disease. Munksgaard, Copenhagen, 112 pp.
Windeler, A. S., and Feldman, G. L. 1970. The isolation and partial structural characterization of some ocular gangliosides. *Biochim. Biophys. Acta* **202**:361–366.
Worthington, C. R. 1971. Structure of photoreceptor membranes. *Fed. Proc.* **30**:57–63.
Yamada, E., and Shikano, S. -J. (eds.) 1973. Electron Microscopic Atlas in Ophthalmology. G. Thieme, Stuttgart, 375 pp.
Young, R. W. 1967. The renewal of photoreceptor cell outer segments. *J. Cell Biol.* **33**:61–72.
Young, R. W. 1974. Biosynthesis and renewal of visual cell outer segment membranes. *Exp. Eye Res.* **18**:215–223.
Young, R. W., and Droz, B. 1968. The renewal of proteins in retinal rods and cones. *J. Cell Biol.* **39**:169–184.
Zimmerman, W. F. 1974. The distribution of proportions of vitamin A compounds during the visual cycle in the rat. *Vision Res.* **14**:795–802.

Skeletal Muscle

KEIZO WAKU

I. *Lipid Composition of Mammalian Skeletal Muscle*

A. *Whole Muscle*

Typical of skeletal muscle is the total lipids of monkey gastrocnemius muscle. It consists of one-third neutral lipid and two-thirds phospholipid. The neutral lipid is 83% triacylglycerol and 16% cholesterol (Masoro *et al.*, 1964) together with much smaller amounts of cholesterol ester, free fatty acid, diacylglycerol, and monoacylglycerol (Boichot and Constanzo, 1968; Fröberg, 1967). The phospholipid compositions of the skeletal muscle of several mammalian species are summarized in Table 1. These are similar to results obtained for liver (Singh and Swartwout, 1972) but show higher levels of plasmalogens, especially of ethanolamine plasmalogen (1-alk-1′-enyl-2-acyl-*sn*-glycero-3-phosphorylethanolamine) (Dawson, 1960; Davenport, 1964).

Comprehensive studies on the fatty acid pattern of phosphatidylethanolamine and phosphatidylcholine in the lipids of rat organs were performed by Kanoh (1967), and no significant departures from the liver pattern were observed. Age-related changes in the phospholipids of human skeletal muscle have been investigated (Bruce and Svennerholm, 1971; Bruce, 1974a,b). The concentration of total phospholipid increased slightly after birth and remained unchanged thereafter at 11 μmoles/g. The ratio of phosphatidylethanolamine to phosphatidylcholine was not changed with age, but diphosphatidylglycerol increased from 3% in the fetal stage to 9% by the end of the first year of life. The proportion of linoleate in phosphati-

---

KEIZO WAKU • Laboratory of Medical Research Institute, Tokyo Medical and Dental University, No. 5. 45, 1-Chome Yushima, Bunkyo-ku, Tokyo, Japan.

TABLE 1

Phospholipid Compositions of Skeletal Muscle of Several Mammals[a]

PA	DPG	PE	EP	PC	CP	Sph	PS	PI	LPC	PG	PL[b]	
							(mole %)					
0.1	5.8	25.8	—	52.4	—	3.6	4.0	6.7	0.6	0.7	3.52	Mouse, legs Simon and Rouser (1969)
3.1 (PA + DPG)		21.9	—	67.4	—	2.2	—	2.5	—	—	0.98	Rat, hind legs Boichot and Costanzo (1968)
0.8	1.4	22.2	—	51.1	—	2.7	3.7	8.9	2.7	0.9	0.88	Rat, legs Simon and Rouser (1969)
—	4	24[c]	—	54	11	—	—	4	—	—	0.51	Rabbit Gray and Macfarlane (1961)
—	7.0	14.7	6.6	39.9	2.3	6.4	3.8	2.6	—	—	1.37	Sheep Dawson (1960)
—	5.8	5.4	13.3	35.2	13.6	3.0	—	2.7	—	—	0.38	Ox, longissimus dorsi Davenport (1964)
0.3	8.9	26.6	—	46.5	—	4.5	4.1	5.6	0.7	0.3	1.00	Ox, neck Simon and Rouser (1969)
—	8.1	23.2	—	54.6	—	3.8	2.4	7.6	—	—	0.93	Monkey, gastrocnemius Masoro *et al.* (1964)
1.2	6.6	26.4	—	48.0	—	4.0	3.3	8.8	Tr	1.0	1.31	Human, legs Simon and Rouser (1969)
1.0	5.0	27.0	—	48.0	—	4.0	4.0	9.0	2.0	—	—	Human Singh and Swartwout (1972)

[a] Abbreviations: PA, Phosphatidic acid; DPG, Diphosphatidylglycerol; PE, Phosphatidylethanolamine; EP, Ethanolamine plasmalogen; PC, Phosphatidylcholine; CP, Choline plasmalogen; Sph, Sphingomyelin; PS, Phosphatidylserine; PI, Phosphatidylinositol; LPC, Lysophosphatidylcholine; PG, Phosphatidylglycerol; PL, Phospholipid.

[b] Phospholipid 25% in wet tissue.

[c] As phosphatidylethanolamine.

dylcholine increased tenfold from the middle of gestation to the age of 1 year. After the age of 1 year, phospholipid and fatty acid composition did not change much.

B. Sarcoplasmic Reticulum

Since the demonstration by Kielley and Meyerhof (1950) that the major lipid component of the membranous fraction of rat skeletal muscle (sarcoplasmic reticulum) is phosphatidylcholine, many researchers have studied the lipid composition of mammalian sarcoplasmic reticulum. The lyophilized sarcoplasmic reticulum of rabbits was found to be 44.5% lipid, of which 88.8% was phospholipid (Waku *et al.*, 1971). These values agree with those of Meissner and Fleischer (1971) and Marai and Kuksis (1973a), although the detailed results on the proportions of individual neutral lipids differ significantly (Table 2). These discrepancies may have arisen from differences in preparation or in analytical methods.

The phospholipid composition data of Table 3 show several noteworthy features. Phosphatidylcholine forms 50–70% of total phospholipid. The plasmalogens accounted for a significant proportion of the total choline and ethanolamine phosphoglycerides of rabbit, over 70% of the ethanolamine phosphoglyceride being in this form (Waku *et al.*, 1971; Marai and Kuksis, 1973a). Furthermore, a significant amount of 1-alkyl-2-acyl-*sn*-glycero-3-phosphorylcholine or ethanolamine (8.4% of choline phosphoglycerides and 7.5% of ethanolamine phosphoglycerides) was present (Waku *et al.*, 1971). In rabbit sarcoplasmic reticulum, some glycosphingolipid (0.613 μg lipid-bound sialic acid/mg protein, 47.5% in G_{M3},* 18.0% in G_{M1},† and 27.0% in G_{D1a}‡ or G_{D1b}§ and 7.5% in G_T||) was found by Narasimhan and Murray (1974).

The fatty acid compositions of the phospholipids of sarcoplasmic reticulum are shown in Table 4. Those occupying positions 1 and 2 of

*G_{M3} = *N*-Acetylneuraminyl-galactosyl-glucosylceramide (NANA-Gal-Glc-Cer).

†G_{M1} = Galactosyl-*N*-acetylgalactosaminyl-[*N*-acetylneuraminyl] galactosyl-glucosylceramide (Gal-NAcGal-Gal-Glc-Cer).
```
                      |
                     NANA
```

‡G_{D1a} = *N*-Acetylneuraminyl-galactosyl-*N*-acetylgalactosaminyl-[*N*-acetylneuraminyl] galactosyl-glucosylceramide (NANA-Gal-NAcGal-Gal-Glc-Cer).
```
                                 |
                                NANA
```

§G_{D1b} = Galactosyl-*N*-acetylgalactosaminyl-[*N*-acetylneuraminyl-*N*-acetylneuraminyl] galactosyl-glucosylceramide (Gal-NAcGal-Gal-Glc-Cer).
```
                            |
                        NANA-NANA
```

||G_T = *N*-Acetylneuraminyl-galactosyl-*N*-acetylgalactosaminyl-[*N*-acetylneuraminyl-*N*-acetylneuraminyl] galactosyl-glucosylceramide (NANA-Gal-NAc-Gal-Gal-Glc-Cer).
```
                                             |
                                         NANA-NANA.
```

TABLE 2
Composition of the Neutral Lipids of Rat or Rabbit Sarcoplasmic Reticulum

Cholesterol esters	Triacyl-glycerols	Free fatty acids	Cholesterol	Diacyl-glycerols	
			% (w/w)		
34.0	18.9	28.4	5.9	12.7	Rat Sanslone *et al.* (1972)
16.2	26.4	11.6	45.5	Tr	Rabbit Waku *et al.* (1971)
Tr	7.6	Tr	92.4	—	Rabbit Marai and Kuksis (1973a)
—	59.7	13.8	26.4	—	Rabbit Meissner and Fleischer (1971)

TABLE 3
Phospholipid Compositions of Sarcoplasmic Reticulum of Several Mammals[a]

PA	DPG	PE	EP	PC	CP	Sph	PS	PI	
						(mole %)			
—	—	15.5	—	69	—	—	4.5	—	Rat Martonosi *et al.* (1968)
—	—	29.4	—	58.2	—	3.2	9.4	—	Rat Fiehn and Peter (1971)
—	—	18.4	—	69.5	—	—	—	9.3	Rat Sanslone *et al.* (1972)
—	2.3	15.2	3.2	60.4	1.7	5.6	2.8	8.8	Rat Marai and Kuksis (1973b)
—	—	13	—	75	—	7	—	—	Rabbit Drabikowski *et al.* (1966)
—	—	12.3	—	65.0	—	—	6.2	—	Rabbit Martonosi *et al.* (1968)
Tr	0.3	5.1	11.4	64.1	6.7	6.2	1.7	2.3	Rabbit Waku *et al.* (1971)
0.2	0.3	13.5	—	72.7	—	1.0	1.8	8.7	Rabbit Meissner and Fleisher (1971)
—	0.1	4.3	13.3	58.9	8.1	3.9	2.1	9.2	Rabbit Owens *et al.* (1972)
—	0.4	5.4	12.0	58.2	7.7	4.7	0.8	10.7	Rabbit Marai and Kuksis (1973b)
0.9	1.6	16.9	—	65.1	—	2.7	2.7	8.7	Human Takagi (1971)
—	2.2	12.4	11.6	43.3	9.7	10.5	1.5	8.6	Human Marai and Kuksis (1973b)

[a]Abbreviations are the same as in Table 1.

TABLE 4

Fatty Chain Compositions of Sarcoplasmic Reticulum Phospholipids of Several Mammals

16:0	18:0	18:1	18:2	20:3	20:4	20:5	22:5	22:6			
							(mole %)				
12.5	11.9	7.6	14.5	0.9	22.6	1.0	3.5	23.2	Rat, total phospholipids Fiehn and Peter (1971)		
26.5	8.4	16.0	17.9	1.1	16.0	2.1	7.3	1.8	Rabbit, total phospholipids Fiehn and Hasselbach (1970)		
22.8	14.1	10.8	28.9	1.5	12.5	—	0.9	1.5	Human, total phospholipids Takagi (1971)		
14.8	59.8	15.2	7.8	—	—	—	—	—	Rabbit, Waku *et al.* (1974)		
									1-position	diacyl	Ethanolamine phosphoglycerides
3.9	—	13.9	28.0	—	37.8	—	8.7	2.8	2-position		
54.3	30.0	8.4	—	—	—	—	—	—	1-position	1-alkenyl-2-acyl	
0.7	—	12.7	4.7	0.5	30.1	8.2	31.3	8.6	2-position		
67.3	21.7	5.3	—	—	—	—	—	—	1-position	1-alkyl-2-acyl	
7.2	2.7	7.6	3.2	—	10.5	3.7	50.7	6.2	2-position		
82.2	5.3	4.3	5.7	—	—	—	—	—	1-position	diacyl	Choline phosphoglycerides
5.0	—	25.2	56.5	—	7.6	1.0	2.8	—	2-position		
84.7	7.1	1.9	—	—	—	—	—	—	1-position	1-alkenyl-2-acyl	
1.7	—	33.7	15.0	—	31.4	2.2	9.6	4.5	2-position		
84.5	7.5	8.0	—	—	—	—	—	—	1-position	1-alkyl-2-acyl	
8.0	1.5	12.1	5. 9	2.6	22.3	7.9	24.4	4.1	2-position		

diacyl-, 1-alk-1′-enyl-2-acyl-, and 1-alkyl-2-acyl-*sn*-glycero-3-phosphorylethanolamine or choline have been analyzed (Waku *et al.,* 1974). In the 1-position of the choline phosphoglyceride fraction, over 80% of the fatty acid was palmitic acid, but in ethanolamine phosphoglyceride there was also a significant proportion of stearic acid. In plasmalogen and in 1-alkyl-2-acyl-*sn*-glycero-3-phosphorylcholine or ethanolamine, there was a higher proportion of polyunsaturated fatty acids in the 2-position. Marai and Kuksis (1973a,b) analyzed the molecular species of diacyl-*sn*-glycero-3-phosphorylcholine and the ethanolamine analog in chickens, rats, rabbits, and humans and found that in phosphatidylcholine 16:0/18:1 or 16:0/18:2 are predominant and in phosphatidylethanolamine the molecular species were rather nonspecific. The physiological significance of the existence of plasmalogens, alkyl glycerolipids, or of specific molecular species in the sarcoplasmic reticulum will not be clarified until the physicochemical nature of protein–lipid interaction is understood.

C. *Sarcolemma and Mitochondria*

Madeira and Antunes-Madeira (1973) isolated sarcolemma from rabbit skeletal muscle and found it to be composed of 63% protein, 17.5% lipid, 3–5% nucleic acid, and a trace amount of carbohydrate. The lipid fraction is composed of 50% neutral lipid and 50% phospholipid. Cholesterol and its ester were found in the neutral lipid fraction, while in the phospholipid 43% was phosphatidylcholine, 24% phosphatidylethanolamine, 23% sphingomyelin, and 10% lysophosphatidylcholine.

Fiehn and Peter (1971) demonstrated that the lipid compositions of sarcolemma, mitochondria, and sarcoplasmic reticulum from rat skeletal muscle were very different. The sarcolemma contained a high proportion of phosphatidylserine and sphingomyelin, and a large amount of cholesterol, cholesterol ester, free fatty acid (FFA), and triacylglycerols (phospholipid/neutral lipid: 48/52 wt/wt%). The mitochondria contained 83% phospholipid and 17% neutral lipid. There was a higher proportion of phosphatidylethanolamine and less phosphatidylcholine than in the other two fractions. The fatty acid content of the neutral lipid in sarcolemma was higher in saturated fatty acids and lower in unsaturated fatty acids than was that of mitochondria or sarcoplasmic reticulum.

II. *Lipid Metabolism*

A. *Biosynthesis*

Although not many experiments on lipid synthesis or metabolism have been carried out using skeletal muscle, the biosynthetic route of phosphatidic acid and phosphatidylcholine has been studied by several researchers. Neptune *et al.* (1963) reported synthesis of the glycerides and phosphatidic acid by incubation of the microsomal fraction of rat skeletal muscle together with CoA, ATP, M_g^{++}, *sn*-glycerol-3-P and [^{14}C]fatty acids in phosphate buffer at pH 8.0. The phosphatidic acid synthesized was identified by silicic acid impregnated paper chromatography. Waku *et al.* (1975) also examined this reaction with a more simplified system, that is, [^{14}C]acyl-CoA, *sn*-glycerol-3-P, M_g^{++}, sarcoplasmic reticulum from male rabbits, and Tris buffer (pH 7.4) at room temperature. The synthesized phosphatidic acid was isolated by thin-layer chromatography, and, for identification, it was hydrolyzed by phosphatidic acid phosphatase and the radioactive diacylglycerol determined by TLC. The extent of reaction was proportional to time and protein concentration. The specific activities, however, were much lower (approximately one-tenth) than those found when liver microsomes were used. The specific activities found when different acyl-CoAs were employed are shown in Table 5.

Phosphatidic acid phosphatase activity in the muscle tissue has not been found experimentally, but CDP-choline: 1, 2-diacyl-*sn*-glycerol phosphorylcholine transferase activity has been demonstrated. A particle fraction (mainly mitochondria and microsomes) obtained from the hind leg muscle of mice was incubated with [Me-^{14}C]choline (Pennington and Worsfold, 1965) and the radioactive phosphatidylcholine was isolated by TLC. Up to 0.14 nmoles of labeled phosphatidylcholine per mg of fraction N were formed in 2 hr at 30°C. The reaction rate was much lower than that obtained using liver fractions. They also demonstrated the same results using CDP-[^{14}C]choline or [^{14}C]choline phosphate as a substrate (Pennington and Worsfold, 1969). Its incorporation required M_g^{++}, M_n^{++}, or C_o^{++}, and externally added diacylglycerol from egg lecithin stimulated the incorporation of CDP-choline. From these results, they concluded that the

TABLE 5
Specific Activities of Acyl-CoA: sn-glycerol-3-P Acyltransferase in Rabbit Sarcoplasmic Reticulum

Acyl-CoA	Glycerol-3-P (nmoles/hr/mg protein)[a]
16:0	1.26
18:0	1.37
18:1	1.93
18:2	2.61
20:4	1.05

[a] This unit shows the nmoles of fatty acid incorporated into phosphatidic acid.

cytidine nucleotide pathway for the biosynthesis of phosphatidylcholine is functioning in skeletal muscle.

Waku (unpublished data) also incubated CDP-[^{14}C-Me]choline and M_g^{++} with sarcoplasmic reticulum (77 mg of protein) in 0.1 M Tris buffer (pH 7.4) for 2hr at 37°C, and 2.6% of the radioactivity was incorporated into the choline phosphoglyceride fraction. No significant difference was observed among the specific activities of diacyl-, 1-alk-1′-enyl-2-acyl-, and 1-alkyl-2-acyl-*sn*-glycero-3-phosphorylcholine. The biosynthesis of fatty acids, cholesterol, or phospholipids other than choline phosphoglycerides in skeletal muscle does not appear to have been studied.

B. Acyltransferase Activities

Webster (1965) determined the activity of acylation of lysophosphatidylcholine with [^{14}C]oleic acid in a rat skeletal muscle homogenate to be 2.0 μmoles/g/hr. The acylation activities of 1-acyl-, 1-alk-1′-enyl-, and 1-alkyl-*sn*-glycero-3-phosphorylcholine determined using rabbit sarcoplasmic reticulum (Waku and Lands, 1968; Waku and Nakazawa, 1970; Waku *et al.,* 1975) are shown in Table 6. From these results it is apparent that sarcoplasmic reticulum has about the same specific activity for the acylation of 1-acyl-*sn*-glycero-3-phosphorylcholine as rat liver microsomes (Lands, 1965), but about one-tenth the activity of 2-acyl-*sn*-glycero-3-phosphorylcholine acyltransferase. However, the rate of hydrolysis of fatty acid esters in sarcoplasmic reticulum is not known.

C. Turnover of Phospholipids

There have been several *in vivo* and *in vitro* experiments on the incorporation of radioactive precursors into phospholipids of skeletal mus-

TABLE 6

Specific Activities of Acyl-CoA:1-acyl-, 2-Acyl-, 1-Alk-1′-enyl-, and 1-Alkyl-GPC Acyltransferases in Rabbit Sarcoplasmic Reticulum[a]

Acyl-CoA	1-Acyl-GPC[b]	2-Acyl-GPC	1-Alk-1′-enyl-GPC	1-Alkyl-GPC
16:0	10	2.2	<1	0.05
18:0	9	0.5	<1	0.02
18:1	13	0.8	<1	0.00
18:2	27	0.6	9.4	0.56
18:3	29	0.05	1.9	0.42
20:4	27	—	4.7	—

[a](nmoles/min/mg protein)
[b]GPC = *sn*-glycero-3-phosphorylcholine

cle or sarcoplasmic reticulum. The incorporation of [^{32}P]P_i into femoral muscle phosphatides of rat after an intraperitoneal injection was studied by Hayasi *et al.* (1961). Incorporation of [^{32}P]P_i into phosphatidylcholine accounted for approximately 50% of the radioactivity incorporated into total phosphatides. The highest specific radioactivity observed was in inositol phosphatides. Olthoff and Kunze (1970) incubated [^{32}P]P_i with guinea pig muscle homogenate. The specific activities of phosphatidylcholine and lysophosphatidylcholine increased with incubation time while that of diphosphatidylglycerol decreased. The specific activities of phosphatidylethanolamine and phosphatidylserine were considerably lower.

Incorporation of the radioactive precursors of phospholipids into sarcoplasmic reticulum was studied by Martonosi *et al.* (1968) by injecting [^{32}P]P_i into rats and analyzing the microsomal phospholipids. The radioactivity appeared in phosphatidylcholine and ethanolamine. They also incubated [^{32}P]P_i or [^{32}P]ATP with sarcoplasmic reticulum *in vitro,* and found incorporation of radioactivity only into the phosphatidylethanolamine fraction and not into phosphatidylcholine. The rate of incorporation of [^{32}P]P_i, [^{14}C]leucine, and [^{14}C]acetate into the phospholipids and proteins of sarcoplasmic reticulum membranes of rat skeletal muscle was estimated after intraperitoneal injection of a single dose of the radioactive precursor (Martonosi and Halpin, 1972). The [^{32}P]P_i was incorporated into phospholipids and the [^{14}C]leucine into microsomal proteins; [^{14}C]acetate was incorporated into both fractions. The half-lives calculated from the decay of radioactivity of proteins and phospholipids were both in the range of 10–18 days. These results may be compared with 78 hr for the half-life of liver phospholipids determined by Omura *et al.* (1967) using acetate incorporation. The similarity in turnover rates of protein and phospholipids in sarcoplasmic reticulum suggests that their metabolism may be coordinated in some way.

Waku and Nakazawa (1973) estimated the rates of incorporation of [^{32}P]P_i, [^{3}H]glycerol, and [^{3}H]acetate, injected as single doses intravenously into male rabbits, into phospholipid fractions of sarcoplasmic reticulum. High specific activities were found in 1-alkyl compounds, especially in ethanolamine phosphoglycerides. The specific activities of 1-alk-1′-enyl compounds were lower than those of 1-acyl compounds. Therefore, it was suggested that 1-alkyl compounds have a high turnover rate in the phospholipids of sarcoplasmic reticulum.

In conclusion, phosphatidylcholine may be synthesized by the following pathway:

$$sn\text{-glycerol-3-P} \xrightarrow{\text{fatty acyl-CoA}} \text{phosphatidic acid} \xrightarrow{\text{phosphatidic acid phosphatase}}$$

$$\text{1,2-diacyl-}sn\text{-glycerol} \xrightarrow{\text{CDP-choline}} \text{1,2-diacyl-}sn\text{-glycero-3-phosphorylcholine}$$

(introductory chapter, Vol. 1, II.A.1, II.B.1, and II.C.1 reactions). The fatty acid composition of phosphatidylcholine is controlled by acyltransferase activity (introductory chapter, II.G.2 reaction). The synthesis and turnover of phospholipids in muscle or sarcoplasmic reticulum are significantly slower than those in liver tissues. The turnover rate of alkyl phospholipids in sarcoplasmic reticulum is much higher than other ethanolamine phosphoglycerides.

III. Lipid Metabolism in Muscular Dystrophy

A. Human Muscular Dystrophy

It has been shown that there is an accumulation of lipid in diseased muscle and this has been considered to represent a replacement of muscle fibers by fat cells (Adams *et al.,* 1962). Kunze and Olthoff (1970) analyzed the composition of lipids in progressive muscular dystrophy and found an increase of triacylglycerols (ten times), FFA, cholesterol, cholesterol esters lysophosphatidylcholine, and sphingomyelin, and a decrease of phosphatidylcholine and phosphatidylethanolamine.

Pennington *et al.* (1966) found no marked abnormality in the composition of fatty acids of dystrophic muscle, while Sudo *et al.* (1966) reported changes in the fatty acid composition of cholesterol esters and phospholipids in Duchenne progressive muscular dystrophy. They found a decrease of stearic and linoleic acids and an increase of oleic and palmitoleic acids. Takagi *et al.* (1968) also analyzed the fatty acid composition of phosphatidylcholine of tissues with the same kind of muscular dystrophy and confirmed the increase of oleic acid and decrease of linoleic acid.

Kunze *et al.* (1971) compared rates of incorporation of ^{14}C-labeled palmitic, stearic, and linoleic acids into lipids of progressively dystrophic and normal human muscle slices. A decreased rate of incorporation of linoleic acid into sphingomyelin, phosphatidylcholine, and triacylglycerols was found for the diseased muscle. Kunze and Olthoff (1972) estimated acyl-CoA:1-acyl-*sn*-glycerol-3-phosphorylcholine acyltransferase activity of a homogenate of the diseased human muscle. This enzyme could control the type of fatty acid in the 2-position of phosphatidylcholine, but no change in activity was observed in the dystrophic muscle, despite a marked change in fatty acid composition of lipids in human muscular dystrophy. The mechanism of this change is not yet resolved.

B. Muscular Dystrophy in Experimental Animals

In the dystrophic mouse, Shull and Alfin-Slater (1958) analyzed the lipid content of several organs and observed an increase of total lipids from

6.5–7.5% to 10.7–11.2% and of free cholesterol from 0.9 mg/g to 1.6 mg/g in skeletal muscle. Comparison of pooled muscle from normal and dystrophic mice revealed that the latter contained relatively less phosphatidylcholine and more sphingomyelin (Hughes and Frais, 1965). The difference was small but statistically significant. Jordan *et al.* (1964) demonstrated a marked increase in total lipids from 2.4% to 37.0% in the muscles of chickens with hereditary dystrophy and observed an increase in cholesterol, cholesterol esters, triacylglycerols, FFA, and phospholipid. However, the rate of incorporation of [1-^{14}C]palmitate or *sn*-[1-^{14}C]glycerol 3-P into triacylglycerols, and phospholipids, estimated in homogenates of skeletal muscle from normal and dystrophic mice, revealed no marked differences in the incorporation of either precursor (Jato-Rodriguez *et al.*, 1974). From these data, it was suggested that no quantitative differences exist in the *de novo* synthesis of triacylglycerols and phospholipids in normal and dystrophic muscle.

A decrease in palmitic acid oxidation by skeletal muscle mitochondria of dystrophic mice was observed by Lin *et al.* (1970) and a lower respiration and oxidative phosphorylation in muscle mitochondria from hamsters with muscular dystrophy was demonstrated by Jacobson *et al.* (1970).

Chio *et al.* (1972) also observed a marked increase in the concentration of both triacylglycerols (from 2.33 to 34.6 mg/g muscle) and phospholipids (9.9 to 15.1 μmole/g muscle) in dystrophic chickens. The rate of incorporation of [2-^{14}C]acetate into muscle lipids, both *in vivo* and *in vitro* was increased in dystrophy. From the turnover rate, they concluded that the increased lipid synthesis is accompanied by slower breakdown, so that an accumulation of lipid results. Susheela (1968) reported an increase in FFA (from 0.241 μmoles to 0.924 μmoles/100 mg tissue) in mice but no change in human dystrophic muscle. The content of ganglioside in the skeletal muscle of dystrophic mice was determined by Max and Brady (1971), and they observed a 20% decrease.

C. *Sarcoplasmic Reticulum*

Takagi (1973) estimated the $\overline{\text{Ca}}^{++}$-uptake activity of human Duchene dystrophy muscle (biopsy sample) and found a decrease from 4.3 to 1.2 μmol Ca^{++}/mg sarcoplasmic reticulum protein/15 min. Under these conditions, no significant difference was observed in ATPase activity, nor in concentration of total lipid, phospholipid, or cholesterol. However, phosphatidylcholine content was decreased from 65% to 50% and phosphatidylserine and sphingomyelin were slightly increased.

Owens and Hughes (1970) analyzed the composition of the muscle microsomes of the dystrophic mouse. They found less phosphatidylcholine and more phosphatidylethanolamine, sphingomyelin, and cholesterol.

There was an increase in stearic and linoleic acid, and a decrease in docosahexaenoic acid in the total phospholipids.

In chicken muscular dystrophy, Hsu and Kaldor (1971) found a significant decrease of Ca^{++}-uptake activity (80% loss). Decreased phosphatidylcholine and increased cholesterol, triacylglycerols, sphingomyelin, phosphatidylethanolamine, and phosphatidylserine were observed. Total phospholipid content was not changed.

As shown above, various changes in lipid compositions have been observed in hereditary muscular dystrophy, but it has not yet been proved whether altered lipid metabolism is one of the causes of this disease, or whether the disease induces such changes. Enzymatic activity related to lipid metabolism, or turnover rate of lipid fractions has been discussed only in a few papers, and it may be necessary to study more precisely lipid metabolism in muscular dystrophy.

IV. Role of Phospholipids in Sarcoplasmic Reticulum

Ebashi (1960, 1961) demonstrated that sarcoplasmic reticulum has a strong Ca^{++} binding activity. The biologically important relaxing action of sarcoplasmic reticulum on contractile elements may be explained on this basis.

Since Kielley and Meyerhof (1950) prepared and studied the ATPase from rat muscle, the role of lipids in this microsomal fraction has been investigated by many researchers. They considered that choline-containing phospholipids were essential for ATPase activity (phosphatidylcholine accounts for 60% of total phospholipid in this fraction) on the grounds that ATPase was inactivated by phospholipase-C *(Cl. welchii)* showing equivalence of phospholipid-splitting and inactivation up to 80%.

Martonosi (1963) reported the rather important finding that addition of phospholipid to the phospholipase-C-treated microsomes resulted in recovery of ATPase and Ca^{++}-uptake activity. Hydrolysis of rabbit sarcoplasmic reticulum by phospholipase-C *(Cl. perfringens)* was accompanied by phospholipid degradation and decreased Ca^{++}-uptake and ATPase activities. The addition of lysophosphatidylcholine or phosphatidylcholine resulted in almost 100% recovery, but phosphatidylserine or phosphatidylethanolamine had little effect.

A scheme for the general mechanism of ATP hydrolysis was shown as follows (Ebashi and Lipmann, 1962; Hasselbach and Makinose, 1962; Makinose, 1969; Yamamoto and Tonomura, 1968):

1. enzyme + ATP → enzyme P + ADP
2. enzyme P + H_2O → enzyme + P_i

According to Martonosi (Martonosi, 1969; Martonosi *et al.*, 1971), phospholipase-C inhibits the hydrolysis of ATP without significant inhibition of the formation of phosphoprotein intermediates. Therefore, a strict requirement for membrane phospholipids in the decomposition of phosphoprotein was suggested.

The inactivation of ATPase and Ca^{++}-uptake activity by treatment of microsomes with phospholipase-C and restoration of the activities by added phospholipids were confirmed by Yu *et al.* (1968), Masoro and Byung (1970), and Fiehn and Hasselback (1970). Yu *et al.* (1968) studied phospholipase-C-treated sarcoplasmic reticulum, which had lost 75% of its total phospholipids, by electron-microscopy. The untreated preparation primarily contained irregular, large membranous vesicles, while the phospholipase-C treated vesicles were smaller, more regularly ovoid, and had thinner walls. In addition, no effect of phospholipase-D on the microsomes was found by Yu *et al.* (1968), Martonosi *et al.* (1968), and Masoro and Byung (1970), and less change was observed in electron micrographs (Yu *et al.*, 1968). However, the results on phospholipase-A treatment of microsomes are conflicting. Yu *et al.* (1968) treated microsomes with phospholipase-A in the presence of 2% of albumin. The rate of Ca^{++}-uptake and the ATPase activity were not affected, and, in the absence of albumin, there was a slight but significant increase in ATPase activity. However, according to Martonosi *et al.* (1971), phospholipase-A treatment in the absence of albumin increased ATPase activity and inhibited Ca^{++}-transport activity. Washing the digested microsomes with serum albumin resulted in nearly complete inhibition of ATPase activity without significant inhibition of phosphoprotein formation. Addition of lysophosphatidylcholine to phospholipase-A-treated and albumin-washed microsomes caused partial restoration of ATPase activity. Fiehn and Hasselbach (1970) also treated the microsomes with phospholipase-A in the absence of albumin. Ca^{++} uptake was abolished but ATPase activity remained unimpaired. The fatty acids and lysophosphatidylcholine were removed from the protein by bovine serum albumin; ATPase and the phosphoryl transfer from ATP to the membrane protein were abolished in this case. Unsaturated fatty acids restored the phosphoryl transfer reaction and ATPase activity. Lysophosphatidylcholine restored the ATPase activity but the phosphoryl transfer reaction was not reactivated. Ca^{++} accumulation was not restored either by unsaturated fatty acids or lysophosphatidylcholine. The and Hasselbach (1973) examined the reconstituting effect of unsaturated fatty acids whose chain length varied between C_6 and C_{24} on delipidated sarcoplasmic reticulum (phospholipase-A treated). The most effective reactivating unsaturated fatty acids are those with a chain length of C_{16}–C_{18} whose double bonds are in the middle of the chain. Meissner and Fleischer (1972) and Carvalho (1972) arrived at the same conclusion from experiments involving phospholipase-

A treatment. Observing the experimental results as above, phospholipase-A treatment and subsequent bovine serum albumin washing may cause the inactivation of ATPase and Ca^{++}-uptake activity.

The results of Tume *et al.* (1973) are of interest in relation to the effect of fatty acid composition in the phospholipids of sarcoplasmic reticulum. Feeding rats on a diet containing safflower oil or hydrogenated coconut oil caused a marked increase in the content of linoleic acid in the sarcoplasmic reticulum phospholipids, but did not affect such properties as the rate of Ca^{++} uptake and ATPase activity.

Boland and Martonosi (1974) observed that developmental changes in the protein, phospholipid, cholesterol, and fatty acid composition of chicken skeletal microsomes were correlated with Ca^{++} transport activity in the embryos and during the early postnatal period. The induction of Ca^{++} transport activity was accompanied by a marked decrease in the palmitate and increase in the linoleate content of the phospholipid in sarcoplasmic reticulum. It may be of inerest to inquire whether this increase of linoleate is directly correlated with the function of sarcoplasmic reticulum.

As shown in Table 3, a significant level of plasmalogens is present in rabbit sarcoplasmic reticulum, but in rat sarcoplasmic reticulum only a small percentage of plasmalogens are found in phosphoglycerides. Therefore, the relationship between the content of plasmalogen and Ca^{++}-uptake or ATPase activity in sarcoplasmic reticulum is rather obscure.

V. Energy Factors in Muscular Exercise

A. Lipids as an Energy Source

There is little doubt that the oxidation of albumin-bound FFAs is an important source of energy in skeletal muscle. To get a quantitative measurement of the combustion of plasma FFA, the inflow of FFA from the plasma and liberation of carbon dioxide have been estimated by using radioactive fatty acids. Fritz *et al.* (1958) studied the metabolism of [1-^{14}C]palmitate by isolated skeletal muscle of rats. Increased concentration of fatty acids led to enhanced conversion of the substrate to [^{14}C]CO_2. Electrical stimulation of muscles under tension resulted in a 60% increase in oxygen consumption and 100% rise in fatty acid oxidation. Eaton and Steinberg (1961) also demonstrated that the rate of oxidation of [1-^{14}C]palmitate to [^{14}C]CO_2 by rat skeletal muscle increased markedly as a function of the concentration of FFA in the medium. Consequently, skeletal muscle blood flow and arterial FFA concentration are considered to be important in FFA uptake. Beatty and Bocek (1971) demonstrated, using skeletal muscle fiber groups obtained from monkey, that palmitate did not affect either glucose uptake, production of [^{14}C]lactate from [^{14}C]glucose, or the tissue glycogen level. When muscle fiber groups were incubated with [^{14}C]palmitate, about 60% of the uptake appeared in the lipid fraction and about 10% in CO_2. Sixteen percent of the total CO_2 was derived from palmitate.

Havel *et al.* (1963) studied the *in vivo* metabolism of [1-^{14}C]palmitate infused intravenously into healthy young men who walked on a treadmill at a rate of approximately 6 km/hr. Of the plasma FFA entering the tissue, 57–87% was oxidized. This accounted for 41–49% of energy metabolism during the exercise. These results suggest that FFA's are the major fuel delivered to working muscle from the blood. Miller *et al.* (1963), Spitzer and Gold (1964), and Issekutz *et al.* (1964) observed almost the same results. Furthermore, Havel *et al.* (1967) conducted experiments in exercising men who were receiving constant intravenous infusion of [1-^{14}C]palmitate during the exercise. They found by analyzing the arterial and femoral venous blood that approximately 60% of the oxidative metabolism was of fatty acids. The plasma FFA supplied approximately half and the fatty acids of triacylglycerols in plasma less than 10% of the total fatty acids burned in the leg during exercise. Therefore, it was suggested that local stores of the lipid would be used in the leg during exercise.

Estimation of local lipid stores during exercise, however, gave conflicting results. Masoro *et al.* (1966) questioned whether the intracellular lipids of monkey skeletal muscle are used as a fuel for contractile activity. The concentration of the various classes of phospholipids and triacylglycerols in the muscle was not affected by contractile activity. Morgan *et al.* (1969) also could not observe a significant reduction in lipid stores in the human muscle after 4–6 weeks exercise.

However, Fröberg *et al.* (1971) showed a marked decrease of triacylglycerols and glycogen during exercise. Fasting healthy volunteers were subjected to exercise on a bicycle to exhaustion (99 min). The concentration of triacylglycerols was decreased from 10.4 μmole/g of wet muscle tissue to 7.8 μmole and glycogen decreased from 10.4 mg to 3.4 mg/g wet tissue. It was estimated that 75% and 25% of the fatty acids oxidized were derived from muscle triacylglycerols, and plasma fatty acids, respectively. They also found using muscle from rats that the concentration of triacylglycerols was reduced, but that of phospholipid was unchanged (Fröberg *et al.*, 1972).

Masoro (1967) found that the amount of phospholipid in muscle remained constant throughout prolonged fasting while triacylglycerols tended to increase in muscle during the early stages of fasting, but were depleted when the period of fasting was prolonged. Thiele *et al.* (1960) estimated the phospholipid content in exercised muscle and observed that the plasmalogen concentration was reduced despite an increase in the concentration of total phosphatides.

These rather conflicting results may have arisen from experimental differences in the conditions of exercising. However, recent studies indicate that triacylglycerols may be consumed as a fuel in exercising muscle, although phospholipid is not (Fröberg *et al.*, 1971, 1972).

Boichot *et al.* (1970) studied the fate of FFA incorporated into lipids of skeletal muscle. Lipids of skeletal muscle (rat hind leg) were analyzed after intravenous perfusion of [11, 12-^{3}H]lauric acid and [1-^{14}C]palmitic acid; triacylglycerols and phosphatidylcholine contained most of the radioactivity. The shortening or lengthening of fatty acid chains was not observed during incorporation into triacylglycerols or phospholipids (Boichot *et al.*, 1973).

B. *Skeletal Muscle Mitochondria*

The mitochondria of rat skeletal muscle are able to oxidize octanoate as well as palmitate in *in vitro* experiments. The oxidation is strongly carnitine dependent and this fact suggests that the acyl-CoA forming enzyme is localized on the outer membrane (Groot and Hülsmann, 1973). Gollnick and King (1969) observed that mitochondria in the skeletal muscle of trained rats increased in numbers, size, and density of cristae compared to those in the sedentary control. Leshkevich and Chagovets (1971) also observed the swelling of mitochondria as an effect of training. However, the results of Terjung *et al.* (1972) are rather conflicting. They evaluated the effect of any disruption of rat skeletal muscle mitochondria associated with exercise to exhaustion. For example, the capacity of pyruvate oxidation of mitochondria in leg muscle cells was not affected by exhaustion and they concluded that significant functional impairment of skeletal muscle mitochondria does not occur during exercise.

ACKNOWLEDGMENTS

The author expresses his gratitude to Dr. S. Ebashi (Department of Pharmacology, Faculty of Medicine, University of Tokyo, Tokyo) and Dr. A. Takagi (Department of Neurology, Faculty of Medicine, University of Tokyo, Tokyo) for advice and helpful discussion.

References

Adams, R. D., Denny-Brown, D., and Pearson, C. M. 1962. Disease of Muscle, 2nd ed. Harper & Row, New York. 360 pp.

Beatty, C. H., and Bocek, R. M. 1971. Interrelation of carbohydrate and palmitate metabolism in skeletal muscle. *Am. J. Physiol.* **220**:1928–1934.

Boichot, J., and Costanzo, G. 1968. Les lipides du muscle squelettique au repos chez le rat. *Bull. Soc. Chim. Biol.* **50**:1711–1722.

Boichot, J., Paris, R., and Clement, J. 1970. Incorporation *in vivo* des acides laurique et palmitique dans les differents lipides du muscle squelettique chez le rat en periode de repos. *Bull. Soc. Chim. Biol.* **52**:1381–1392.

Boichot, J., Didier, J. P., Boucrot, P., and Klepping, J. 1973. Incorporation d'acide érucique[14^{14}C] dans les lipides du muscle squelettique isolé, au repos et excité. *Biochimie* **55**:1153–1157.

Boland, R., and Martonosi, A. 1974. Developmental changes in the composition and function of sarcoplasmic reticulum. *J. Biol. Chem.* **249**:612–623.

Bruce, Å. 1974a. Skeletal muscle lipids. II. Changes in phospholipid composition in man from fetal to middle age. *J. Lipid Res.* **15**:103–108.

Bruce, Å. 1974b. Skeletal muscle lipids. III. Changes in fatty acid composition of individual phosphoglycerides in man from fetal to middle age. *J. Lipid Res.* **15**:109–113.

Bruce, Å., and Svennerholm, L. 1971. Skeletal muscle lipids I. Changes in fatty acid composition of lecithin in man during growth. *Biochim. Biophys. Acta* **230**:393–400.

Carvalho, A. P. 1972. Binding and release of cations by sarcoplasmic reticulum before and after removal of lipid. *Eur. J. Biochem.* **27**:491–502.

Chio, L. F., Peterson, D. W., and Kratzer, F. H. 1972. Lipid composition and synthesis in the muscles of normal and dystrophic chickens. *Can. J. Biochem.* **50**:1267–1272.

Davenport, J. B. 1964. The phospholipids of pigeon and ox skeletal muscle. *Biochem. J.* **90**:116–122.

Dawson, M. C. 1960. A hydrolytic procedure for the identification and estimation of individual phospholipids in biological samples. *Biochem. J.* **75**:45–53.

Drabikowski, W., Dominas, H., and Dabrowska, M. 1966. Lipid patterns in microsomal fractions of rabbit skeletal muscle. *Acta Biochim. Pol.* **13**:11–24.

Eaton, P., and Steinberg, D. 1961. Effects of medium fatty acid concentration, epinephrine, and glucose on palmitate-1-C^{14} oxidation and incorporation into neutral lipids by skeletal muscle *in vitro*. *J. Lipid Res.* **2**:376–382.

Ebashi, S. 1960. Calcium binding and relaxation in the actomyosin system. *J. Biochem.* **48**:150–151.

Ebashi, S. 1961. Calcium binding activity of vesicular relaxing factor. *J. Biochem.* **50**:236–244.

Ebashi, S., and Lipmann, F. 1962. Adenosine triphosphate-linked contraction of calcium ions in a particulate fraction of rabbit muscle. *J. Cell Biol.* **14**:389–400.

Fiehn, W., and Hasselbach, W., 1970. The effect of phospholipase A on the calcium transport and the role of unsaturated fatty acids in ATPase activity of sarcoplasmic vesicles. *Eur. J. Biochem.* **13**:510–518.

Fiehn, W., and Peter, J. B. 1971. Lipids and fatty acids of sarcolemma, sarcoplasmic reticulum, and mitochondria from rat skeletal muscle. *J. Biol. Chem.* **246**:5617–5620.

Fritz, I. B., Davis, D. G., Holtrop, R. H., and Dundee, H. 1958. Fatty acid oxidation by skeletal muscle during rest and activity. *Am. J. Physiol.* **194**:379–386.

Fröberg, S. O. 1967. Determination of muscle lipids. *Biochim. Biophys. Acta* **144**:83–93.

Fröberg, S. O., Carlson, L. A., and Ekelund, L. G. 1971. Local lipid stores and exercise. *Adv. Exp. Med. Biol.* **11**:307–313.

Fröberg, S. O., Östman, I., and Sjöstrand, N. O. 1972. Effect of training on esterified fatty acids and carnitine in muscle and on lipolysis in adipose tissue *in vitro*. *Acta Physiol. Scand.* **86**:166–174.

Gollnick, P. D., and King, D. W. 1969. Effect of exercise and training on mitochondria of rat skeletal muscle. *Am. J. Physiol.* **216**:1502–1509.

Gray, G. M., and Macfarlane, M. G. 1961. Composition of phospholipids of rabbit, pigeon and trout muscle and various pig tissues. *Biochem. J.* **81**:480–488.

Groot, P. H. E., and Hülsmann, W. C. 1973. The activation and oxidation of octanoate and palmitate by rat skeletal muscle mitochondria. *Biochim. Biophys. Acta.* **316**:124–135.

Hasselbach, W., and Makinose, M. 1962. ATP and active transport. *Biochem. Biophys. Res. Commun.* **7**:132–136.

Havel, R. J., Naimark, A., and Borchgrevink, C. F. 1963. Turnover rate and oxidation of free fatty acids of blood plasma in man during exercise: Studies during continuous infusion of palmitate-1-C^{14}. *J. Clin. Invest.* **42**:1054–1063.

Havel, R. J., Pernow, B., and Jones, N. L. 1967. Uptake and release of free fatty acids and other metabolites in the legs of exercising men. *J. Appl. Physiol.* **23**:90–99.

Hayashi, K., Kanoh, T., and Yamazoe, S. 1961. Studies on the incorporation of P^{32}-labelled orthophosphate into the phosphatides of heart and femoral muscle of rat. *J. Biochem.* **50**:202–209.

Hsu, Q-S., and Kaldor, G. 1971. Studies on the lipid composition of the fragmented sarcoplasmic reticulum of normal and dystrophic chickens. *Proc. Soc. Exp. Biol. Med.* **138**:733–737.

Hughes, B. P., and Frais, F. F. 1965. Muscle phospholipids: Thin-layer chromatographic studies on normal and diseased tissue. *Biochem. J.* **96**:6p.

Issekutz, B. Jr., Miller, H. I., Paul, P., and Rodahl, K. 1964. Source of fat oxidation in exercising dogs. *Am. J. Physiol.* **207**:583–589.

Jacobson, B. E., Blanchaer, M. C., and Wrogemann, K., 1970. Defective respiration and oxidative phosphorylation in muscle mitochondria of hamster of hereditary muscular dystrophy. *Can. J. Biochem.* **48**:1037–1042.

Jato-Rodriguez, J. J., Hudson, A. J., and Strickland, K. P., 1974. Triglyceride metabolism in skeletal muscle from normal and dystrophic mice. *Biochim. Biophys. Acta* **348**:1–13.

Jordan, J. P., Kratzer, F. H., and Zargham, N. S., 1964. Lipid composition of the pectoral muscles of chickens with inherited muscular dystrophy. *Proc. Soc. Exp. Biol. Med.* **116**:243–246.

Kanoh, H., 1967. Fatty acid composition on the 1-(α-) and 2-(β-)position of phosphatidylcholine and phosphatidylethanolamine of various organs of rats. *Sapporo Med. J.* **32**:197–204.

Kielley, W. W., and Meyerhof, O., 1950. Studies on adenosinetriphosphatase of muscle III. The lipoprotein nature of the magnesium-activated adenosinetriphosphatase. *J. Biol. Chem.* **183**:391–401.

Kunze, D., and Olthoff, D. 1970. Der Lipidgehalt menschlicher Skelettmuskulatur bei primären und sekundären Myopathien. *Clin. Chim. Acta.* **29**:455–462.

Kunze, D., and Olthoff, D. 1972. Acylation of lysolecithin with long-chain fatty acids by normal and dystrophic human muscle. *Clin. Chim. Acta.* **36**:564–565.

Kunze, D., Olthoff, D., and Bönsch, G. 1971. Der Einbau von [^{14}C]Fettsäuren in Muskellipide bei progressiver Muskeldystrophie (Duchenne). *Clin. Chim. Acta* **33**:373–378.

Lands, W. E. M. 1965. Effects of double bond configuration on lecithin biosynthesis. *J. Am. Oil Chem. Soc.* **42**:465–467.

Leshkevich, L. G., and Chagovets, N. R. 1971. Content of phospholipid fractions in mitochondria of skeletal muscles during work and rest. *Ukr. Biokhim. Zh.* **36**:712–716.

Lin, C. H., Hudson, A. J., and Strickland, K. P., 1970. Palmitic acid-1-^{14}C oxidation by skeletal muscle mitochondria of dystrophic mice. *Can. J. Biochem.* **48**:566–572.

Madeira, V. M. C., and Antunes-Madeira, M. C. 1973. Chemical composition of sarcolemma isolated from rabbit skeletal muscle. *Biochim. Biophys. Acta* **298**:230–238.

Makinose, M. 1969. The phosphorylation of the membrane protein of the sarcoplasmic vesicles during active calcium transport. *Eur. J. Biochem.* **10**:74–82.

Marai, L., and Kuksis, A. 1973a. Molecular species of glycerolipids of adenosine triphosphatase and sarcotubular membranes of rabbit skeletal muscle. *Can. J. Biochem.* **51**:1248–1261.

Marai, L., and Kuksis, A. 1973b. Comparative study of molecular species of glycerolipids in sarcotubular membranes of skeletal muacle of rabbit, rat, chicken, and man. *Can. J. Biochem.* **51**:1365–1379.

Martonosi, A. 1963. The activating effect of phospholipids on the ATPase activity and C_a^{++} transport of fragmented sarcoplasmic reticulum. *Biochem. Biophys. Res. Commun.* **13**:273–278.

Martonosi, A. 1969. Sarcoplasmic reticulum. VII. Properties of a phosphoprotein intermediate implicated in calcium transport. *J. Biol. Chem.* **244**:613–620.

Martonosi, A., and Halpin, R. A. 1972. Sarcoplasmic reticulum. XVII. The turnover of proteins and phospholipids in sarcoplasmic reticulum membranes. *Arch. Biochem. Biophys.* **152**:440–450.

Martonosi, A., Donley, J., and Halpin, R. A. 1968. Sarcoplasmic reticulum. III. The role of phospholipids in the adenosine triphosphatase activity and C_a^{++} transport. *J. Biol. Chem.* **243**:61–70.

Martonosi, A., Donley, J. R., Pucell, A. G., and Halpin, R. A. 1971. Sarcoplasmic reticulum. XI. The mode of involvement of phospholipids in the hydrolysis of ATP by sarcoplasmic reticulum membranes. *Arch. Biochem. Biophys.* **144**:529–540.

Masoro, E. J., 1967. Skeletal muscle lipids. III. Analysis of the functioning of skeletal muscle lipids during fasting. *J. Biol. Chem.* **242**:1111–1114.

Masoro, E. J., and Byung, P. Y. 1970. The functioning of the lipids and lipoproteins of sarcotubular membranes in calcium transport. *Lipids.* **6**:357–368.

Masoro, E. J., Rowell, L. B., and McDonald, R. M. 1964. Skeletal muscle lipids. I. Analytical method and composition of monkey gastrocnemius and soleus muscles. *Biochim. Biophys. Acta.* **84**:493–506.

Masoro, E. J., Rowell, L. B., and McDonald, R. M., 1966. Intracellular muscle lipids as energy sources during muscular exercise and fasting. *Fed. Proc.* **25**:1421–1424.

Max, S. R., and Brady, R. O., 1971. Alteration of the ganglioside composition of skeletal muscle in murine muscular dystrophy. *Nature (London) Biol.* **233**:55–56.

Meissner, G., and Fleischer, S. 1971. Characterization of sarcoplasmic reticulum from skeletal muscle. *Biochim. Biophys. Acta* **241**:356–378.

Meissner, G., and Fleischer, S., 1972. The role of phospholipid in C_a^{++}-stimulated ATPase activity of sarcoplasmic reticulum. *Biochim. Biophys. Acta* **255**:19–33.

Miller, H., Issekutz, B. Jr., and Rodahl, K. 1963. Effect of exercise on the metabolism of fatty acids in the dog. *Am. J. Physiol.* **205**:167–172.

Morgan, T. E., Short, F. A., and Cobb, L. A. 1969. Effect of long-term exercise on skeletal muscle lipid composition. *Am. J. Physiol.* **216**:82–86.

Narasimhan, R., and Murray, R. K., 1974. Presence of glycosphingolipids in the sarcoplasmic reticulum fraction of rabbit sarcoplasmic muscle. *FEBS Lett.* **43**:23–26.

Neptune, Jr. E. M., Sudduth, H. C., Brigance, W. H., and Brown, J. D. 1963. Lipid glyceride synthesis by rat skeletal muscle. *Am. J. Physiol.* **204**:933–938.

Olthoff, D., and Kunze, D., 1970. Untersuchungen über *in vitro*- einbau von ^{32}P in die Phosphatide der Skelettmuskulatur. *Acta. Biol. Med. Germ.* **25**:805–812.

Omura, T., Siekevitz, P., and Palade, G. E. 1967. Turnover of constituents of the endoplasmic reticulum membranes of rat hepatocytes. *J. Biol. Chem.* **242**:2389–2396.

Owens, K., and Hughes, B. P. 1970. Lipids of dystrophic and normal mouse muscle: whole tissue and particulate fractions. *J. Lipid Res.* **11**:486–494.

Owens, K., Ruth, R. C., and Weglicki, W. B. 1972. Lipid composition of purified fragmented sarcoplasmic reticulum of the rabbit. *Biochim. Biophys. Acta* **288**:479–481.

Pennington, R. J., and Worsfold, M. 1965. Lecithin biosynthesis in skeletal muscle. *Biochem. J.* **96**:5p.

Pennington, R. J., and Worsfold, M. 1969. Biosynthesis of lecithin by skeletal muscle. *Biochim. Biophys. Acta* **176**:774–782.

Pennington, R. J., Park, D. C., and Freeman, C. P., 1966. The fatty acid composition of infiltrating fat in muscle from a case of muscular dystrophy. *Clin. Chim. Acta* **13**:399–400.

Sanslone, W. R., Bertrand, H. A., Yu, B. P., and Masoro, E. J. 1972. Lipid components of sarcotubular membranes. *J. Cell. Physiol.* **79**:97–102.

Shull, R. L., and Alfin-Slater, R. B., 1958. Tissue lipids of dystrophia muscularis, a mouse with inherited muscular dystrophy. *Proc. Soc. Exp. Biol. Med.* **97**:403–405.

Simon, G., and Rouser, G. 1969. Species variations in phospholipid class distribution of organs: II. Heart and skeletal muscle. *Lipids.* **4**:607–614.

Singh, E. J., and Swartwout, J. R. 1972. The phospholipid composition of human placenta, endometrium and amniotic fluid: a comparative study. *Lipids.* **7**:26–29.

Spitzer, J. J., and Gold, M. 1964. Free fatty acid metabolism by skeletal muscle. *Am. J. Physiol.* **206**:159–163.

Sudo, M., Kuroda, T., and Usui, T. 1966. On the fatty acid composition of lipids of progressive muscular dystrophy. *Jap. J. Clin. Pathol.* **14**:535–538.

Susheela, A. K. 1968. Free fatty acid concentrations in normal and diseased human muscle and in blood sera from patients with neuromuscular disease. *Clin. Chim. Acta* **22**:219–222.

Takagi, A., 1971. Lipid composition of sarcoplasmic reticulum of human skeletal muscle. *Biochim. Biophys. Acta* **248**:12–20.

Takagi, A., 1973. Sarcoplasmic reticulum in Duchenne muscular dystrophy. *Arch. Neurol.* **28**:380–384.

Takagi, A., Muto, Y., Takahasi, Y., and Nakao, K. 1968. Fatty acid composition of lecithin from muscles in human progressive muscular dystrophy. *Clin. Chim. Acta.* **20**:41–42.

Terjung, R. L., Baldwin, K. M., Mole, P. A., Klinkerfuss, G. H., and Holloszy, J. O. 1972. Effect of running to exhaustion on skeletal muscle mitochondria: a biochemical study. *Am. J. Physiol.* **223**:549–554.

The, R., and Hasselbach, W., 1973. Unsaturated fatty acids as reactivators of the calcium-dependent ATPase of delipidated sarcoplasmic membranes. *Eur. J. Biochem.* **39**:63–68.

Thiele, O. W., Schröder, H., and Berg, W. V. 1960. Über die Phosphatide der Muskulatur mit besonderer Berücksichtigung der Plasmalogene. *Z. Physiol. Chim.* **322**:147–157.

Tume, R. K., Newbold, R. P., and Horgan, D. J., 1973. Changes in the fatty acid composition of sarcoplasmic reticulum lipids and calcium uptake activity. *Arch. Biochem. Biophys.* **157**:485–490.

Waku, K., and Lands, W. E. M. 1968. Acyl coenzyme A:1-alkenylglycero-3-phosphorylcholine acyltransferase action in plasmalogen biosynthesis. *J. Biol. Chem.* **243**:2654–2659.

Waku, K., and Nakazawa, Y. 1970. Acyltransferase activity to 1-0-alkyl-glycero-3-phosphorylcholine in sarcoplasmic reticulum. *J. Biochem.* **68**:459–466.

Waku, K., and Nakazawa, Y. 1973. The rate of incorporation of inorganic orthophosphate, glycerol, and acetate into phospholipids of rabbit sarcoplasmic reticulum. *J. Biochem.* **73**:497–504.

Waku, K., Uda, Y., and Nakazawa, Y. 1971. Lipid composition in rabbit sarcoplasmic reticulum and occurrence of alkyl ether phospholipids. *J. Biochem* **69**:483–491.

Waku, K., Ito, H., Bito, T., and Nakazawa, Y. 1974. Fatty chains of acyl, alkenyl and alkyl phosphoglycerides of rabbit sarcoplasmic reticulum. *J. Biochem.* **75**:1307–1312.

Waku, K., Hayakawa, F., and Nakazawa, Y. 1975. On the phospholipid synthesis in rabbit sarcoplasmic reticulum. *Proceeding of the 95th annual meeting of Japan Pharmaceutical Society, Biochemistry. 5M10-3.*

Webster, G. R. 1965. The acylation of lysophosphatides with long-chain fatty acids by rat brain and tissues. *Biochim. Biophys. Acta* **98**:512–519.

Yamamoto, T., and Tonomura, Y. 1968. Reaction mechanism of the C_a^{++}-dependent ATPase of sarcoplasmic reticulum from skeletal muscle. *J. Biochem.* **62**:558–575.

Yu, B. P., DeMartinis, F. D., and Masoro, E. J., 1968. Relation of lipid structure of sarcotubular vesicles to C_a^{++}transport activity. *J. Lipid Res.* **9**:492–500.

Skin

M. R. GRIGOR

I. *Structure of Skin*

Before discussing lipid metabolism in skin, it is necessary first to consider the heterogeneous nature of this tissue. Skin is made up of two main layers, the epidermis and the underlying dermis. The epidermis itself consists of both living and dead tissue and contains several sublayers. Mitotic activity is confined to the innermost zone, the stratum germinativum, from which the cells migrate to the stratum granulosum, where they undergo a degenerative process becoming filled with keratohyalin granules and then dehydrated to form the horny layer of dead cells, the stratum corneum. A fourth sublayer, the stratum lucidum, is sometimes observed between the strata granulosum and corneum. The production and subsequent degeneration of the epidermal cells is a continuous process and dead cells are constantly desquamated from the free surface.

The dermis consists of dense connective tissue, through which blood capillaries and nerves run. The epidermis has no direct blood supply, and the cells of the epidermis obtain their nutrients by diffusion from the dermis. The dermis merges into the hypodermis, a layer of rather looser connective tissue, interspersed with adipose cells. The hypodermis is not generally regarded as a part of skin.

Several specialized structures occur in skin. These include the hair follicles from which the hairs grow, the associated sebaceous glands, and two types of sweat glands.

Although both the hair follicles and the sebaceous glands are embedded in the dermis, the follicular and sebaceous cells both originate from the stratum germinativum. Sebaceous cells arise at the periphery of the glands,

---

M. R. GRIGOR • Department of Biochemistry, University of Otago, Dunedin, New Zealand.

and these enlarge as they fill with lipid to eventually rupture forming the sebaceous secretion, sebum, in a typical holocrine manner. The sebum is secreted through the sebaceous duct into the follicular canal and onto the surface of the skin.

Among mammals humans are unusual in having two types of hair. Dense coarse hairs are found on the scalp, in the armpit, and genital regions, whereas finer vellous hairs are more sparsely distributed over much of the body. This has consequences in the distribution of sebaceous glands in humans. Highest densities are found on the scalp and forehead, while none are found on either the soles of the feet or palms of the hands. Variable densities are found in other locations. In addition a number of large sebaceous glands, which do not appear to be associated with hair follicles, are found in humans. These include the Meiobium glands of the eyelid, the glands of the buccal mucosa, the lips, the nipples, and the prepuce.

Animals with a dense pelage have a more even distribution of sebaceous glands. In rats, for example, only the soles of he feet are free from sebaceous glands. In contrast, other mammals such as pigs have only sparsely distributed hairs and sebaceous glands, while whales and porpoises have neither hair nor sebaceous glands.

A number of very large discrete sebaceous structures have been found in a variety of mammalian species. Some of these have proved useful to study because they can be excised free from other tissue. They include the preputial glands of rodents, the harderian glands of rabbits, the brachial glands of lemurs. The secretion of some of these specialized structures has been shown to contain pheromones (Mykytowycz and Goodrich, 1974). The biochemistry of some of these glands will be discussed in section VIII.

Two types of sweat glands are found in human skin, although these are absent from the skin of most other mammals. The eccrine are the most common and are found over much of the body. They open directly onto the surface of the skin. In contrast, the aprocrine glands are much fewer and confined to certain locations only. These open into the follicular canals. Both types of glands arise from the epidermis, although they are found deep in the dermis or sometimes in the hypodermis.

This chapter will be concerned mainly with the lipid-producing sebaceous glands. They are metabolically very active and produce a lipid with particularly unusual composition when compared to most tissue lipids (see section III). Lipid metabolism in the epidermis and dermis will also be considered insofar as these fractions can be distinguished from other skin fractions. The sweat glands do not appear to produce lipids (Montagna, 1962) and will not be discussed further.

Montagna (1962) gives a detailed description of the anatomy of skin. Comprehensive reviews of the general biochemistry and physiology of skin are provided by Rothman (1954) and Montagna and Lobitz (1964). The biology of the sebaceous gland has been described in considerable detail in the proceedings of two symposia (Montagna *et al.*, 1963, 1974).

II. *Sampling Skin Lipids*

Nicolaides and Kellum (1965) have thoroughly reviewed the problems of sampling skin lipids. Because of the ease in obtaining the skin surface lipids, they have been most extensively studied. The possibility of external contamination is very real but procedures are available to minimize this. Also there is the possibility of modification of the surface lipids by oxidation or microbial action. For human surface lipids, direct sampling with solvents, swabbing the skin with solvent-soaked cotton wool, and extraction of hair fats have been used. The vernix caseosa, the wax on the fetal surface, has been used as a source of sterile human sebum (Downing 1963, 1965; Nicolaides and Ray, 1965; Kärkkäinen *et al.*, 1965). For rats and mice a method involving their partial immersion in a solvent bath has proved very useful in sample collection (Nikkari and Haahti, 1964). Nicolaides and Kellum (1965) used hexane-impregnated cotton balls to swab the backs of shaved rats to obtain the surface lipid. The hair fats have been used to study the surface lipids of a number of larger animals.

While the surface lipid of animals with dense fur or hair is almost entirely of sebaceous origin, that from animals, including humans, which have a more sparse distribution of hair, may be either sebaceous or epidermal in origin. The relative contribution of these sources is determined by the anatomical site sampled and the solvent used for sampling. The surface lipid from areas free of sebaceous glands must be of epidermal origin. However, that from areas rich in sebaceous glands contains mainly sebaceous lipid (Green *et al.*, 1970). The use of relatively mild lipid solvents serves to minimize the extraction of epidermal lipids; hexane, diethyl ether, and acetone have been regularly used for this purpose. Stronger solvent systems, mainly choloroform : methanol mixtures, are required to extract the lipids from whole skin samples or fractions derived from them.

A number of physical, chemical, and enzymic methods exist for separating the epidermis from the dermis in skin samples (see Nicolaides, 1965 for review). A recent development has been to expose the skin to 2 M $CaCl_2$ (Kellum, 1966). While these methods yield satisfactory results in terms of the morphology of the fractions obtained, they are not necessarily suitable for subsequent metabolic studies, and there is a possibility of modifying the lipids present in the two fractions. Wilkinson and Walsh (1974) used four such methods to separate the epidermis from the dermis in newborn mouse skin samples and found that the ability of the fractions to incorporate acetate into lipid was severely impaired when compared to whole skin.

Kellum (1966, 1967) has used epidermal fractions to obtain preparations of isolated sebaceous glands by using microdisection. Subsequently Summerly and Woodbury (1971), Lutsky *et al.* (1974), and Im and Hoopes (1974) have all reported experiments involving lipid or carbohydrate metabolism performed with sebaceous glands isolated by a microdissection procedure.

III. Composition of Skin Lipids

A number of comprehensive reviews of the composition of skin surface lipids have been published (Nicolaides, 1965; Nicolaides *et al.,* 1968, 1970; Nikkari, 1974; Nicolaides, 1974). These will be summarized briefly in this chapter.

A. Nature of Skin Surface Lipids

Skin surface lipids are characterized by the presence of some unusual lipids. These include squalene, a monoester fraction made up of both steryl esters (often several different sterols) and wax esters, and a diester fraction containing two main molecular forms. Triacylglycerols are not common components and form the major proportions of the surface lipid of only a few species, one being humans. In addition the fatty acids obtained from many of these fractions frequently contain significant amounts of odd carbon number and branched chains (in some cases well over 50% are branched). An unusual pattern of unsaturation has also been reported.

Squalene has long been recognized as a component of human skin surface lipid and amounts to between 5% and 20% of the total lipids (Haahti, 1961; Nicolaides 1965; Greene *et al.* 1970; Downing and Strauss, 1974). Other hydrocarbons have been reported in human skin surface lipid, but the fatty chains present in this lipid class suggest that they might be contaminants (Downing and Strauss, 1974). Squalene is a minor component of the surface lipids of other species (Nikkari, 1965) and small amounts of other hydrocarbons have also been reported in the hair fats of other species, including sheep (Downing *et al.,* 1960).

The wax esters appear to be constituents of the surface lipid of practically every species, and in many cases they form the major components (Nicolaides *et al.,* 1968, Nikkari, 1974). The distribution of steryl esters is quite variable, being very minor components of adult human skin surface lipids (Downing *et al.,* 1969), major components in fetal and newborn human samples (Karkkainen *et al.*, 1965; Nicolaides *et al.*, 1972), and major components of a number of other species including the rat (Nikkari, 1965; Nikkari and Haahti, 1964), mouse (Wilkinson and Karasek, 1966), and sheep (Downing *et al.*, 1960).

The presence of 2-hydroxy acids (Horn *et al.*, 1954; Weitkamp, 1945) and alkane-1, 2-diols (Horn and Hougen, 1953) in hydrolysates of wool wax suggested the possible occurrence of diester compounds. Nikkari and Haahti (1964) showed that the same compounds occurred in hydrolysates of rat skin surface lipids. Nikkari (1965) obtained from the surface lipids of rat skin two fractions that were more polar than monoesters, but not as polar as triacylglycerols. The first of these on hydrolysis yielded fatty alcohols, unsubstituted fatty acids, and hydroxysubstituted fatty acids, while the second yielded as major products alkane diols and unsubstituted fatty acids. These were labeled diesters type I and type II and together comprised 20% of the rat skin surface lipid. The structures of the type I and II diester compounds are shown below.

Type I wax diester [(O-acyl)-2-hydroxy fatty acid alkyl ester]:

$$\begin{array}{l} CO—O—CH_2—R \\ | \\ CH—O—CO—R \\ | \\ CH_2 \\ | \\ R \end{array}$$

Type II wax diester [1,2-diacyl alkane diol]:

$$\begin{array}{l} CH_2—O—CO—R \\ | \\ CH—O—CO—R \\ | \\ CH_2 \\ | \\ R \end{array}$$

The presence of these diester compounds in the surface lipids of many species has subsequently been reported (Nikkari and Haahti, 1968; Nikkari, 1969; Nicolaides *et al.*, 1970; Nikkari, 1974). Of the nine mammalian species reported by Nicolaides *et al.* (1970), only the rat was shown to produce significant quantities of both type I and type II diesters. The cow, rabbit, and cat contain mainly type I, while the dog, mouse, guinea pig, baboon, and humans (vernix caseosa) contain mainly the type II diester. Nikkari (1974) reports that sheep sebum also contains both types of diesters, consistent with the earlier observations of both hydroxy acids and alkane diols in wool wax. In many species the diesters are the most abundant lipid class, amounting to 50% or more for the rabbit, cat, mouse, and dog. They appear to be absent or present in only trace amounts in adult human skin lipids, while they comprise around 10% of the lipids of the vernix caseosa (Nicolaides *et al.*, 1970).

Triacyglycerols and the products of their hydrolysis, partial glycerides, and free fatty acids, appear to be important constituents of the surface lipids of humans and also of the pig (Nicolaides *et al.*, 1968). The vernix caseosa contains triglycerides but no free fatty acids (Karkkainen *et al.*, 1965; Nicolaides *et al.*, 1970).

Free sterols are present in all samples reported to date. They amount to from 1.5% to 4% of the surface lipids of humans to around 13% in the mouse (Nikkari, 1974).

Small amounts of polar material are observed on thin-layer chromatograms of skin surface lipids of most species (Nicolaides *et al.*, 1968). This material is possibly phospholipid either extracted from the epidermis or present in membrane fragments resulting from the destruction of the sebaceous cell during sebum excretion.

1. *Nature of Sterols Present in Skin Surface Lipids*

Several different sterols have been identified in the surface lipids of a variety of species. These include as well as the C27 sterols, cholesterol, and lathosterol (cholesta-7-en-3β-ol), lanosterol, and 24,25-dihydrolanosterol, and several C28 and C29 compounds. These sterols appear to be imtermediates in a pathway from lanosterol to cholesterol, the significance of which will be discussed in section VII. Apart from the rat, the surface lipids of every species reported contains cholesterol as the major sterol. It accounts for 90% or more of the total sterols in humans, the guinea pig, and rabbit (Nikkari, 1974). The next most abundant sterol is lathosterol. The C28, C29, and C30 sterols are present in lesser amounts, almost entirely in the esterified form. In most species cholesterol is the only sterol found in the free sterol fraction.

2. *Nature of Fatty Chains in Skin Surface Lipid*

Weitkamp (1945) was first to demonstrate the presence of unusual fatty chains in surface lipids when he obtained four homologous series of fatty acid methyl esters from wool wax. These were normal unsubstituted fatty acids, 2-hydroxysubstituted normal acids, an iso-branched series with an even carbon number, and an anteiso-branched series with an odd carbon number. The term "iso-" referred to a single methyl branch on the penultimate carbon of the chain, and the term "anteiso-" referred to a single methyl branch on the antepenultimate carbon. These structures were determined by studying the physical properties of binary mixtures with known esters. The structures of the iso- and anteiso-branched chains are shown below:

$CH_3—CH(CH_3)—CH_2—(CH_2)_X—CO—OH$ (iso)

$CH_3\text{-}CH_2CH(CH_3)—(CH_2)_X—CO—OH$ (anteiso)

James and Wheatley (1956) and Wheatley and James (1957) applied gas chromatography to the study of the fatty acids of the surface lipid of a number of species and confirmed the presence of branched-chain acids and odd carbon number acids. Downing *et al.* (1960) investigated the nature of the fatty chains found in fatty acids, fatty alcohols, alkane diols, and 2-hydroxy fatty acids of wool wax. All classes contained iso- and anteiso-branches. The branched-chain compounds made up over 80% of the total unsubstituted fatty acids, monohydric alcohols, and diols.

Branched-chain fatty acids were first shown to be present in skin surface lipids of humans by James and Wheatley (1956). Subsequent analyses by Boughton and Wheatley (1959), Haahti (1961), and Nicolaides *et al.* (1972) confirmed their presence and showed that they contained the iso- and anteiso-branching.

The branched-chain compounds in the vernix caseosa have been estensively studied (Downing, 1963, 1965; Karkkainen *et al.*, 1965; Nicolaides, 1971; Nicolaides *et al.*, 1972). The fatty acids, fatty alcohols, 2-OH acids, and alkane diols all contain large amounts (50% or more) of iso- and anteiso-branched chains. In addition the presence of small amounts of methyl-branched fatty acids, other than iso- or anteiso-branching, have been reported. These contain branches at even carbons along the chain and, sometimes, more than one branch point (Nicolaides, 1971). The proportion of branched-chain acids is not uniform throughout all the lipid classes. The steryl esters and the diesters contain the highest proportions of branched acids, the wax esters somewhat less, while the fatty acids of the triacylglycerols contain only a low proportion of branched chains (Karkkainen *et al.*, 1965; Nicolaides *et al.*, 1972).

Nikkari and Haahti (1964) investigated the fatty chains found in the rat skin surface lipids. The branched-chain compounds made up 30% each of the fatty acids and monohydric alcohols and 17–20% of the hydroxy acids and alkane diols. The chain lengths of the acids ranged from C14 to C34, the alcohols from C14 to C26, while the hydroxy acids and diols had chains from C14 to C18 in length. Nikkari (1965) extended these observations and showed the presence of small amounts of branched chain monoenoic acids. The branched-chain acids were present in all lipid classes, although the type I diesters had lower proportions than the wax esters, steryl esters, or type II diesters. The alcohols found in the type I diesters, however, had higher proportions of branch chains than those of the monoester waxes. Grigor *et al.* (1970) detected small amounts of anteiso-branched acids with an even number of carbons in the surface lipids of rats fed a diet containing a supplement of isoleucine (see section V).

Branched-chain fatty acids are found in varying proportions in the surface lipids of all species thus far reported. In some like the dog they amount to 60% of the total fatty acids (Nikkari, 1969: Nicolaides *et al.*, 1970), while in the mouse and guinea pig the proportion is much lower (10–20%) (Nikkari, 1974).

Weitkamp *et al.* (1947) studied the mono-enoic acids found in the nonesterified fatty acids of human hair fat. On degradation these were shown to have a $\Delta 6$ structure or to be derived from a $\Delta 6$ structure by addition or removal of C2 units. This contrasts with the $\Delta 9$ structure commonly found in fatty acids from animal tissues. These results were confirmed by Nicolaides *et al.* (1964). Subsequently it has been shown that the monounsaturated alcohols also had or were based on the $\Delta 6$ structure (Nicolaides, 1967) as were the esterified fatty acids (Nicolaides *et al.*, 1972). The monounsaturated fatty acids of the vernix caseosa contain both $\Delta 6$ and $\Delta 9$ type acids. The sterol ester and diester fatty acids are mainly $\Delta 9$, while the wax ester fatty acids are mainly $\Delta 6$ (Nicolaides *et al.*, 1972). The monounsaturated fatty acids of the total surface lipids of the rat and mouse both have the basic $\Delta 9$ structures (Nicolaides and Ansari, 1968; Wilkinson, 1970a).

Polyunsaturated acids are only very minor components of surface lipids. Small amounts of an octadecadienoic acid are found in rat surface lipids (Nikkari, 1965) and human surface lipids (Nicolaides *et al.*, 1972).

B. Nature of Epidermal Lipids

Reinertson and Wheatley (1959) have studied the composition of human epidermal lipids. In contrast to the surface lipids, significant amounts of phospholipids occur (10% of the total lipids) in living epidermis, but the amount is somewhat lower in the stratum corneum. Significant amounts of free cholesterol were also observed, but again these levels were decreased in the stratum corneum.

Nicolaides (1965) examined the lipid composition of several different samples of human epidermis. The polar lipids amounted to between 13% and 22% of the total lipids for leg samples and 66% for a sole sample. The nonpolar lipids were made up of triacylglycerols, free sterols, and free fatty acids; significant amounts of steryl esters (around 10%) were also found (Nicolaides, 1974). The polar lipids of human epidermis have been investigated by Gerstein (1963), and these contained as major components phosphatidylcholine, phosphatidylethanolamine, phosphatidylserine, and sphingomyelin. Signigicant amounts (10%) of ethanolamine plasmalogens were also present.

Carruthers and Davis (1961) and Carruthers (1962) have examined the lipids from epidermal and dermal fractions of mouse skin. The dermis contained 90% triacylglycerols and 10% phospholipids and only very low amounts of cholesterol or cholesterol esters (less than 1%), while the epidermis contained 12–14% phospholipids, 80% triacylglycerols, and 2–3% each of cholesterol and cholesteryl esters. The major fatty acids of both the dermal and epidermal fractions were octadecenoic acid (55–60%), palmitic acid (20%), and octadecadienoic acid (10%). Different patterns were observed for the fatty acids of phospholipids that contained much lower amounts of the octadecenoic acid. Wilkinson and Karasek (1966) compared the mouse epidermal lipids with the mouse surface lipids. The epidermal lipids contained 15% phospholipids against none reported for the surface lipids, and some 40% triacylglycerols. This lipid class comprises only 8% of the surface lipids, where the major components are the diester waxes (66%). The epidermal lipids were shown to contain 25% diesters, but these may represent contaminants from the sebum. Carruthers and Davis (1961) did not report the presence of any epidermal diesters, although it is probable that had they been there they would have been included with the triacylglycerol fraction.

Christie *et al.* (1972) have described the lipids of pig skin. Only triacylglycerols and phospholipids were detected in any significant amounts. Of the fatty acids, around 50% was an octadecenoic acid.

C. *Nature of Sebaceous Gland Lipids*

Because of the difficulty in obtaining isolated sebaceous glands from skin, very little has been reported specifically on the lipids of these glands. Nicolaides (1965) examined the lipids from the sebaceous glands of the human scalp and observed a pattern similar to human surface lipids. It was not possible to conclude whether steryl esters were present. Kellum (1967) published thin-layer chromatograms of human sebaceous gland lipids showing an absence of both free fatty acids and steryl esters. Small amounts of free sterol were present, and wax esters, triacylglycerols, and squalene were all readily detectable. The absence of steryl esters suggests that the steryl esters found in human surface lipids are derived mainly from the epidermal lipids. This is confirmed by finding higher proportions of free and esterified sterols in the surface lipids with low concentrations of sebaceous glands (Greene *et al.*, 1970). It would appear that the human sebaceous gland differs from those of other species in two major aspects: It cannot readily convert squalene to sterols, and it produces large quantities of triacylglycerols.

A number of reports have been published on the lipid composition of specialized sebaceous structures from several species. These will be discussed in section VIII.

IV. Factors Affecting Rate of Sebum Production

A number of factors have been shown to affect the rate of sebum production. While considerable variation of sebum excretion rates is observed between individuals, differences related to age and sex have been observed. Levels are low before puberty, but increase four- to five-fold at puberty. With increasing age there is a decrease to a level approaching the prepuberty value for women, but not men (Pochi and Strauss, 1974). These changes all seem to be related to endocrine changes within the body.

The effects of hormones on secretion by sebaceous glands have been reviewed by Pochi and Strauss (1974), Ebling (1974), and Shuster and Thody (1974). General patterns appear to exist in all species. That is, there is a stimulation of sebum secretion rate following androgen treatment, while estrogens decrease the secretion rate. Progesterone does not appear to have any effect, although certain synthetic steroids with progestational activity can cause an increase in sebum production. Both gonadal androgens (testosterone) and adrenal androgens (dehydroepiandrosterone or Δ4-androstenedione) appear to affect the increase in sebum secretion. Thyroidectomy decreases sebum production, which may be restored through thyroxine treatment.

A number of anterior pituitary hormones including adrenocorticotrophin (ACTH), thyroid-stimulating hormone (TSH), and gonadotrophin also affect sebum production, but their action is mediated through their respective target organs. On the other hand, melanocyte-stimulating hormone (MSH), produced by the intermediate lobe of the pituitary, may have a direct effect on the sebaceous gland (Shuster and Thody, 1974).

Ebling (1974) distinguished between hormones that alter the rate of cell division in the sebaceous gland and those that also alter the rate of lipid production. The androgens appear to act at both levels, while the estrogens act only at the latter, decreasing lipid synthesis but not the mitotic rate.

The rate of sebum production is also affected by the nutritional state of the animal. Fasting results in a decrease in sebum production in human subjects (Downing and Strauss, 1974). Nikkari (1965) has shown that the rate of sebum production in rats is higher when a fat-free diet is fed than when fat is included in the diet. The rate is insensitive to the nature of the fat included. Feeding rats a diet low in protein (5% compared with 20%) decreases the yield of the surface lipids (Grigor *et al.*, 1970).

V. *Factors Affecting the Composition of Skin Surface Lipids*

No change in the composition of human surface lipids can be detected relative to the site sampled until areas of very low or zero sebaceous densities are involved (Greene *et al.*, 1970). Here the surface lipid tends to reflect the epidermal pattern rather than the sebaceous distribution. Changes in composition, however, do occur with age up to puberty. Ramasastry *et al.* (1970) have shown a decrease in the proportion of total cholesterol along with an increase in the proportion of wax esters near puberty. An interesting change occurs in the first 2 years of life. In newborn samples, the cholesterol and wax ester contents are close to the adult levels and change to give the juvenile pattern only slowly over this 2-year period.

Nikkari (1965) could detect only minor differences in the composition of the surface lipids of intact rats treated with various hormones even though significant differences in sebum production rates were observed. Hypothysectomized rats, however, presented a different pattern showing an increase in the type I diester. Treatment with testosterone restored the normal pattern with a shift from the type I diesters to the wax monoesters (Nikkari and Valavaara, 1970). These authors were also able to show a correlation between the C16/C18 fatty acid ratio and the rate of sebum production.

Downing *et al.* (1972) showed a consistent change in the composition of surface lipids in humans during fasting. As the sebum production rate decreased, the rate of squalene production appeared unaltered. As a consequence the proportion of squalene increased at the expense of the triacylglycerols and free fatty acids.

A number of reports have described experiments investigating the effect of dietary lipids on the composition of the skin surface lipids. Boughton *et al.* (1959) and Nikkari (1965) could not detect any significant change in the fatty acid composition of mice or rats fed lipid supplements. Grigor (1969a) observed a modest increase in the proportion of linoleic acid (from less than 3% to over 10%) in rats fed soya bean oil.

The lipid composition of the skin surface in rats is markedly dependent on the level of dietary protein (Grigor, 1969b; Grigor *et al.*, 1970). As the protein content of the diet is increased from 5% to 40% there is a parallel increase in the proportion of the branched-chain fatty acids. These tended to decrease when diets of higher protein content were fed. It was also observed that when supplements of each of the branched amino acids (valine, leucine, and isoleucine) were fed there were specific increases in the proportions of particular classes of branched fatty acids, viz., valine—even carbon number iso-branched acids, leucine—odd carbon number iso-branched acids, and isoleucine—odd carbon number anteiso-branched acids. Feeding leucine also increased the proportions of the odd carbon number normal acids. Both the L- and D-isomers of valine were effective. These observations will be discussed further in section VI.

Skin surface lipids, because of their exposure to the atmosphere, are prone to oxidation and possible modification due to microbial action. Downing *et al.* (1960) found that the sterols on the outer two-thirds of the fleece from a sheep had been extensively oxidized when compared with those on the inner one-third.

The action of skin micro-organisms appears to be of major importance only in species where triacylglycerols are present, which includes humans. While freshly secreted sebum and the vernix caseosa contain triacylglycerols and only very low, if any, amounts of free fatty acids, considerable quantities of free fatty acids are often found in the surface lipids of adult human skin (Nicolaides, 1963; Nicolaides *et al.,* 1968; Downing and Strauss, 1974; Nikkari, 1974). Furthermore, the combined content of triacylglycerols and free fatty acids for a number of samples appears to be very similar. There is now considerable evidence that the free fatty acids arise from lipolysis of the triacylglycerols by surface bacteria (Shalita, 1974). Marples *et al.* (1970, 1971, 1972) investigated the relation between the microflora of the skin surface and the production of free fatty acids. Using combinations of antibiotics, it was possible to show that under physiological conditions *Corynebacterium acnes* alone was responsible for the hydrolysis of the triacylglycerols. Lipases from this organism have been isolated and partially characterized (Freinkel and Shen, 1969; Hassing, 1971; Pablo *et al.,* 1974).

VI. Lipogenesis in Skin

Because skin is not amenable to ready fractionation or to the preparation of cell-free suspensions, only few enzyme studies have been reported. Rather, more work has been done using isolated skin samples—either whole skin or skin fractions. In addition, studies using perfused skin and some *in vivo* studies have been reported.

In much of the work reported it is not always possible to distinguish between the activities of the various skin fractions. While it would be attractive to consider the sebaceous gland as the major site of lipid synthesis, this is not always so, since the epidermis can actively catalyze lipid synthesis. The dermis is comparatively inactive with respect to lipogenesis.

Several questions should be considered at this stage. These relate to the carbon sources involved, the nature of the fatty acid synthase activities, the source of the reducing equivalents, the pathways by which the unusual compounds found in skin lipids arise, and the control of lipogenesis. Knowledge of the nature of the fatty acid synthase has important consequences, since it allows one to consider what intermediates are required and their location within the cell. The question of control in a holocrine gland is an intriguing one. In this respect there are marked similarities between the sebaceous cells and those of the epidermis. Both have limited life-spans, after which they go through a degenerative process, the one becoming filled with lipid and the other with protein.

Several reports have appeared in the literature describing the incorporation of radioactive acetate into lipids by skin samples. While it is unlikely that acetate provides the normal carbon source for skin cells, these studies have been important in showing that all the various components found in skin lipids can be synthesized locally (Nicolaides *et al.*, 1955; Patterson and Griesemer, 1959; Hsia *et al.*, 1966; Wilkinson, 1970b; 1973; Wheatley *et al.*, 1971; Summerly and Woodbury, 1971).

Glucose is also readily incorporated into lipids and the observations of Wheatley *et al.* (1971) and Wheatley (1974) suggest that it might be a major carbon source. Besides glucose, intermediates of glycoloysis, notably pyruvate and lactate, were readily incorporated into lipids, and ability to incorporate the label from a number of precursors including acetate, pyruvate, lactate, some amino acids, and tritiated water were all enhanced in the presence of glucose. Certain amino acids, including alanine and particularly the branched amino acids, were also readily incorporated into lipids by both perfused dog skin and isolated guinea pig skin preparations (Wheatley *et al.* 1967, 1971). It is difficult to estimate the relative quantitative importance of the amino acids as a carbon source for lipogenesis. Wheatley (1974) has concluded that they play a comparatively minor and rather specialized role compared to glucose.

Little is known as to whether the circulating lipids contribute significantly to skin lipids. In an experiment where rats were fed [^{3}H]stearic acid and [^{14}C]linoleic acid, the skin surface lipids were labeled with both isotopes. While very little of the ^{3}H was present as the parent stearic acid, some 30% of the ^{14}C found in the unsubstituted fatty acids was as the octadecanoic acid; some 0.2% of the ingested ^{14}C was recovered in the skin surface lipids (Grigor, unpublished observations). These data suggest that while there is limited evidence of direct incorporation of fatty acids from the circulation into the skin surface lipids, the circulating fatty acids can certainly act as a carbon source for lipogensis in skin. Adachi *et al.* (1967b) measured the activity of β-hydroxyacyl-CoA dehydrogenase in skin and showed that the sebaceous gland in particular has significant activities of this enzyme. This is consistent with the ability of skin to degrade fatty acids.

Skin has the ability to convert glucose very efficiently to lactate (Freinkel, 1960; Mier, 1969), and high activities of lactate dehydrogenase are found in skin fractions (Im and Hoopes, 1974). These observations and those of Wheatley *et al.*, (1971) indicating that skin can use lactate for lipogenesis raise the possibility that lactate might be an important carbon source for lipid synthesis. To have a tissue both actively producing lactate and then using it subsequently might appear contradictory. However, lactate production is confined mainly to the epidermis (Halprin and Ohkawara, 1966), where oxygen levels are low. Lipogenesis occurs primarily in the sebaceous glands, which, because of their location in the dermis, are likely to be able to catalyze the further oxidation of lactate. Mier (1969) has suggested that the lactate produced in the epidermis enters the circulation and is taken up by the liver. It is highly probable that the bulk of the lactate does follow this path with only small amounts, if any, used by the sebaceous glands. An additional possibility exists where, under conditions of low blood glucose, lactate might be taken up by the sebaceous glands from the circulation. The quantitative importance of lactate as a carbon source for lipid synthesis in skin and the potential metabolic interactions between the epidermis and the sebaceous gland remain to be clarified.

The observation that lipogenesis in guinea pig skin was more active in buffers containing bicarbonate than those containing phosphate (Wheatley *et al.,* 1971) provided indirect evidence that fatty acid synthesis in skin is catalyzed by a fatty acid synthase complex similar to that found in other tissues. Recently Wilkinson (1974) has shown that a particle free fraction from newborn mouse skin can catalyze the formation of saturated fatty acids from malonyl-CoA consistent with the operation of such a synthase. This study also showed that the particulate fractions could desaturate and elongate saturated fatty acids. Previously Wilkinson (1970b) has shown that isolated epidermal cells could elongate existing fatty acids by the addition of C2 units from acetate. It is probable that the sebaceous cells also have more than one synthase. If one examines the distribution of saturated fatty acids in the surface lipids of the rat (Fig. 1), the normal saturated fatty acids have a bimodal distribution. The peaks at C16 and C17 would correspond to the end products of *de novo* synthesis, while those at C24 and C25 would appear to result from the elongation of the former acids by some four C2 units.

The operation of the *de novo* synthase has certain implications. It requires extramitochondrial acetyl-CoA as substrate and the production of NADPH to supply the reducing equivalents. Acetyl-CoA is formed in the

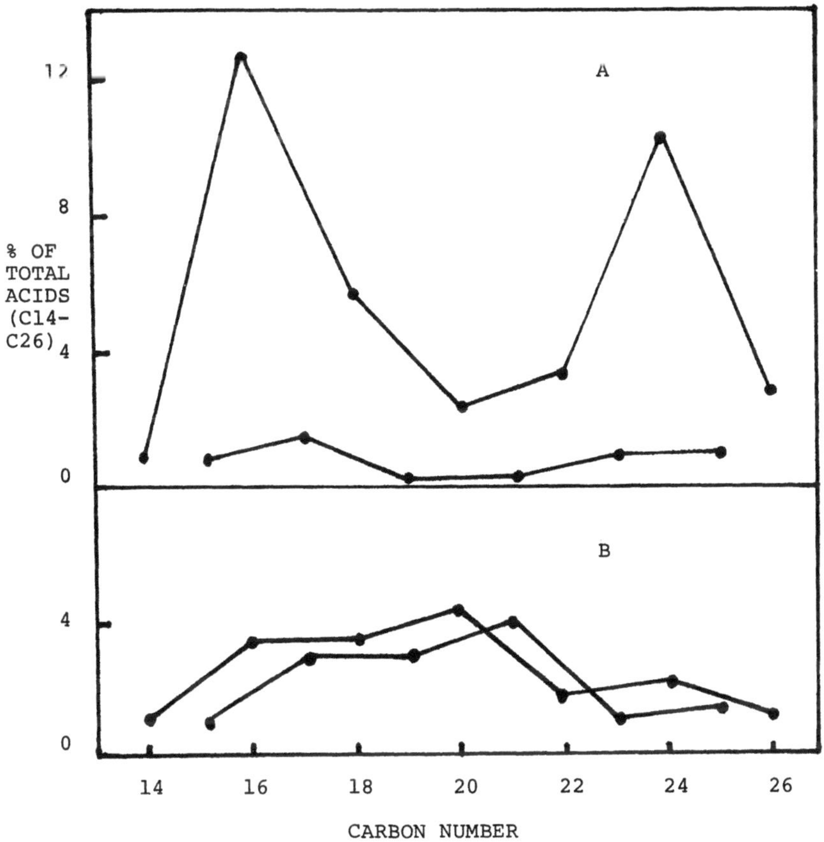

FIGURE 1. Fatty acid composition of pooled rat skin surface lipids. (A) normal fatty acids, (B) branched iso- and anteiso-acids (Grigor, unpublished data).

mitochondria either from pyruvate by the action of pyruvate dehydrogenase or by β-oxidation of fatty acids. The availability of extramitochondrial acetyl-CoA involves a translocation process. Wheatley *et al.* (1973) have shown that the major route in skin involves the translocation of citrate followed by the action of the citrate cleavage enzyme to produce acetyl-CoA and extra mitochondrial oxaloacetate.

Three possible sources exist for the production of NADPH. These are the operation of the hexose monophosphate pathway (HMP), the NADP-linked isocitrate dehydrogenase, and the malate cycle. The HMP pathway has been shown to be active in human skin (Freinkel, 1960; Yardley and Godfrey, 1963), rat skin (Ziboh *et al.*, 1970), and guinea pig ear skin (Wheatley *et al.*, 1973). Im and Hoopes (1974) have assayed key enzymes of the HMP pathway in fractions from human skin and showed that the sebaceous gland in particular had significant activities of glucose 6-phosphate dehydrogenase. The epidermis also had significant but somewhat lower activities. The enzymes of the malate cycle, and the NADP-linked isocitrate dehydrogenase were also present in human skin (Im and Hoopes, 1974; Cruickshank *et al.*, 1958). Wheatley *et al.* (1973) have shown that both the malate cycle and isocitrate dehydrogenase were active in ear skin of guinea pigs. There is little information available concerning the relative contributions of these three sources for NADPH in skin. As a result of inhibition studies, Ziboh *et al.* (1970) have suggested that the HMP pathway provides only about half the NADPH required for lipid synthesis in rat skin. This is somewhat less than the figures suggested for other tissues.

The utilization of alanine, aspartate, and glutamate as carbon sources for lipogenesis (Wheatley *et al.*, 1971; 1973) is consistent with the earlier observations of Adachi *et al.* (1967a,c) showing the presence of aspartate and alanine transaminases and glutamate dehydrogenase in human skin. Each of these enzymes tended to be localized in the sebaceous gland rather than the other skin fractions. The incorporation of branched amino acids into lipid is of particular significance. In 1945, Velick and English first suggested that the branches of the iso- and anteiso-acids might originate from the branched amino acids, valine, leucine, and isoleucine. Wheatley *et al.* (1961, 1967) showed that these amino acids were readily incorporated into lipids by isolated perfused dog skin. Isoleucine was preferentially incorporated into a C15 branched acid (Wheatley *et al.*, 1961). Grigor (1969b) and Grigor *et al.* (1970) showed that the proportions of each of the different homologous series of branched fatty acids in surface lipids in rat skin could be altered by feeding dietary supplements of one the branched amino acids. Furthermore, intraperitoneal injections of the radioactive amino acids resulted in the specific labeling of the corresponding series of fatty acids. These results were consistent with the operation of the pathways shown in Fig. 2. A single transaminase is active for the L-isomer of all three amino acids, and a single decarboxylase appears to be active for the three ketoacids. Hodgins and Wheatley (1973) confirmed the pathway for valine in mouse ear. They also degraded the iso-C16 acid and found that 76% of the activity in this acid was in the branched portion. Labeled isobutyrate was also incorporated into long-chain iso-branched acids (Grigor *et al.*, 1970).

To date nothing is known about the enzyme systems involved in the production of the $\Delta 6$ bond, nor is there any direct evidence concerning the production of fatty alcohols, hydroxy fatty acids, or alkane diols in normal sebaceous glands. A number of mammalian tissues including the mouse preputial gland tumor can reduce activated fatty acids to fatty alcohols in the presence of $NADPH_2$ (Snyder and Malone, 1970). The production of 2-

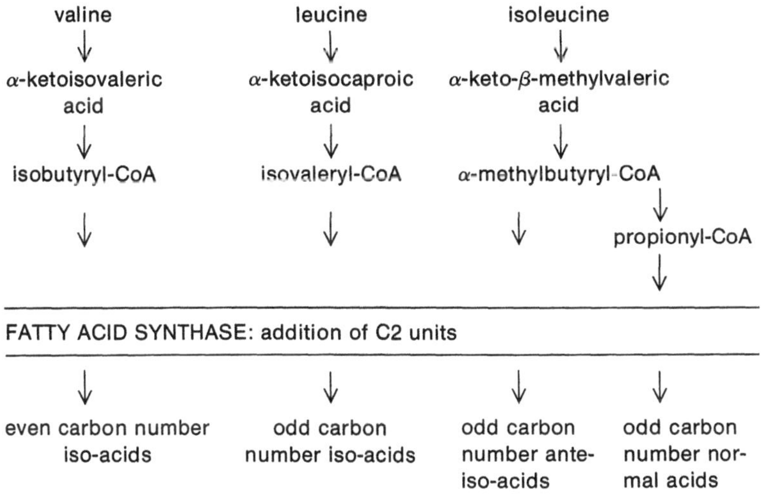

FIGURE 2. Pathways for the formation of branched and odd carbon number fatty acids in skin.

hydroxy fatty acids in brain has been studied (Bowen and Radin, 1968), but nothing is known about the snythesis of alkane-1,2-diols. It is possible that the diols are formed by reduction of the 2-hydroxy fatty acids, since there is a similarity in the chain lengths of these two classes (Nikkari, 1965). Wax esters are formed in the preputial gland tumor by an acyltransferase using an acyl-CoA as acyl donor (Snyder and Malone, 1971), but again there is no information about the enzyme systems involved in the formation of diesters.

Control of lipogenesis is achieved in most tissues by (a) the activities, usually proportional to the amounts, of the fatty acid synthase and, possibly more important, the acetyl-CoA carboxylase, (b) the availability of substrates, and (c) the availability of reducing equivalents. In skin, direct assays of the acetyl-CoA carboxylase or the fatty acid synthase have not been performed. The ability of the skin to use glucose, fatty acids, and possibly lactate as carbon sources, and the relative unimportance of the hexose monophosphate pathway for the production of NADPH have already been mentioned. These would tend to make lipogenesis in skin less susceptible to dietary control than in most tissues involved in lipogenesis.

There is conflicting evidence about the role of insulin in glucose uptake by skin. Wheatley *et al.* (1971) showed that guinea pig skin was insensitive to insulin, while Kahlenberg and Kalant (1966) have reported moderate increases in glucose uptake by human skin, particularly diabetic skin, in the presence of insulin. Ziboh *et al.* (1971) reported that normal rat skin responded readily to insulin.

It was shown earlier that androgens increase both the mitotic rate and the production of lipids within the sebaceous gland (Ebling, 1974). The metabolic events following testosterone administration in the normal sebaceous gland are not known, but Sansone *et al.* (1971) report an increase in the RNA/DNA ratios along with an increase in the rate of synthesis of all lipid classes in mouse preputial glands.

VII. Sterol Metabolism in Skin

The presence of unusual sterols in the surface lipid of a number of species has already been mentioned. Kandutsch and Russell (1960a,b), working with mouse preputial gland tumors, demonstrated that these sterols were intermediates in a pathway for the conversion of lanosterol to cholesterol. This pathway is outlined in Fig. 3, where it is contrasted with that believed to occur in liver (Holloway, 1970). Of particular note in the Kandutsch-Russell pathway are the initial saturation of the 24,25 double

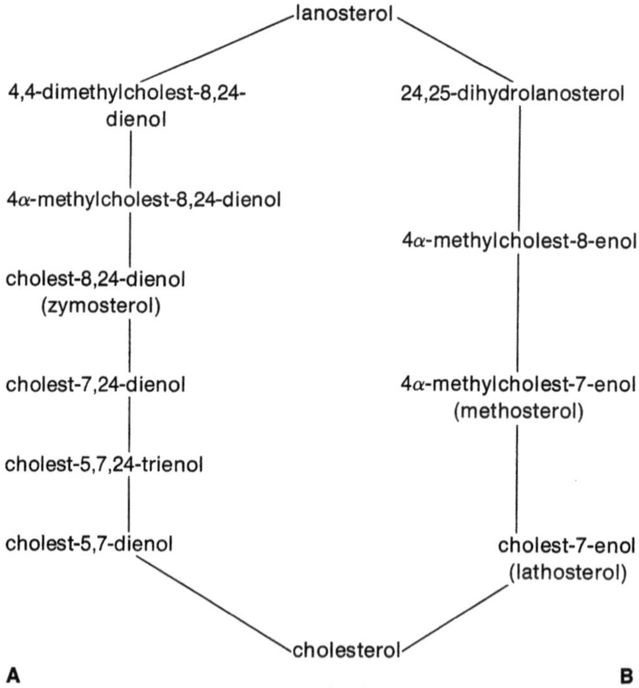

FIGURE 3. Pathways for the conversion of lanosterol to cholesterol in liver (A) and the Kandutsch–Russell pathway in skin (B).

bond, and the retention of the 4α-methyl to allow for the formation and accumulation of 4α-methylcholest-7-enol (methosterol). An additional point not apparent in the figure is that the intermediates of the Kandutsch–Russell pathway tend nearly always to be found in the esterified form. Thus, while cholesterol usually makes up the entire free sterol fraction of skin lipids, the steryl ester fraction contains several different sterols (Frantz *et al.*, 1957; Wilson, 1963; Nikkari, 1965). While Wilson (1963) has suggested that the ester linkage might be essential for the operation of the Kandutsch–Russell pathway, the observations of Brady and Gaylor (1971) suggest rather that these intermediates are trapped as esters. The rate of esterification for the methylsterols in skin is several times that of the demethylation, and the two reactions compete for the available sterol. In liver, at least, the esterified sterols were not substrates for the demethylation enzymes.

Based on the results of studies done with rat skin and rat preputial gland tumors, Wilson (1963) has concluded that both pathways operate in skin, and that the sebaceous gland uses the Kandutsch–Russell pathway preferentially. Testosterone treatment was shown to enhance the synthesis of the esterified sterols in both skin and the preputial glands, while estradiol markedly inhibited both free and esterified sterol production in skin, but not in the preputial glands.

Gaylor (1963a,b) and Gaylor *et al.* (1966) have studied sterol synthesis in rat skin and detected the formation of a highly labeled intermediate, lanosta-7,24-dien-3β-ol. This does not appear to fit either the "liver" or Kandutsch-Russell pathway. A probable explanation for this observation is that there is a general lack of specificty of the various demethylation enzymes for the substrate sterol. Thus, there is the possibility of several pathways between lanosterol and cholesterol. Clayton *et al.* (1963) described some 15 different sterols in whole rat skin, some of which do not fit either pathway shown in Fig. 3.

Inhibition studies using triparanol (an inhibitor of the 24,25 saturation) result in the accumulation of both lanosterol and desmosterol (cholesta-7,24-dien-3β-ol) in the preputial gland (Wilson, 1963) and in rat skin (Horlick and Avigan, 1963). Clayton *et al.* (1963) performed a similar experiment, but reported in a somewhat different pattern of sterols in rat skin. All these data are consistent with the presence of at least two and probably more pathways for the conversion of lanosterol to cholesterol in skin.

The presence of esterified sterols in skin lipids and the production of steryl esters by isolated preputial glands (Wilson, 1963) suggest that the skin contains an active sterol esterfying system. Freinkel and Aso (1971) examined cholesteryl ester synthesis in guinea pig epidermis and showed that it was catalyzed by a fatty acyl-CoA : cholesterol acyltransferase. Brady and Gaylor (1971) reported that the esterification of sterols in rat skin required CoA and ATP, implicating an acyltransferase. Grigor *et al.* (1972) described a similar system in the mouse preputial gland tumor.

For a long time skin has been implicated as the site of vitamin D synthesis, since small quantities of provitamin D (cholesta-5,7-dien-3β-ol) have been observed in skin. While in humans these appear to be confined to the epidermal cells (Wheatley and Reinertson, 1958), in the rat they are found primarily in the sebaceous cells (Gaylor and Sault, 1964). This has important consequences in the mode by which the vitamin D is made available to the animal. For humans it would appear that back diffusion from the epidermis would be important. In the rat, however, the vitamin D synthesized appears to be taken in orally. The ingestion of skin surface lipids by the rat is well established (Grigor *et al.*, 1971).

Because of its size, the skin appears to be the major target organ for androgens (Sansone-Bazzano and Reisner, 1974). As well as metabolizing testosterone to products of low androgenicity, skin, like the prostate gland (Bruchovsky and Wilson, 1968) can also modify testosterone to form 5α-dihydrotestosterone, a much more potent androgen (Gomez and Hsia, 1968; Sansone *et al.*, 1971). Both rat and mouse preputial glands also produce 5α-androgens (Sansone-Bazzano and Reisner, 1974), suggesting that the 5α-reductase may be primarily sebaceous. Thus not only is skin a target organ of androgens, it appears also to have the ability to amplify the effectiveness of these hormones.

VIII. *Biochemistry of Specialized Sebaceous Structures*

A number of quite large specialized sebaceous structures exist. Because they can frequently be dissected out free from nonsebaceous tissue, they provide attractive model systems for studying sebaceous gland metabolism. However, some caution is needed in interpreting the results obtained from these experiments. Frequently the lipid composition of these glands is quite different from that of the normal sebaceous gland of the same animal. Although these glands often respond to hormones like the normal gland, one can never be sure to what extent they can be used as satisfactory models.

The preputial gland of rodents has been the most thoroughly studied. Lipids from both rat and mouse preputial glands have a composition quite unlike the sebum. The gland does not contain any diester waxes of the type found in the sebum, but rather contains ether-linked diesters, alkyldiacylglycerols, and alk-1-enyldiacylglycerols (Sansone and Hamilton, 1969; Snyder and Blank, 1969). Furthermore, unlike the surface lipids of these species where, for the rat in particular, there are large quantities of branched chain fatty acids, the preputial gland lipids do not appear to contain any such acids. Nevertheless, the preputial gland has been used extensively to study hormone action (Burgess and Wilson, 1962), hormone metabolism (Sansone-Bazzano and Reisner, 1974), and sterol synthesis (Wilson, 1963).

A transplantable tumor of the mouse preputial gland exists and this has been studied extensively. The Kandutsch–Russell pathway for cholesterol synthesis was discovered in this tumor (Kandutsch and Russell, 1960b). Snyder and co-workers have used the tumor to study aspects of the synthesis of ether-linked lipids (Snyder *et al.*, 1969, 1970), the production of fatty alcohols (Snyder and Malone, 1970), and wax esters (Snyder and Malone, 1971). Grigor *et al.* (1972) examined the synthesis of cholesteryl esters in this tumor.

The harderian gland near the eye of the rabbit represents a specialized sebaceous gland. This also contains ether-linked glycerides including two very unusual classes (see the chapter The Eye), one containing isovaleric acid exclusively in position three (Blank *et al.*, 1972), and one containing a hydroxy-substituted alkyl chain (Kasama *et al.*, 1973) (see the chapter Harderian Gland).

The lipid of the meiboium gland in the human eyelid has been studied by Andrews (1971). Wax esters and cholesteryl esters form the major components of their lipids. This contrasts with the presence of triacylglycerols and only low amounts of cholesteryl esters in normal human sebum.

The composition of the secretions of a number of other mammalian skin glands has been reviewed by Mykytowycz and Goodrich (1974). Some, but not all, of these can be regarded as sebaceous structures. The secretions contain a variety of unusual lipid components, many of which act as pheromones.

In conclusion it is interesting to compare the sebaceous system of mammals with the equivalent glands in birds. Birds have a single sebaceous structure—the uropygial or preen gland found near the base of the tail. Two main lipid patterns appear to exist: In aquatic birds wax esters, made up of a straight-chain alcohol and a highly branched acid, are the major components (Odham, 1967). Typical of the branched-chain acids is a 2,4,6,8-tetramethyldecanoic acid. Here the branching portion does not come from the amino acids, but, rather, the efficient incorporation of propionate into these acids suggests the presence of a fatty acid synthase utilizing methylmalonyl-CoA (Noble *et al.*, 1963). The preen gland lipid of the second group of birds, which includes mainly non-water-birds, contains a diester wax which yields on hydrolysis an alkane-2, 3-diol, rather than the alkane-1, 2-diol found for mammalian diesters (Haahti and Fales, 1967; Sawaya and Kolattukudy, 1973).

References

Adachi, K., Lewis, C., and Hershey, F. B. 1967a. Enzymes of amino acid metabolism in normal human skin I. Glutamate dehydrogenase. *J. Invest. Dermatol.* **48**:226–229.

Adachi, K., Lewis, C., and Hershey, F. B. 1967b. β-Hydroxyacyl-CoA dehydrogenase activities in various parts of normal human skin. *J. Invest. Dermatol.* **49**:97–99.

Adachi, K., Lewis, C., and Hershey, F. B. 1967c. Enzymes of amino acid metabolism in normal human skin II. Alanine and aspartate transaminases. *J. Invest. Dermatol.* **49**:240–245.

Andrews, J. S. 1971. Human tear film lipids I. Composition of the principal non-polar component, *Exp. Eye Res.* **10**:223–227.

Blank, M. L., Kasama, K., and Snyder, F. 1972. Isolation and identification of an alkyldiacylglycerol containing isovaleric acid. *J. Lipid Res.* **13**:390–395.

Boughton, B., and Wheatley, V. R. 1959. The fatty acid composition of the skin surface fat ("sebum") of normal human subjects. *J. Invest. Dermatol.* **33**:49–55.

Boughton, B., MacKenna, R. M. B., Wheatley, V. R., and Wormall, A. 1959. The fatty acid composition of the surface skin fats ("sebum") in acne vulgaris and seborrheic dermatitis. *J. Invest. Dermatol.* **33**:57–64.

Bowen, D. M., and Radin, N. S. 1968. Hydroxy fatty acid metabolism in brain. *Adv. Lipid Res.* **6**:255–272.

Brady, D. R., and Gaylor, J. L. 1971. Enzymic formation of esters of methyl sterol precursors of cholesterol. *J. Lipid Res.* **12**:270–276.

Burgess, T. L., and Wilson, J. D. 1962. Studies on hormonal regulation of squalene synthesis in preputial gland and skin of the rat. *Proc. Soc. Exp. Biol. Med.* **113**:747–750.

Bruchovsky, N., and Wilson, J. D. 1968. The conversion of testosterone to 5α-androstan-17β-ol-3-one by rat prostrate *in vivo* and *in vitro*. *J. Biol. Chem.* **243**:2012–2021.

Carruthers, C. 1962. The fatty acid composition of dermal and epidermal triglycerides and phosphatides in mouse skin during normal and abnormal growth. *Cancer Res.* **22**:294–298.

Carruthers, C., and Davis, B. 1961. The lipid composition of epidermis and dermis of mice undergoing carcinogenesis by methylcholanthrene. *Cancer Res.* **21**:82–85.

Christie, W. W., Jenkinson, D. M., and Moore, J. H. 1972. Variation in lipid composition through the skin and subcutaneous adipose tissue of pigs. *J. Sci. Food Agric.* **23**:1125–1129.

Clayton, R. B., Nelson, A. N., and Frantz, I. D. 1963. The skin sterols of normal and triparanol treated rats. *J. Lipid Res.* **4**:166–178.

Cruickshank, C. N. D., Hershey, F. B., and Lewis, C. 1958. Isocitrate dehydrogenase activity of human epidermis. *J. Invest. Dermatol.* **30**:33–37.

Downing, D. T. 1963. The fatty acid composition of vernix caseosa. *Aust. J. Chem.* **16**:679–682.

Downing, D. T. 1965. Composition of the unsaponifiable matter of vernix caseosa. *Aust. J. Chem.* **18**:1287–1291.

Downing, D. T., and Strauss, J. S. 1974. Synthesis and composition of surface lipids of human skin. *J. Invest. Dermatol.* **62**:228–244.

Downing, D. T., Kranz, Z. H., and Murray, K. E. 1960. Studies in waxes XIV. An investigation of the aliphatic constituents of hydrolysed wool wax by gas chromatography. *Aust. J. Chem.* **13**:80–94.

Downing, D. T., Strauss, J. S., and Pochi, P. E. 1969. Variability in the chemical composition of human skin surface lipids. *J. Invest. Dermatol.* **53**:322–327.

Downing, D. T., Strauss, J. S., and Pochi, P. E. 1972. Changes in skin surface lipid composition induced by severe caloric restriction in man. *Am. J. Clin. Nutr.* **25**:365–367.

Ebling, F. J. 1974. Hormonal control and methods of measuring sebaceous gland activity. *J. Invest. Dermatol.* **62**:161–171.

Frantz, I. D., Dulit, E., and Davidson, A. G. 1957. The state of esterification of the sterols of rat skin. *J. Biol. Chem.* **226**:139–144.

Freinkel, R. K. 1960. Metabolism of glucose-^{14}C by human skin *in vitro*. *J. Invest Dermatol.* **34**:37–42.

Freinkel, R. K., and Aso, K. 1971. Esterification of cholesterol by epidermis. *Biochim. Biophys. Acta* **239**:98–102.

Freinkel, R. K., and Shen, Y. 1969. The origin of free fatty acids in sebum II. Assay of the lipases of the cutaneous bacteria and effects of pH. *J. Invest. Dermatol.* **53**:422–427.

Gaylor, J. L. 1963a. Biosynthesis of skin sterols III. Conversion of squalene to sterols by rat skin. *J. Biol Chem.* **238**:1643–1648.

Gaylor, J. L. 1963b. Biosynthesis of skin sterols IV. Identification of labelled lanosta-7,24-dien-3β-ol from rat skin. *J. Biol. Chem.* **238**:1649–1655.

Gaylor, J. L., and Sault, F. M. 1964. Localization and biosynthesis of 7-dehydrocholesterol in rat skin. *J. Lipid Res.* **5**:422–431.

Gaylor, J. L., Delwiche, C. V., Brady, D. R., and Green, A. J. 1966. Preparation and properties of a cell-free system from rat skin that catalyzes sterol biosynthesis. *J. Lipid Res.* **7**:501–510.

Gerstein, W. 1963. The phospholipids of normal and psoriatic skin. *J. Invest. Dermatol.* **40**:105–109.

Gomez, E. C., and Hsia, S. L. 1968. *In vitro* metabolism of testosterone-4-^{14}C and Δ4-androsten-3,17-dione-4-^{14}C in human skin. *Biochemistry* **7**:24–32.

Greene, R. S., Downing, D. T., Pochi, P. E., and Strauss, J. S. 1970. Anatomical variation in the amount and composition of human skin surface lipid. *J. Invest. Dermatol.* **54**:240–247.

Grigor, M. R. 1969a. The effect of dietary content on the rat skin surface fatty acids. *Proc. Univ. Otago Med. Sch.* **47**:46–47.

Grigor, M. R. 1969b. Lipogenesis in rat skin from the branched amino acids. *Proc. Univ. Otago Med. Sch.* **47**:48–49.

Grigor, M. R., Dunckley, G. G., and Purves, H. D. 1970. The synthesis of the branched-chain fatty acids of rat skin surface lipid. *Biochim. Biophys. Acta* **218**:389–399.

Grigor, M. R., Dunckley, G. G., and Purves, H. D. 1971. The branched chain fatty acids of rat faecal lipids: The contribution of undigested sebaceous lipid. *Biochim. Biophys. Acta* **231**:264–269.

Grigor, M. R., Pratt, R. D., and Snyder, F. 1972. Acylation of cholesterol by microsomal enzymes from mouse preputial gland tumors. *Arch. Biochem. Biophys.* **150**:371–375.

Haahti, E. 1961. Major lipid constituents of human skin surface with special reference to gas chromatographic methods. *Scand. J. Clin. Lab. Invest.* Suppl. 59, **13**:1–108.

Haahti, E. O. A., and Fales, H. M. 1967. The uropygiols: Identification of the unsaponifiable constituent of a diester wax from chicken preen glands. *J. Lipid Res.* **8**:131–137.

Halprin, K. M., and Ohkawara, A. 1966. Lactate production and lactate dehydrogenase in the human epidermis. *J. Invest. Dermatol.* **47**:222–226.

Hassing, G. S. 1971. Inhibition of *Corynebacterium acnes* lipase by tetracycline. *J. Invest. Dermatol.* **56**:189–192.

Hodgins, L. T., and Wheatley, V. R. 1973. Cutaneous lipogenesis III. The incorporation of valine carbons into the branched-chain fatty acids of rodent ear skin lipids. *Biochim. Biophys. Acta* **316**:173–179.

Holloway, P. W. 1970. Steroid Metabolism, pp. 371–429. *In* S. J. Wakil (ed.). Lipid Metabolism. Academic Press, New York.

Horlick, L., and Avigan, J. 1963. Sterols of skin in the normal and triparanol-treated rat. *J. Lipid. Res.* **4**:160–165.

Horn, D. H. S., and Hougen, F. W. 1953. Wool wax Pt. IV. The constitution of the aliphatic diols. *J. Chem. Soc.* **1953**:3533–3538.

Horn, D. H. S., Hougen, F. W., von Rudloff, E., and Sutton, D. A. 1954: Wool wax Pt. V. The constitution of the α-hydroxy-acids derived therefrom. *J. Chem. Soc.* **1954**:177–180.

Hsia, S. L., Sofer, G., and Lane, B. 1966. Lipid metabolism in human skin I. Lipogenesis from acetate-1-^{14}C. *J. Invest. Dermatol.* **47**:437–442.

Im, M. J. C., and Hoopes, J. E. 1974. Enzymes of carbohydrate metabolism in normal human sebaceous glands. *J. Invest. Dermatol.* **62**:153–160.

James, A. T., and Wheatley, V. R. 1956. Studies of sebum, 6. The determination of the component fatty acids of human forearm sebum by gas–liquid chromatography. *Biochem J.* **63**:269–273.

Kahlenberg, A., and Kalant, N. 1966. The effect of insulin and diabetes on glucose metabolism in human skin. *Can. J. Biochem.* **44**:801–808.

Kandutsch, A. A., and Russell, A. E. 1960a. Preputial gland tumor sterols II. The identification of 4α-methyl-Δ8-cholesten-3β-ol. *J. Biol. Chem.* **235**:2253–2255.

Kandutsch, A. A., and Russell, A. E. 1960b. Preputial gland tumor sterols III. A metabolic pathway from lanosterol to cholesterol. *J. Biol. Chem.* **235**:2256–2261.

Kärkkäinen, J., Nikkari, T., Ruponen, S., and Haahti, E. 1965. Lipids of vernix caseosa. *J. Invest. Dermatol.* **44**:333–338.

Kasama, K., Rainey, W. T., Jr., and Snyder, F. 1973. Chemical identification and enzymatic synthesis of a newly discovered lipid class—hydroxyalkylglycerols. *Arch. Biochem. Biophys.* **154**:648–658.

Kellum, R. E. 1966. Isolation of human sebaceous glands. *Arch. Dermatol.* **93**:610–612.

Kellum, R. E. 1967. Human sebaceous gland lipids: Analysis by thin-layer chromatography. *Arch. Dermatol.* **95**:218–220.

Lutsky, B. N., Casmer, C., and Koziol, P. 1974. In vitro incorporation of ^{14}C-acetate into lipids of hamster flank organ (costovertebral organ) sebaceous glands and epidermis. *Lipids* **9**:43–48.

Marples, R. R., Kligman, A. M., Lantis, L. R., and Downing, D. T. 1970. The role of the aerobic microflora in the genesis of fatty acids in human surface lipids. *J. Invest. Dermatol.* **55**:173–178.

Marples, R. R., Downing, D. T., and Kligman, A. M. 1971. Control of free fatty acids in human surface lipids by *Corynebacterium acnes*. *J. Invest. Dermatol.* **56**:127–131.

Marples, R. R., Downing, D. T., and Kligman, A. R. 1972. Influence of Pityrosporum species in the generation of free fatty acids in human surface lipids. *J. Invest. Dermatol.* **58**:155–159.

Mier, P. D. 1969. The carbohydrate metabolism of skin. *Br. J. Dermatol.* (Suppl 2) **81**:14–17.

Montagna, W. 1962. The Structure and Function of Skin. Academic Press, New York.

Montagna, W., and Lobitz, W. C. (eds.). 1964. The Epidermis. Academic Press, New York.

Montagna, W., Ellis, R. A., and Silver, A. F. (eds.). 1963. Advances in the Biology of Skin, Vol. 4. The Sebaceous Glands. Pergamon Press, Oxford. 260 pp.

Montagna, W., Bell, M., and Stauss, J. S. 1974. Proceedings of the 22nd annual symposium on the biology of skin: Session on sebaceous glands and *Acne vulgaris*. *J. Invest. Dermatol.* **62**:120–339.

Mykytowycz, R., and Goodrich, B. S. 1974. Skin glands as organs of communication in mammals. *J. Invest. Dermatol.* **62**:124–131.

Nicolaides, N. 1963. Human skin surface lipids—origin, composition and possible function, pp. 167–187. *In* W. Montagna, R. A. Ellis, and A. F. Silver (eds.), Advances in Biology of Skin, Vol. 4. The Sebaceous Glands. Pergamon Press, Oxford.

Nicolaides, N. 1965. Skin lipids II. Lipid class composition of samples from various species and anatomical sites. *J. Am. Oil Chem. Soc.* **42**:691–702.

Nicolaides, N. 1967. The monoene and other wax alcohols of human skin surface lipid and their relation to fatty acids of this lipid. *Lipids* **2**:266–275.

Nicolaides, N. 1971. The structures of the branched fatty acids in wax esters of Vernix caseosa. *Lipids* **6**:901–905.

Nicolaides, N. 1974. Skin lipids: Their biochemical uniqueness. *Science* **186**:19–26.

Nicolaides, N., and Ansari, M. N. A. 1968. Fatty acids of unusual double-bond positions and chain lengths found in rat skin surface lipids. *Lipids* **3**:403–410.

Nicolaides, N., and Kellum, R. E. 1965. Skin Lipids. I. Sampling problems of the skin and its appendages. *J. Am. Oil Chem. Soc.* **42**:685–690.

Nicolaides, N., and Ray, T. 1965. Skin lipids III. Fatty chains in skin lipids. The use of vernix caseosa to distinuish between endogenous and exogenous components of human skin surface lipid. *J. Am. Oil Chem. Soc.* **42**:702–707.

Nicolaides, N., Reis, O. K., and Langdon, R. G. 1955. Studies of the *in vitro* lipid metabolism of the human skin. 1. Biosynthesis in scalp skin. *J. Am. Chem. Soc.* **77**:1535–1538.

Nicolaides, N., Kellum, R. E., and Wooley, P. V. 1964. The structure of the free unsaturated fatty acids of human skin surface fat. *Arch. Biochem. Biophys.* **105**:637–639.

Nicolaides, N., Fu, H. C., and Rice, G. R. 1968. The skin surface lipids of man compared with those of eighteen species of animals. *J. Invest. Dermatol.* **51**:83–89.

Nicolaides, N., Fu, H. C., and Ansari, M. N. A. 1970. Diester waxes in surface lipids of animal skin. *Lipids* **5**:299–307.

Nicolaides, N., Fu, H. C., Ansari, M. N. A., and Rice, G. 1972. The fatty acids of wax esters and sterol esters from vernix caseosa and from human skin surface lipid. *Lipids* **7**:506–517.

Nikkari, T. 1965. Composition and secretion of the skin surface lipids of the rat; effects of dietary lipids and hormones. *Scand. J. Clin. Lab. Invest.* Suppl. 85 **17**:1–140.

Nikkari, T. 1969. The occurrence of diester waxes in human vernix caseosa and in hair lipids of common laboratory animals. *Comp. Bioch. Physiol.* **29**:795–803.

Nikkari, T. 1974. Comparative chemistry of sebum. *J. Invest. Dermatol.* **62**:257–267.

Nikkari, T., and Haahti, E. 1964. The composition of rat skin surface lipids. Analysis of the saponifiable and nonsaponifiable fractions. *Acta Chem. Scand.* **18**:671–680.

Nikkari, T., and Haahti, E. 1968. Isolation and analysis of two types of diester waxes from the skin surface lipids of the rat. *Biochim. Biophys. Acta* **164**:294–305.

Nikkari, T., and Valavaara, M. 1970. The influence of age, sex, hypophysectomy and various hormones on the composition of the skin surface lipids of the rat. *Br. J. Dermatol.* **83**:459–472.

Noble, R. E., Stjernholm, R. L., Mercier, D., and Lederer, E. 1963. Incorporation of propionic acid into a branched chain fatty acid of the preen gland of the goose. *Nature* **199**:600–601.

Odham, G. 1967. Studies of feather waxes of waterfowl. *Ark. Kem.* **27**:295–307.

Pablo, G., Hammons, A., Bradley, S., and Fulton, J. E. 1974. Characteristics of the extracellular lipases from *Corynebacterium acnes* and *Staphylococcus epidermidis*. *J. Invest. Dermatol.* **63**:231–238.

Patterson, J. F., and Griesemer. 1959. Lipogenesis in human skin. *J. Invest. Dermatol.* **33**:281–285.

Pochi, P. E., and Strauss, J. S. 1974. Endocrinologic control of the development and activity of the human sebaceous gland. *J. Invest. Dermatol.* **62**:191–201.

Ramasastry, P., Downing, D. T., Pochi, P. E., and Strauss, J. S. 1970. Chemical composition of human surface lipids from birth to puberty. *J. Invest. Dermatol.* **54**:139–144.

Reinertson, R. R., and Wheatley, V. R. 1959. Studies on the chemical composition of human epidermal lipids. *J. Invest. Dermatol.* **32**:49–59.

Rothman, S. 1954. Physiology and Biochemistry of the Skin. University of Chicago Press, Chicago.

Sansone, G., and Hamilton, J. G. 1969. Glyceryl ether, wax ester and triglyceride composition of the mouse preputial gland. *Lipids* **4**:435–440.

Sansone-Bazzano, G., and Reisner, R. M. 1974. Steroid pathways in sebaceous glands. *J. Invest. Dermatol.* **62**:211–216.

Sansone, G., Davidson, W., Cummings, B., and Reisner, R. M. 1971. Sebaceous gland lipogenesis induced by testosterone: Early metabolic events. *J. Invest. Dermatol.* **57**:144–148.

Sawaya, W. N., and Kolattukudy, P. E. 1973. Enzymatic esterification of alkane-2,3-diols by the microsomes of the uropygial glands of ring-necked pheasants. (Phasianus colchicus.) *Arch. Biochem. Biophys.* **157**:309–319.

Shalita, A. R. 1974. Genesis of free fatty acids. *J. Invest. Dermatol.* **62**:332–335.

Shuster, S., and Thody, A. J. 1974. The control and measurement of sebum secretion. *J. Invest. Dermatol.* **62**:172–190.

Snyder, F., and Blank, M. L. 1969. Relationships of chain lengths and double bond locations in O-alkyl, O-alk-1-enyl, acyl, and fatty alcohol moieties in preputial glands of mice. *Arch. Biochem. Biophys.* **130**:101–110.

Snyder, F., and Malone, B. 1970. Enzymic interconversion of fatty alcohols and fatty acids. *Biochem. Biophys. Res. Commun.* **41**:1382–1387.

Snyder, F., and Malone, B. 1971. The biosynthesis of wax esters: Enzymic interconversions of fatty acids and fatty alcohols. *J. Am. Oil Chem. Soc.* **48**:85A.

Snyder, F., Malone, B., and Wykle, R. L. 1969. The biosynthesis of alkyl ether bonds in lipids by a cell-free system. *Biochem. Biophys. Res. Commun.* **34**:40–47.

Snyder, F., Malone, B., and Blank, M. L. 1970. Enzymic synthesis of O-alkyl bonds in glycerolipids. *J. Biol. Chem.* **245**:1790–1799.

Summerly, R., and Woodbury, S. 1971. The *in vitro* incorporation of ^{14}C-acetate into the isolated sebaceous glands and appendage-freed epidermis of human skin. *Br. J. Dermatol.* **85**:424–431.

Velick, S. F., and English, J. 1945. The synthesis and configuration of d-14-methylpalmitic acid and its identity with the natural fatty acid from wool fat. *J. Biol. Chem.* **160**:473–480.

Weitkamp, A. W. 1945. The acidic constituents of degras. A new method of structure elucidation. *J. Am. Chem. Soc.* **67**:447–454.

Weitkamp, A. W., Smiljanic, A. M., and Rothman, S. 1947. The free fatty acids of human hair fat. *J. Am. Chem. Soc.* **69**:1936–1939.

Wheatley, V. R. 1974. Cutaneous lipogenesis. Major pathways of carbon flow and possible interrelationships between the epidermis and sebaceous glands. *J. Invest. Dermatol.* **62**:245–256.

Wheatley, V. R., and James, A. T. 1957. Studies of sebum 7: The composition of the sebum of some common rodents. *Biochem. J.* **65**:36–42.

Wheatley, V. R., and Reinertson, R. P. 1958. The presence of vitamin D precursors in human epidermis. *J. Invest. Dermatol.* **31**:51–54.

Wheatley, V. R., Chow, D. C., and Keenen, F. D. 1961. Studies on the lipids of dog skin II. Observations of the lipid metabolism of perfused surviving dog skin. *J. Invest. Dermatol.* **36**:237–239.

Wheatley, V. R., Lipkin, G., and Woo, T. H. 1967. Lipogenesis from amino acids in perfused isolated dog skin. *J. Lipid Res.* **8**:84–89.

Wheatley, V. R., Hodgins, L. T., Coon, W. M., Kumarisiri, M., Berenzweig, H., and Feinstein, J. M. 1971. Cutaneous lipogenesis: Precursors utilized by guinea pig skin for lipid synthesis. *J. Lipid Res.* **12**:347–360.

Wheatley, V. R., Kumarisiri, M., and Brind, J. L. 1973. Cutaneous lipogenesis. IV. Role of the pentose phosphate pathway during lipogenesis in guinea pig ear skin. *J. Invest. Dermatol.* **61**:357–365.

Wilkinson, D. I. 1970a. Monounsaturated fatty acids of mouse skin surface lipids. *Lipids* **5**:148–149.

Wilkinson, D. I. 1970b. Incorporation of acetate-1-^{14}C into fatty acids of isolated epidermal cells. *J. Invest. Dermatol.* **54**:132–138.

Wilkinson, D. I. 1973. Some factors affecting C^{14}-acetate incorporation into polyunsaturated fatty acids of skin. *J. Invest. Dermatol.* **60**:188–192.

Wilkinson, D. I. 1974. Biosynthesis of fatty acids by subcellular fractions of newborn mouse skin. *J. Invest. Dermatol.* **63**:350–355.

Wilkinson, D. I., and Karasek, M. A. 1966. Skin lipids of a normal and a mutant (asebic) mouse strain. *J. Invest. Dermatol.* **47**:449–455.

Wilkinson, D. I., and Walsh, J. T. 1974. Effect of various methods of epidermal–dermal separation on the distribution of ^{14}C-acetate-labelled polyunsaturated fatty acids in skin compartments. *J. Invest. Dermatol.* **62**:517–521.

Wilson, J. D. 1963. Studies on the regulation of cholesterol synthesis in the skin and preputial gland of the rat. pp. 148–166. *In* W. Montagna, R. A. Ellis, and A. F. Silver (eds.). Advances in the Biology of Skin, Vol. 4. The Sebaceous Glands. Pergamon Press, Oxford.

Yardley, H. J., and Godfrey, G. 1963. Direct evidence for the hexose monophosphate pathway of glucose metabolism in skin. *Biochem. J.* **86**:101–103.

Ziboh, V. A., Dreize, M. A., and Hsia, S. L. 1970. Inhibition of lipid synthesis and glucose 6-phosphate dehydrogenase in rate skin by dehydroepiandrosterone. *J. Lipid Res.* **11**:346–354.

Ziboh, V. A., Wright, R., and Hsia, S. L. 1971. Effects of insulin on the uptake and metabolism of glucose by rat skin *in vitro*. *Arch. Biochem. Biophys.* **146**:93–99.

Calcified Tissues

THOMAS R. DIRKSEN

I. Introduction

Studies that indicate dentin, enamel, and bone contain lipids of diverse types generally fall into three categories: (a) the employment of histochemical methodology for localization of lipids within calcified tissues, (b) the extraction of bulk quantities of pulverized tissues for more exact identification and quantitation of lipid classes, and (c) the utilization of radioactive substrates to study lipid synthesis by calcified tissues under both *in vitro* or *in vivo* conditions. The mechanism whereby such tissues normally undergo calcification while other tissues of similar organic composition do not is still an unresolved problem. Recent work on calcified tissues suggests that lipids play a role in this process. Histologic evidence of a "lipid-like" material at sites of active calcification, coupled with observations that lipid extracts of bone, dentin, and enamel differ both quantitatively and qualitatively, depending upon whether samples were demineralized before extraction, strengthens their importance in biological mineralization.

II. Anatomy and Physiology of Teeth and Bone

A. Macroscopic and Microscopic Organization of Teeth

Teeth are anatomically divided into the crown and the root portion at the cervical line or neck. A diagram of a tooth is shown in Fig. 1. The root portion is embedded within alveolar bone and is held there by periodontal membrane fibers. These fibers extend from bone to tooth and suspend the

THOMAS R. DIRKSEN • Departments of Oral Biology and Cell and Molecular Biology, Schools of Dentistry and Medicine, Medical College of Georgia, Augusta, Georgia 30902.

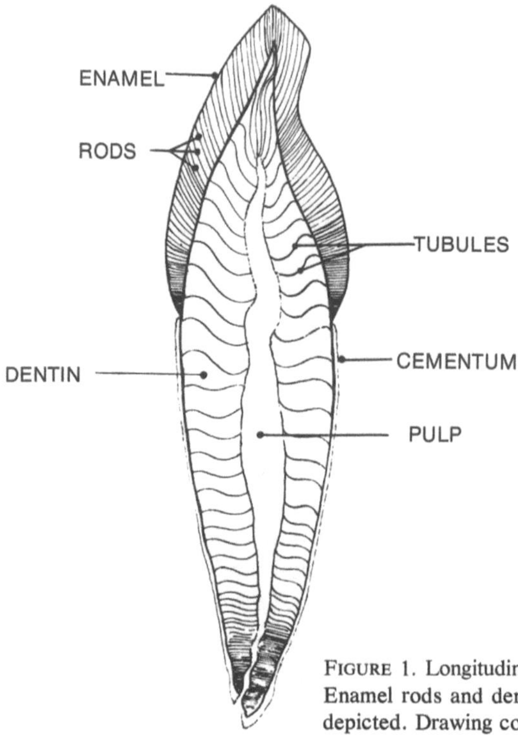

FIGURE 1. Longitudinal view of an upper central human incisor. Enamel rods and dentinal tubules are far more numerous than depicted. Drawing courtesy of St. Paul Gaffney, Jr.

tooth within its socket in a manner somewhat analogous to the supporting ropes of a hammock.

The main bulk of both crown and root is dentin. Dentin gives the tooth its general form and elastic strength. It is partially transparent and yellowish in color. Normal dentin is both avascular and acellular. Its formative cells, the odontoblasts, line the pulp chamber and are capable of further matrix synthesis should dental caries occur. Their protoplasmic extensions penetrate to the dentin enamel junction through small canals or dentinal tubules.

A central cavity within dentin contains pulp, a soft connective tissue with blood vessels and nerves. This chamber communicates with the periodontal membrane through a small foramen at the apex of the root. The anatomical root is covered by a thin layer of cementum, a cellular calcified material which approximates the composition of bone. It is within cementum that the periodontal membrane fibers are embedded. The anatomical crown is covered by enamel, the only dental tissue of epithelial origin. Cementum and dentin are of mesodermal origin. In some species, cementum also covers the enamel, which complicates chemical analysis.

Mature enamel is acellular and consists of prismatic rods with interprismatic substance. It is the hardest animal substance and is quite brittle when not supported by the more elastic dentin. The ameloblasts, once enamel matrix has been elaborated, degenerate before the tooth erupts into the oral cavity.

B. *Macroscopic and Microscopic Organization of Bone*

Two specialized connective tissues, bone and cartilage, are the predominant components of the skeletal system. While bones may be classified as to their shape—long, short, flat, and irregular—their chemical composition is quite similar.

1. *Anatomical Parts of Long Bones*

Each of the long bones of the body (femur, radius, etc.) consists of the following anatomical parts:

a. Diaphysis. The main portion or shaft of a bone. The outer compact or cortical bone provides strength, the central core or marrow cavity precludes excess weight.

b. Metaphysis. The wider part at the end of the diaphysis and adjacent to the epiphysis.

c. Epiphysis. The bulbous portions at each end of the metaphysis and diaphysis, which include the growth centers of long bones. A microscopic view of the epiphysis is shown in Fig. 2.

d. Articular Cartilage. A thin layer of hyaline cartilage covering the articular surfaces of the epiphyses.

e. Periosteum. A dense fibrous membrane covering the outer surfaces of bone, except at the joints. Fibers from both ligaments and bone (Sharpey's fibers) entermesh with fibers of the periosteum.

f. Marrow Cavity. An inner cavity which runs the entire length of the diaphysis. In the young, blood-forming elements are present, but red marrow is replaced by yellow marrow (high in triacylglycerols) as the individual matures.

g. Endosteum. A membrane which lines the marrow cavity and Haversian systems.

Compact bone is a connective tissue in which the intercellular substance is calcified and arranged in concentric layers or osteones. At the center of each osteon are nutrient vessels within Haversian canals. In histologic section, these cylindrical structures appear as rings, somewhat analogous to the growth lines of a tree, as can be seen in Fig. 2. Numerous blood vessels from the periosteum penetrate bone through Volkmann's canals to connect with blood vessels in the Haversian system.

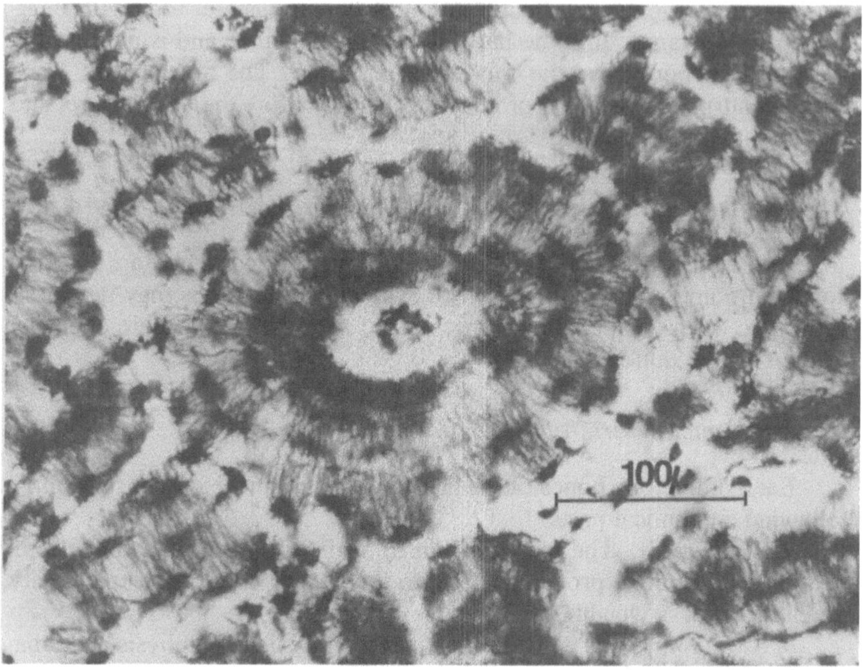

FIGURE 2. A ground section of human compact bone. The central canal surrounded by concentric layers of lamellae of bone is called a Haversian system or osteon. The lacunae in which the osteocytes reside appear as dark flattened structures. The fine lines connecting the lacunae with the Haversian canal are canaliculi. Photo courtesy of Dr. F. T. Lake.

2. *Cells of Bone*

a. Osteoprogenerator cells. Bone cells arise through differentiation of osteoprogenerator cells into one of the three cell types of bone. Each type is associated with specific functions: osteoblasts with the formation of new bone material, osteocytes with the maintenance of bone as a living tissue, and osteoclasts with the resorption of bone. During active bone growth and remodeling, frequent transformation from one cell type to another may occur.

b. Osteoblasts. The deeper layers of the periosteum and endosteum contain a layer of osteogenic cells called osteoblasts. They are flattened, spindle-shaped cells that form a continuous layer such that bone is encased within a membrane. Large amounts of endoplasmic reticulum are evident during active synthesis of bone matrix. These cells enable bone to grow by apposition.

c. Osteocytes. The osteocyte may be viewed as an osteoblast that has become surrounded by a calcified interstitial matrix. Osteocytes occupy spaces, or lacunae, between the lamellae and small microscopic canals, or canaliculi, and radiate in all directions to provide a continuous flow of fluid and nutrients.

d. Osteoclasts. Giant cells with variable numbers of nuclei are a prominent feature in patients with hyperparathyroidism or following administration of parathyroid hormone. These osteoclasts may be observed around bone spicules or within cups known as Howship's lacunae. Electromicrographs show a brush border appearance of the cell wall adjacent the site of active resorption and small fragments of calcified collagen are frequently observed within the cells.

C. Matrix of Teeth and Bone

1. Inorganic Composition

The hardness of bone and teeth results from mineral being deposited within an extracellular organic matrix. When incinerated, enamel is found to contain some 96% ash as compared to approximately 70% for dentin, cementum, and bone. The crystalline structure is generally considered that of calcium hydroxyapatite, whose formula is $Ca_{10}(PO_4)_6(OH)_2$. The composition of this biological material is modified from true hydroxyapatite by ion substitution; in particular, carbonate may replace phosphate and fluoride the hydroxyl group. In addition to apatite, amorphous calcium phosphate has been observed in dentin and bone. It is likely that amorphous calcium phosphate matures into apatite as the structure ages.

2. Organic Composition

Decalcified dentin and bone consists primarily of collagen, a protein widely distributed throughout the mammalian body. Under the electron microscope, collagen fibers present a characteristic 640-Å banding pattern. In mineralizing bone and dentin, crystals of inorganic salt may be seen deposited between the bands. This close relationship has suggested to some the possible role of collagen in initiating calcification. Collagen as first secreted by fiberblasts or osteoblasts is relatively soluble. As the collagen fibers mature, both inter- and intramolecular crosslinks increase in number such that collagen becomes highly insoluble. As with other tissues, the collagen fibers lie within a ground substance, the organic portion of which consists mainly of glycosaminoglycans. Glycogen and lipid are other organic constituents of cartilage, bone, and teeth or their cells.

The large content of mineral, coupled with difficulties in isolation and purification, has precluded many studies on the organic matrix of enamel. Many enamel proteins or amelogenins and other matrix components appear to be lost during calcification in a manner not yet explained. Collagen is not present in enamel.

D. *General Metabolism of Teeth and Bone*

1. *Formation of Teeth*

Teeth develop from tooth buds, which consist of three parts: (1) an enamel organ that is derived from the oral ectoderm, (2) a dental papilla that is derived from mesenchyme, and (3) a dental sac that is also derived from mesenchyme. Tooth formation is initiated by a thickening of the oral epithelium, the dental lamina, which proliferates and penetrates into mesenchyme to form the cap-like enamel organ. The cells on the concavity of the cap represent the inner enamel epithelium and differentiate into ameloblasts.

Under the organizing influence of the enamel organ, mesenchyme adjacent to the inner enamel epithelium proliferates and condenses to form the dental papilla. This is the formative organ of the dentin and dental pulp. Before the inner enamel epithelium begins to produce enamel, the peripheral cells of the dental papilla differentiate into odontoblasts. The boundary between the enamel organ and dental papilla determines the shape of the tooth since it constitutes the future dental enamel junction.

Dentin formation precedes that of enamel although ameloblasts are first to differentiate. An uncalcified organic matrix, the predentin, is first formed and subsequently mineralizes. From specific growth centers, matrix synthesis and calcification spread along the dental enamel junction. As the functioning odontoblasts recede from the dental enamel junction, they leave behind a cytoplasmic process that becomes embedded in the predentin matrix. When 10–20μm of predentin are formed, mineralization of the layers closest the dental enamel junction begins. The process advances pulpward at a rate approximating that of predentin formation. The earliest crystal deposition occurs in the ground substance and on the surface of collagen fibrils. Globular islands of mineralization are present onto which crystal deposition occurs.

Enamel formation or amelogenesis occurs first through matrix formation and then by maturation. After the first evidence of odontogenesis, enamel matrix is elaborated by ameloblasts as they move away from the dental enamel junction. The first activity is observed at growth centers and spreads laterally along the dental enamel junction toward the future cervical line. The formation of enamel involves extracellular secretion of a matrix that is partially mineralized. At a later stage, the final deposition of mineral occurs with loss of organic material from the area.

2. *Formation of Bone*

There are two general processes by which bone is formed: (1) endochondral, and (2) intramembranous. The long bones of the body form and grow through endochondral ossification. An example of the growth center of a long bone is shown in Fig. 3. During embryonic life, centers of

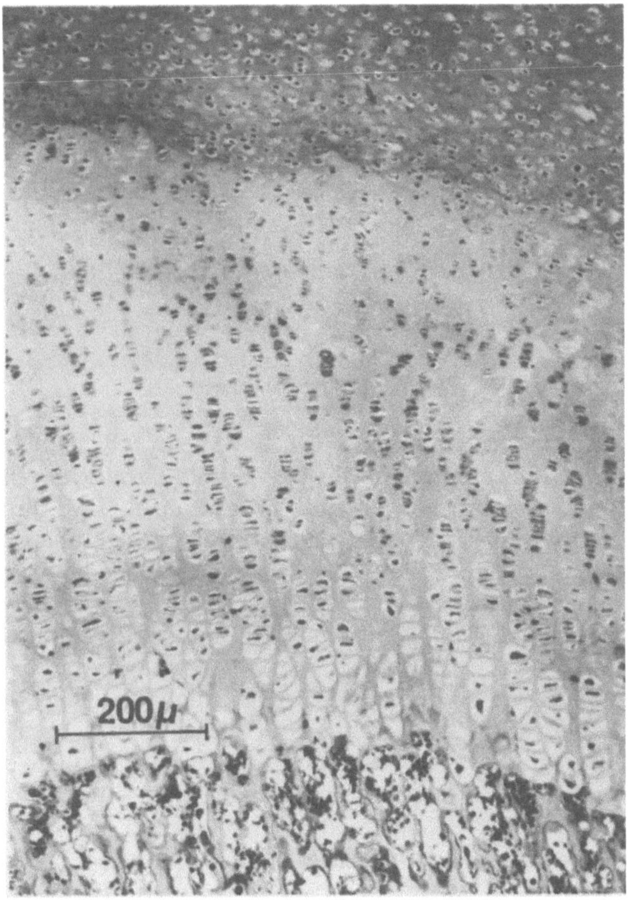

FIGURE 3. Longitudinal section of fetal bovine metatarsal showing the different zones of cells of the epiphyseal plate. A zone of resting cartilage is seen at the top, below which are zones of proliferating cartilage, maturing cartilage, calcifying cartilage, and developing bone of the metaphysis. Photo courtesy of Dr. F. T. Lake.

ossification appear in growing cartilage at the ends of long bones. Ossification is initiated first by maturation and then by the subsequent destruction of cartilage cells.

Several zones of cartilage cells that merge one into the other can be observed in the epiphyses. From epiphysis to diaphysis, there is first a zone of resting cartilage consisting mainly of primitive hyaline cartilage. In the next zone of proliferating cartilage, mitosis occurs with hypertrophy of the cells. They appear to stack or pile one on the other so as to form columns. The third cartilage layer contains cells in various stages of maturation, i.e., the zone of maturing cartilage. They are also arranged in columns of increasing cell size with the absence of mitosis. As this zone merges with the zone of calcified cartilage, the intercellular substance becomes increasingly calcified. This results in cell death, since cartilage depends upon permeation of nutrients through the matrix. The zone of calcified cartilage is quite thin and the transverse septa between cells are lost. Osteogenic cells of the diaphysis differentiate into osteoblasts, which then elaborate bone matrix onto that portion of calcified cartilage that persists.

III. Histochemical Observations on Lipids of Calcified Tissues

A. General Histochemical Technology

Histochemical tests for lipids usually employ oil-soluble dyes that dissolve in lipid. The criterion for lipid involves the ability to remove such dyes with an appropriate solvent and then restore the dye, provided the lipid is not removed in the process. Sudan black B is highly sensitive and is frequently used for studies of intracellular structures and lipid droplets. Oil red O and Sudan IV also give good results. The dyes are generally prepared as a saturated solution in 70% alcohol, so some risk of lipid extraction is always present.

Acidic lipids can be distinguished with Nile blue. Acidic lipids stain blue, whereas nonacidic lipids stain pink. Other tests have been developed for phospholipids, cholesterol, and unsaturation, but they have not been utilized to any appreciable extent for lipids in calcified tissues.

B. Histochemical Demonstration of Lipids in Calcified Tissues

A recent review has covered some of the early work dealing with histochemical observation on lipids in mineralizing tissues (Shapiro, 1973).

During the nineteenth century, von Bibra (1884) demonstrated the presence of lipids in bone, enamel, and dentin. Both Debreuil (1910) and Hill (1939) described lipids in bone cells, and this was confirmed by Conklin *et al.* (1965), who also found lipids in bone matrix.

Cartilage has been found to contain both cellular and extracellular lipids. Lipid components of cartilage have been reviewed by Stockwell (1965, 1967a,b), who studied lipids in articular cartilage from 35 humans ranging in age from 2 days to 89 years. Lipids were detected in the matrix of the superficial zone in subjects below the age of 20. From the third decade on, however, a sudanophilic layer became evident from which stain was lost following hot pyridine extraction. Nile blue and hematein denoted the presence of both neutral lipids and phospholipids. No sudanophilia was observed in osteoarthritic cartilage. In rabbits, only phospholipids were demonstrated in newborn cartilage cells, but extracellular phospholipid became evident with increasing age.

The early works of Euler and Meyer (1927), Weber (1930), and Hatton (1930) indicated traces of fat in dentinal tubules of teeth affected by dental decay or abrasion. These findings were corroborated by Bodecker (1931), who stained sections of teeth with Sudan IV. On the basis of his finding and the analytical work of von Bibra (1884), which indicated 0.2 percent fat in enamel and 0.4 percent fat in dentin, he formulated a theory where lipid might contribute to making teeth impermeable to acid and thereby greatly reduce the incidence of dental decay. Bodecker (1938) further suggested that the inclusion of soap in dentifrice, while keeping teeth clean, would increase their susceptibility to acid attack by removing the protective lipids.

The so called "dead tracts" beneath incisal margins of teeth showing attrition were interpreted by Fish (1932) as a defense mechanism on the part of odontoblasts to peripheral injury. Rushton (1940) observed a similar phenomenon in carious teeth. Numerous tracts were found to stain with Sudan black B and were, therefore, considered to contain large quantities of fat. Adjacent sound dentin showed only minute amounts of the stained material. Rushton concluded that either fat itself entered the tubules or a foreign substance from the oral cavity entered and promoted the conversion of tubular contents into stainable fat. Tubules containing a material which stained with Sudan black B were also reported to occur below carious lesions by Willner (1926).

Wislocki *et al.* (1948) demonstrated lipid in the odontoblastic processes of monkey, guinea pig, and fetal and adult human teeth. The sudanophilic positive area in the tubules of carious dentin was attributed to the presence of bacteria (Sognnaes and Wislocki, 1950). The prism sheaths in the developing enamel of humans, rats, and monkeys were also found to stain with Sudan black B (Wislocki and Sognnaes, 1950).

Lindemann (1957) reported the presence of dark-blue granules in paraffin and celloidin sections of decalcified bone and teeth stained with hematoxylin and eosin. She suggested that these granules consisted largely of lipid material.

Histochemical methods were used to observe the staining characteristics of carious dentin (Miller and Massler, 1962; Miller, 1969). Fat stains were found to localize in the surface or necrotic layer of carious dentin and in the upper half of the destructive layer. This distribution corresponded to that observed in Gram-stained sections and was attributed to lipids of bacteria. Deeper staining below the bacterial layer was thought to be related to an unmasking process in peritubular dentin.

Stewart and co-workers (Stewart, 1961, 1962, 1963; Stewart *et al.*, 1965) used a variety of stains to demonstrate lipid granules within odontoblastic cytoplasm and odontoblastic extensions within dentinal tubules. Granules were frequently associated with mitochondria in the Golgi zone. Assuming this zone as their origin, the granules were thought to be discharged from the cell into the predentin area among the Korff's fibers. Large number of granules were also seen in areas subjacent the odontoblastic processes immediately following experimental cavity preparation. They were believed to represent dentin maturation.

Opdyke (1962) observed a lipid-staining material in both natural and artificially induced lesions of enamel and dentin. The stain was removed by extraction with fat solvents and no dye uptake was observed when sections were extracted of lipids before staining. In early carious lesions, enamel rods stained in a linear fashion with a periodicity of approximately 4 μm. In the deep decay of advanced lesions, dentin tubules were filled with fat. This was believed to arise from destroyed enamel matrix, which preceded the advancing carious front. Exposure of sound tooth surfaces to sterile dextrose agar containing lactic acid showed, sequentially over a 20-day period, staining of calcium, later protein, and finally lipid. This, plus the periodic staining arrangement in both early artificial and natural carious lesions, indicated the endogenous origin of lipid laid down as part of the organic matrix during enamel formation.

In studies on sound dentin, Allred (1966a, 1968a,b,c) found that only dentinal tubule contents and uncalcified peritubular dentin stained with acetone-Sudan black B, acetone-acetylated Sudan black B, osmium tetroxide vapor, Nile blue sulphate, phosphomolbdyic acid, or Luxol fast blue. When various unmasking techniques were utilized (ranging from a brief acid etching to boiling in acetic acid), a changing pattern of dye uptake by previously nonstainable tissue was revealed. The most recently calcified peritubular dentin matrix adjacent to the predentin was the first to show lipid staining. As the intensity of unmasking increased, the peritubular matrix in the more peripheral dentin was stained. The positive results obtained with various stains suggested the presence of choline-containing phospholipids in these areas. When the specimens became completely decalcified, almost all lipid-staining characteristics were lost. It was assumed that lipids were in such intimate contact with both organic and inorganic fractions that some degree of tissue disorganization (acid treatment) was requisite for staining. At the junction of sound and carious dentin, Allred (1966b, 1969) observed that peritubular matrix staining increased in density toward the more advanced portions of the carious lesion until the intertubular and peritubular matrices were equally stained. Further progression of the carious process resulted in decreased staining, a condition similar to that observed when sound dentin was completely decalcified by unmasking procedures.

It would appear that the carious process first unmasks lipids of the peritubular matrix, thus allowing their release from the most advanced portions of the lesion. Further, since staining characteristics of carious dentin follows a pattern similar to that observed with progressive unmasking of sound dentin, much of carious dentin lipid, aside from that in bacteria, is assumed of intrinsic origin.

The distribution of lipids within compact bone has been studied by Enlow and Conklin (1964) and Enlow *et al.* (1965). Long bones of dogs, monkeys, and rats were sectioned and stained with a variety of lipid stains, both before and after organic solvent extraction. A lipid-positive material was observed in cortical bone, diffusely spread throughout the intercellular matrix. Dye was not taken up by chloroform–methanol extracted bones. Intracellular staining of lipid droplets was observed in hypertrophied chondrocytes within the zone of calcified cartilage as well as in cytoplasm of occasional osteocytes. Virtually all vascular channels contained lipid droplets, and the osteocytes, which demonstrated lipids, were generally found adjacent to these vascular canals.

Traumatic loading of incisors in rats has been shown to alter the lipid-staining characteristics of alveolar bone (Ingervall *et al.,* 1972). Incisal planes were fastened to mandibular teeth in such a manner as to cause abnormal biting pressures. Animals were sacrificed 10–27 days following application of the planes, and the bones were studied histochemically. No changes in lipid content of the alveolar bone were observed in young rats (6 weeks), even though destructive forces were sufficient to disrupt the enamel organs. Older rats (52 weeks), however, showed histochemical reactions for phospholipids in circular bone lamella. Results were attributed to either unmasking of bone lipids or deposition of material into the area as associated with vascular injury.

Lipid accumulation within the periodontal membranes of children, 12–

14 years of age, has been demonstrated following application of light orthodontic forces to teeth (Buck *et al.,* 1973). A force of 75*g* over a 21-day period was sufficient to cause extracellular lipid deposition in the cell-free layer of the periodontal membrane on the compression side of the teeth. Both the origin and fate of this lipid was unknown, but the effect was believed to be transitory in nature, since lipid deposition has not been observed around teeth when orthodontic forces are relieved.

C. The Sudanophilic Reaction

1. The Sudanophilic Reaction at Sites of Calcification

Irving (1958a,b, 1959, 1960, 1962, 1963) and Irving and Wuthier (1961, 1968) have reported a method for staining those sites within tissues that are rapidly undergoing active calcification. Essentially, the procedure involves formalin–saline fixation to retain fine detail, pyridine or ethanol-benzene extraction, decalcification, gelatin embedding, sectioning on a freezing microtome, and Sudan black B staining. In developing bone, the matrix surrounding hypertrophic cells of epiphyseal cartilage stained blue. In developing teeth, the junction of dentin and predentin and the enamel, where it became acid soluble, was found to become sudanophilic. Fully calcified tissues remained unstained, and it appears that the sudanophilic material was associated with the onset of calcification. The validity of the method has been demonstrated in a variety of separate studies, but the nature of the sudanophilic material remains to be established.

The localization of lipid components in the developing femurs of normal and tetracycline-treated chick embryos has been described by Rolle (1965). An alcohol-extractable lipid (probably neutral fat) was found in the cytoplasm of osteoblasts and osteocytes and also in osteoid. Phospholipids were detected in the cytoplasm of chondrocytes. A sudanophilic lipid, readily unmasked by ethanol but removed by pyridine, was found in the calcified bone of both normal and experimental animals as well as in the calcified cartilage matrix of experimental femurs. Another sudanophilic lipid, resistant to extraction with pyridine, was found in both rapidly calcifying bone and cartilage matrix. The amount of this material appeared reduced in the older, less rapidly calcifying bone. A similar pyridine-resistant, sudanophilic, bound lipid was found in cytoplasmic granules of the chondrocytes and those osteocytes situated in a calcifying matrix. Rolle (1965) suggested a direct relationship between both the size and number of sudanophilic granules and calcification of the surrounding matrix.

2. *Rickets and Scurvy and the Sudanophilic Reaction*

Further confirmation of the association between sudanophilia and calcification has come from experiments where bones from rats with severe rickets failed to demonstrate the staining pattern typical of normal growth and mineralization (Irving, 1960). When given vitamin D, however, the Sudan black B positive material appeared at the epiphysis, which also became positive to the von Kossa reaction for calcium.

Somewhat similar results were seen in scorbutic guinea pigs (Irving and Durkin, 1965). In this condition, an area of inactive osteoblasts develops below epiphyseal cartilage that does not stain with Sudan black B until some 72 hours following vitamin C administration. By that time, new bone development initiates and the sudanophilic material becomes evident.

3. *Calciphylaxis and the Sudanophilic Reaction*

Calciphylaxis is an ectopic form of calcification that can be induced in experimental animals. Irving (1965) studied the condition induced in young rats by dihydrotachysterol and 5-HT (serotonin creatine sulfate) administration. Mineral deposition occurred within soft tissue such as the salivary glands, stomach, heart, and kidneys. When observed by the Sudan black B technique, all tissues found to exhibit calciphylaxis also contained adjacent Sudan black B positive material.

4. *Identification of the Sudanophilic Material*

Extraction of histologic sections with 0.3% HCl in ethanol will remove the sudanophilic material from calcified tissues (Irving and Wuthier, 1968). As will be discussed in section IV.C, certain lipid classes are not removed from bone and teeth unless acidified solvents are employed. Such lipid extracts contain a very high percentage of those acidic phospholipids (phosphatidylserine, phosphatidylinositol) known to complex with calcium (Cotmore *et al.,* 1971). Thus, the correlation between acidic phospholipids, sudanophilia, and calcification is too strong to be discarded lightly.

Not all Sudan black B staining material is removed from calcified tissue sections by acidified chloroform–methanol, however, which suggests a material being present other than lipid (Shapiro *et al.,* 1966). Fincham *et al.* (1972) observed that bovine enamel matrix is partially soluble in organic solvents. The lipid content was quantitated and felt to be too low to explain sudanophilia on the basis of lipoprotein formation. Certain enamel matrix proteins of a hydrophobic nature were isolated and considered to play a role in the sudanophilic reaction. The suggestion that Sudan black B stains components in enamel other than lipid has also been made by Irving (1973a).

IV. Lipid Extraction Procedures for Calcified Tissues

A. Sample Preparation

A variety of procedures have been employed to separate calcified tissues from contaminants. In studying the lipid composition of bone, periosteal and endosteal surfaces are generally removed under a water spray with various abrasive disks or wheels. Since contents of the Haversian canals are not removed by these procedures, results must be interpreted in light of contaminate blood vessels and other connective tissue elements.

Separation of pure enamel and dentin under a dissecting microscope is difficult at best. A more rapid method for separation of powdered tooth material, based upon differences in density, was developed by Manley and Hodge (1939). Appropriate amounts of acetone in bromoform provide a solution of sufficient density such that enamel will sink, while dentin and cementum float when placed in a centrifugal field. Further addition of acetone allows separation of dentin and cementum. While quite suitable for studies of inorganic components of teeth, this method is totally unsatisfactory for lipid studies, since lipids are removed during the process. A somewhat similar procedure of differential flotation was established by Prout and Shutt (1970b) using cadmium tungstoborate solutions. The suitability of their method was established in that thin layer chromatograms of cadmium tungstoborate solutions demonstrated no trace of lipids and enamel and dentin samples of high purity were obtained.

B. Demineralization of Calcified Tissues

Until recently, the process of extracting lipids from calcified tissues was no different from procedures utilized for soft tissues. Samples were reduced to small particle size and then washed with the appropriate solvents. Of recent importance to studies on calcified tissues, however, is the observation that lipid extracts differ both quantitatively and qualitatively if samples are first demineralized. Dirksen (1963) and Dirksen and Ikels (1964) noted that additional lipids may be extracted from previously extracted dentin if samples are demineralized before the second extraction.

Their method of demineralization employed removal of inorganic salts from ground dentin matrix with 15% EDTA at pH 7.5, a procedure that appears to have no deleterious effects upon lipids. As discussed later, these results have been verified many times, and, qualitatively, the lipid extracts obtained postdemineralization always contain a higher percentage of acidic phospholipids.

C. *Acidified Solvents and Solvent Artifacts*

Some lipids are so tightly bound to the organic matrix of calcified tissues that they require acidified solvents for their extraction. Shapiro *et al.* (1966) found additional lipids over and above those already obtained with chloroform–methanol before and subsequent to EDTA demineralization of enamel and dentin. A word of caution should be introduced, however, since potentially high levels of methyl esters of fatty acids may be produced through transesterification at the expense of other lipid classes.

V. *Quantitative and Qualitative Anaylsis of Mineralized Tissues*

A. *Lipid Analysis of Teeth*

The studies cited in section III utilized almost exclusively histochemical methodology for the localization of lipid components in calcified tissues. As indicated, fat stains are frequently nonspecific and more exact identification requires isolation of lipids from bulk quantities of calcified tissues.

1. *Total Lipids of Enamel and Dentin*

Pincus (1938) isolated a fatty substance from sound enamel that amounted to approximately 1% of the original tissue. While the nature of this material was not elucidated, he felt sterols were a possible constituent. Krasnow (1934) determined the phosphorus content of an alcohol–ether soluble material from 15 teeth. The lipid phosphorus per 100 g of enamel and dentin was reported to range from 0.5 to 2.8 mg and from 0.1 to 1.7 mg, respectively.

Leopold *et al.* (1951) examined the cholesterol content of human root dentin following acidic and basic hydrolysis of pulverized tooth material. Yields of cholesterol as high as 24 mg/100 g dentin were found. Soyenkoff *et al.* (1951) extracted 300 g of mixed dental tissue (enamel, dentin, and cementum) and found about 300 mg of a chloroform-soluble material. The presence of sterols and fatty acids was established, as well as an aminophospholipid.

Hess *et al.* (1956) found that human enamel contained 0.008% cholesterol. Total lipids exclusive of cholesterol amounted to 0.60% dry calcified material and had saponification and iodine numbers of 161 and 20, respectively. Phosphorus analysis of enamel lipids indicated that 12.5% was phospholipid or approximately 0.075% enamel. Total lipids in dentin were 0.26–0.36% calcified material and had saponification and iodine numbers of 215 and 46. Phosphorus determinations indicated that 3.06% of the total lipids was phospholipid and 0.014% dentin.

Development of paper chromatographic methods for lipid analysis led to the identification of specific lipid classes in mineralized tissues. Using silicic acid impregnated papers, Dirksen (1963) studied the lipids of sound and carious dentin. The presence of cholesterol esters, triacylglycerols, fatty acids, cholesterol, diacylglycerols, monoacylglycerols (tentative), and various phospholipids was demonstrated. Sound dentin was found to contain phosphatidylinositol, sphingomyelin, phosphatidylcholine, phosphatidylethanolamine, lysophosphatidylethanolamine, and three unidentified phosphatides which may have included polyglycerol phosphatides and/or phosphatidic acid. Decayed material further contained phosphatidylserine and other phosphatides, which had undergone various degrees of degradation.

Of particular importance to the study of lipids in calcified tissues was the observation that certain lipid classes are not generally extracted from dentin unless samples are demineralized, either by the pathological process of caries formation (Dirksen, 1963) or by less degradatory laboratory procedures such as with EDTA (Dirksen and Ikels, 1964). As seen in section V.A.2, this finding has been applied to several studies of calcified tissues. Results obtained with dentin, enamel, and bone are shown in Table 1. In all instances, additional quantities of lipids were extracted as samples were demineralized before solvent addition.

2. *Phospholipids of Enamel and Dentin*

Shapiro *et al.* (1966) quantitated the different phospholipids in fetal bovine enamel and dentin. Extracting lipids both before and after demineralization, plus the additional use of acidified solvents, provided three different lipid samples for analysis. Average results calculated from their data are shown in Tables 2 and 3. Approximately 20 mg/100 g dry dentin was extracted before EDTA demineralization. Another 29 mg/100 g dry demineralized matrix was extractable after treatment of dentin with EDTA. A third extract of the same dentin matrix as obtained with acidified chloroform–methanol provided some 6 mg/100 g of residual matrix. Since the lipid

TABLE 1
Lipid Components of Dentin, Enamel, and Bone (mg/100 g)

Lipid class		Human dentin[a]	Bovine dentin[b]	Human dentin[c]	Bovine enamel[b]	Human bone[b]
Cholesterol	*d*	2.89	0.59	5.1	0.45	9.8
esters	*e*	—	0.36	—	—	6.6
	f	4.14	0.95	—	—	16.4
Triacylglycerol	*d*	1.59	2.16	28.7	.41	1626.0
	e	—	2.68	—	—	27.0
	f	1.61	4.84	—	—	1653.0
Fatty acids	*d*			36.5	—	—
Cholesterol	*d*	3.42	6.08	3.3	.81	14.1
	e	—	1.30	—	—	7.4
	f	6.53	7.38	—	—	21.5
Phospholipid	*d*	2.04	5.62	12.2	.30	21.5
	e	—	1.49	—	—	4.6
	f	4.94	7.11	—	—	26.1
TOTAL	*d*	11.14	14.45	90.4	1.97	1929.0
	e	—	5.83	—	—	75.9
	f	176.60	20.83	—	—	2004.9

[a]Dirksen and Ikels (1964).
[b]Dirksen and Marinetti (1970).
[c]Rabinowitz *et al.* (1967).
[d]Lipids extracted before demineralization.
[e]Lipids extracted after demineralization.
[f]Combined lipid extract.

yields are based upon calcified material in the first instance and demineralized matrix for the latter two values, it is somewhat difficult to ascertain the exact lipid yield for dentin from these experiments.

Fetal enamel, subjected to the same treatment, gave approximately 24 mg lipid per 100 g dry enamel before demineralization. The postdemineralization experiment yielded 271 mg/100 g matrix material. It must be remembered, however, that enamel contains far less matrix per gram of tissue than

TABLE 2
Total Phospholipids Extracted from Bovine Fetal Dentin and Enamel (mg/100 g)

	Before demineralization[a]	After demineralization[b]	Acid extraction[c]
Enamel	24.75	271.35	24.3
Dentin	20.35	29.15	5.85

[a] Based upon dry, nondemineralized tissue weight.
[b] Based upon demineralized tissue weight.
[c] Based upon demineralized, extracted tissue weight.

TABLE 3
Individual Phospholipids of Dentin and Enamel Expressed as a Percent of Total Phospholipids Extracted[a]

Lipid class	Tissue[b]	Before demineralization	After demineralization	Acid extraction	Total
Phosphatidylcholine	D	50.0	4.0	1.5	56.0
	E	29.0	4.0	2.2	33.2
Phosphatidylethanolamine	D	10.1	6.0	1.5	17.5
	E	5.0	29.0	—	34.0
Sphingomyelin	D	7	4.0	—	11.0
	E	8	—	0.9	8.9
Cardiolipin	D	1.0	—	—	1.0
	E	—	6.0	—	6.0
Phosphatidylserine	D	—	8.0	1.0	9.0
	E	—	8.0	0.8	8.8
Phosphatidylinositol	D	—	3.0	1.0	4.0
	E	—	6.4	0.4	6.8
Phosphatidic acid	D	—	1.0	1.0	2.0
	E	—	2.0	—	2.0

[a] Shapiro *et al.* (1966).
[b] D = dentin; E = enamel.

does any other tissue in the body, thus accounting for the high lipid yield when calculated on the basis of decalcified matrix. The third extraction using acidified solvents yielded another 24 mg lipid/100 g matrix.

The distribution of individual phospholipids among the three extracts of enamel and dentin is shown in Table 3. The values in the table represent the percentage that each lipid class contributes to the total phosphorus extracted. It is obvious that both phosphatidylcholine and phosphatidylethanolamine are present in highest concentrations in enamel and dentin. Further, with exception of the phosphatidylethanolamine content in enamel, the bulk of these two lipids is extracted before demineralization. Cardiolipin was the only acidic phospholipid to be extracted from dentin before demineralization. All other acidic phospholipids required EDTA demineralization and/or utilization of acidified solvents for their removal from the matrix.

Similar results have been obtained with both human and rat teeth where the calcified tissue was lipid extracted both before and after acid demineralization. Pulverized dentin and enamel were separated by density flotation in cadmium-tungstoborate solutions. In rat teeth (Prout *et al.*, 1973; Odutuga and Prout, 1973) some 66% of dentin lipid and 57% of enamel lipid was extracted before demineralization. A total of 333–350 mg lipid/100 g dentin and 540–570 mg lipid/100 g enamel was obtained. The total amount of individual lipids obtained before and after demineralization is reported in Table 4. With human enamel and dentin (Odutuga and Prout,

TABLE 4
Lipids of Rat Enamel and Dentin (mg/100 g Sound Tissue)[a]

	Total lipid	Cholesterol	Cholesterol esters	Triacylglycerols	Fatty acids and mono- and diacylglycerol	Phosphatidylcholine	Sphingomyelin, phosphatidylserine, plus phosphatidylinositol	Phosphatidylethanolamine	Phosphatidic acid plus cardiolipin
Incisor dentin	333	80.1	45.6	83.8	11.1	48.1	24.6	23.7	16.3
Incisor enamel	539	113	107	134	30.2	63.8	33.5	30.8	25.9
Molar dentin	344	68.6	56.8	75.9	10.3	57.9	26.8	29.5	13.2
Molar enamel	557	100	132	167	22.8	56.9	36.2	28.5	15.2

[a] Lipid extracts obtained before and after demineralization were combined before analysis calculated from Odutuga and Prout (1973).

1974a), 330 mg and 510 mg lipid/100 g dentin and enamel, respectively, were obtained. Before demineralization, 68% and 60% of dentin and enamel total lipids were extracted.

3. *Fatty Acids of Dental Tissues*

a. Total Fatty Acid of Teeth. Total fatty acids of dental tissues were examined by Schubert (1962), who extracted noncarious human dentin and subjected the lipid extracts to gas–liquid chromatography. While palmitic, stearic, oleic, and linoleic acids were identified, column retention times indicated that most fatty acids were of the unsaturated and branched chain type.

Das and Harris (1970a) examined lipids and total fatty acids in teeth from different animal species. Dentin and enamel were not separated. Total lipids ranged from 0.19% to 0.38% of calcified tissue and total fatty acids from 0.19% to 0.24%. Palmitic, stearic, and oleic acids were present in largest amounts. The ratios of these three fatty acids in the different lipid fractions were as follows: cholesterol esters, 62/38/trace; phospholipids, 39/16/22; and triacylglycerols, 52/39/trace. Total fatty acids of fossil teeth, 100,000 to 230,000,000 years old, were also investigated (Das and Harris, 1970b). Anhydrous total tooth samples, exclusive of pulpal tissue, contained 0.69–1.39% chloroform soluble material of which 0.38–0.88% were fatty acids. As in modern animal teeth, over 60% of the total fatty acids was palmitic, stearic, and oleic acids.

Total and individual fatty acids in human dentin and enamel have been studied by Prout and Shutt (1970a). Freshly extracted teeth were dissected with a dental bur and extracted for lipid after EDTA demineralization. In enamel, 0.01% of the tissue was fatty acids, whereas dentin contained 0.012–0.016% on a dry weight basis. The distribution of fatty acids in enamel and dentin showed a similar pattern: palmitic acid (19–25%), stearic acid (14–17%), and oleic acid (20–21%). Low levels of linoleic and linolenic acids were present in both tissues, but enamel contained more than dentin.

In human dentin Rabinowitz *et al.* (1967) found that 40% of the total lipid extract was free fatty acids. The procedures used during sample preparation were shown to result in the hydrolysis of 30% of the sterol esters and 27% of triacylglycerols. Lipid hydrolysis and denaturation during sample preparation were also investigated by Alam *et al.* (1973a). However, in this investigation the addition of small amounts of exogenous lipid to ground tooth material and subsequent extraction with acidified solvents resulted in only minor amounts of free fatty acids being released.

The distribution of individual fatty acids within different lipid classes of rat enamel and dentin was reported by Odutuga and Prout (1973). A combined lipid extract, obtained both before and after acid demineralization, was purified and separated by chromatography into several lipid classes. In the neutral lipid fractions, fatty acids of 12–20 carbons were obtained, with the major fatty acid components being palmitic, stearic, and oleic acids (See Table 5.). Linoleic acid was observed in all neutral lipid classes except cholesterol esters. Palmitic, stearic, and oleic acids were also the major fatty acids of both enamel and dentin phospholipids. Generally, higher levels of linoleic and linolenic acids were observed in phospholipids than in neutral lipids.

b. Effect of Dietary Alterations on the Fatty Acid Composition of Teeth. The effects of dietary alterations on the lipid components of rat teeth have been the subject of a series of investigations. Alam and Harris (1972) compared the fatty acid composition of depulped rat teeth from animals fed a diet containing either 67% starch or 67% sucrose for 8 weeks. Fatty acids extracted with chloroform–methanol from EDTA demineralized teeth, as well as those obtained by acid refluxing the methanolic phase from the first extraction, provided two samples for analysis. Tooth fatty acid composition was essentially the same whether rats were fed either starch or sucrose. Palmitic, stearic, oleic, linoleic, and arachidonic acids were present in the first extract, with 37–40% of total fatty acids being palmitic. Teeth from rats fed the starch diet contained only a slightly higher percentage of palmitic acid and less arachidonic acid when compared to the sucrose fed rats (37% versus 40% and 10% versus 8%). In the acidified extracts, stearic acid was the major fatty acid observed (44%).

The effects of food restriction (65% of control group food intake) formed the basis of a study by Alam *et al.* (1973a). Weanling rats were fed a cariogenic diet for 3 weeks or 6 weeks and the lipids analyzed. Enamel

TABLE 5

Fatty Acid Composition of Neutral Lipids and Phospholipids of Rat Enamel and Dentin (Percent of Total Fatty Acids)[a]

	Enamel		Dentin	
	Molar	Incisor	Molar	Incisor
Neutral lipids				
$C_{16:0}$	35	33	46	48
$C_{18:0}$	26	24	21	18
$C_{18:1}$	25	26	20	21
Phospholipids				
$C_{16:0}$	33	29	46	34
$C_{18:0}$	24	22	23	28
$C_{18:1}$	21	24	18	20

[a]Odutuga and Prout, 1973

and dentin were separated from pulp and cementum by density centrifugation using cadmium tungstoborate. Tooth material was extracted with chloroform–methanol both before and after EDTA demineralization and a third extract was obtained using acidified solvents. As expected, food restriction resulted in less body weight gain and also a lowered caries incidence in the experimental group. Lower quantities of fatty acids were extracted from teeth of the diet-restricted groups before tooth demineralization, which is probably a reflection of the increased caries experience (pathological decalcification) observed in the control group. Phospholipids and free fatty acids accounted for most of the total fatty acids in the teeth. Teeth from the experimental group contained somewhat lower concentrations of palmitic, stearic, and oleic acids when compared to the control group.

In another study on the effects of food restriction on lipids of rat teeth, Das and Harris (1975) fed experimental animels 65% of the control animal diet (57% corn starch, 20% casein, 15% vegetable shortening, 5% sucrose, plus salts vitamins and cystine) throughout a 78-day postweaning period. Pooled tooth powers were extracted after demineralization and the lipids analyzed. Tooth lipids from restricted animals consisted mainly of phospholipids and cholesterol whereas control tooth lipids were predominantly triacylglycerols and phospholipids. In contrast to previous findings (Das and Harris, 1970a), palmitic acid contributed a larger percentage of total fatty acids in control animals (80% vs approximately 50%). In the restricted group, some 60% of total fatty acids was palmitic, whereas 32–36% was oleic acid (compared to 11–25%) in the control group. The authors suggest that phospholipids and cholesterol are structural in nature, whereas triacylglycerols constitute a lipid store for energy production, which was consequently reduced in experimental animals.

Caries incidence in rats fed a cariogenic diet containing one of four different oils or fats at a 9% level was generally lower than in rats fed the same oil at a 3% level (Alam *et al.*, 1973b). Weanling rats received a standard cariogenic diet beginning at 19 days of age for a total period of 3 weeks with soy bean oil, cottonseed oil, vegetable shortening, or butter added at the expense of cellulose. Dentin and enamel were separated from residual pulp and cementum by centrifugation in cadmium tungstoborate, and lipids were extracted from the dentin-enamel samples with chloroform–methanol both before and after EDTA decalcification. The aqueous methanolic phase was acidified to obtain a third extract, and fatty acids were subjected to gas–liquid chromatography after methylation. While a lower caries score was generally observed with the 9% lipid diet, differences were not significant except when cottonseed oil was fed. The major fatty acids of molar teeth were found to be palmitic, stearic, oleic, linoleic, and arachidonic. Butter-fed animals had the lowest levels of linoleic acid but the proportions of palmitic, stearic, and oleic acids were not appreciably changed owing to the type of lipid included within the diet.

A 14-week high-sucrose diet, with or without 1.5% corn oil to supply essential fatty acids, produced alteration in the fatty acid patterns of rat enamel and dentin when compared to a control diet [Prout and Atkin (nee Shutt), 1973]. This was particularly true for incisors, which, in the rat, are continuously erupting and were replaced entirely during the period of essential fatty acid deficiency. The major fatty acids of normal enamel and dentin were palmitic, stearic, and oleic acids. Incisor enamel from rats on the stock diet differed from the corn-oil-supplemented group in that the oleic and linoleic acid content was increased. Further, the latter group had no hexadecadieonic or linolenic acids in molar enamel. The high-sugar, fat-free diet resulted in an appreciable increase in the quantity of longer-chain fatty acids, particularly docosahexenoic acid, a fatty acid not observed in the other two groups.

It is interesting that no significant differences in the relative proportions of various lipid classes within teeth were observed between corn oil supplemented (5% corn oil for 14 weeks) and essential fatty acid deficient rats (Prout and Odutuga, 1974a). However, rats fed an essential fatty acid deficient diet for 19 weeks, followed by 5 weeks supplementation with 5% corn oil, had teeth whose lipid content was intermediate to that of control animals and essential fatty acids deficient animals (Prout and Odutuga, 1974b). As can be seen in Table 6, incisor dentin of rehabilitated rats contained 0.34% total lipids compared to 0.37% and 0.33% for deficient and control animals, respectively. Corresponding values of molar dentin and molar and incisor enamel are also given. Since the proportions of various lipids obtained from rehabilitated rats were similar to and sometimes fell between those of the deficient and supplemented rats, it was considered that changes produced in teeth by essential fatty acid deficiencies are reversible.

TABLE 6

Total Lipid of Rat Enamel and Dentin as Affected by Essential Fatty Acid Deficiency (Percent of Dry Weight)

	Enamel		Dentin	
	Molar	Incisor	Molar	Incisor
Deficiency[a]	0.58	0.59	0.38	0.37
Rehabilitated[a]	0.57	0.57	0.36	0.34
Control or supplemented[b]	0.56	0.53	0.32	0.33

[a] Rats were fed an essential fatty acid deficient diet for 26 weeks. Rehabilitated rats received 5% corn oil the last 5 weeks (Prout and Odutuga, 1974b).

[b] Rats were fed an essential fatty acid deficient diet for 14 weeks (Prout and Odutuga, 1974a).

Further study indicated that the fatty acid composition of selected lipid classes also became altered. Two groups of rats were placed on a high-sugar synthetic diet, deficient in essential fatty acids, or the same diet supplemented with 5% corn oil (53% linoleic and 1.5% linolenic acids) for 20 weeks (Odutuga and Prout, 1974b). The overall effect of essential fatty acid deficiency on neutral lipid composition was to lower the proportion of palmitic and to raise the content of stearic, oleic, and longer-chain fatty acids. This was true for all neutral lipids except the free fatty acid fraction plus the mono- and diacylglycerol fraction. Essential fatty acid deficiency resulted in the appearance of docosahexaenoic acid in phospholipids with a definite decrease in the proportion of linoleic acid. The phosphatidylethanolamine and phosphatidic acid plus cardiolipin fractions of enamel and dentin from the deficient rats contained up to 20% of the docosahexaenoic acid, a fatty acid not observed at all in corn-oil-fed animals.

Some 0.48% and 0.63% lipid was extracted from carious dentin and enamel, respectively, compared to 0.35% and 0.58% for sound tissues (Odutuga and Prout, 1975). The appearance of fatty acids having odd numbers of carbon atoms in carious tissues was considered to be due to the presence of bacteria.

Further evidence that the changes produced by essential fatty acid deficiencies are reversible was derived from experiments utilizing [^{14}C]linoleic acid. Control and essential fatty acid deficient rats received injections of [^{14}C]linoleic acid (Prout and Odutuga, 1974c); ^{14}C-activity was incorporated into lipids, particularly acidic phospholipids, of incisor enamel and dentin and molar dentin. Only a small amount of radioactivity was observed in molar enamel. This was to be expected, since no molar enamel was formed during the experimental period.

B. *Lipid Analysis of Bone*

1. *Total Lipids of Bone*

Other than the few studies mentioned in section III.A, a comprehensive analysis of bone lipids has not been conducted until comparatively recent times. The cholesterol content of compact bone has been assayed by Pikular (1955) after demineralization of the sample material with hydrochloric acid. Cholesterol was found to range from 0.0028 to 0.0040 g/100 g dry ash-free protein. Leach (1958) determined the lipid content of compact, ox leg bone which he rinsed with acetone during the cleaning procedures; no demineralization step was used. Lipid extracts were found to contain triacylglycerols, cholesterol, cholesterol esters, phospholipids, and possibly a beta-carotene. Dry ox femur contained 0.0673% lipid, and other bones studied contained 0.0629% as determined by a cold acetone extraction procedure. The various lipid components assayed and their amounts (mg/

100 mg dry bone) were triacylglycerols (53.5), free cholesterol (8.9), cholesterol esters (1.1), and phospholipids (1.5).

Other lipid classes have been demonstrated in human bone by paper chromatography (Dirksen and Marinetti, 1970). In addition to the above lipid classes, fatty acids, and mono- and diacyl-glycerols were observed with no alterations in the neutral lipid patterns whether sample materials had been demineralized or not. Phosphatidylcholine, lysophosphatidylcholine, sphingomyelin, phosphatidylethanolamine, phosphatidylserine, phosphatidic acid, and cardiolipin were all extracted before demineralization. However, the major portion of acidic phospholipids obtained before demineralization was believed to have originated from cell membranes and other vascular elements rather than from the matrix itself. Shapiro (1970a) observed similar results with mature bovine cortical bone except that phosphatidylserine, phosphatidylinositol, and phosphatidic acid were not extractable until after demineralization.

Quantitative analysis of human bone, obtained following autopsy or after surgery, is shown in Table 1 (Dirksen and Marinetti, 1970). The major portion of total lipids was found to be extracted before demineralization and consisted primarily of triacylglycerols (85%). With the exception of triacylglycerols, the neutral lipid values reported in Table 1 are less than those of mature bovine bone (Table 7). Shapiro (1971) obtained a total of

TABLE 7

Lipid Composition of Bovine Bone (mg/100 g of Calcified Tissue)

Lipid class	Pre-demineralization	Post-demineralization	Acid hydrolysis	Total lipids
Triacylglycerols[a]	691.04	67.20	19.04	777.28
Cholesterol esters[a]	8.96	116.48	51.52	176.96
Cholesterol[a]	8.96	42.56	8.96	60.48
Fatty acids[a]	7.84	63.84	20.16	91.84
Mono- and diacylglycerol[a]	11.20	—	—	11.20
Hydrocarbons[a]	2.24	—	—	2.24
Sphingomyelin[b]	5.49	1.82	0.08	7.37
Phosphatidylcholine[b]	4.75	1.07	0.19	6.01
Phosphatidyl-ethanolamine[b]	2.56	1.12	—	3.69
Phosphatidylserine[b]	—	0.85	0.14	0.99
Phosphatidylinositol[b]	—	0.27	0.14	0.41
Phosphatidic acid[b]	—	0.27	0.06	0.33
Cardiolipin[b]	3.00	—	—	3.00
Total neutral lipids	730.24	290.08	99.68	1120.00
Total phospholipids	13.40	5.40	0.60	19.40

[a]Calculated from the data of Shapiro (1971).
[b]Calculated from the data of Shapiro (1970a).

1120 mg neutral lipid/100 g calcified tissue. Some 777.3 mg of this was triacylglycerol, 88% of which was extracted before EDTA demineralization. The phospholipids of bovine bone are also reported in Table 6. Of the acidic phospholipids, only cardiolipin was extracted before demineralization.

The highest lipid yields for human bone as reported in Table 1 (up to 5 g lipid/100 g bones), were obtained from subjects who suffered localized vascular blockage with resultant limb loss. On the basis of fatty acid analysis (Table 8), it would appear that much of this neutral lipid was due to fatty marrow infiltration of the Haversian systems. Lund *et al.* (1962) found that palmitic and oleic acids constituted some 26% and 46%, respectively, of the total fatty acids of triacylglycerols in bone marrow from subjects ranging in age from 1 to 78 years. This is almost identical to the results for bone, where these two fatty acids comprised approximately 25% and 50% of fatty acids of human cortical bone, the triacylglycerol fraction of bone lipids, and bone marrow of the same sample. Rat costal cartilage demonstrated a somewhat different fatty acid spectrum (Penton, 1968); only 33% of total fatty acids was oleic and 13% palmitoleic. Administration of sodium fluoride to rats resulted in increased levels of fatty acids, particularly palmitic and oleic acids.

2. *Phospholipids of the Epiphyseal Apparatus*

Development of new paper chromatographic techniques enabled Wuthier (1966, 1968) to quantitate the individual phospholipids of bone and cartilage. The epiphyseal apparatus (section II.B) of leg bones from full-term calf fetuses were dissected into several fractions: resting cartilage, proliferative cartilage, hyaline cartilage, calcified cartilage, cancellous

TABLE 8
Fatty Acids of Human Bone, Bone Triacylglycerols, and Bone Marrow (Percent)[a]

Fatty acid	Bone total lipids	Bone triacylglycerols	Bone marrow
Lauric	0.34	—	0.64
Myristic	2.89	1.40	2.23
Palmitic	22.53	25.70	25.66
Palmitoleic	5.69	5.11	5.89
Stearic	6.06	4.60	5.42
Oleic	46.90	54.19	49.29
Linoleic	10.41	9.01	10.68

[a] Dirksen and Marinetti (1970).

TABLE 9
Lipids of Calf Epiphyseal Apparatus[a]

Zone	Total lipids[b]	Lipid phosphorus[b] Before EDTA	After EDTA	Acidified solvents	Total phosphorus
Resting cartilage	1000	7.2	6.7	0.5	14.3
Proliferating cartilage	4860	61.5	3.9	0.8	66.2
Hypertrophic cartilage	7700	52.6	6.6	1.7	60.9
Calcified cartilage	8500	100.0	9.2	2.6	111.9
Cancellous bone	610	5.2	0.4	0.5	6.1
Compact bone	980	9.4	2.4	1.3	13.2

[a] Wuthier (1968).
[b] mg/100 g demineralized dry weight.

bone, and compact bone. Lipids were extracted from these samples both before and after EDTA demineralization as well as with acidified solvents. The total lipids, as well as total lipid phosphorus, from the different zones are reported in the Table 9. It is obvious that the lipid content increased progressively from the resting cartilage zone through the calcified cartilage zone. Phosphatidylcholine constituted some 46% of total lipid phosphorus extracted before demineralization, but only 31% and 26% of that extracted after EDTA demineralization or with acidified solvents. While phosphatidylserine and its lyso-derivative constituted only 4% of lipid phosphorus extracted before demineralization, these values rose to 24% and 16%, respectively, after EDTA treatment and employment of acidified solvents.

Quite similar findings were obtained (Eisenberg *et al.* 1970) from studies on the uptake of radioactive phosphate into lipids of the epiphyses of growing chickens. The percent of total lipid phosphorus extracted from different epiphyseal zones before and after decalcification was as follows: phosphatidylcholine (46–49% and 13–37%), phosphatidylethanolamine (15–24% and 3–13%), sphingomyelin (8–16% and 6–16%), phosphatidylserine (2–6% and 5–23%), and phosphatidylinositol (3–6% and 16–17%). The pattern of lipid distribution in the second extract (that obtained after demineralization) was similar to that of the first except more lyso-phospholipids were observed.

A variety of radioactive precursors have been utilized to study the metabolism of lipids in the epiphysis. These include [^{14}C]glucose, [^{14}C]palmitate, [^{14}C]acetate, and [^{14}C]citrate (Wolinksy and Guggenheim, 1970, 1971), [^{3}H]serine (Wuthier and Cummins, 1974), and [^{32}P]phosphate (Cruess and Hong, 1973; Eisenberg *et al.*, 1970; and Havivi, 1971). These experiments have demonstrated an active lipogenic system within both bone and cartilage of growing animals. The degree of activity seems dependent upon the uniqueness of the epiphyseal zone under study as well as the choice of substrate. For example, resting cartilage has a relatively sluggish phospholipid metabolism when compared to other epiphyseal zones and liver phosphatidylserine turnover is quite slow when compared to that of proliferating and calcified cartilage (Eisenberg *et al.*, 1970).

3. *Alterations in Bone Lipid Composition*

a. Vitamin A and Vitamin D. Alterations in the lipids of bone due to hypervitaminosis A and D have been reported by Cruess and Clark (1965). Toxic amounts of vitamin A (15,000 IU/day) over a 44-day period were found to decrease the total fatty acid content of the long bones of rats. (4.21 vs 4.09 mg/g) whereas toxic amounts of vitamin D (30,000 IU/day) increased triacylglycerols (1.05 vs 2.13 mg/g), esterified cholesterol (0.34 vs 0.55 mg/g), and in particular the phospholipids (1.54 vs 4.42 mg/g). Vitamin A alone did not alter lipid content. When toxic amounts of both vitamins were administered, the alterations in bone lipids caused by vitamin D alone were generally prevented. A number of experiments have been conducted to determine the effects of vitamin D deficiency upon lipids of bone and cartilage. The following variable results can best be explained at present on the basis of the length of vitamin D deprivation, the tissue zone, the species, and the methods used for lipid extraction.

Howell *et al.* (1965) found that costal cartilage in rachitic calves contained one-tenth to one-half the phospholipid content of normal cartilage; 0.9 mg% vs 1.29 mg% dry weight basis for combined resting and proliferating cartilage zones and 0.32 vs 0.72 mg% for combined hypertrophic and osteoid zones. The experimental animals in this study had frank clinical signs of rickets. The proportion of phosphatidylethanolamine (percent of total phospholipids) was less in rachitic tissue, whereas no difference existed for phosphatidylcholine. Phospholidylserine was not reported, which may have been due to tissues being extracted without demineralization.

Different results were obtained by Havivi and Bernstein (1969), who quantitated lipids extracted from undemineralized rat epiphyseal cartilage. Cartilage from rats fed a rachitogenic diet for 1–2 weeks contained comparatively more total lipids (12.08 mg vs 8.15 mg/100 mg dry weight) as well as DNA and RNA. Accompanying the increase in total lipids was an increase in total phospholipid (13.4 vs 6.9 mg/mg DNA), free fatty acids (1.13 vs 0.88 mg/mg DNA), and total cholesterol (2.80 vs 0.13 mg/mg DNA). Triacylglycerol values were affected very little. Metabolic differences were also present in that uptake of [^{14}C]palmitic acid by rachitic cartilage was significantly greater than that of control cartilage. Fractionation of lipid showed 65–75% of label in free fatty acids, 14–19% in triacylglycerols, 9–13% in phospholipids, and 3–6% in cholesterol and its esters.

In contrast to the above, elevated epiphyseal phospholipid values were not observed by Cruess and Clark (1971). A vitamin-D-free, semisynthetic diet was fed throughout a 60-day period to Holtzman rats with control animals receiving 75 IU of vitamin D_2 three times each week. Lipids were extracted from long bones both before and after EDTA demineralization as well as with acidified chloroform–methanol to provide three lipid extracts for comparison. An increase in the amount of organic matter was noted in all regions of bones analyzed, with the greatest change being seen in the metaphyseal region. Of the lipids extracted from bone before demineralization, vitamin-D-deficient rats had greater quantities of total lipids in the epiphysis (120 vs 83.9 mg/g organic matrix) and diaphysis (16.4 vs 7.6 mg/g organic), but less total lipids in the metaphysis (51.2 vs 71.1 mg/g organic). While diaphyseal phospholipids, cholesterol, total fatty acids, and triacylglycerols were elevated in vitamin-D-deficient animals, phospholipids of the metaphysis and epiphysis were unchanged. There were no striking differences in the content of individual phospholipids, but sample sizes were too small for statistical analysis.

The most comprehensive study to date on the effects of vitamin D deficiency in bone is that of Wuthier (1971). Both chickens and pigs were used and seven different zones of the epiphysis and bone were taken for analysis. Lipids were extracted both before and after demineralization as well as with acidified solvents for individual phospholipid analysis. Both increased and decreased levels of lipids due to hypovitaminosis D were noted, depending upon the epiphyseal zone under question as well as the type of lipid. Calcified cartilage showed the least effect, whereas alterations in phospholipid composition were quite pronounced in proliferating and hypertrophic cartilage and cancellous bone. The results suggest that synthesis of phosphalidylserine, phosphatidylglycerol, and sphingomyelin increased in rachatic epiphyseal cartilage with increased synthesis or reduced breakdown of the first two in cancellous bone. There was also an apparent decreased synthesis of phosphatidylinositol and chondrolipid in cartilage, but an increased synthesis in rachitic bone. Rickets also reduced the extractability of certain lipids from mineralizing zones, which was interpreted to indicate localized formation of a calcium phospholipid phosphate complex that was unable to participate in the normal mineralization process.

b. Cortisone. Two studies have been conducted to discern the effects of cortisone on bone lipids. In one, four groups of rats were utilized: (1) controls, (2) adrenalectomized, (3) cortisone treated (2 mg cortisone acetate/100 g body weight daily for 10 days), and (4) adrenalectomized and cortisone treated (Sakai and Cruess, 1967). Total lipids of cortical bone decreased in adrenalectomized animals and significantly increased in cortisone-treated and adrenalectomized–cortisone-treated rats. The triacylglycerol fraction of bone lipids was most dramatically affected by the experimental procedures, whereas phospholipids did not change. Only 50–60% of the total lipids isolated could be accounted for as triacylglycerols, phospholipids, and cholesterol or its esters.

Similar experiments were conducted (Cruess and Sakai, 1972) to determine cortisone effects upon phospholipid synthesis in bone. Groups of animals were treated as above and portions of metaphysis and diaphysis incubated with radioactive phosphate. Only phosphatidylcholine and phosphatidylethanolamine incorporated significant amounts of radioactivity. Adrenalectomy increased incorporation of [^{32}P]phosphate into both lipids in metaphysis and diaphysis, whereas cortisone injection had little effect upon intact or adrenalectomized rats. In terms of phospholipid content, adrenalectomy increased the amount of both the choline and ethanolamine phosphoglycerides in metaphysis and diaphysis, and cortisone treatment caused a decrease.

c. Growth Hormone. Observations that growth hormone alters systemic lipid metabolism as well as producing changes in bones of growing animals prompted Sakai *et al.* (1968) to study the effect of this hormone on bone matrix lipids. Groups of rats (control, hypophysectomized, and hypophysectomized–growth hormone treated (4 mg/rat)) were maintained for 7 days and the cortical bone of long bones taken for lipid analysis. Hypophysectomized rats lost weight following the operation, but regained some of the weight if given growth hormone. Hypophysectomy resulted in significant increases in all lipid classes analyzed; total lipids, phospholipids, triaçylglycerols, total fatty acids, and cholesterol. Administration of growth hormone to hypophysectomized rats reduced the elevated values to normal.

d. Estrogen. Studies by Sakai *et al.* (1970) have demonstrated the effects of estrogen upon lipids of rat diaphysis, metaphysis, and epiphysis. Both male and female rats were administered estradiol benzoate (0.8 mg/100 g body weight twice a week for 3 weeks) beginning 14 days after castration. Both normal and castrated rats were used for comparison. Some sex-related differences were noted. Control female rats had less phospholipids, more diaphyseal triacylglycerols, and less epiphyseal cholesterol in all regions of bone than males. Castration in the male caused little change from the norm, whereas oophorectomy resulted in female values approaching those of the male. Estrogen increased both total lipids and triacylglycerols in most zones of male and female castrates. Further, some sex-related differences were noted in response to estrogen when other lipid classes were considered.

Both *in vitro* and *in vivo* systems have been used to examine the effects of estrogen on bone phospholipid synthesis (Cruess *et al.*, 1973). Control and ovariectomized rats were injected with 17-estradiol benzoate/100 g body weight for 2 weeks. No difference in the total phospholipid content of metaphysis was observed be ween groups. However, experimental rats had elevated values in the diaphysis (0.66 vs 0.81 mg/g bone). Use of [^{32}P]orthophosphate showed that the effect of estrogen on bone was due primarily to increased phosphatidylcholine and sphingomyelin, with no measurable differences in lysophosphatidylcholine, phosphatidylserine, or phosphatidylethanolamine content.

3. *Parathyroid Hormone*. The success of the above experiments in demonstrating an estrogen effect upon bone prompted Tanaka and Hollander (1971) to study the influence of parathyroid hormone on osseous lipid. Parathyroidectomized rats received either the vehicle or 1.0 IU parathyroid hormone/g body weight for periods ranging from 76 to 4 hr before sacrifice. Portions of long bone were incubated with [^{32}P]orthophosphate or other phospholipid precursors. Following 3 days of daily injections, experimental animal bones showed no differences in ash, organic matter, or total extractable lipid when compared to bones from vehicle injected animals. Total phospholipids extractable from bone was greater in hormone-treated animals (1.95 mg/g vs 2.59 mg/g dry bone). Further, [^{32}P]orthophosphate incorporation was almost double control values (310 dpm vs. 665 dpm/10 mg bone) with radioactive serine, ethanolamine, and glycerol giving much the same results. This stimulation was apparently not a general systemic one as [^{32}P]orthophosphate incorporation into liver was not affected.

f. *Age*. The speculation that lipids are involved in calcification prompted an investigation to correlate various lipid classes with bone calcium and phosphate in rats ranging from 10 to 50 days of age (Sakai *et al.,* 1969). Total organic content of epiphysis decreased from 85% to approximately 50% at 50 days of age. That of metaphysis and diaphysis decreased only slightly and averaged some 35% and 30%, respectively. In a like manner, total lipids, phospholipids, cholesterol, total fatty acids, and triacylglycerols also decreased (expressed in mg/g organic material). The calcium and phosphorus content of the epiphysis and diaphysis increased with age, but no such changes were noted in the metaphysis.

Lipid changes in the epiphyseal cartilage of rat tibia in animals ranging from 10 days to 100 days of age have been reported by Foldes *et al.* (1971). Total lipid content decreased from 16.1% to 3.9%, whereas the phospholipid/total lipid ratio increased with age.

VI. *In Vitro Lipid Metabolism by Bone and Bone Cell Cultures*

Until the last decade, little information has been available on lipid synthesis by bone, particularly under *in vitro* conditions. Although various studies have suggested the possible importance of lipids in bone, they provide little information regarding mechanisms of lipid synthesis or breakdown. Furthermore, the various functions of lipids other than storage or structural purposes have only been surmised. It is chiefly because of this lack of information that lipid synthesis in bone has been presumed to follow pathways similar to those described for other tissues. Diagrams depicting lipid synthesis in biological systems have been presented in the introductory chapter. A number of recent studies have been conducted using newborn rat calvaria as an experimental model for the study of lipid biosynthesis *in vitro*. These bones at birth are of comparable thickness to a tissue slice system and are rapidly undergoing calcification. Their lack of marrow elements makes them ideally suited for studies of this nature.

A. *Experimental Methodology*

Under *in vitro* conditions, calvaria are incubated with various radioactive substrates in buffer, the bones demineralized with EDTA, extracted, and the lipids chromatographed on silicic acid papers and localized by autoradiography: Bone cell cultures can be studied by digesting calvaria with collagenase and growing cells in culture as originally described by Peck *et al.* (1964). Such studies have demonstrated that both bone and bone cell cultures have the ability to synthesize lipids from various radioactive precursors (Peck and Dirksen, 1966). An analysis of the results allows one to speculate on the importance of the various metabolic pathways described in the introductory chapter.

B. *Incorporation of [^{14}C]Glycerol into Lipids of Rat Calvaria and Bone Cell Cultures*

When rat calvaria and bone cell cultures were incubated in media containing [^{14}C]glycerol, the radioactivity of all identified lipid components generally increased throughout the incubation period (Dirksen *et al.*, 1970a). Triacylglycerols and phosphatidylcholine were the two major lipid classes labeled (approximately 50% and 20% of total incorporated ^{14}C, respectively). Phosphatidylserine, cholesterol, sphingomyelin, and fatty acids did not become labeled with glycerol. Of the various tissues studied by Marinetti *et al.* (1964), only liver and kidney were able to convert glycerol into lipids to any appreciable extent. With bone and bone cells, the rapid incorporation of [^{14}C]glycerol, as well as [^{32}P]phosphate (see section VI.D), suggests that bone possesses an active glycerol kinase.

Exogenous ATP was found to stimulate neutral lipid synthesis from [^{14}C]glycerol, but retarded phospholipid synthesis at low glucose concentrations (0 or 18 mg%). At higher glucose levels (100 mg%), decreased lipid labeling from radioactive glycerol resulted, probably through an isotope dilution effect. At higher levels of ATP (0.01 M ATP as opposed to 0.004 M ATP) and "cold" glucose, varied results were obtained, which suggests that critical levels of both substrates are required for optimum lipogenesis. CTP (0.01 M) stimulated synthesis of di- and triacyl-glycerols and phosphatidylcholine, but inhibited that of phosphatidylethanolamine.

C. *Incorporation of [^{14}C]Glucose into Lipids of Rat Calvaria and Bone Cell Cultures*

The labeling of lipids from [^{14}C]glucose by calvaria and cell cultures was similar to the pattern obtained with [^{14}C]glycerol (Dirksen *et al.*, 1970a). Radioactive carbon from glucose, like that of glycerol, was found almost exclusively in the glycerol moiety of the various lipids, thus demonstrating conversion of glucose to *sn*-glycerol-3-P. This presumes that glucose is metabolized by the Embden–Meyerhoff system to yield glyceraldehyde-P and dihydroxyacetone-P, the latter compound being converted to *sn*-glycerol-3-P. No significant radioactive cholesterol or fatty acids were observed in studies with [^{14}C]glucose and bone, showing that glucose did not yield sufficient acetate for lipogenesis during the time period of the experiments. The synthesis of lipid from [^{14}C]glucose was generally inhibited by 0.01 M ATP. The allosteric effect of phosphofructokinase must be considered in this inhibition.

D. *Incorporation of [^{32}P]Phosphate into Lipids of Rat Calvaria and Bone Cell Cultures*

[^{32}P]Orthophosphate was readily incorporated into lipid by both bone and bone cell cultures with over 50% of the incorporated radioactivity migrating with phosphatidylcholine (Dirksen *et al.* 1970b). Phosphatidylinositol, phosphatidylethanolamine, and cardiolipin contained most of the other ^{32}P. For all practical purposes, sphingomyelin and phosphatidylserine did not incorporate [^{32}P]phosphate in these experiments. Supplementation of media with 0.01 M ATP results in decreased labeling of lipids from [^{32}P]phosphate. This is to be expected if phosphate must first be incorporated into ATP before its appearance in lipid.

E. *Incorporation of [^{14}C]Acetate into Lipids of Rat Calvaria and Bone Cell Cultures*

That bone has the ability to synthesize fatty acids as well as cholesterol has been demonstrated in studies employing [^{14}C]acetate as a labeled precursor (Dirksen, 1969, 1971b). [^{14}C]Acetate was incorporated into the fatty acids of several different lipid classes. Krebs–Ringer phosphate buffer was better for lipogenesis than a carbonate buffer. As with [^{14}C]glycerol and [^{14}C]glucose, triacylglycerols and phosphatidylcholine were the two most heavily labeled lipid classes with some 30% of incorporated counts found in each of these two fractions. An unidentified component, which migrated between cholesterol esters and triacylglycerols during chromatography, also contained radioactivity. Utilization of [^{14}C]palmitic acid and [^{14}C]oleic acid also resulted in the labeling of a compound with similar chromatographic characteristics. The unknown may be fatty acid methyl esters.

Saponification of bone and bone cell culture lipid extracts following incubation with [^{14}C]acetate gave radioactive hexane-soluble materials which migrated during paper chromatography with mobilities characteristic of fatty acids, squalene, and cholesterol (Schuster *et al.*, 1975). Other radioactive lipids were also observed by thin-layer chromatography. On the basis of R_f and R_c values, dihydrocholesterol, squalene, lanosterol, lathosterol, and demosterol were identified. The presence of 7-dehydrocholesterol in bone and bone cell cultures was not confirmed.

Different types of fatty acids synthesized by bone from [^{14}C]acetate have been investigated in collaboration with Dr. John Coniglio (unpublished observations). Following incubation of calvaria with [^{14}C]acetate, C-16:0 comprised some 40–50% of synthesized radioactive fatty acids; C-14:0, 21–30%; and C-18:0, up to a maximum of 16%. It is suggested that in view of the large proportion of radioactive palmitic acid, an active fatty acid synthase system is operative within bone.

F. Incorporation of Radioactive Bases into Lipids of Rat Calvaria

Calvaria of newborn rats were incubated with [^{14}C]serine, [^{14}C]choline, [^{14}C]ethanolamine, [^{14}C-methyl]methionine, and [^{3}H]inositol for more positive identification of lipids and to further study bone lipid metabolism.

1. [^{14}C]Serine

Since radioactive phosphatidylserine was not observed in extracts of calvaria or bone cell cultures incubated with radioactive glycerol, glucose, or phosphate, experiments were conducted with [3-^{14}C]serine to determine whether bone had the capacity to synthesize this lipid. Labeled serine was incorporated not only into phosphatidylserine, but also into sphingomyelin, lysophosphatidylserine and phosphatidylethanolamine. Other unidentified components also became radioactive (Dirksen *et al.*, 1970b).

Addition of 0.01 M ATP to the incubation media stimulated the incorporation of [^{14}C]serine into phosphatidylserine some 50% over control values. A more dramatic effect was obtained with 0.01 M CTP in that experimental values ranged up to 550% over control values. Hubscher *et al.* (1959) noted that both CMP and CTP stimulated [^{14}C]serine incorporation into phospholipids of rat liver mitochondria. They did not, however, separate phosphatidylserine from other phospholipids. A base exchange mechanism has been documented in rat liver for phosphatidylserine synthesis, but this does not explain the role of ATP or CTP observed in bone. A CDP-diglyceride pathway is known for bacteria, but has not yet been observed in mammalian systems (see the introductory chapter, Vol. 1).

The amount of radioactivity observed in phosphatidylethanolamine when [^{14}C]serine was used as the radioactive precursor far surpassed that obtained with radioactive glycerol, glucose, or phosphate. One must therefore consider the decarboxylation of phosphatidylserine to form phosphatidylethanolamine as an active pathway within bone.

Migration of phosphatidylserine during paper chromatography of [^{14}C]serine labeled lipids depended upon the methodology used during lipid extraction. Phosphatidylserine obtained from bone before decalcification with EDTA streaked and formed an elongated spot. Calvaria demineralized with EDTA before lipid extraction yielded a discrete phosphatidylserine area, thus suggesting formation of a phosphatidylserine-calcium salt when bone mineral was readily available. The major bulk of phosphatidylserine was obtained from bone only after EDTA demineralization.

2. [^{14}C]*Ethanolamine*

[^{14}C]Ethanolamine was employed to more positively identify phosphatidylethanolamine and its lyso-derivative (Dirksen, 1975). Some 90% of the incorporated ^{14}C was observed in the parent compound at the end of a 4-hr incubation period. No radioactivity from this precursor appeared in phosphatidylcholine.

3. [*Methyl*-^{14}C]*Methionine*

The ability of bone to synthesize phosphatidylcholine through progressive methylation of phosphatidylethanolamine by S-adenosyl-methionine appears lacking since radioactivity from [^{14}C]ethanolamine failed to appear in phosphatidylcholine, and the methyl group of [methyl-^{14}C]methionine failed to become incorporated into lipid by bone. Comparable experiments conducted with newborn rat liver rapidly utilized the methyl group of methionine for lipid synthesis (Dirksen, 1975).

4. [^{14}C]Choline

Phosphatidylcholine was the most heavily labeled lipid synthesized from [^{14}C]choline by calvaria. Some 90% of incorporated counts were usually found in this particular lipid. Radioactive lysophosphatidylcholine and sphingomyelin were also synthesized, as was phosphatidylethanolamine on occasion (Dirksen, 1975).

5. Double Isotope Studies

Utilization of a double labeling technique which employed either [^{14}C]choline, [^{14}C]ethanolamine, or [^{14}C]serine with [^{3}H]glycerol suggests that different rates of base exchange occur in regard to lipid synthesis within bone (Dirksen, 1975). Virtually all phosphatidylserine synthesis in bone appears to result from base exchange as suggested by the high [^{14}C]serine/[^{3}H]glycerol ratios, and preliminary observations suggest that high calcium levels in the medium facilitates serine incorporation. Phosphatidylethanolamine synthesis also occurs by base exchange, but *de novo* synthesis as well as phosphatidylserine decarboxylation also play a role. The greater portion of phosphatidylcholine labeling from [^{14}C]choline appears to be due to *de novo* synthesis, probably by the CDP-choline pathway.

G. In Vitro Incorporation of [^{14}C]Palmitic and [^{14}C]Oleic Acids into Rat Calvaria

Studies were conducted to discern the capability of rat calvaria to utilize exogenous fatty acids for lipid synthesis under *in vitro* conditions and also to investigate the identity of the unknown lipid component, which migrated chromatographically between the cholesterol esters and triacylglycerols (see section VI.E).

The unknown was previously observed to become radioactive when bones were incubated with [^{14}C]acetate but not with [^{14}C]glycerol or [^{14}C]glucose. The radioactivity incorporated into lipid from [^{14}C]palmitic acid by rat calvaria increased throughout a 4-hr incubation period as can be seen in Fig. 4 (Burnett and Dirksen, 1972). Similar results were observed with [^{14}C]oleic acid and both of the precursors labeled the unknown lipid. The overall distribution of radioactivity among the different lipid classes was somewhat similar to that observed with [^{14}C]acetate except that over 50% of incorporated counts was found in free fatty acids. This high percentage of radioactivity in free fatty acids was partially attributed to binding of the labeled substrate by calcium of the mineral phase (soap

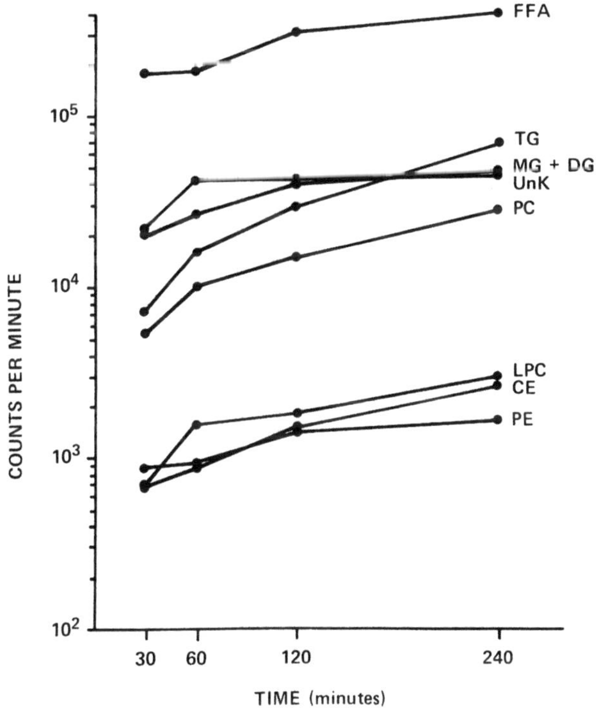

FIGURE 4. Incorporation of [14C]palmitic acid into lipids of rat calvaria over a 4-hr period. Four groups of approximately 15 calvaria each were incubated for varying periods of time in 4 ml Krebs–Ringer bicarbonate buffer (pH 7.2) containing 5 μCi [14C]palmitic acid. The radioactivity of most lipid classes increased with incubation time. Abbreviations: FFA, free fatty acids; TG, triacylglycerols; MG + DG, mono- and diacylglycerols; UnK, unknown; PC, phosphatidylcholine; LPC, lysophosphatidylcholine; CE, cholesterol esters; PE, phosphatidylethanolamine.

formation). Most of the remaining label was observed in triacylglycerols, phosphatidylcholine, and the unknown lipid.

While identity of the unknown lipid remains to be established, methyl esters of fatty acids remain a likely candidate. This was suggested by experiments in which treatment of radioactive lipids with sodium methoxide resulted in formation of labeled methyl esters having chromatographic mobilities identical to the unknown. Such methyl esters of fatty acids may result as solvent artifacts, since extracts of frozen–thawed bones incubated with [14C]palmitic acid included the labeled unknown lipid, particularly when using acidified solvents during extraction procedures. No other lipid class in extracts from frozen–thawed bones was radioactive, except for free fatty acids.

H. Lipid Synthesis by Frozen–Thawed Bones

The studies (described in section VI.F) with newborn rat calvaria suggested that labeling of lipids from radioactive precursors may occur by *de novo* synthesis or by base exchange, the former being energy dependent while the latter is not. To further support this assumption, lipid synthesis by frozen–thawed bones was studied using a variety of metabolic precursors (Dirksen *et al.* 1974). Groups of bones were frozen–thawed three times to the temperature of dry ice–acetone and incubated along with control groups with one of the following substrates: [^{14}C]acetate, [^{14}C]glycerol, [^{14}C]choline, [^{14}C]ethanolamine, [^{14}C]palmitic acid, [^{14}C]serine, or [^{3}H]inositol.

The incorporation of [^{14}C]choline into phosphatidylcholine, lysophosphatidylcholine, and sphingomyelin by experimental bones was less than 20% of control values, thus supporting the concept that base exchange is of minor consequence for phosphatidylcholine synthesis in bone. [^{3}H]Inositol, [^{14}C]palmitic acid, and [^{14}C]acetate incorporation into lipid was also greatly diminished. Labeling of phosphatidylethanolamine and phosphatidylserine from [^{14}C]ethanolamine and [^{14}C]serine, respectively, was significantly higher in frozen–thawed bones as compared to control bones. Experimental counts were approximately 120–130% of control values, which support the importance of base exchange for synthesis of these lipids by bone.

I. Effect of Metabolic Inhibitors on Bone Lipid Synthesis

The effects of iodoacetate, sodium cyanide, and dinitrophenol, and in some instances, fluoride upon incorporation of various radioactive precursors into lipids by calvaria have been examined.

1. [^{14}C]Acetate

Newborn rat calvaria were incubated with and without various metabolic inhibitors for 4 hr in the presence of [^{14}C]acetate and the lipids extracted. [^{14}C]CO_2 and lactic acid production were also monitored. The labeling of all lipids was significantly reduced in the presence of iodoacetate at both 10^{-3} M and 10^{-4} M concentrations. Experimental values were less than 13% of control values, with the exception of fatty acids, which were 45% of control values at 10^{-4} M. Synthesis of phosphatidylinositol and sphingomyelin were reduced to the point where these lipids were not detected on autoradiograms (Dirksen and McPherson, 1973).

Dinitrophenol and sodium cyanide at 10^{-5} concentration had no effect upon [^{14}C]acetate incorporation into lipids by bone throughout a 4-hr incubation period. Lipid labeling was diminished in the presence of 10^{-4} M and 10^{-3} M dinitrophenol, but not to the extreme extent as that observed with iodoacetate. Except for monoacylglycerols, diacylglycerols, and cholesterol, where no effects were observed, 10^{-4} M dinitrophenol reduced lipid radioactivity for all other identified lipid classes to approximately one-half that of control bones. In contrast to the effects observed with iodoacetate and dinitrophenol, sodium cyanide at 10^{-4} M resulted in enhanced lipid radioactivity. Experimental values ranged from 150% to 200% of control values for all lipid classes. This was true for both 10^{-3} M and 10^{-4} M sodium cyanide. Fluoride, at 100 ppm or more was required to cause decreased lipid labeling by bone during a 4-hr incubation period. The presence of significant amounts of mineral in the calvaria may have protected bone cells from the deleterious effects of fluoride by precipitation of calcium fluoride (Dirksen, 1972).

Elevated lactic acid production was noted with dinitrophenol, whereas sodium cyanide had no effect at the concentration employed. Iodoacetate at 10^{-4} M resulted in significantly less lactic acid production when compared to control bones.

A concentration of 50 ppm fluoride, while not affecting lipid synthesis, resulted in decreased [^{14}C]CO_2 production. Iodoacetate reduced [^{14}C]CO_2 production to less than half that of control bones. The other two metabolic inhibitors had no effect upon this parameter of metabolism.

2. [^{14}C]*Glycerol*

Iodoacetate, a potent inhibitor of anaerobic glycolysis, resulted in low levels of both radioactive lipids and [^{14}C]CO_2 production from radioactive glycerol by rat calvaria. Experimental lipid radioactivities were 15% and 26% of control values, respectively, with 10^{-3} M and 10^{-4} M iodoacetate. Dinitrophenol at 10^{-3} M and 10^{-4} M resulted in decreased lipid radioactivity, but not to the same extent as that observed with iodoacetate (18–48% of control values). [^{14}C]CO_2 production in the presence of this metabolic inhibitor was also decreased (Dirksen and McPherson, 1974).

Sodium cyanide at 10^{-4} M had little effect upon lipid radioactivity, whereas 10^{-3} M values were 65% of control values: [^{14}C]CO_2 production was depressed at both concentrations. The observed effects with sodium cyanide are in contrast to those of the previous section, where this metabolic inhibitor stimulated lipogenesis from [^{14}C]acetate above control values.

3. *Radioactive Bases*

Calvaria were incubated with [^{14}C]choline, [^{14}C]ethanolamine, [^{14}C]serine, and [^{3}H]inositol in the presence of various metabolic inhibitors, including iodoacetate, sodium cyanide, and dinitrophenol (Dirksen *et al.* 1975c). Iodoacetate at 10^{-3} M and 10^{-4} M strongly inhibited incorporation of both [^{14}C]choline and [^{3}H]inositol into their respective phospholipids, had somewhat less of an effect upon [^{14}C]ethanolamine utilization, and almost no effect upon [^{14}C]serine incorporation. Somewhat similar results were obtained from experiments employing cyanide and dinitrophenol at similar concentrations. The results suggest that choline and inositol incorporation into calvaria lipids occurs by energy-requiring reactions, that enthanolamine labeling occurs by both base exchange and *de novo* lipid synthesis, and that serine incorporation is almost exclusively by base exchange.

The results of the preceding experiments, as well as those of Tanaka and Hollander (1971), suggest that lipid synthesis within calcified tissues depends quite heavily upon an active glycolytic pathway for lipid synthesis. Inhibitors of the Embden–Meyerhof pathway significantly retard lipogenesis whereas those exerting their action on oxidative phosphorylation have little effect.

J. *In Vitro Bone Lipid Synthesis in Air or Nitrogen*

While several investigators have studied oxygen tension and its effect on bone resorption, oxygen partial pressure and lipid synthesis have only recently been reported (Dirksen, 1971a). In this regard, calvaria were incubated for 4 hr with radioactive substrates in a 5% CO_2/95% air or nitrogen atmosphere. The radioactivity of almost all [^{14}C]acetate-, [^{14}C]ethanolamine-, and [^{14}C]serine-labeled lipids was enhanced when calvaria were incubated in the nitrogen atmosphere and some elevation was observed with [^{14}C]choline. Labeling of lipids with [^{14}C]glucose and [^{14}C]glycerol as precursors was considerably less in the nitrogen-incubated bones. Whether these results were due to decreased lipolysis or enhanced lipogenesis remains to be determined, although the former seems a more plausible explanation.

K. *Effect of Exogenous Lipids on Lipid Synthesis from [^{14}C]Acetate by Bone and Bone Cell Cultures*

Studies were conducted to determine the effects of exogenous lipids on the pattern of lipid synthesis by rat calvaria and bone cell cultures (Schuster *et al.*, 1975). Bone or bone cell cultures were incubated with [^{14}C]acetate in the presence or absence of 5 mg% palmitic acid (complexed with albumin) or 10% fetal bovine serum. The majority of incorporated radioactivity by calvaria from [^{14}C]acetate was found in phosphatidylcholine, free fatty acids, cholesterol, and mono- and diacylglycerols. These different lipid classes represented 70% of total ^{14}C incorporated. The inclusion of 10% serum in the incubation media resulted in decreased radioactivity for all lipid classes to approximately 40% of control counts. The greatest decrease was observed with free fatty acids. In bone cell cultures, a different distribution of [^{14}C]acetate among the different lipid classes was observed. Some 40% of total incorporated radioactivity was found in the mono- and diacylglycerols plus cholesterol fraction. As with calvaria, bone cells utilized less [^{14}C]acetate for lipogenesis when grown in the presence of serum. Substitution of serum with 5 mg% palmitic acid bound to albumin also resulted in decreased radioactivity for all lipid classes in both calvaria and bone cell systems.

The production of [$^{14}C]CO_2$ by calvaria from [^{14}C]acetate in media with and without added serum was also monitored throughout a 4-hr incubation period. In three separate studies, experimental values ranged from 80% to 106% of control values. In only one experiment were results significantly different to indicate that serum had an effect upon [$^{14}C]CO_2$ production by calvaria from [^{14}C]acetate.

L. *Effect of Beta-Aminoproprionitrile (BAPN) on Lipid Synthesis from [^{14}C]Acetate by Calvaria and Bone Cell Cultures*

The lathyrogenic compound BAPN has been used to study various aspects of connective tissue production, especially that of collagen. The possible close association between collagen and extracellular lipid as suggested by Sudan black B staining (Melcher, 1966, 1969) prompted the investigation of lipid synthesis by bone and bone cells as affected by BAPN (Dirksen *et al.*, 1975b). Newborn rat calvaria were incubated with or without 1–100 mM BAPN for 1 to 24 hr before a 4-hr incubation period with [^{14}C]acetate; [$^{14}C]CO_2$ was collected in hyamine. [^{14}C]Acetate incorporation into lipids was found to be dependent upon BAPN concentration and incubation time. Generally, exposure of bones to 1, 5, or 10 mM BAPN throughout a 5-hr incubation period resulted in little effect upon lipid synthesis when compared to that of control bones. Higher BAPN concentrations from 20 to 100 mM during this short incubation period resulted in enhanced lipid radioactivity for many of the identified lipid classes (with a major exception of neutral lipids). A longer preincubation period of 20 hr and low BAPN concentrations also resulted in enhanced lipid radioactivity. At BAPN concentrations above 20 mM, lipogenesis decreased to where some lipid classes failed to become labeled after the 20-hr preincubation period. [$^{14}C]CO_2$ production, as affected by BAPN concentration and incubation time generally followed a pattern similar to that of lipid radioactivity.

Bone cells harvested by collagenase digestion of rat calvaria and grown in culture were also studied with BAPN (Erbland *et al.*, 1974). Cell morphologic alterations were monitored throughout the incubation by standard histological techniques. BAPN at 20 and 40 mM decreased lipid labeling from [^{14}C]acetate and 5 mM BAPN enhanced lipid labeling regardless of preincubation time. Results with 10 mM BAPN were variable. Distinct cytoplasmic vacuoles became evident within cells at 2 hr after addition of 5 or 10 mM BAPN. Their number became maximal at 4 hr and remained so throughout 24 hr. Electron microscopy revealed dilatation of cytoplasmic vacuoles and blebbing of the endoplasmic reticulum.

Both [^{3}H]acetate and [^{14}C]acetate were utilized for autoradiographic studies of bone cells incubated with or without BAPN. Calvaria incubated with 5 mM BAPN showed more radioactivity than did controls. Lipid histochemistry demonstrated the presence of sudanophilic and oil red-O positive material in the osteoblasts and osteoid tissue. The bone cells showed vaculation after BAPN incubation with continued presence of lipid droplets. Fine structural studies of the bone cells indicated the vacuoles to be ballooning endoplasmic reticulum with cytoplasmic pseudopod formation. The morphological data correlated well with the biochemical data of increased lipid radioactivity in BAPN-treated bone and bone cells.

M. Effect of Chlorophenoxyisobutyrate (CPIB) on Incorporation of Radioactive Precursors into Lipids of Calvaria

CPIB is a commonly employed drug used in the treatment of hyperlipemia. Its exact mechanism of action remains to be documented, but reduced mobilization from adipose tissue and decreased synthesis of liver cholesterol have been suggested. To test the effects of CPIB on bone and bone cell lipid metabolism, concentrations ranging from 0.1 mg/ml to 1.0 mg/ml of the agent were added to media containing radioactive substrates (Dirksen *et al.*, 1975a).

The results demonstrated a direct effect of CPIB on tissues over and above that which might result at the liver or adipose tissue level. CPIB had little effect on calvaria lipid metabolism at 0.1 mg/ml, but bone cell cultures demonstrated stimulation of [^{14}C]acetate incorporation into triacylglycerols. Synthesis of phospholipids and cholesterol by cell cultures was lower at 1.0 mg/ml CPIB. Incorporation of [^{14}C]serine, [^{14}C]ethanolamine, [^{14}C]glucose, [^{14}C]glycerol, and [^{3}H]inositol into calvaria lipids was reduced in the presence of 1.0 mg/ml CPIB, but [^{14}C]acetate and [^{14}C]choline utilization was affected very little.

N. Effect of Triton WR-1339 on Lipid Synthesis by Calvaria and Bone Cell Cultures

The hyperlipemic action of Triton WR-1339 in whole animals as well as its effect on organs removed from injected animals has been known for years. Lipid biosynthesis by rat calvaria and bone cell cultures was studied *in vitro* by incubation in media containing [^{14}C]acetate, with or without 250 μg/ml Triton (McPherson *et al.*, 1975). A preincubation period of 1 or 20 hr was employed before addition of the labeled precursor.

The studies demonstrated an *in vitro* effect of the agent in that it stimulated synthesis of most lipid classes. Many experimental values ranged from 1.5 to 10 times those of control values. Cell cultures incubated for 20 hr appeared to thin out and more floating cells were observed in the presence of Triton. This suggested a toxic effect of the agent which heretofore has not been reported at the concentration employed.

O. *Lipid Biosynthesis from [^{14}C]Acetate by Rat Calvaria of Different Ages*

Bone at the age of newborn rat calvaria is characterized by rapid growth and mineralization. A study was therefore conducted to determine whether lipid synthesis from [^{14}C]acetate by calvaria of different ages might in some manner be related to the degree of mineralization (Dirksen and O'Dell, 1972). Calvaria from rats 1, 7, and 14 days of age were incubated with [^{14}C]acetate for 4 hr and the lipids extracted. Wet weight, dry weight, ash weight, and calcium content were determined in bones of similar ages to those used in lipid experiments.

Calvaria in the three age groups were found to differ with regard to lipid synthesis from [^{14}C]acetate under *in vitro* conditions. Newborn calvaria converted 3.89% of media radioactivity into lipids compared to only 0.67% for the oldest bones. Free fatty acids, some 32% of incorporated ^{14}C, constituted the most heavily labeled lipid class, followed in turn by the mono- and diacylglycerols plus cholesterol, and phosphatidylcholine. While total lipid radioactivity decreased with age, the percent of total ^{14}C incorporated into triacylglycerols increased. From birth to 14 days of age, bone water decreased as total solids increased. The percent ash of total dry weight did not vary with age, although the calcium content of ash did increase.

VII. *The Role of Lipids in Calcification*

Theories on the mechanism of calcification abound and their number is virtually as extensive as there are investigators in the field. Seemingly every tissue component has at one time or another been proposed as being responsible for the initial precipitation of mineral. This includes organic phosphate esters, collagen, chondroitin, sulfate, glycoseaminoglycans, lipids, enzymes, and various cell organelles. While much has been learned, the problem is far from resolved (Irving, 1973b).

The requirements for biological mineralization include an organic matrix and mechanisms sufficiently unique to concentrate ions in the appropriate location. Under conditions of high calcium flux such as found in the hypertrophic zone of the epiphysis, mitochondria accumulate both calcium and phosphate in the form of granules (Matthews *et al.*, 1970). As mentioned previously, the inorganic fraction of mammalian hard tissue belongs to the apatite family although both dentin and bone contain measurable amounts of amorphous calcium phosphate. The latter appears to mature into the former with increasing age. The concomitant loss of sudanophilia and mitochondrial granules in the rachitic growth plate (Irving, 1960, Matthews *et al.*, 1970), coupled with specific lipid classes not being extracted from bone, except through use of acidified solvents, has supported the role of lipids in the mineralization process. The attraction between lipid and calcium is quite evident to any person who has attempted to use soap in hard water, and Johnson (1960) suggested that calcification might result through formation of calcium salts of fatty acids.

A. *Binding of Calcium and Phosphate by Lipids*

In view of the high lipid content in cell membranes, as well as the importance of ionized calcium in various membrane phenomena, much interest has developed in the binding characteristics of this ion with both artificial and natural membranes as well as with lipids in solution. As an example, Seimiya and Ohki (1972) studied the accessibility of calcium ions to anionic sites of lipid membranes. Lipids studied were phosphatidylcholine, phosphatidylethanolamine, stearic acid, and phosphatidylserine. With the first two lipids, the site of interaction was the phosphate group. The latter two bound calcium through the carboxyl group, and in addition, phosphatidylserine by the phosphate group. The degree to which association occurred was influenced by lipid concentration, since increased molecular packing masked the phosphate group. Thus, both phosphatidylcholine and phosphatidylethanolamine demonstrated an inverse correlation between concentration and ion binding capacity as the phosphate group became progressively inaccessible.

Calcium can also influence the organization of lipids within membranes (Omnishi and Ito, 1973). When the ion was added to artificial membranes composed of phosphatidylcholine and phosphatidylserine, rapid formation of calcium phosphatidylserine aggregates occurred leaving clusters of phosphatidylcholine. While these experiments were conducted to examine membrane permeability and excitability, they also allow one to consider the sequestering of calcium by membranes rich in acidic phospholipids and also help to explain why certain lipids are so difficult to extract from tissues rich in inorganic salts.

Variations in the attraction of different phospholipids for the mineral phase of bone has been investigated by Shapiro (1970b). Phospholipids, extracted from kidney, spleen, and liver, were mixed with powdered inorganic bone. The bone was then eluted sequentially with chloroform, chloroform–methanol, and chloroform–methanol–HCl. Approximately 80% of the lipid was removed during the first chloroform extraction, and only 4% required acidified solvents for removal. The neutral phospholipids were extracted primarily with chloroform or chloroform–methanol, whereas the acidified solvents were necessary for extraction of the acid phospholipids.

In other types of experiments, the transport of calcium from one aqueous phase through an organic solvent to another aqueous layer has been shown to be facilitated by the presence of dissolved lipids (Woolley and Campbell, 1962). In addition to calcium, lipids can also attract inorganic phosphate, though the amount is influenced by the presence of cations (Bader, 1962). Treatment of lipid extracts containing inorganic [^{32}P]phosphate with aqueous 0.1 N $CaCl_2$ left 60 times more radionucleotide than extracts treated with 0.1 N KCl.

The observation of membrane-bound vesicles (section VII.B) in epiphyseal plates of growing animals prompted Cotmore *et al.* (1971) to study calcium and phosphate binding by phospholipids. At physiological pH, each molecule of phosphatidylinositol and phosphatidylserine complexed slightly less than 0.5 molecules of calcium. Phosphatidylethanolamine was less effective, whereas phosphatidylcholine failed to bind calcium. Addition of phosphate to the mixture greatly enhanced the attraction of lipid for ions. With phosphatidylserine, a complex formed which had the calcium–phosphate–lipid molecular ratio of 12.12/6.74/1.0. The complex was easily visible by electron microscopy and existed in the form of spherules with an average diameter of 175–280 Å. A typical lipid bilayer with calcium and phosphate attracted to lipid polar groups was assumed.

A complex consisting of 2.3% calcium, 1.4% inorganic phosphate, and 80% phospholipid (by weight) has been isolated from bone by Boskey *et al.* (1975). The undissociated complex would not migrate during chromatography and required dissolution in formic acid for lipid analysis. The lipid portion consisted mainly of phosphatidic acid and phosphatidylserine.

B. Matrix Vesicles

As mentioned in section VII.A, electron-micrographic studies have demonstrated the presence of small (1,000 Å diameter), membrane-bound, extracellular vesicles in tissues undergoing calcification. They are considered the site of initial mineral deposition since the first deposits of apatite recognizable by electron microscopy are associated with these structures. Matrix vesicles have been observed in epiphyseal cartilage, bone, dentin, and other calcifying structures which include aorta, algae, oral bacteria, fish scales, and deer antlers. A recent review by Anderson (1976) includes numerous references for these studies. In cartilage, the origin of matrix vesicles would appear to be the chondrocyte through processes not yet well understood, although budding from cells seems a likely possibility (Rabinovich and Anderson, 1976). While these structures have not yet been demonstrated in enamel, it might be that this structure also relies upon vesicles for mineralization. Crystal growth by apposition onto dentin vesicles could spread across the dentin–enamel junction into the enamel matrix, thus explaining the observation that dentin mineralization always precedes that of enamel.

Kashiwa and Mukai (1971) reported both cytoplasmic and extracellular spherules in and around chondrocytes of developing bones. These structures had staining characteristics of lipid with Sudan black B and were also positive for calcium and phosphate. Acidified chloroform–methanol removed the lipid-staining characteristics.

Through the use of an electron probe, Ali (1976) determined the calcium and phosphate content of matrix vesicles distributed throughout a single septum of rabbit epiphyseal plate from the resting cartilage zone through the calcified zone. Matrix vesicles appeared to be secreted into the proliferative zone without either calcium or phosphate. As analysis progressed toward the calcification front, the vesicle content of first calcium and then phosphate was found to increase progressively.

It was previously mentioned (section III.C.2) that the sudanophilic reaction is lost from the epiphysis during rickets, but is rapidly restored following addition of Vitamin D to the diet (Irving, 1960, 1976). Simon *et al.* (1973) noted that matrix vesicles in the rachitic rat contained no crystallites throughout the entire epiphyseal zone. During healing, however, mineral was deposited in vesicles of the hypertrophic zone 12 hr following administration of phosphate and vitamin D. This time period corresponds quite favorably to that recorded by Irving (1960) for the vitamin-D-induced return of sudanophilia to the calcification zone. Aspiration of fluid from healing rachitic rat cartilage has also demonstrated the presence of matrix vesicles. When they were added to metastable solutions of calcium and phosphate, mineral formed. The mineral-forming agent was resistant to acid demineralization, but was destroyed by freezing and thawing, heating, sonication, and phospholipase A at 10^{-5} M but not at 10^{-7} M (Howell *et al.*, 1976).

Vesicles have been isolated in relatively pure form for both chemical analysis and enzymatic assay. Bovine fetal cartilage, digested with collagenase, was partitioned into several fractions by differential centrifugation (Ali *et al.*, 1970). Extracellular fractions containing matrix vesicles in addition to apatite crystals were found to have the highest specific activity of any cellular fraction for alkaline phosphatase, pyrophosphatase,

ATPase, and 5′-AMPase. The former has been recognized for years as being quite characteristic of mineralized tissue formation. Studies on matrix vesicles isolated from chick epiphyseal cartilage suggest that both magnesium-dependent ATPase and pyrophosphatase activities are due to a single enzyme, alkaline phosphatase (Majeska and Wuthier, 1975).

A phosphatase of cartilage, active at neutral pH, has a wide specificity for different substrates. These include nucleotide triphosphates, diphosphates, monophosphates, and other phosphate-containing substrates. No acid phosphatase, β-glucuronidase, or cathepsin D activities were found in the same fraction, thus demonstrating the difference between matrix vesicles and lysozomes (Ali, 1976).

Under *in vitro* conditions, ATP has been shown to stimulate the uptake of ^{45}Ca by matrix vesicles (Ali and Evans, 1973; Ali, 1976), and pyrophosphate has the same effect (Anderson and Reynolds, 1973). On this basis, it would appear that matrix vesicles are able to effectively concentrate calcium and phosphate by a calcium pump and pyrophosphate hydrolysis. Majeska and Wuthier (1975) reported, however, that high levels of calcium were inhibitory to ATPase activity, and Sajdera *et al.* (1976) observed that the ATPase of matrix vesicles was not stimulated by calcium ions. Further, vesicles do not appear to be able to metabolize glucose.

The nature of lipids from matrix vesicles has been studied by Peress *et al.* (1974). Vesicles from fetal calf epiphysis contained 0.54 mg lipid/mg protein as compared to only 0.21 mg/mg protein for chondrocytes from the same source. Vesicles were enriched in phospholipid, cholesterol, and glycolipid. The percentage distribution of phospholipids for matrix vesicles is shown in Table 10. It is quite evident that both phosphatidylserine and

TABLE 10
Lipid Composition of Fetal Calf Epiphyseal Chondrocytes and Matrix Vesicles (mg Lipid/100 mg Protein)[a]

Lipid	Chondrocytes	Matrix vesicles
Phosphatidylcholine	9.58	18.25
Phosphatidylethanolamine	5.82	11.27[b]
Sphingomyelin	1.53	8.05[b]
Phosphatidylinositol	1.53	4.48
Phosphatidylserine	1.41	7.13[b]
Phosphatidic acid	0.25	1.40
Cardiolipin	0.23	1.57
Cholesterol	4.00	11.00
Glycolipids	5.00	18.00

[a]Peress *et al.* (1974).
[b]$P > 0.05$.

sphingomyelin comprise a much larger portion of vesicle total phospholipid, whereas phosphatidylcholine and phosphatidylethanolamine values are of a lesser magnitude when compared to chondrocytes. Wuthier (1976) studied the lipid composition of cartilage cells and vesicle fractions from both calves and chickens. Generally, the phospholipid content in the cell fractions from both species were comparable. A comparison of lipids extracted from tissues after demineralization with those of isolated matrix vesicles suggests that the sudanophilic material seen by Irving and the extracellular matrix lipid of mineralizing cartilage is indeed the lipid of matrix vesicles.

C. *Release of Lipid-Bound Calcium*

It has been suggested that owing to the strong association between phosphatidylserine and calcium at body pH, the calcium of matrix vesicles would not be available for formation of hydroxyapatite or amorphous calcium phosphate (Peress *et al.*, 1974). Yet, release may not be necessary if the calcium ions bound to lipid serve only to orientate inorganic phosphate during the initial stages of mineral aggregation.

Another suggestion for calcium release is offered by the work of Wuthier (1973). Lyso-derivatives of phospholipids are common to lipid extracts of calcified tissues, and Wuthier, using tissue slices from chicken epiphysis under *in vitro* conditions, observed the presence of phospholipase activity within the different zones. Phosphatidylserine turnover was uniquely high in the zone of calcification. While further investigation is required, it is possible that lipid-bound calcium might be released through enzymatically controlled reactions, although Wuthier suggested that the phospholipase activity may be related to maturation of amorphous calcium phosphate.

D. *Induction of Calcification by Lipids*

The observation that apatite formation in the microorganisms *Bacterionema matruchotti* is lipid dependent (Vogel and Ennever, 1971) has been cited as another form of evidence to implicate lipids as being responsible for biological mineralization. Such experiments generally involve placing lipid or an extracted matrix in a calcium phosphate solution. Ennever *et al.* (1974) used bone powder prepared from long bones of marmoset; the material was decalcified and suspended in a metastable solution of calcium and phosphate along with samples that had been extracted with acidified chloroform–methanol. The lipid extracts were fractionated into neutral lipids and phospholipids and also tested in the calcification solution. Results demonstrated that the matrix not extracted of lipids was able to induce the precipitation of mineral having an X-ray diffraction pattern similar to apatite. The phospholipids extracted from bone also induced apatite formation, whereas results with the lipid-extracted matrix and neutral lipids were negative. Essentially the same experiments have been conducted with oral calculus (Ennever *et al.,* 1973) and a lysozyme–phosphatidylinositol complex (Vogel *et al.,* 1973).

Lipids extracted from human teeth, rat teeth, rat liver, and rat kidney were tested by Odutuga *et al.* (1975) for their ability to induce precipitation of hydroxyapatite. Weighed quantities of the different lipid extracts as well as individual lipids obtained commercially were added to metastable solutions of calcium phosphate or carbonate. The precipitates, if any, were then studied by X-ray analysis. Total lipid extracts from each of the tissues were able to induce precipitation of apatite or calcite, respectively, but this ability was due to the phospholipids, not neutral lipids. Up to 2% by weight of the lipid apatite complex was lipid, which could not be extracted unless a demineralization step was employed. The proportion of phospholipids associated with apatite was similar to that observed in matrix vesicles by Peress *et al.* (1974). When individual phospholipids were tested (150 mg lipid/100 ml solution), phosphatidylcholine, sphingomyelin, and phosphatidylethanolamine caused the least precipitation (3.6–6.1 mg), with little of these lipids being bound in the mineral (0.9–1.9% of lipid recovered from precipitate). Phosphatidylserine, phosphatidylinositol, phosphatidic acid, and cardiolipins precipitated some 7.8–8.5 mg mineral, with 2.7% of lipid being recovered in the complex. These studies demonstrate the ability of pure lipid compounds to induce the formation of apatite under *in vitro* conditions, but further work is necessary before the exact *in vivo* mechanism is known.

VIII. Summary

Evidence to support the presence of lipids in calcified tissues has been in existence for a long period of time. Extraction of bulk quantities of bone or tooth components has allowed more exact identification and quantitation, and it was first noted with human dentin that additional lipids may be extracted if sample material is demineralized before extraction. The existence of lipids quite firmly bound to the matrix of calcified tissues has been demonstrated through the use of acidified solvents.

Analysis of different zones of the epiphysis demonstrated the intimate association between acidic phospholipids and calcium deposition. This has its counterpart histochemically in the sudanophilic reaction at sites of active calcification.

The first recognizable deposits of apatite during endochondrial ossification are associated with extracellular matrix vesicles whose lipid and enzymatic qualities have been studied. Their characteristics support the deposition of calcium and phosphate ions as mineral through one or a combination of mechanisms: (1) increased localized concentration of phosphate by hydrolysis of organic phosphate esters (alkaline phosphatase) or pyrophosphate (pyrophosphatase); (2) increased calcium concentration through ion exchange involving phosphatidylserine or other acidic phospholipid molecules in the vesicular membrane; (3) increased calcium concentration in matrix vesicles by a calcium pump (ATPase); (4) destruction of pyrophosphate, a known inhibitor of crystal growth (pyrophosphatase); (5) release of lipid-bound calcium through phospholipase activity; (6) orientation of ions in space by membranes.

While the vesicles have yet to be demonstrated in enamel, their existence has been confirmed in cartilage, dentin, and bone. Their enrichment in acidic phospholipids makes them a prime candidate for the nucleation of calcium for the formation of mineralized tissues in mammals.

References

Alam, S. Q., and Harris, R. S. 1972. Effects of nutrition on the composition of tooth lipids and fatty acids in rats: I. The effect of dietary carbohydrate on fatty acid composition of rat teeth. *J. Dent. Res.* **51**:1474–1477.

Alam, S. Q., Alvarez, C. J., and Harris, R. S. 1973a. Effects of nutrition on the composition of tooth lipids and fatty acids in rats: II. Effects of restriction of a cariogenic diet on caries and lipid composition of molars and incisors. *J. Dent. Res.* **52**:229–235.

Alam, S. Q., Alvarez, C. J., and Harris, R. S. 1973b. Effects of nutrition on the composition of tooth lipids and fatty acids in rats: III. Effects of feeding different oils and fats on caries and on fatty acid composition of teeth. *J. Dent. Res.* **52**:236–241.

Ali, S. Y. 1976. Analysis of matrix vesicles and their role in the calcification of epiphyseal cartilage. *Fed. Proc.* **35**:135–142.

Ali, S. Y., and Evans, L. 1973. The uptake of [^{45}Ca]calcium ions by matrix vesicles isolated from calcifying cartilage. *Biochem. J.* **134**:647–650.

Ali, S. Y., Sajdera, S. W., and Anderson, H. C. 1970. Isolation and characterization of calcifying matrix vesicles from epiphyseal cartilage. *Proc. Natl. Acad. Sci.* **67**:1513–1520.

Allred, H. 1966a. Histochemical observations on the lipids of sound human dentine. *Nature* **210**:646–47.

Allred, H. 1966b. Histochemical observations on the lipids of carious human dentin. *Nature* **210**:748.

Allred, H. 1968a. The differential staining of pertubular and intertubular matrices in human dentin. *Arch. Oral Biol.* **13**:1–11.

Allred, H. 1968b. The staining of lipids in human dentin matrix. *Arch. Oral Biol.* **13**:433–444.

Allred, H. 1968c. Investigations into the relationship between the lipid and other components of human dentine. *Arch. Oral Biol.* **13**:1077–1093.

Allred, H. 1969. The staining of the lipids in carious human dentin. *Arch. Oral Biol.* **14**:271–276.

Anderson, H. C. 1976. Matrix vesicle calcification. Introduction. *Fed. Proc.* **35**:105–108.

Anderson, H. C., and Reynolds, J. J. 1973. Pyrophosphate stimulation of calcium uptake into cultured embryonic bones. Fine structure of matrix vesicles and their role in calcification. *Dev. Biol.* **34**:211–227.

Bader, H. 1962. The uptake of inorganic phosphate by the lipid extract of rat liver. *Biochim. Biophys. Acta* **65**:178–180.

Bibra, E. von. 1884. Chemische Untersuchungen über die Knochen and Zähne des Menschen und der Wirbeltiere, Schweinfurt, Germany, p. 114.

Bodecker, C. F. 1931. The lipin content of dental tissues in relation to decay. *J. Dent. Res.* **11**:278–84.

Bodecker, C. F. 1938. A research conference on the cause and prevention of dental caries, p. 13. The Good Teeth Council for Children, Inc., Chicago, Illinois.

Boskey, A. L., Galperin, D., Ronner, and Posner, A. S. 1975. A calcium-phospholipid complex in bone. Paper presented at the 53rd General Session of the International Association for Dental Research, London, England. Abstract No. L316.

Buck, D. L., Griffith, D. A., and Mills, M. J. 1973. Histologic evidence for lipids during human tooth movement. *Am J. Orthod.* **1973**:619–624.

Burnett, G. W., and Dirksen, T. R. 1972. *In vitro* incorporation of ^{14}C-palmitic acid into rat calvaria. Paper presented at the 50th General Session of the International Association for Dental Research, Las Vegas, Nevada. Abstract No. 665.

Conklin, J. L., Enlow, D. H., and Bang, S. 1965. Methods for the demonstration of lipid applied to compact bone. *Stain Technol.* **40**:183–191.

Cotmore, J. M., Nichols, G., Jr., and Wuthier, R. E. 1971. Phospholipid-calcium complex: Calcium migration in the presence of phosphate. *Science* **172**:1339–1341.

Cruess, R. L., and Clark, I. 1965. Alterations in the lipids of bone caused by hypervitaminosis A and D. *Biochem. J.* **96**:262–65.

Cruess, R. L., and Clark, I. 1971. The effect of vitamin D deficiency on the lipids of bone matrix. *Proc. Soc. Exp. Biol. Med.* **136**:415–419.

Cruess, R. L., and Hong, K. C. 1973. Synthesis rates of phospholipids derived from bone matrix of the rat. *Calcif. Tissue Res.* **13**:305–310.

Cruess, R. L., and Sakai, T. 1972. Effect of cortisone upon synthesis rates of some components of rat bone matrix. *Clin. Orthop. Related Res.* **86**:253–259.

Cruess, R. L., Hong, K. C., and Iida, K. 1973. Effect of estrogen on the phospholipid metabolism of rat bone. *Endocrinology* **92**:961–963.

Das, S. K., and Harris, R. S. 1970a. Fatty acids in the tooth lipids of 16 animal species. *J. Dent. Res.* **49**:119–125.

Das, S. K., and Harris, R. S. 1970b. Lipids and fatty acids in fossil teeth. *J. Dent. Res.* **49**:119–125.

Das, S. K., and Harris, R. S. 1975. Effects of dietary restriction on the composition of lipids in rat teeth. *Arch. Oral Biol.* **20**:131–135.

Debreuil, G. 1910. Vacuoles A lipoides des osteoblastes, des cellules osseuses et des osteoclastes. *C. R. Soc. Biol.* **69**:189–190.

Dirksen, T. R. 1963. Lipid components of sound and carious dentin. *J. Dent. Res.* **42**:128–132.

Dirksen, T. R. 1969. The *in vitro* incorporation of acetate-^{14}C into the lipids of new born rat calvaria. *Arch. Biochem. Biophys.* **134**:603–609.

Dirksen, T. R. 1971a. *In vitro* bone lipid synthesis in air and nitrogen. *Biochem. Biophys. Acta* **231**:458–464.

Dirksen, T. R. 1971b. Bone lipids: *in vitro* synthesis in bicarbonate and phosphate buffers. *J. Dent. Res.* **50**:640–642.

Dirksen, T. R. 1973. *In vitro* effect of fluoride on lipid synthesis by rat calvaria. *Arch. Oral Biol.* **17**:55–59.

Dirksen, T. R. 1975. Incorporation of radioactive bases into calvaria of the new-born rat, *Rattus Norvegicus*. *Comp. Biochem. Physiol.* **50B**:345–349.

Dirksen, T. R., and Ikels, K. G. 1964. Quantitative determination of some constituent lipids in human dentin. *J. Dent. Res.* **43**:246–251.

Dirksen, T. R., and Marinetti, G. V. 1970. Lipids of bovine enamel and dentin and human bone. *Calcif. Tissue Res.* **6**:1–10.

Dirksen, T. R., and McPherson, J. C. 1973. Effect of metabolic inhibitors on bone lipid synthesis from acetate-^{14}C. *Int. J. Biochem.* **4**:102–106.

Dirksen, T. R., and McPherson, J. C. 1974. Effect of metabolic inhibitors on bone lipid synthesis from [^{14}C]glycerol. *Int. J. Biochem.* **5**:321–324.

Dirksen, T. R., and O'Dell, N. L. 1972. *In vitro* lipid biosynthesis from [^{14}C]acetate by *Rattus Norvegicus* calvaria of different ages. *Int. J. Biochem.* **3**:151–154.

Dirksen, T. R., Marinetti, G. V., and Peck, W. A. 1970a. Lipid metabolism in bone and bone cells. I. The *in vitro* incorporation of [^{14}C]glycerol and [^{14}C]glucose into lipids of bone and bone cell cultures. *Biochim. Biophys. Acta* **202**:67–79.

Dirksen, T. R., Marinetti, G. V., and Peck, W. A. 1970b. Lipid metabolism in bone and bone cells. II. The *in vitro* incorporation of [^{32}P]orthophosphate and [^{14}C]serine into lipids of bone and bone cell cultures. *Biochim. Biophys. Acta* **202**:80–90.

Dirksen, T. R., Schuster, G. S., Bustos, S. E., and McKinney, R. V. 1974. Lipid synthesis by frozen-thawed bones. Paper presented at the 52nd General Session of the International Association for Dental Research, Atlanta, Georgia. Abstract No. 318.

Dirksen, T. R., Schuster, G. S., McPherson, J., Bustos, S. E., and McKinney, R. V. 1975a. The effect of CPIB on lipid synthesis by bone and bone cells. Paper presented at the 53rd General Session of the International Association for Dental Research, London, England. Abstract No. L315.

Dirksen, T. R., Schuster, G. S., McKinney, R. V., and Bustos, S. E. 1975b. Effect of β-aminopropionitrile on *in vitro* bone lipid synthesis. *Lab. Invest.* **32**:133–139.

Dirksen, T. R., Schuster, G. S., McKinney, R., and Bustos, S. E. 1975c. Effect of metabolic inhibitors on bone lipid synthesis from radioactive bases. *J. Dent. Res.* **54**:1009–1014.

Eisenberg, E., Wuthier, R. E., Frank, R. B., and Irving, J. T. 1970. Time study of *in vivo* incorporation of ^{32}P orthophosphate into phospholipids of chicken epiphyseal tissues. *Calcif. Tissue Res.* **6**:32–48.

Enlow, D. H., and Conklin, J. L. 1964. A study of lipid distribution in compact bone. *Anat. Rec.* **148**:279.

Enlow, D. H., Conklin, J. L., and Bang, Seong. 1965. Observations on the occurrence and the distribution of lipids in compact bone. *Clin. Orthop. Related Res.* **38**:157–169.

Ennever, J., Vogel, J. J., and Benson, L. A. 1973. Lipid and calculus matrix calcification *in vitro*. *J. Dent. Res.* **52**:1056–1059.

Ennever, J., Vogel, J. J., and Levy, B. M. 1974. Lipid and bone matrix calcification *in vitro*. *Proc. Soc. Exp. Biol. Med.* **145**:1386–1388.

Erbland, J. F., McKinney, R. V., Dirksen, T. R., Schuster, G. S., and Bustos, S. E. 1974. β-aminopropionitrile effect on acetate uptake by bone. Paper presented at the 52nd General Session of the International Association for Dental Research, Atlanta, Georgia. Abstract No. 776.

Euler, H., and Meyer, W. 1927. Pathohistologie der Zähne mit bsonderer Berücksichtigung der Pathologie. J. F. Bergmann, Munich. 353 pp.

Fincham, A. G., Burkland, G. A., and Shapiro, I. M. 1972. Lipophilia of enamel matrix. A chemical investigation of the neutral lipids and lipophilic proteins of enamel. *Calcif. Tissue Res.* **9**:247–259.

Fish, E. W. 1932. An experimental investigation of enamel, dentin, and the dental pulp. John Bale Sons and Danielsson, London.

Foldes, I., Baczi, L. Jr., and Modis, L. 1971. The role of lipids in endochondral ossification. *Acta Biol. Acad. Sci. Hung.* **22**:9–18.
Hatton, E. H. 1930. Pulp pathology from the standpoint of the clinician. *J. Am. Dent. Assoc.* **17**:2262–65.
Havivi, E. 1971. Incorporation of P^{32} orthophosphate into phospholipid of epiphyseal cartilage. *Lipids* **6**:314–317.
Havivi, E., and Bernstein, D. S. 1969. Lipid metabolism in normal and rachitic rat epiphyseal cartilage. *Proc. Soc. Exp. Biol. Med.* **131**:1300–1304.
Hess, W. C., Lee, C. Y., and Peckham, S. C. 1956. The lipid content of enamel and dentin. *J. Dent. Res.* **35**:273–275.
Hill, J. C. 1939. The cytology and histochemistry of osteoblasts grown *in vitro*. *Arch. Exp. Zellforsch. Gewebezucht.* **18**:496–511.
Howell, D. S., Marquez, J. F., and Pita, J. C. 1965. The nature of phospholipids in normal and rachitic costochondral plates. *Arthritis Rheum.* **8**:1039–1046.
Howell, D. S., Pita, J. C., and Alvarez, J. 1976. Possible role of extracellular matrix vesicles in initial calcification of healing rachitic cartilage. *Fed. Proc.* **35**:122–126.
Hubscher, G., Dils, R. R., and Pover, W. F. R. 1959. Studies on the biosynthesis of phosphatidyl serine. *Biochim. Biophys. Acta* **36**:518–528.
Ingervall, B., Freden, H., and Heyden, G. 1972. A histochemical study of the lipid content of rat alveolar bone after traumatic loading of the teeth. *Scand. J. Dent. Res.* **80**:453–456.
Irving, J. T. 1958a. A histochemical stain for newly calcified tissues. *Nature* **181**:704–5.
Irving, J. T. 1958b. Sudanophil inclusions in ameloblasts, odontoblasts and cells of the oral epithelium. *Nature* **181**:569–70.
Irving, J. T. 1959. A histological staining method for sites of calcification in teeth and bone. *Arch. Oral Biol.* **1**:89–96.
Irving, J. T. 1960. Histochemical changes in the early stages of calcification. *Clin. Orthop.* **17**:92–102.
Irving, J. T. 1962. Sudan black method for staining sites of calcification. *Nature* **194**:390.
Irving, J. T. 1963. The sudanophil material at sites of calcification. *Arch. Oral Biol.* **8**:735–45.
Irving, J. T. 1965. Lipids and calciphylaxis. *Arch. Oral Biol.* **19**:189–190.
Irving, J. T. 1973a. The pattern of sudanophilia in developing rat molar enamel. *Arch. Oral Biol.* **18**:137–140.
Irving, J. T. 1973b. Theories of mineralization of bone. *Clin. Orthop. Related Res.* **97**:225–236.
Irving, J. T. 1976. Interrelations of matrix lipids, vesicles, and calcification. *Fed. Proc.* **35**:109–111.
Irving, J. T., and Durkin, J. F. 1965. A comparison of the changes in the mandibular condyle with those in the upper tibial epiphyses during the onset and healing of scurvy. *Arch. Oral Biol.* **10**:179.
Irving, J. T., and Wuthier, R. E. 1961. Further observations on the sudan black stain for calcification. *Arch. Oral Biol.* **5**:323–324.
Irving, J. T., and Wuthier, R. E. 1968. Histochemistry and biochemistry of calcification with special reference to the role of lipids. *Clin. Orthop.* **56**:237–260.
Johnson, L. C. 1960. Mineralization of turkey leg tendon. I. Histology and histochemistry of mineralization. *Publ. Am. Assoc. Sci.* **64**:117–128.
Kashiwa, H. K., and Mukai, C. D. 1971. Lipid-calcium-phosphate spherules in chondrocytes of developing long bones. *Clin. Orthop. Related Res.* **78**:223–229.
Krasnow, F. 1934. Cholesterol and lecithin in teeth and saliva. *J. Dent. Res.* **14**:226–227, 480.
Leach, A. A. 1958. The lipids of ox compact bone. *Biochem. J.* **69**:429–432.
Leopold, R. S., Hess, W. C., and Carter, W. J. 1951. Dentinal protein: bound cholesterol. *J. Dent. Res.* **30**:837–839.

Lindeman, G. 1957. A lipid material in bone and teeth in experimental chronic fluorosis. *Nature* **180**:926.

Lund, P. K., Abadi, D. M., and Mathies, J. C. 1962. Lipid composition of normal human bone marrow as determined by column chromatography. *J. Lipid Res.* **3**:95–98.

Majeska, R. J., and Wuthier, R. E. 1975. Studies on matrix vesicles isolated from chick epiphyseal cartilage. Association of pyrophosphatase and ATPase activities with alkaline phosphatase. *Biochim. Biophys. Acta* **391**:51–60.

Manly, R. S., and Hodge, H. C. 1939. Density and refractive index studies of dental hard tissue. *J. Dent. Res.* **18**:133–141.

Marinetti, G. V., Erbland, J. F., and Brossard, M. 1964. Biosynthesis of phospholipids and glycerides, pp. 71–93. *In* R. M. C. Dawson and D. N. Rhodes (eds.). Metabolism and Physiological Significance of Lipids. John Wiley and Sons, London.

Matthews, J. L., Martin, J. H., Sampson, H. W., Kunin, A. S., and Roan, J. H. 1970. Mitochondrial granules in the normal and rachitic rat epiphysis. *Calcif. Tissue Res.* **5**:91.

McPherson, J. C., Schuster, G. S., and Dirksen, T. R. 1975. The *in vitro* effects of triton WR-1339 on lipid synthesis by bone cells. *Proc. Soc. Exp. Biol. Med.* **149**:172–177.

Melcher, A. H. 1966. On sudan black positive material in connective tissue. *Nature* **211**:593–95.

Melcher, A. H. 1969. Histologically demonstrable bound lipid apparently associated with relatively stable, mature collagen fibres. *Gerontologia* **15**:217–232.

Miller, W. A. 1969. Fat staining in carious dentin. *J. Dent. Res.* **48**:109–113.

Miller, W. A., and Massler, M. 1962. Permeability and staining of active and arrested lesions in dentine. *Br. Dent. J.* **112**:187–197.

Odutuga, A. A., and Prout, R. E. S. 1973. Fatty acid composition of neutral lipids and phospholipids of enamel and dentine from rat incisors and molars. *Arch. Oral Biol.* **18**:689–697.

Odutuga, A. A., and Prout, R. E. S. 1974a. Lipid analysis of human enamel and dentine. *Arch. Oral Biol.* **19**:729–731.

Odutuga, A. A., and Prout, R. E. S. 1974b. Effect of essential fatty acid deficiency on the fatty acids composition of individual lipids from enamel and dentine of the rat. *Arch. Oral Biol.* **19**:911–920.

Odutuga, A. A., and Prout, R. E. S. 1975. Fatty acid composition of carious molar enamel and dentine from rats deficient in essential fatty acids. *Arch. Oral Biol.* **20**:49–51.

Odutuga, A. A., Prout, R. E. S., and Hoare, R. J. 1975. Hydroxyapatite precipitation *in vitro* by lipids extracted from hard and soft tissues. *Arch. Oral Biol.* **20**:311–316.

Ohnishi, S., and Ito, T. 1973. Clustering of lecithin molecules in phosphatidylserine membranes induced by calcium ion binding to phosphatidylserine. *Biochem. Biophys. Res. Commun.* **51**:132–138.

Opdyke, D. L. J. 1962. The histochemistry of dental decay. *Arch. Oral Biol.* **7**:207–219.

Peck, W. A., and Dirksen, T. R. 1966. The metabolism of bone tissue *in vitro*. *Clin. Orthop. Related Res.* **48**:243–265.

Peck, W. A., Birge, S. J., Jr., and Fedak, S. A. 1964. Bone cells: Biochemical and biological studies after enzymatic isolation. *Science* **146**:1476–77.

Penton, Z. G. 1968. Some effects of administration of fluoride on calcifying cartilage in rat. *Proc. Soc. Exp. Biol. Med.* **129**:978–981.

Peress, N. S., Anderson, H. C., and Sajdera, S. W. 1974. The lipids of matrix vesicles from bovine fetal epiphyseal cartilage. *Calcif. Tissue Res.* **14**:275–281.

Pikular, A. T. 1955. Quoted by A. A. Leach. *Ukr. Biokhem. Zh.* **27**:517.

Pincus, P. 1938. Enamel Protein. *Proc. R. Soc. Med.* **32**:513–518.

Prout, R. E. S., and Atkin (nee Shutt), E. R. 1973. Effect of diet deficient in essential fatty acid on fatty acid composition of enamel and dentine of the rat. *Arch. Oral Biol.* **18**:583–589.

Prout, R. E. S., and Odutuga, A. A. 1974a. Short communications: Lipid composition of dentine and enamel of rats maintained on a diet deficient in essential fatty acids. *Arch Oral Biol* **19**:725–728.
Prout, R. E. S., and Odutuga, A. A. 1974b. The effect on the lipid composition of enamel and dentine of feeding a corn oil supplement to rats deficient in essential fatty acids. *Arch. Oral Biol.* **19**:955–958.
Prout, R. E. S., and Odutuga, A. A. 1974c. *In vivo* incorporation of [1-^{14}C]-linoleic acid into the lipids of enamel and dentine of normal and essential fatty acid deficient rats. *Arch. Oral Biol.* **19**:1167–1170.
Prout, R. E. S., and Shutt, E. R. 1970a. Analysis of fatty acids in human root dentine and enamel. *Arch. Oral Biol.* **15**:281–286.
Prout, R. E. S., and Shutt, E. R. 1970b. Separation of enamel and dentine using cadmium-tungstoborate solution. *Arch. Oral Biol.* **15**:559–561.
Prout, R. E. S., Odutuga, A. A., and Tring, F. C. 1973. Lipid analysis of rat enamel and dentine. *Arch. Oral Biol.* **18**:373–380.
Rabinovitch, A. L., and Anderson, H. C. 1976. Biogenesis of matrix vesicles in cartilage growth plates. *Fed. Proc.* **35**:112–116.
Rabinowitz, J. L., Luddy, F. E., Barford, R. A., Herb, S. F., Orlean, S. L., and Cohen, D. W. 1967. Lipid determination in powdered dentin by thin-layer and gas–liquid chromatography. *J. Dent. Res.* **46**:1086–1089.
Rolle, G. K. 1965. Histochemistry of bone formation in normal and tetracycline-treated chick embryos-I. *Arch. Oral Biol.* **10**:393–405.
Rushton, M. A. 1940. Observations on Fish's "dead tracts" in dentine. *Br. Dent. J.* **68**:11–13.
Sajdera, S. W., Franklin, S., and Fortuna, R. 1976. Matrix vesicles of bovine fetal cartilage: Metabolic potential and solubilization with detergents. *Fed. Proc.* **35**:154–156.
Sakai, T., and Cruess, R. L. 1967. Effect of cortisone on the lipids of bone matrix in the rat. *Proc. Soc. Exp. Biol. Med.* **124**:490–93.
Sakai, T., Yoshinari, T., and Cruess, R. L. 1968. Effect of growth hormone upon the lipids of bone matrix. *Endocrinology* **83**:51–55.
Sakai, T., Cruess, R. L., and Iida, K. 1969. The effect of age upon the lipids of the long bones of the rat. *Proc. Soc. Exp. Biol. Med.* **132**:100–104.
Sakai, T., Cruess, R. L., Yoshinari, T., and Iida, K. 1970. Effect of estrogen upon the lipids of various regions of the long bones of the rat. *Endocrinology* **86**:167–171.
Schubert, F. W., Jr. 1962. Lipid composition of dentin. *Dent. Stu. Mag.* **40**:239–42.
Schuster, G. S., Dirksen, T. R., and Harms, W. S. 1975. Effect of exogenous lipid on lipid synthesis by bone and bone cell cultures. *J. Dent. Res.* **54**:131–139.
Seimiya, T., and Ohki, S. 1972. Accessibility of calcium ions to anionic sites of lipid monolayers. *Nat. New Biol.* **239**:26–27.
Shapiro, I. M. 1970a. The phospholipids of mineralized tissues. I. Mammalian compact bone. *Calcif. Tissue Res.* **5**:21–29.
Shapiro, I. M. 1970b. The association of phospholipids with anorganic bone. *Calcif. Tissue Res.* **5**:13–20.
Shapiro, I. M. 1971. The neutral lipids of bovine bone. *Arch. Oral Biol.* **16**:411–421.
Shapiro, I. M. 1973. Biological Mineralization. Wiley, Interscience, New York. 20 pp.
Shapiro, I. M., Wuthier, R. E., and Irving, J. T. 1966. A study of the phospholipids of bovine dental tissues—I. Enamel matrix and dentine. *Arch. Oral Biol.* **11**:501–512.
Simon, D. R., Irwin, B., and Howell, D. S. 1973. Relationship of extracellular matrix vesicles to calcification in normal and healing rachitic epiphyseal cartilage. *Anat. Rec.* **176**:167–180.
Sognnaes, R. F., and Wislocki, G. B. 1950. Observations on enamel and dentin undergoing carious destruction. *Oral Surg.* **3**:1238–1996.

Soyenkoff, B. C., Friedman, B. K., and Newton, M. 1951. The lipids of dental tissues. *J. Dent. Res.* **30**:599.

Stewart, J. M. 1961. Odontoblasts: vacuoles and inclusions. *Science* **133**:1011–1012.

Stewart, J. M. 1962. Life cycle of odontoblasts. *Int. Assoc. Dent. Res. Preprinted Abs.* **40**:1.

Stewart, J. M. 1963. Immediate response to odontoblasts to injury. Paper presented at the 41st General Meeting of the International Association of Dental Research, Pittsburgh, Pennsylvania. Abstract No. 15.

Stewart, J. M., Claibourne, P. A., and Luikart, G. A. 1965. A histologic and histochemical study of lipids in human odontoblasts. *J. Dent. Res.* **44**:608–613.

Stockwell, R. A. 1965. Lipid in the matrix of ageing articular cartilage. *Nat.* **207**:427–428.

Stockwell, R. A. 1967a. The lipid and glycogen content of rabbit articular hyaline and non-articular hyaline cartilage. *J. Anat.* **102**:87–94.

Stockwell, R. A. 1967b. Lipid content of human costal and articular cartilage. *Ann. Rheum. Dis.* **26**:481–486.

Tanaka, T., and Hollander, V. P. 1971. Effect of parathyroid extract on bone phospholipid. *Proc. Soc. Exp. Biol. Med.* **136**:174–177.

Vogel, J. J., and Ennever, J. 1971. The role of a lipoprotein in the intracellular hydroxyapatite formation in bacterionema matruchotii. *Clin. Orthop. Related Res.* **78**:218–222.

Vogel, J. J., Campbell, N. M., and Ennever, J. 1973. Calcification of a lysozymeinositol phosphatide. *Proc. Soc. Exp. Biol. Med.* **143**:677.

Weber, R. 1930. Neue Untersuchungen über das Auftreten von Felt in Zahn. *Deutsche Monatschr. Zahnh.* **48**:1489.

Willner, H. 1926. Untersuchunger über das Vorkommen von Felt in der Zahnpulpa. *Z. Stomatol.* **24**:1084–1099.

Wislocki, G. B., and Sognnaes, R. F. 1950. Histochemical reactions of normal teeth. *Am. J. Anat.* **87**:239–275.

Wislocki, G. B., Singer, M., and Waldo, C. M. 1948. Some histochemical reactions of mucopolysaccharides, glycogen, lipids and other substances in teeth. *Anat. Rec.* **101**:487–514.

Wolinsky, I., and Guggenheim, K. 1970. Lipid metabolism of chick epiphyseal bone and cartilage. *Calcif. Tissue Res.* **6**:113–119.

Wolinsky, I., and Guggenheim, K. 1971. Effect of fluoride on the lipid metabolism of rat bone. *Isr. J. Med. Sci.* **7**:527–28.

Woolley, D. W., and Campbell, N. K. 1962. Tissue lipids as ion exchangers for cations and the relationship to physiological processes. *Biochim. Biophys. Acta* **57**:384–385.

Wuthier, R. E. 1966. Two-dimensional chromatography on silica gel-loaded paper for the microanalysis of polar lipids. *J. Lipid Res.* **7**:544–50.

Wuthier, R. E. 1968. Lipids of mineralizing epiphyseal tissues in the bovine fetus. *J. Lipid Res.* **9**:68–78.

Wuthier, R. E. 1971. Zonal analysis of phospholipids in the epiphyseal cartilage and bone of normal and rachitic chickens and pigs. *Calcif. Tissue Res.* **8**:36–53.

Wuthier, R. E. 1973. The role of phospholipids in biological calcification. *Clin. Orthop. Related Res.* **90**:191–200.

Wuthier, R. E. 1976. Lipids of matrix vesicles. *Fed. Proc.* **35**:117–121.

Wuthier, R. E., and Cummins, J. W. 1974. *In vitro* incorporation of ^{3}H serine into phospholipids of proliferating and calcifying epiphyseal cartilage and liver. *Biochim. Biophys. Acta* **337**:50–59.

Cancer Cells

TEN-CHING LEE AND FRED SNYDER

I. Introduction

Cancer cells and healthy cells metabolize lipids and sterols similarly, although in some instances quantitative differences in levels and enzymatic activities have been observed. Some studies have also established that cancer cells inadequately regulate specific enzymatic steps in lipid biosynthesis, but the mechanisms responsible for most of the reported differences are still obscure. In this chapter, we summarize the current status of metabolic information on glycerolipids, glycolipids, long-chain aliphatic moieties, and sterols in cancer cells. Since books edited by Carroll (1975) and Wood (1973) contain in-depth reviews of most of these areas, our approach is deliberately brief, the emphasis being on how lipid and sterol metabolism differs in normal and neoplastic growth.

II. Fatty Acids

A. Biosynthesis

The synthesis of fatty acids is thought to be qualitatively the same in both healthy and cancer cells. *De novo* fatty acid synthesis in normal cells requires the sequential action of two enzyme systems (see the introductory chapter), the first of which catalyzes the carboxylation of acetyl-CoA to malonyl-CoA by acetyl-CoA carboxylase. The synthesis of palmitic acid from acetyl-CoA and malonyl-CoA is then catalyzed by the multienzyme complex, fatty acid synthase. Most tumors appear to synthesize fatty acids

TEN-CHING LEE AND FRED SNYDER • Medical and Health Sciences Division, Oak Ridge Associated Universities, P. O. Box 117, Oak Ridge, Tennessee 37830.

de novo through the same pathway, but the rate at which this occurs may vary widely. This statement is supported by the fact that the incorporation of glucose and acetate into fatty acids by Ehrlich ascites cells is stimulated by the addition of bicarbonate to the incubation medium and is dependent on an NADPH-generating system (Kimura *et al.*, 1964; Pedersen *et al.*, 1972). Furthermore, in Morris hepatomas 7777 and 9618A, and Ehrlich ascites cells, the acetyl-CoA carboxylase activity was shown to depend on ATP, acetyl-CoA, Mg^{++}, citrate, and CO_2, and fatty acid synthase required acetyl-CoA, malonyl-CoA, and NADPH for enzyme activity (Majerus *et al.*, 1968; McGee and Spector, 1974).

Long-term regulation of fatty acid biosynthesis in normal tissue is manifested over a period of 2–3 days and is reflected by changes in the amounts of both acetyl-CoA carboxylase and fatty acid synthase. For example, the increase or decrease in the level of carboxylase in refed or diabetic animals occurs because of the rise or fall in the rate of enzyme synthesis, whereas the decrease in the level of carboxylase in fasted animals is due to both diminished enzyme synthesis and accelerated enzyme degradation (Majerus and Kilburn, 1969; Nakanishi and Numa, 1970). Similarly, fatty acid synthase has been shown to undergo accelerated enzyme degradation and decreased enzyme synthesis during fasting and increased enzyme synthesis during fat-free diet refeeding (Volpe *et al.*, 1973). Also, the marked reduction in activities of fatty acid synthase and carboxylase in livers of adult rats given glucagon and in adipose tissue of adult rats given glucocorticoid is caused by a decrease in the synthesis of fatty acid synthase (Volpe and Marasa, 1975). Another illustration of long-term regulation is the apparent induction of fatty acid synthase in cell cultures upon the removal of serum for 24–48 hr, and the addition of insulin to the culture media. These results are explained on the basis of increased levels of enzyme, rather than by factors affecting the activation of existing enzyme (Alberts *et al.*, 1974).

Unlike normal tissues, dietary manipulation of host animals did not affect the rate of acetate incorporation into fatty acids by tissue slices from ten different transplantable hepatomas (Sabine *et al.*, 1966, 1968; Zuckerman *et al.*, 1970; Elwood and Morris, 1968). Parallel experiments have demonstrated that neither acetyl-CoA carboxylase nor fatty acid synthase activities of two hepatomas were changed by altering the diet (Majerus *et al.*, 1968). Comparative studies done with Morris hepatoma 7777 and normal rat liver show that acetyl-CoA carboxylase was essentially identical for both tissues with respect to heat inactivation, affinities for acetyl-CoA and ATP, and activation and aggregation by citrate. Both partially purified acetyl-CoA carboxylase and crude fatty acid synthase from hepatoma 7777 were equally inhibited by palmitoyl-CoA, which compares with results from healthy livers. Similarly, acetyl-CoA carboxylase of Ehrlich ascites cells was also inhibited by stearoyl-CoA (McGee and Spector, 1974). These data indicate that defective long-term control of fatty acid biosynthesis in tumors does not result from structural alterations in the enzyme. Recently, Halperin *et al.* (1975) found low levels of acid-insoluble acyl-CoAs in hepatomas 7777 (180 pmoles/mg protein), and the level remained essentially unchanged in the fasted state, whereas the acid-insoluble CoA levels in livers increased from 720 pmoles/mg protein in fed rats to 1050 pmoles/mg protein during fasting. In addition, the K_m for the mitochondrial citrate transport is about one-third less in the hepatoma than the value for host livers. The high levels of acyl-CoA increase the K_m for citrate transport causing citrate levels in the cytosol to fall, resulting in a decreased activation of acetyl-CoA carboxylase. Furthermore, the activity of acetyl-CoA carboxylase is also inhibited by the elevated acyl-CoA levels. The coordinated effect results in a profound decrease in the rate of fatty acid synthesis. The regulation of citrate transport by altered levels of fatty acid metabolites does not occur in hepatomas, and this impaired regulatory mechanism could be one of the factors accounting for the constant rate of fatty acid synthesis in hepatomas during different nutritional states.

In short-term regulation, modification of the rate of fatty acid biosynthesis can occur very rapidly in the absence of changes in the enzyme levels. For example, a marked decrease in the rate of fatty acid synthesis in liver occurs within 2 hr after rats are fed a high-fat diet or after animals are fasted 12 and 18 hr (Hill *et al.*, 1960; Lyon *et al.*, 1952; Bortz *et al.*, 1963). In addition, stimulation of acetate incorporation into lipids occurs within hours after removing serum from cell culture media (Jacobs *et al.*, 1973; Jacobs and Majerus, 1973; Howard *et al.*, 1974). A number of possible modulators for acetyl-CoA carboxylase have been proposed as mediators of this short-term control in healthy tissues. *In vitro,* acetyl-CoA carboxylase is subject to activation by citrate, isocitrate, and magnesium (Waite and Wakil, 1962; Martin and Vagelos, 1962; Greenspan and Lowenstein, 1968). It is inhibited by long-chain acyl-CoA derivatives, free fatty acids (Numa *et al.*, 1965a; Korchak and Masoro, 1964), and 1 mM N^6,$O^{2\prime}$-dibutyryl cAMP (Allred and Roehrig, 1973). Furthermore the enzyme can be inactivated after phosphorylation catalyzed by an ATP-dependent, cAMP-independent kinase and can be reactivated by a Mg^{++}-requiring phosphatase (Carlson and Kim, 1973).

However, conflicting views are held on whether the intracellular level of citrate in rat liver is an important modifier of acetyl-CoA carboxylase activity *in vivo*. Even when the citrate concentration in the fat-fed group was three times that of the starved animal, the malonyl-CoA concentration and rate of fatty acid synthesis was low in both groups (Guynn *et al.*, 1972). In addition, the liver content of citrate in neonatal chicks remains unchanged during times of widely varying rates of fatty acid synthesis

(Goodridge, 1973a). In contrast, the citrate content of the isolated chick hepatocytes was positively correlated with fatty acid synthesis under all incubation conditions (Goodridge, 1973b). Therefore, the possibility exists that different pools of citrate could explain the conflicting *in vivo* data, e.g., significant, but undetected changes in citrate concentrations could have occurred in the extramitochondrial compartment. It is also possible that the isolated cells do not represent the *in vivo* physiological state. Although cAMP or dibutyryl cAMP, when added to rat liver slices incubated with labeled acetate, was capable of greatly diminishing acetate incorporation into fatty acids (Bricker and Levey, 1972a) and decreasing the activity of acetyl-CoA carboxylase, the concentrations of nucleotides required for inhibition were generally high (10^{-3}–10^{-4} M) compared with the known range of cAMP concentrations in liver tissue (10^{-5}–10^{-7} M) (Exton *et al.*, 1971). cAMP at a concentration of 5 mM markedly depressed fatty acid biosynthesis but had no effect at 0.5 mM.The rate of fatty acid synthesis in the intact perfused rat liver also appears to be totally independent of actively induced changes in intracellular cAMP concentrations (Raskin *et al.*, 1974). Such data indicate that cAMP may not be an important regulator in short-term control of fatty acid synthesis *in vivo*.

There is also controversy over the role of free fatty acids and long-chain acyl-CoAs in the regulation of fatty acid synthesis because these lipids appear to be nonspecific in their mode of inhibition (Taketa and Pogell, 1966; Dorsey and Porter, 1968; Parvin and Dakshinamurti, 1970) and their inhibitory effects could be due to their detergent properties. The nonspecific inhibitory effects of palmitoyl-CoA on several lipogenic and nonlipogenic enzymes can be eliminated if the free concentration of palmitoyl-CoA is maintained at a low level by using albumin (Goodridge, 1972). Inhibition of acetyl-CoA carboxylase by palmitoyl-CoA *in vitro* is also reversible by albumin. Goodridge (1973a) observed an inverse relation between fatty acid synthesis and the acyl-CoA content of livers from neonatal and growing chicks. Yet, no change in total cellular long-chain acyl-CoA levels was found when the rate of acetate incorporation into fatty acids was decreased by switching human skin fibroblast cultures from lipid-deficient medium to one supplemented with free fatty acids (Jacobs *et al.*, 1973). Recent findings by Howard *et al.* (1974) suggest that acetyl-CoA synthetase may be an additional control point in the regulation of acetate incorporation into lipids of cultured cells. This mechanism also appears to be at the level of enzyme activity rather than via induction or repression of enzyme synthesis. Evidence also indicates that free fatty acids in the incubation medium can regulate the *de novo* rate of fatty acid synthesis by altering the composition of the cellular long-chain acyl-CoA pool (McGee and Spector, 1975). When the total activities of acetyl-CoA carboxylase and fatty acid synthase increased before or after hatching (Goodridge,

1973a), the level of free stearate and stearoyl-CoA in the liver decreased. In this context, it should be noted that stearoyl-CoA is a more potent inhibitor of purified rat liver acetyl-CoA carboxylase (Numa *et al.*, 1965b) than either palmitoyl-CoA or oleoyl-CoA (K_i = 0.53 μM for stearoyl-CoA, 0.91 μM for palmitoyl-CoA, and 0.67 μM for oleoyl-CoA).

In neoplastic tissue, short-term regulation of fatty acid biosynthesis appears to be modulated by extracellular free fatty acids. Some striking similarities have been observed in short-term control of fatty acid synthesis in Ehrlich ascites cells, hepatocytes, and skin fibroblasts (Jacobs *et al.*, 1973; Goodridge, 1973b; McGee and Spector, 1974). The relative ability of free fatty acids in the incubation medium to inhibit fatty acid biosynthesis in these cells is stearate > palmitate > oleate > linoleate. Furthermore, as shown for the liver enzyme (Numa *et al.*, 1965b), various long-chain acyl-CoAs inhibit the purified acetyl-CoA carboxylase from Ehrlich ascites cells to a different extent (McGee and Spector, 1975). For example, 2 μM stearoyl-CoA almost totally inhibits the enzyme, whereas 2 μM palmitoyl-CoA produces only about 50% inhibition and 2 μM linoleoyl-CoA produces essentially no inhibition.

As mentioned earlier for healthy tissues, cAMP has also been implicated as a possible negative mediator of short-term fatty acid regulation. When tissue slices from either fast-growing (3924A) or slow-growing (9121) Morris hepatomas were incubated with cAMP (5×10^{-3} M) or dibutyryl cAMP (3×10^{-4} M), the incorporation of acetate into fatty acids was not suppressed (Bricker and Levey, 1972b). These data correlate well with the finding that cAMP-binding protein was not detected in HTC hepatoma cells, but was present in normal liver cells (Granner, 1972). The physiological importance of this observation is still unknown.

B. *Desaturation*

There are at least three enzymatic desaturation systems present in rat liver microsomes. The Δ9-desaturase converts stearic acid to oleic acid and palmitic acid to palmitoleic acid by placing a double bond between carbon atoms 9 and 10 of the aliphatic chain. The Δ6-desaturase introduces a new double bond between carbon atoms 6 and 7 of oleic acid ($18{:}1\Delta^{9}$), linoleic acid ($18{:}2\Delta^{9,12}$), or α-linoleic acid ($18{:}3\Delta^{9,12,15}$) to form octadeca-6,9-dienoic acid, γ-linolenic acid ($18{:}3\Delta^{6,9,12}$), and octadeca-6,9,12,15-tetraenoic acid, respectively. The Δ5-desaturase inserts a double bond in fatty acids containing 20 carbon atoms, i.e., $20{:}2\Delta^{8,11} \rightarrow 20{:}3\Delta^{5,8,11}$ and $20{:}3\Delta^{8,11,14} \rightarrow 20{:}4\Delta^{5,8,11,14}$. The existence of these three distinct desaturase systems is based primarily on the fact that the three types of reactions are catalyzed at somewhat different rates under different experimental conditions. Although all three desaturase activities decrease in fasted normal and alloxanic diabetic rats, increasing the level of dietary protein stimulates only the Δ5- and Δ6-desaturation (Castuma *et al.*, 1972; Inkpen *et al.*, 1969; Brenner, 1974), but has no effect on Δ9-desaturation. In contrast, carbohydrate diets inhibit the desaturation of linoleic acid at the 6,7-position and stimulate the desaturation of stearic acid at the 9,10-position. Also, the conversion of oleic and linoleic acids to the corresponding Δ6 acids was increased in rats fed fat-free diets, whereas desaturation of eicosadienoic acid at the 5,6-position was not changed (Castuma *et al.*, 1972). In addition, no competitive effects with regard to fatty acid composition of liver lipids were observed when linoleate and eicosa-8,11-dienoate were fed to rats (Sprecher, 1974).

All three desaturases require NADPH or NADH and molecular oxygen. The multicomponent nature of the enzyme systems has been studied most extensively with stearoyl-CoA desaturation. It is now well established that this reaction consists of at least four components: (1) NADH-cytochrome b_5 reductase, (2) NADPH-dependent flavoprotein, (3) cytochrome b_5, and (4) a cyanide-sensitive terminal desaturase (Oshino and Sato, 1972). The terminal component with a rather short half-life of 3–4 hr probably regulates the overall desaturation sequence (Oshino and Sato, 1972). The significant response of the stearoyl-CoA desaturase system to fasting and refeeding appears to be due to the amount of this terminal desaturase, and the control mechanism for this probably is mediated by protein synthesis and not by activation of the pre-existing terminal protein component. It has been assumed, but not proved, that the electron transport systems for the Δ5- and Δ6-desaturation reactions are similar to that described for Δ9-desaturase. It has been shown that the Δ9-desaturase activity is present in ss1K fast-growing hepatomas, ss1H slow-growing hepatomas of mice, and Morris hepatomas 5123, 7800, and 7777 of rats (Chiappe *et al.*, 1974; Raju, 1974; Lee *et al.*, 1974). The activity of Δ6-desaturase, although less than in host liver, was also present in two fast- and slow-growing mouse hepatomas, ss1K and ss1H, respectively (Chiappe *et al.*, 1974). Heteroploid or transformed cells (mouse fibrosarcoma L929 and L2071 transformed mouse fibroblasts) do not desaturate linoleate (Δ6-desaturase, $18{:}2\Delta^{9,12} \rightarrow 18{:}3\Delta^{6,9,12}$), but they readily convert dihomo-γlinolenate ($20{:}3\Delta^{8,11,14}$) to arachidonate ($20{:}4\Delta^{5,8,11,14}$) when these fatty acids are supplemented to the new medium (Dunbar and Bailey, 1975).

Different results in studies of the effects of diet on the regulation of desaturase activities in tumors have been obtained. When host animals were kept for several months on a fat-free diet before and after the tumor transplants, or when tumor-bearing rats were fasted for 48 hr and then fed the fat-free diet for 48 hr, stearoyl-CoA desaturase was increased in Fischer R-3259 tumors (Lee *et al.*, 1973) and in Morris 5123C and 7777 hepatomas

(Lee *et al.,* 1974). On the other hand, the desaturase activities in Morris hepatomas 5123C and 7800 were not altered when the host animals were on fat-free diets for a week (Raju, 1974).

C. Oxidation

Very little enzymatic information is available on fatty acid oxidation in tumors. It is known that, as in other tissues, fatty acid oxidation occurs in the mitochondria of hepatomas and is closely linked to the tricarboxylic acid cycle (Brown *et al.,* 1956; Emmelot and Bos, 1955). In hepatomas, an inverse relation appears to exist between glucose utilization and fatty acid oxidation (Bloch-Frankenthal *et al.,* 1965). Those tumors, which grow slowly and lack the capability to phosphorylate glucose readily oxidize fatty acids, whereas the high glycolyzing tumors do not. There does not appear to be any consistent pattern in the effects of dietary manipulation on fatty acid oxidation by tumors (Sabine *et al.,* 1968).

III. Fatty Alcohols

Fatty alcohols, the precursors of ether-linked aliphatic chains in glycerolipids, are synthesized from acyl-CoAs via fatty aldehydes by oxidoreductases in mammalian cells. Reduction of these aliphatic moieties appears to be most prominent in those tissues that contain significant quantities of alkyl and/or alk-1-enyl lipids such as brain (Tabakoff and Erwin, 1970), heart (Johnson and Gilbertson, 1972; Kawalek and Gilbertson, 1973), and cancer cells (Snyder and Malone, 1970; Snyder *et al.,* 1971). Fatty aldehydes formed in liver by catabolic reactions involving alkylglycerols and sphingolipids can also be enzymatically reduced to alcohols (Tietz *et al.,* 1964; Pfleger *et al.,* 1967; Stoffel *et al.,* 1970; Ferrell and Kessler, 1971). The specificity for the reduced pyridine nucleotides (NADPH or NADH) required for the reaction appears to depend on the source of the enzyme. In *Euglena gracilis,* the reduction of fatty acids to fatty alcohols seems to be closely coupled with fatty acid synthase (Khan and Kolattukudy, 1973). Only the overall conversion of acids to alcohols has been investigated in cell-free preparations of cancer cells (Snyder and Malone, 1970), whereas in brain and heart (Tabakoff and Erwin, 1970; Johnson and Gilbertson, 1972; Kawalek and Gilbertson, 1973) the step involving the reduction of the aldehyde to alcohol has been studied separately (see the chapter Brain and the chapter Cardiac Muscle).

Experiments with mammalian tissue preparations have also been reported for the reverse reactions, i.e., the oxidation of long-chain aliphatic moieties. In such reactions, fatty aldehydes (Erwin and Deitrich, 1966; Nilsson, 1970) and fatty alcohols (Snyder and Malone, 1970; Snyder *et al.*, 1971) are converted to the corresponding fatty acids by the oxidoreductases (dehydrogenases). In brain and tumor systems, the oxidation of long-chain aldehydes and alcohols required NAD.

Acyl-CoA oxidoreductase and fatty aldehyde oxidoreductase activities have been found in both particulate and soluble protein fractions of mammalian cells. Although many questions remain to be answered about the properties, quantitative distribution in tissues and organelles, and the reaction mechanism of the acyl-CoA and fatty aldehyde oxidoreductases, it is clear that these enzymes are of primary importance in the biosynthesis of the ether-linked glycerolipids in cancer cells.

IV. Glycerolipids

The thorough review (Snyder and Snyder, 1975) of glycerolipids and their metabolism in cancer cells and the specific enzymes involved in the metabolism of the ether-linked lipids (Wykle and Snyder, 1976) makes it unnecessary to explore this area in similar detail. Other earlier chapters on compositional and metabolic studies (Haven and Bloor, 1956; Snyder, 1971) and a bibliographical index (Snyder, 1970) are also available on the broader aspects of lipids and cancer. Here, we will simply amplify those aspects of glycerolipid metabolism that appear to be particularly associated with cancerous growth.

A. Biosynthesis

A number of the investigations of glycerolipids in cancer cells indicate differences in levels of specific lipid classes and/or their aliphatic moieties when comparisons are made with healthy tissues. However, unless comparisons are made with the same cells before and after transformation, differences must be interpreted with extreme caution. Evaluation of the available literature on glycerolipids in neoplastic growth does not permit one to make any broad generalizations, except to point out the biosynthesis of the ether-linked aliphatic moieties in lipids is closely associated with most cancers. As pointed out earlier, the relatively high levels of ether lipids in a variety of tumors have been documented by a number of investigators (Snyder and Wood, 1968, 1969; Howard *et al.*, 1972; Steele and Jenkin, 1972; see reviews by Snyder, 1972; Snyder and Snyder, 1975).

Pathways for the synthesis of glycerolipids require *sn*-glycerol-3-P or dihydroxyacetone-P for the glycerol portion, and acyl-CoAs or fatty alcohols for the aliphatic portions (cf. the introductory chapter). In tumors, the dihydroxyacetone phosphate pathway is very prominent (Pollock *et al.*, 1975) and appears to be directly related to the characteristically high levels of ether lipids in such cells. The acyldihydroxyacetone-P formed by the initial acylation of dihydroxyacetone-P (Hajra, 1968; Hajra and Agranoff, 1968) serves as a substrate in the biosynthesis of alkyl lipids. In this unique reaction, the fatty acid of acyldihydroxyacetone-P is replaced by a long-chain fatty alcohol (Hajra, 1970; Wykle *et al.*, 1972a). The enzyme responsible for this step, as with all the biosynthetic steps in the ether-lipid pathway, is located in the microsomal fraction. Although the mechanism for the alcohol substitution is unknown, it has been shown with hexadecanol labeled with ^{18}O, ^{3}H, and ^{14}C that the entire alcohol grouping (RCH_2O^-) forms the ether-linked hydrocarbon chain of alkyldihydroxyacetone-P (see Wykle and Snyder, 1976 for review). Recently, this enzyme has been solubilized from Ehrlich ascites cells without loss of activity when cholic acid or Triton were used as the detergents (Rock and Snyder, 1976).

The second step in the pathway forms 1-alkyl-*sn*-glycerol-3-P and is catalyzed by an NADPH-linked oxidoreductase (Snyder *et al.*, 1970a; Wykle and Snyder, 1970; Chae *et al.*, 1973a; LaBelle and Hajra, 1974). Alternately, the 1-alkyl-*sn*-glycerol-3-P can be synthesized from 1-alkyl-*sn*-glycerols by an ATP-linked phosphotransferase (Chae *et al.*, 1973b; Rock and Snyder, 1974). Under certain conditions, relatively high levels of NADH can substitute for NADPH in the oxidoreductase reaction (LaBelle and Hajra, 1972; Wykle *et al.*, 1972a; Chae *et al.*, 1973a). The oxidoreductase has been solubilized and purified approximately fivefold (LaBelle and Hajra, 1974).

Next in the biosynthetic sequence is the reaction catalyzed by acyltransferases that utilize the alkylglycerol-P and acyl-CoAs to form the alkyl analog of phosphatidic acid, 1-alkyl-2-acyl-*sn*-glycerol-3-P, (Wykle and Snyder, 1970). The subsequent reactions involving phosphatidate phosphohydrolase to produce alkyldiacylglycerols (Blank *et al.*, 1974) and choline- or ethanolamine-phosphotransferases to produce alkyl phospholipids (Snyder *et al.*, 1970b) are analogous to those responsible for the biosynthesis of the diacyl glycerolipids, and in fact the enzymes might be identical for the corresponding ester- and ether-linked lipid substrates.

Plasmalogens, another important class of ether-linked glycerolipids in cancer cells, are formed from 1-alkyl-2-acyl-*sn*-glycero-3-phosphoethanolamines (Wykle *et al.*, 1972b; Paltauf and Holasek, 1973). The conversion of the alkyl to alk-1-enyl moiety at the 1-*sn*-glycerol position is known to occur only on the intact phospholipid, i.e., both the acyl and phosphoethanolamine groups must be present at the other two positions of the glycerol moiety. The microsomal alkyl desaturase that catalyzes the reaction requires a reduced pyridine nucleotide, oxygen, cytochrome b_5, and a cyanide-sensitive factor (Wykle *et al.*, 1972b; Paltauf and Holasek, 1973;

Paltauf *et al.,* 1974). The components of this mixed-function oxidase complex are similar to those required for acyl-CoA desaturases; however, it differs in that the alkyl desaturase in tumors, unlike the acyl desaturase, does not respond to dietary regulation (Lee *et al.,* 1973). This indicates that the terminal oxidase in these two systems must be different.

Howard and co-workers (1972) reported a close association between increased levels of ether-linked glycerolipids and decreased activities of glycerol-3-P dehydrogenase in six different Morris hepatomas and five different types of cell cultures. Their results suggest that the low glycerol-3-P dehydrogenase activities might be responsible for high levels of dihydroxyacetone-P, a precursor required for ether-lipid synthesis. However, this does not always appear to be consistent, since Rao and Abraham (1973) found the cellular concentration of glycerol-3-P was 12 times greater than dihydroxyacetone-P in a mouse mammary tumor that contained only 5% of the glycerol-3-P dehydrogenase activities present in the normal gland.

In contrast to the ether-linked glycerolipid analogs, there does not appear to be any unique pattern in the biosynthesis of triacylglycerols and diacyl types of phospholipids in cancer cells (Snyder and Snyder, 1975). Hillyard and Abraham (1972) have shown the incorporation of ^{14}C-labeled choline into phosphatidylcholine of normal, preneoplastic, and neoplastic mammary glands is closely correlated with membrane proliferation, but the rate of incorporation was not a useful indicator of cancerous growth.

B. Degradation

Available information about lipases that degrade ester-linked aliphatic moieties in glycerolipids of cancer cells is scarce. On the basis of studies with lysophosphatidylcholine, Stein and Stein (1967) proposed that phospholipase A_1 is localized at the cell surface of the Landschutz ascites tumor. Lumb and Allen (1976) obtained kinetic data for phospholipase A_2 in microsomes from Novikoff hepatomas indicating that the properties of A_2 in this tumor are significantly different from those in normal liver. One important difference is that the A_2 from the hepatoma microsomes required calcium for activity, whereas the A_2 from the liver did not. Phospholipase C has not been investigated in tumor preparations. Phospholipase D activity was thought to occur only in plant tissue, but it has recently been detected in rat brain (Saito and Kanfer, 1975). Lysophospholipase D has also been reported to occur in brain (Wykle and Schremmer, 1974) and liver, and to a much lesser extent in Fischer rat sarcomas (Wykle, unpublished data). The significance of the lysophospholipase D in the tumors and the other mammalian tissues is not known, but it could play an important role in preventing lysophospholipids from building up to toxic levels in cells.

Enzymes are also known to cleave the aliphatic moieties that are ether-linked at the number 1 carbon atom in glycerolipids. The alk-1-enyl grouping can be cleaved enzymatically (plasmalogenase) in liver (Warner and Lands, 1961) and brain (Ansell and Spanner, 1965), and also nonenzymatically by a ferrous-ascorbate complex isolated from brain (Yavin and Gatt, 1972a,b). Alk-1-enyl cleavage enzyme activities have not yet been studied in cancerous tissues. The alkyl cleavage enzyme, which requires tetrahydropteridine as a cofactor, has been detected in a variety of healthy tissues and its activity is highest in liver (Teitz *et al.*, 1964; Pfleger *et al.*, 1967). In contrast, most tumor cells are essentially devoid of the alkyl cleavage enzyme activities; only a very slow-growing Morris hepatoma had a cleavage activity similar to that found in normal liver (Soodsma *et al.*, 1970). In general, the absence of the alkyl cleavage enzyme appears to be closely correlated with the higher levels of alkyl glycerolipids.

V. Glycolipids

Reviews by Hakomori (1973, 1975) and by Brady and Fishman (1974) describe the most current views of glycolipid metabolism in cancer cells. The significance of glycolipids in neoplastic growth is that they are located almost exclusively at the surface of cells, a strategic position for interactions between cells. During cell transformation, there is an incomplete synthesis of glycolipids (Hakomori and Murakami, 1968). Hakomori (1975) believes that this change in glycolipid metabolism in cancer cells could account for associated alterations in antigenicity, loss of growth control, and weakened intercellular linkages. Brady and Fishman (1974) also stress in their review on the biosynthesis of glycolipids in virus-transformed cells that the decreased glycosyltransferase activities in the virally infected cells appear to be associated with the transformation process.

VI. Cholesterol

When an animal consumes cholesterol in the diet, suppression of hepatic cholesterol biosynthesis occurs. It is well established that this feedback control mechanism acts specifically at the conversion of 3-hydroxy-3-methyl-glutaryl CoA (HMG-CoA) to mevalonic acid, catalyzed by HMG-CoA reductase (Dempsey, 1974). By using a specific antiserum to purified HMG-CoA reductase, Higgins and Rudney (1973) demonstrated that HMG-CoA reductase is under more than one control by cholesterol feeding. The immediate effect of dietary cholesterol is independent of protein synthesis and the slower effect is caused by a decrease in the synthesis of new enzyme molecules (Higgins and Rudney, 1973). Feedback regulation of cholesterol synthesis in response to dietary cholesterol is consistently absent in all primary and transplantable hepatomas of animals and humans that have been studied (Sabine, 1975). The defective dietary feedback control in hepatomas is associated with HMG-CoA reductase.

Recently, a comparison was made between the properties of HMG-CoA reductase after solubilization and partial purification from microsomes of normal rat liver and Morris hepatoma 9121 (Brown *et al.,* 1974). The solubilized enzyme from the two sources behaved identically with respect to substrate kinetics and heat or cold inactivation. Furthermore, the two enzymes also have similar sizes as measured by agarose gel filtration and similar charges as measured by polyacrylamide gel electrophoresis. This evidence suggests that the lack of suppression of HMG-CoA reductase in hepatomas is not due to the production of an altered form of HMG-CoA reductase, but is instead a result of a cholesterol-induced suppression of enzyme synthesis. However, the possibility of an inactivation factor (Sabine, 1975) also needs to be considered since such a factor has recently been found in the soluble fraction of human fibroblasts and rat liver (Brown *et al.,* 1975). Microsomal HMG-CoA reductase prepared from either rat liver or human fibroblasts can be inactivated *in vitro* by this factor(s) in the presence of ATP or ADP and Mg^{++} or Mn^{++}. The physiological role that the inactivation factor may play in the regulation of HMG-CoA reductase in normal tissues and in hepatomas remains to be explored.

The complications in interpreting the mechanisms responsible for effects of exogenous cholesterol are illustrated by the contradictory results obtained for the control of HMG-CoA reductase by cholesterol in minimal deviation 7288C hepatomas and the same cells grown in culture (Kirsten and Watson, 1974). It appears that the phenotypic regulatory response of cells to serum lipoproteins (extracellular cholesterol) is still retained in hepatoma 7288C cultured *in vitro,* i.e., HMG-CoA reductase activity in cells grown in medium deficient in serum lipoproteins is higher than those activities in cells grown in medium containing whole serum.

ACKNOWLEDGMENTS

This work was supported by the Energy Research and Development Administration, the American Cancer Society (Grant No. BC-70G), and the National Cancer Institute (Grant No. CA 11949-06).

References

Alberts, A. W., Ferguson, K., Hennessy, S., and Vagelos, P. R. 1974. Regulation of lipid synthesis in cultured animal cells. *J. Biol. Chem.* **249**:5241–5249.

Allred, J. B., and Roehrig, K. L. 1973. Inhibition of rat liver acetyl coenzyme A carboxylase by $N^6,O^{2'}$-dibutyryl cyclic adenosine 3′:5′-monophosphate *in vitro. J. Biol. Chem.* **248**:4131–4133.

Ansell, G. B., and Spanner, S. 1965. The magnesium-ion-dependent cleavage of the vinyl ether linkage of brain ethanolamine plasmalogen. *Biochem. J.* **94**:252–258.

Blank, M. L., Wykle, R. L., Alper, S., and Snyder, F. 1974. Microsomal synthesis of the ether analogs of triacylglycerols. Acyl CoA:alkylacylglycerol and acyl CoA:alk-1-enylacylglycerol acyltransferases in tumors and liver. *Biochim. Biophys. Acta* **348**:397–403.

Bloch-Frankenthal, L., Langan, J., Morris, H. P., and Weinhouse, S. 1965. Fatty acid oxidation and ketogenesis in transplantable liver tumors. *Cancer Res.* **25**:732–736.

Bortz, W., Abraham, S., and Chaikoff, I. L. 1963. Localization of the block in lipogenesis resulting from feeding fat. *J. Biol. Chem.* **238**:1266–1272.

Brady, R. O., and Fishman, P. H. 1974. Biosynthesis of glycolipids in virus-transformed cells. *Biochim. Biophys. Acta* **355**:121–148.

Brenner, R. R. 1974. The oxidative desaturation of unsaturated fatty acids in animals. *Mol. Cell. Biochem.* **3**:41–52.

Bricker, L. A., and Levey, G. S. 1972a. Evidence for regulation of cholesterol and fatty acid synthesis in liver by cyclic adenosine 3′,5′-monophosphate. *J. Biol. Chem.* **247**:4914–4915.

Bricker, L. A., and Levey, G. S. 1972b. Autonomous cholesterol and fatty acid synthesis in hepatomas: Deletion of the adenosine 3′,5′-cyclic monophosphate control mechanism of normal liver. *Biochem. Biophys. Res. Commun.* **48**:362–365.

Brown, G. W., Jr., Katz, J., and Chaikoff, I. L. 1956. The oxidative metabolic pattern of mouse hepatoma C954 as studied with C^{14}-labeled acetates, propionate, octanoate, and glucose. *Cancer Res.* **16**:509–519.

Brown, M. S., Dana, S. E., and Siperstein, M. D. 1974. Properties of 3-hydroxy-3-methylglutaryl coenzyme A reductase solubilized from rat liver and hepatoma. *J. Biol. Chem.* **249**:6585–6589.

Brown, M. S., Brunschede, G. Y., and Goldstein, J. L. 1975. Inactivation of 3-hydroxy-3-methylglutaryl coenzyme A reductase *in vitro:* An adenine nucleotide-dependent reaction catalyzed by a factor in human fibroblasts. *J. Biol. Chem.* **250**:2502–2509.

Carlson, C. A., and Kim, K-H. 1973. Regulation of hepatic acetyl coenzyme A carboxylase by phosphorylation and dephosphorylation. *J. Biol. Chem.* **248**:378–380.

Carrol, K. K., (ed.). 1975. Lipids and Tumors, Vol. 10 (Progress in Biochemical Pharmacology). S. Karger, Basel, Switzerland.

Castuma, J. C., Catala, A., and Brenner, R. R. 1972. Oxidative desaturation of eicosa-8,11-dienoic acid to eicosa-5,8,11-trienoic acid: Comparison of different diets on oxidative desaturation at the 5,6 and 6,7 positions. *J. Lipid Res.* **13**:783–789.

Chae, K., Piantadosi, C., and Snyder, F. 1973a. Reductase, phosphatase, and kinase activities in the metabolism of alkyldihydroxyacetone phosphate and alkyldihydroxyacetone. *J. Biol. Chem.* **248**:6718–6723.

Chae, K., Piantadosi, C., and Snyder, F. 1973b. An alternate enzymic route for the synthesis of the alkyl analog of phosphatidic acid involving alkylglycerol. *Biochem. Biophys. Res. Commun.* **51**:119–124.

Chiappe, L. E., De Tomás, M. E., and Mercuri, O. 1974. In vitro activity of Δ6 and Δ9 desaturases in hepatomas of different growth rates. *Lipids* **9**:489–490.

Dempsey, M. E. 1974. Regulation of steroid biosynthesis. *Annu. Rev. Biochem.* **43**:967–990.

Dorsey, J. A., and Porter, J. W. 1968. The effect of palmityl coenzyme A in pigeon liver fatty acid synthetase. *J. Biol. Chem.* **243**:3512–3516.

Dunbar, L. M., and Bailey, J. M. 1975. Enzyme deletion and essential fatty acid metabolism in cultured cells. *J. Biol. Chem.* **250**:1152–1153.

Elwood, J. C., and Morris, H. P. 1968. Lack of adaptation in lipogenesis by hepatoma 9121. *J. Lipid Res.* **9**:337–341.

Emmelot, P., and Bos, C. J. 1955. Factors influencing the fatty acid oxidation of tumor mitochondria with special reference to changes in spontaneous mouse hepatomas. *Experientia* **11**:353–354.

Erwin, V. G., and Deitrich, R. A. 1966. Brain aldehyde dehydrogenase: Localization, purification, and properties. *J. Biol. Chem.* **241**:3533–3539.

Exton, J. H., Robison, G. A., Sutherland, E. W., and Park, C. R. 1971. Studies on the role of adenosine 3′,5′-monophosphate in the hepatic actions of glucagon and catecholamines. *J. Biol. Chem.* **246**:6166–6177.

Ferrell, W. J., and Kessler, R. J. 1971. Enzymic relationship of free fatty acids, aldehydes and alcohols in mouse liver. *Physiol. Chem. Phys.* **3**:549–558.

Goodridge, A. G. 1972. Regulation of the activity of acetyl coenzyme A carboxylase by palmitoyl coenzyme A and citrate. *J. Biol. Chem.* **247**:6946–6952.

Goodridge, A. G. 1973a. Regulation of fatty acid synthesis in the liver of prenatal and early postnatal chicks: Hepatic concentrations of individual free fatty acids and other metabolites. *J. Biol. Chem.* **248**:1939–1945.

Goodridge, A. G. 1973b. Regulation of fatty acid synthesis in isolated hepatocytes: Evidence for a physiological role for long chain fatty acyl coenzyme A and citrate. *J. Biol Chem.* **248**:4318–4326.

Granner, D. K. 1972. Protein kinase: Altered regulation in a hepatoma cell line deficient in adenosine 3′,5′-cyclic monophosphate-binding protein. *Biochem. Biophys. Res. Commun.* **46**:1516–1522.

Greenspan, M. D., and Lowenstein, J. M. 1968. Effects of magnesium ions, adenosine triphosphate, palmitoylcarnitine, and palmitoyl coenzyme A on acetyl coenzyme A carboxylase. *J. Biol. Chem.* **243**:6273–6280.

Guynn, R. W., Veloso, D., and Veech, R. L. 1972. The concentration of malonyl-coenzyme A and the control of fatty acid synthesis *in vivo*. *J. Biol. Chem.* **247**:7325–7331.

Hajra, A. K. 1968. Biosynthesis of phosphatidic acid from dihydroxyacetone phosphate. *Biochem. Biophys. Res. Commun.* **33**:929–935.

Hajra, A. K. 1970. Acyl dihydroxyacetone phosphate: Precursor of alkyl ethers. *Biochem. Biophys. Res. Commun.* **39**:1037–1044.

Hajra, A. K., and Agranoff, B. W. 1968. Acyl dihydroxyacetone phosphate. Characterization of a ^{32}P-labeled lipid from guinea pig liver mitochondria. *J. Biol. Chem.* **243**:1617–1622.

Hakomori, S.-i. 1973. Glycolipids of tumor cell membrane. *Adv. Cancer Res.* **18**:265–315.

Hakomori, S-i. 1975. Fucolipids and blood group glycolipids in normal and tumor tissue. *Prog. Biochem. Pharmacol.* **10**:167–196.

Hakomori, S-i., and Murakami, W. T. 1968. Glycolipids of hamster fibroblasts and derived malignant-transformed cell lines. *Proc. Natl. Acad. Sci. USA* **59**:254–261.

Halperin, M. L., Cheema-Dhadli, S., Taylor, W. M., and Fritz, I. B. 1975. Role of the citrate transporter in the control of fatty acid synthesis. *Adv. Enzyme Regul.* **13**:435–445.

Haven, F. L., and Bloor, W. R. 1956. Lipids in cancer. *Adv. Cancer Res.* **4**:237–314.

Higgins, M., and Rudney, H. 1973. Regulation of rat liver β-methylglutaryl-CoA reductase activity by cholesterol. *Nature (New Biol.)* **246**:60–61.

Hill, R., Webster, W. W., Linazasoro, J. M., and Chaikoff, I. L. 1960. Time of occurrence of changes in the liver's capacity to utilize acetate for fatty acid and cholesterol synthesis after fat feeding. *J. Lipid Res.* **1**:150–153.

Hillyard, L. A., and Abraham, S. 1972. Membrane proliferation and phosphatidylcholine synthesis in normal, preneoplastic, and neoplastic mammary gland tissues in C3H mice. *Cancer Res.* **32**:2834–2841.

Howard, B. V., Morris, H. P., and Bailey, J. M. 1972. Ether-lipids, α-glycerol phosphate dehydrogenase, and growth rate in tumors and cultured cells. *Cancer Res.* **32**:1533–1538.

Howard, B. V., Howard, W. J., and Bailey, J. M. 1974. Acetyl coenzyme A synthetase and the regulation of lipid synthesis from acetate in cultured cells. *J. Biol. Chem.* **249**:7912–7921.

Inkpen, C. A., Harris, R. A., and Quackenbush, F. W. 1969. Differential responses to fasting

and subsequent feeding by microsomal systems of rat liver: 6- and 9-desaturation of fatty acids. *J. Lipid Res.* **10**:277–282.
Jacobs, R. A., and Majerus, P. W. 1973. The regulation of fatty acid synthesis in human skin fibroblasts: Inhibition of fatty acid synthesis by free fatty acids. *J. Biol. Chem.* **248**:8392–8401.
Jacobs, R. A., Sly, W. S., and Majerus, P. W. 1973. The regulation of fatty acid biosynthesis in human skin fibroblasts. *J. Biol. Chem.* **248**:1268–1276.
Johnson, R. C., and Gilbertson, J. R. 1972. Isolation, characterization, and partial purification of fatty acyl coenzyme A reductase from bovine cardiac muscle. *J. Biol. Chem.* **247**:6991–6998.
Kawalek, J. C., and Gilbertson, J. R. 1973. Enzymic reduction of free fatty aldehydes in bovine cardiac muscle. *Biochem. Biophys. Res. Commun.* **51**:1027–1033.
Khan, A. A., and Kolattukudy, P. E. 1973. A microsomal fatty acid synthetase coupled to acyl-CoA reductase in *Euglena gracilis*. *Arch. Biochem. Biophys.* **158**:411–420.
Kimura, Y., Niwa, T., Wada, E., and Komeiji, T. 1964. Incorporation of labelled glucose carbon into different fractions of Ehrlich ascites tumor cells, with special reference to lipogenesis from glucose. *Jap. J. Exper. Med.* **34**:267–291.
Kirsten, E. S., and Watson, J. A. 1974. Regulation of 3-hydroxy-3-methylglutaryl coenzyme A reductase in hepatoma tissue culture cells by serum lipoproteins. *J. Biol. Chem.* **249**:6104–6109.
Korchak, H. M., and Masoro, E. J. 1964. Free fatty acids as lipogenic inhibitors. *Biochim. Biophys. Acta* **84**:750–753.
LaBelle, E. F., Jr., and Hajra, A. K. 1972. Enzymatic reduction of alkyl and acyl derivatives of dihydroxyacetone phosphate by reduced pyridine nucleotides. *J. Biol. Chem.* **247**:5825–5834.
LaBelle, E. F., Jr., and Hajra, A. K. 1974. Purification and kinetic properties of acyl and alkyl dihydroxyacetone phosphate oxidoreductase. *J. Biol. Chem.* **249**:6936–6944.
Lee, T-c., Wykle, R. L., Blank, M. L., and Snyder, F. 1973. Dietary control of stearly CoA and alkylacylglycerophosphorylethanolamine desaturases in tumors. *Biochem. Biophys. Res. Commun.* **55**:574–579.
Lee, T-c., Stephens, N., and Snyder, F. 1974. Dietary effects on stearyl coenzyme A desaturase in Morris hepatomas. *Cancer Res.* **34**:3270–3273.
Lumb, R. H., and Allen, K. F. 1976. Properties of microsomal phospholipase A_1 and A_2 in rat liver and hepatoma. *Biochim. Biophys. Acta* **450**:175–184.
Lyon, I., Masri, M. S., and Chaikoff, I. L. 1952. Fasting and hepatic lipogenesis from C^{14}-acetate. *J. Biol. Chem.* **196**:25–32.
Majerus, P. W., and Kilburn, E. 1969. Acetyl coenzyme A carboxylase: The roles of synthesis and degradation in regulation of enzyme levels in rat liver. *J. Biol. Chem.* **244**:6254–6262.
Majerus, P. W., Jacobs, R., Smith, M. B., and Morris, H. P. 1968. The regulation of fatty acid biosynthesis in rat hepatomas. *J. Biol. Chem.* **243**:3588–3595.
Martin, D. B., and Vagelos, P. R. 1962. The mechanism of tricarboxylic acid cycle regulation of fatty acid synthesis. *J. Biol. Chem.* **237**:1787–1892.
McGee, R., and Spector, A. A. 1974. Short-term effects of free fatty acids on the regulation of fatty acid biosynthesis in Ehrlich ascites tumor cells. *Cancer Res.* **34**:3355–3362.
McGee, R., and Spector, A. A. 1975. Fatty acid biosynthesis in Ehrlich cells: The mechanism of short term control by exogenesis free fatty acids. *J. Biol. Chem.* **250**:5419–5425.
Nakanishi, S., and Numa, S. 1970. Purification of rat liver acetyl coenzyme A carboxylase and immunochemical studies on its synthesis and degradation. *Eur. J. Biochem.* **16**:161–173.
Nilsson, A. 1970. Conversion of dihydrosphingosin to palmitaldehyde and palmitic acid with cell-free preparations of guinea pig intestinal mucosa. *Acta Chem. Scand.* **24**:598–604.

Numa, S., Bortz, W. M., and Lynen, F. 1965a. Regulation of fatty acid synthesis at the acetyl-CoA carboxylation step. *Adv. Enzyme Regul.* **3**:407–423.

Numa, S., Ringelmann, E., and Lynen, F. 1965b. The inhibition of acetyl-coenzyme A-carboxylase by fatty acid co-enzyme A compounds. *Biochem. Z.* **343**:243–257.

Oshino, N., and Sato, R. 1972. The dietary control of the microsomal stearyl CoA desaturation enzyme system in rat liver. *Arch. Biochem. Biophys.* **149**:369–377.

Paltauf, F., and Holasek, A. 1973. Enzymatic synthesis of plasmalogens. Characterization of the 1-*O*-alkyl-2-acyl-*sn*-glycero-3-phosphorylethanolamine desaturase from mucosa of hamster small intestine. *J. Biol. Chem.* **248**:1609–1615.

Paltauf, F., Prough, R. A., Masters, B. S. S., and Johnston, J. M. 1974. Evidence for the participation of b_5 in plasmalogen biosynthesis. *J. Biol. Chem.* **249**:2661–2662.

Parvin, R., and Dakshinamurti, K. 1970. Inhibition of gluconeogenic enzymes by free fatty acids and palmitoyl coenzyme A. *J. Biol. Chem.* **245**:5773–5778.

Pederson, B. N., Gromek, A., and Daehnfeldt, J. L. 1972. Extramitochondrial fatty acid synthesis in Ehrlich ascites tumor cells propagated *in vitro* and *in vivo*. *Proc. Soc. Exp. Biol. Med.* **141**:506–509.

Pfleger, R. C., Piantadosi, C., and Snyder, F. 1967. The biocleavage of isomeric glyceryl ethers by soluble liver enzymes in a variety of species. *Biochim. Biophys. Acta* **144**:633–648.

Pollock, R. J., Hajra, A. K., and Agranoff, B. W. 1975. The relative utilization of the acyl dihydroxyacetone phosphate and glycerol phosphate pathways for synthesis of glycerolipids in various tumors and normal tissues. *Biochim. Biophys. Acta* **380**:421–435.

Raju, P. K. 1974. Altered stearoyl-CoA desaturase activities in Morris hepatomas 5123C and 7800. *Lipids* **9**:795–797.

Rao, G. A., and Abraham, S. 1973. αGlycerophosphate dehydrogenase activity and levels of glyceride-glycerol precursors in mouse mammary tissues. *Lipids* **8**:232–234.

Raskin, P., McGarry, J. D., and Foster, D. W. 1974. Independence of cholesterol and fatty acid biosynthesis from cyclic adenosine monophosphate concentration in the perfused rat liver. *J. Biol. Chem.* **249**:6029–6032.

Rock, C. O., and Snyder, F. 1974. Biosynthesis of 1-alkyl-*sn*-glycero-3-phosphate via adenosine triphosphate:1-alkyl-*sn*-glycerol phosphotransferase. *J. Biol. Chem.* **249**:5382–5387.

Rock, C. O., and Snyder, F. 1976. Stimulation of alkyldihydroxyacetone phosphate synthesis by detergents. *Fed. Proc.* **35**:1724,abstr. 1861.

Sabine, J. R. 1975. Defective control of lipid biosynthesis in cancerous and precancerous liver. *Progr. Biochem. Pharmacol.* **10**:269–307.

Sabine, J. R., Abraham, S., and Chaikoff, I. L. 1966. Lack of feedback control of fatty acid synthesis in a transplantable hepatoma. *Biochim. Biophys. Acta* **116**:407–409.

Sabine, J. R., Abraham, S., and Morris, H. P. 1968. Defective dietary control of fatty acid metabolism in four transplantable rat hepatomas: Numbers 5123C, 7793, 7795, and 7800. *Cancer Res.* **28**:46–51.

Saito, M., and Kanfer, J. 1975. Phosphohydrolase activity in a solubilized preparation from rat brain particulate fraction. *Arch. Biochem. Biophys.* **169**:318–323.

Snyder, F. 1970. Lipids and Cancer. A Bibliography: 1947–1970. USAEC Report—ORAU 111.

Snyder, F. 1971. Glycerolipids in the neoplastic cell: Methodology, metabolism, and composition. *Methods Cancer Res.* **6**:399–436.

Snyder, F. 1972. Ether-linked lipids and fatty alcohol precursors in neoplasms, pp. 273–295. *In* F. Snyder (ed.) Ether Lipids: Chemistry and Biology. Academic Press, New York.

Snyder, F., and Malone, B. 1970. Enzymic interconversion of fatty alcohols and fatty acids. *Biochem. Biophys. Res. Commun.* **41**:1382–1387.

Snyder, F., and Snyder, C. 1975. Glycerolipids and cancer. *Progr. Biochem. Pharmacol.* **10**:1–41.

Snyder, F., and Wood, R. 1968. The occurrence and metabolism of alkyl and alk-1-enyl ethers of glycerol in transplantable rat and mouse tumors. *Cancer Res.* **28**:972–978.

Snyder, F., and Wood, R. 1969. Alkyl and alk-1-enyl ethers of glycerol in lipids from normal and neoplastic human tissues. *Cancer Res.* **29**:251–257.

Snyder, F., Blank, M. L., and Malone, B. 1970a. Requirement of cytidine derivatives in the biosynthesis of *O*-alkyl phospholipids. *J. Biol. Chem.* **245**:4016–4018.

Snyder, F., Malone, B., and Blank, M. L. 1970b. Enzymic synthesis of *O*-alkyl bonds in glycerolipids. *J. Biol. Chem.* **245**:1790–1799.

Snyder, F., Blank, M. L., and Wykle, R. L. 1971. The enzymic synthesis of ethanolamine plasmalogens. *J. Biol. Chem.* **246**:3639–3645.

Soodsma, J. F., Piantadosi, C., and Snyder, F. 1970. The biocleavage of alkyl glyceryl ethers in Morris hepatomas and other transplantable neoplasms. *Cancer Res.* **30**:309–311.

Sprecher, H. 1974. Feeding studies designed to determine whether competitive reactions between acids of the oleate and linoleate families for desaturation chain elongation or incorporation regulate the fatty acid composition of rat liver lipids. *Biochim. Biophys. Acta* **369**:34–44.

Steele, W., and Jenkin, H. M. 1972. The effect of two isomeric octadecenoic acids on alkyl diacyl glycerides and neutral glycosphingolipids of Novikoff hepatoma cells. *Lipids* **7**:556–559.

Stein, O., and Stein, Y. 1967. Utilization of lysolecithin by Landschütz ascites tumor *in vivo* and *in vitro*. *Biochim. Biophys. Acta* **137**:232–239.

Stoffel, W., LeKim, D., and Heyn, G. 1970. Sphinganine (dihydrosphingosine), an effective donor of the alk-1′-enyl chain of plasmalogens. *Hoppe-Seyler's Z. Physiol. Chem.* **351**:875–883.

Tabakoff, B., and Erwin, V. G. 1970. Purification and characterization of a reduced nicotinamide adenine dinucleotide phosphate-linked aldehyde reductase from brain. *J. Biol. Chem.* **245**:3263–3268.

Taketa, K., and Pogell, B. M. 1966. The effect of palmityl coenzyme A on glucose 6-phosphate dehydrogenase and other enzymes. *J. Biol. Chem.* **241**:720–726.

Tietz, A., Lindberg, M., and Kennedy, E. P. 1964. A new pteridine-requiring enzyme system for the oxidation of glyceryl ethers. *J. Biol. Chem.* **239**:4081–4090.

Volpe, J. J., and Marasa, J. C. 1975. Hormonal regulation of fatty acid synthetase, acetyl-CoA carboxylase and fatty acid synthesis in mammalian adipose tissue and liver. *Biochim. Biophys. Acta* **380**:454–472.

Volpe, J. J., Lyles, T. O., Roncari, D. A. K., and Vagelos, P. R. 1973. Fatty acid synthetase of developing brain and liver content, synthesis, and degradation during development. *J. Biol. Chem.* **248**:2502–2513.

Waite, M., and Wakil, S. J. 1962. Studies on the mechanism of fatty acid synthesis. XII. Acetyl coenzyme A carboxylase. *J. Biol. Chem.* **237**:2750–2757.

Warner, H. R., and Lands, W. E. M. 1961. The metabolism of plasmalogen: Enzymatic hydrolysis of the vinyl ether. *J. Biol. Chem.* **236**:2404–2409.

Wood, R. (ed.) 1973. Tumor Lipids: Biochemistry and Metabolism. American Oil Chemists' Society Press, Champaign, Illinois.

Wykle, R. L., and Snyder, F. 1970. Biosynthesis of an *O*-alkyl analogue of phosphatidic acid and *O*-alkylglycerols via *O*-alkyl ketone intermediates by microsomal enzymes of Ehrlich ascites tumor. *J. Biol. Chem.* **245**:3047–3058.

Wykle, R. L. and Schremmer, J. M. 1974. A lysophospholipase D pathway in the metabolism of ether-linked lipids in brain microsomes. *J. Biol. Chem.* **249**:1742–1746.

Wykle, R. L., and Snyder, F. 1976, Microsomal enzymes involved in the metabolism of ether-linked glycerolipids and their precursors in mammals. pp. 87-117. *In* A. Martonosi (ed.). The Enzymes of Biological Membranes, Vol. 2. Plenum Press, New York.

Wykle, R. L., Piantadosi, C., and Snyder, F. 1972a. The role of acyldihydroxyacetone phosphate, NADH, and NADPH in the biosynthesis of *O*-alkyl glycerolipids by microsomal enzymes of Ehrlich ascites tumor. *J. Biol. Chem.* **247**:2944–2948.

Wykle, R. L., Blank, M. L., Malone, B., and Snyder, F. 1972b. Evidence for a mixed-function oxidase in the biosynthesis of ethanolamine plasmalogens from 1-alkyl-2-acyl-*sn*-glycero-3-phosphorylethanolamine. *J. Biol. Chem.* **247**:5442–5447.

Yavin, E., and Gatt, S. 1972a. Oxygen-dependent cleavage of the vinyl-ether linkage of plasmalogens. 1. Cleavage by rat-brain supernatant. *Eur. J. Biochem.* *25:431–436.*

Yavin, E., and Gatt, S. 1972b. Oxygen-dependent cleavage of the vinyl-ether linkage of plasmalogens. 2. Identification of the low molecular-weight active component and the reaction mechanism. *Eur. J. Biochem.* **25**:437–446.

Zuckerman, N. J., Nardella, P., Morris, H. P., and Elwood, J. C. 1970. Lack of adaptation in lipogenesis by hepatomas 9098, 7794A, and 9618A. *J. Nat. Cancer Inst.* **44**:79–83.

Harderian Gland

CHARLES O. ROCK

I. *Introduction*

The harderian gland is a tublo-alveolar organ located on the posterior aspect of the eyeball in animals that possess a nictitating membrane. The main function ascribed to the gland is secretion of an oily substance that facilitates the movement of the third eyelid (Cohn, 1955; Grafflin, 1942; Kennedy, 1970). Recently, the harderian gland has also been implicated as an extraretinal photoreceptor in neonatal rats (Wetterberg *et al.*, 1970a,b,c). These investigators show that the circadian rhythm of pineal serotonin was maintained in blinded rats, but was abolished in animals with their harderian glands excised. The glands have a high porphyrin content (for review see Kennedy, 1970) and the enzymes involved in porphyrin biosynthesis are quite active in this organ (Tomio and Grinstein, 1968; Tomio *et al.*, 1968; Margolis, 1971). This led Wetterberg and his co-workers (1970a) to speculate that the porphyrins may be involved in some way in a phototransduction process. However, the hypothesis that the harderian gland functions as a photoreceptor is not supported by Reiter and Klein (1971), who observed that the glands were not necessary for the development of reproductive changes associated with long-term exposure to light, or regulating the activities of pineal *N*-acetyltransferase and hydroxyindole-*O*-methyltransferase activity in rats. Another possibility is that the hydroxyin-

CHARLES O. ROCK • Medical and Health Sciences Division, Oak Ridge Associated Universities, Oak Ridge, Tennessee.

dole-*O*-methyltransferase activity present in the harderian gland (Vlahakes and Wurtman, 1972; Cardinali and Wurtman, 1972) is mediated by light. However, maintenance of rats in seven different light environments, which markedly affect pineal and retinal hydroxyindole-*O*-methyltransferase, had no effect on the harderian gland enzyme (Cardinali *et al.*, 1972). Demonstration of an unambiguous role for the harderian gland in phototransduction, the function of the abundant quantities of porphyrins, and the relation between the pineal gland and harderian gland await further study.

Macroscopically, harderian glands of rodents exhibit considerable diversity. In the mouse and rat the gland is horseshoe-shaped, reddish in appearance, and closely attached to the posterior aspect of the eyeball (Cohn, 1955; Grafflin, 1942; Woodhouse and Rhodin, 1963). Sexual dimorphism has been noted in the hamster; the gland in the female is highly pigmented, whereas the gland in males is almost devoid of porphyrins (Christensen and Dam, 1953; Woolley and Worley, 1954; Paule *et al.*, 1955). These characteristics can be modulated by light, visual function, and endocrine gland secretion (Hoffman, 1971; Wetterberg, 1972). The rabbit has a large ovoid gland (ca. 1.0 g) that can be easily separated into two lobes: a large pink lobe (ca. 0.7 g) and a smaller white lobe (ca. 0.3 g) (Björkman *et al.*, 1960; Kühnel, 1971).

The ultrastructure of the harderian gland has been examined in the rat (Tsutsumi *et al.*, 1966), mouse (Woodhouse and Rhodin, 1963), rabbit (Björkman *et al.*, 1960; Kühnel, 1971), and hamster (Bucana and Nadakavukaren, 1972a,b). Basically, the gland has a single layer of columnar epithelial cells surrounding the lumen. These cells are characterized by large basally located nuclei, well-developed golgi zones, abundant mitochondria, and many large vacuoles. The electron micrographs show the vacuoles as large holes, indicating that they are storage depots for the neutral lipids, abundantly found in the harderian gland. Electron microscopy has also shown the myoepithelial cells of the mouse (Chiquoine, 1958), rabbit (Kühnel, 1971), and hamster (Bucana and Nadakavukaren, 1972c) harderian gland to be associated with nerve endings. Tashiro *et al.* (1940) reported that injection of rats with the neurotransmitter acetylcholine will produce, in a matter of minutes, copious excretion of "bloody tears" from the harderian gland. Chiquoine (1958) suggested that the myoepithelial cells may respond to acetylcholine by contraction and squeeze out the contents of the secretory cells into the lumen.

Extensive histochemical characterization of the harderian gland of rabbits has been carried out (Kühnel and Wrobel, 1966a,b; Kühnel and Wrobel, 1968; Kühnel, 1966). The most striking feature is the presence of numerous, large sudanophilic granules indicative of the large lipid storage vacuoles present in the gland (Schneir and Hayes, 1951; Kühnel, 1971).

These lipid droplets were found to be nonosmiophilic (Schneir and Hayes, 1951), indicating the presence of saturated lipid components.

II. *Lipid Composition*

The lipid composition of only three harderian gland systems has been examined: the normal harderian gland from the rat and rabbit and a transplantable lipid-producing tumor originating in the mouse harderian gland (Itoh and Kasama, 1970). Hais *et al.* (1968) reported that the lipids of the rat harderian gland are readily labeled by injection of [1-^{14}C]acetate into the bloodstream. About 83% of the activity in the lipid extract migrated with the sterol ester band in their thin-layer chromatographic system (petroleum ether:ethyl ether:acetic acid; 90/10/1, v/v). The unsaponifiable material gave a single radioactive band that the authors considered to be a sterol; however, it should be mentioned that wax esters and sterol ester cochromatograph in the solvent system used. The possibility that the labeled unsaponifiable material was an alkylglycerol for fatty alcohol was not investigated. Recently, Murawski and Jost (1974) identified the major lipid in the rat harderian gland to be a mixture of wax esters. Both the fatty acid and fatty alcohol moieties were determined to be a homologous series of monoenes. The major fatty acid was identified as $C_{20:1}$ and the major fatty alcohol as $C_{24:1}$. Mass spectroscopy placed the double bond position in both fatty acids and fatty alcohols as *n*-7.

Kasama *et al.* (1970) found that the transplantable tumor of the mouse harderian gland contained predominantly alkyldiacylglycerols. At least six other unidentified components were also detected. The alkyl chains were primarily saturated, but the acyl moieties contained appreciable quantities of $C_{18:1}$ and $C_{20:1}$ chains. The position of the double bond in these fatty acids is unknown. It is not known whether the lipid composition of the mouse harderian gland tumor reflects the lipid composition of the normal mouse harderian gland.

The lipid composition of the pink portion of the rabbit harderian gland has been the most thoroughly examined. The total lipids, 93% of which are neutral, comprise 58% of the tissue dry weight. Kasama *et al.* (1973) demonstrated that the major lipid component (ca. 80% of the total lipids) yields hydroxy-substituted alkylglycerols upon chemical reduction. Mass spectroscopy placed the hydroxyl groups at C-10 or C-11 of the hexadecyl moieties and C-11 or C-12 of the octadecyl moieties. Chromatographic and infrared spectroscopic evidence indicates that all hydroxyl groups are esterified in the native lipid class (Fig. 1). On the basis of limited chromatographic data, Jost (1974) has incorrectly identified the major lipid class in the glands as alkyldiacylglycerols containing unsaturated acyl chains.

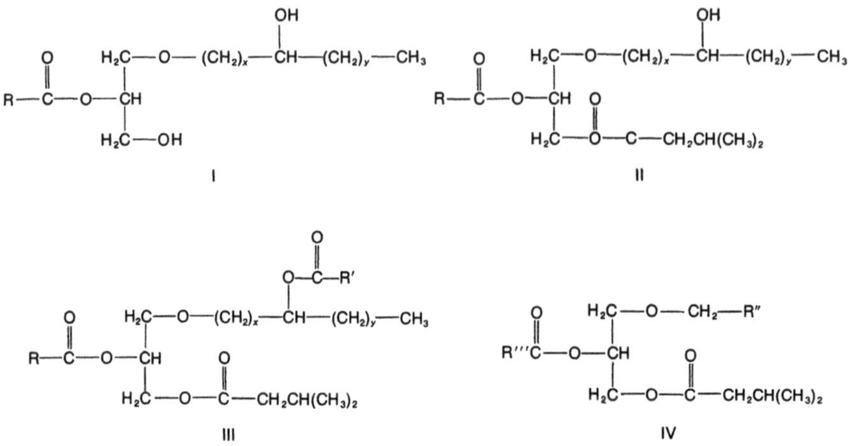

FIGURE 1. Structures of the major neutral lipid classes found in the pink portion of the rabbit harderian gland. (I) 1-hydroxyalkyl-2-acyl-*sn*-glycerol (monoacyl-HAG); (II) 1-hydroxyalkyl-2,3-diacyl-*sn*-glycerol (diacyl-HAG); (III) 1-(*O*-acyl)hydroxyalkyl-2,3-diacyl-*sn*-glycerol (triacyl-HAG); and (IV) 1-alkyl-2,3-diacyl-*sn*-glycerol.

Rock and Snyder (1975a) examined the acyl group composition and positional distribution in the 1-(*O*-acyl)hydroxyalkyl-2,3-diacyl-*sn*-glycerols (triacyl-HAG, Fig. 1). The long-chain acyl moieties consist primarily of hexadecanoic acid and octadecanoic acid in a ratio of 2:1. Only minute amounts of unsaturated fatty acids were detected. Triacyl-HAG also contains a short-chain acid identified as isovaleric acid. There is 1 mole of isovaleric acid for every 2 moles of long-chain fatty acid. The isovaleric acid occupies the 3-position of this lipid class with the long-chain acids occupying the other two positions. Two minor lipid components were also detected in this study and identified as 1-hydroxyalkyl-2,3-diacyl-*sn*-glycerols (diacyl-HAG, Fig. 1) and 1-hydroxyalkyl-2-acyl-*sn*-glycerols (monoacyl-HAG, Fig. 1). The long-chain acyl groups in both classes are almost exclusively hexadecanoic acid and the diacyl-HAG also has isovaleric acid esterified exclusively in the 3-position.

The remaining lipid class (ca. 13% of the total lipids) has been identified by Blank *et al.* (1972) as 1-alkyl-2,3-diacyl-*sn*-glycerols (Fig. 1). The 3-position of this lipid class is also exclusively esterified with isovaleric acid and the long-chain acyl groups at the 2-position are primarily saturated (60% hexadecanoic acid and 30% octadecanoic acid). The alkyl groups are also saturated; but, in contrast to the acyl moieties, they do contain significant quantities of $C_{15:0}$ and $C_{17:0}$ branched chains (ca. 30%) in addition to the $C_{16:0}$ and $C_{18:0}$ carbon chains (ca. 50%).

The presence of short-chain acids in mammalian glycerides is not very common. When short-chain acids do occur, they tend to reside in positions 1 or 3 of the glycerol backbone. For instance, 95% of the short-chain acids in bovine milk fat triacylglycerols are attached to the 3-position of glycerol (Pitas *et al.*, 1967; Breckenridge and Kuksis, 1968), and in porpoise acoustic tissues isovaleric acid occupies the 1-and 3-positions (Varanasi and Malins, 1970, 1972; Varanasi *et al.*, 1973). The isovaleric acid containing glycerides in the pink portion of the rabbit harderian gland also conform to this general rule.

The white portion of the rabbit harderian gland contains a preponderance of neutral lipids (Kasama *et al.*, 1973). A major lipid component in the white portion of the rabbit harderian gland is a mixture of 2-(*O*-acyl)hydroxy fatty acid esters (Fig. 2; Rock *et al.*, 1976). The fatty acid moieties in this lipid class are exclusively saturated and range in chain length from $C_{14:0}$ to $C_{22:0}$, with $C_{16:0}$ being the major component (65%). The fatty alcohol moieties are a simpler mixture, and mass spectroscopy revealed the alcohols to be composed of $C_{20:0}$, $C_{21:0}$, and $C_{22:0}$ chains. The 2-hydroxy fatty acids consist of $C_{14:0}$, $C_{15:0}$, and $C_{16:0}$ chains, and mass spectroscopy and chemical techniques placed the hydroxyl groups at the 2-carbon.

2-(*O*-acyl) Hydroxy fatty acid esters have been identified in surface lipids from a variety of mammals and the vernix caseosa (cf. the chapter Skin). In general, the fatty acids and 2-hydroxy fatty acids from these sources are completely saturated and composed predominantly of $C_{14:0}$ and $C_{16:0}$ chains with smaller amounts of acids with odd numbers of carbon atoms. The fatty alcohols are generally of higher molecular weight, having chain lengths between $C_{20:0}$ and $C_{28:0}$. The 2-(*O*-acyl)hydroxy fatty acid esters from the white portion of the harderian gland conform to the general structural features of 2-(*O*-acyl)hydroxy fatty acid esters obtained from other sources. Note that the hydroxyl group in the fatty acids of the white portion of the harderian gland are at the 2-carbon, whereas the hydroxyl group in the alkyl chains of the pink portion of the gland are in the middle of the chain. Thus, hydroxylation enzymes having different specificities are anticipated to be present in the two portions of the harderian gland.

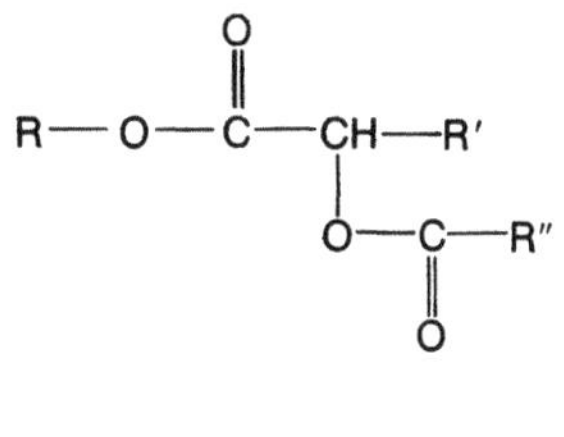

FIGURE 2. 2-(*O*-acyl)Hydroxy fatty acid alkyl esters, a major lipid component of the white portion of the rabbit harderian gland.

III. Lipid Metabolism

1-Alkyl-*sn*-glycero-3-phosphate (alkylglycerol-P) is an important intermediate in the biosynthesis of ether-linked glycerolipids and is normally formed in tissues by the reduction of alkyldihydroxyacetone phosphate by microsomal enzymes in the presence of reduced pyridine nucleotides (cf. the chapter Cancer Cells). However, several *in vivo* experiments have demonstrated that intact alkylglycerols can enter the mainstream of the biosynthetic pathway, indicating that alkylglycerol-P could be formed by an alternate enzymatic route (Thompson, 1968; Blank *et al.*, 1970; Paltauf, 1971). Chae *et al.* (1973) showed the mitochondrial supernatant fraction from mouse preputial gland tumors catalyzed the formation of the alkyl analog of phosphatidic acid. Only minute amounts of alkylglycerol-P were detected, and it was not entirely clear whether phosphorylation preceded or followed the acylation step.

Rock and Snyder (1974) have found the pink portion of the rabbit harderian gland to be a rich source of ATP:1-alkyl-*sn*-glycerol phosphotransferase. This enzyme, associated with the microsomal fraction of the gland, exhibits a pH optimum of 7.1, and ATP and Mg^{2+} are absolute requirements for activity. Manganese partially substituted for Mg^{2+}, but the measured rate in the presence of Mn^{2+} was only half the rate determined in the presence of Mg^{2+}. In incubations supplemented with CoA, 1-alkyl-2-acyl-*sn*-glycero-3 phosphate was identified as the major product. The transfer of the terminal phosphate of ATP to position 3 of 1-alkyl-*sn*-glycerol was stereoselective; other isomeric alkylglycerols, as well as alkylethyleneglycol and S-alkylglycerol, were not appreciably phosphorylated by the phosphotransferase. Attempts were made to show the phosphorylation of 1-acylglycerols, but, owing to the high lipase activity in harderian gland microsomes, results were inconclusive.

The occurrence of isovaleric acid exclusively in the 3-position of the ether-linked glycerolipids of the pink portion of the rabbit harderian gland led to the investigation of the biosynthesis of isovaleric acid and the enzyme(s) responsible for its specific placement in the 3-position of the alkylglycerolipids of the gland. Recently, a microsomal enzyme from Morris hepatoma and rat liver has been described that acylates the 3-position of 1-alkyl-2-acyl-*sn*-glycerols with long-chain acyl-CoAs to form 1-alkyl-2,3-diacyl-*sn*-glycerols (Blank *et al.*, 1974). Rock and Snyder (1975b) demonstrated the existence of a similar enzymatic activity in the microsomal fraction of the pink portion of the rabbit harderian gland, except in this gland the acyltransferase specifically utilitzes isovaleryl-CoA and other closely related short-chain acyl-CoAs.

Isovaleryl-CoA effectively stimulates synthesis of alkyldiacylglycerols in harderian gland microsomes from labeled 1-alkyl-2-acyl-*sn*-glycerols. In contrast, additions of ATP, CoA, Mg^{2+}, and isovaleric acid did not yield substantial quantities of alkyldiacylglycerols. These data indicate that microsomes from this gland do not possess butyrate : CoA ligase (AMP) [EC 6.2.1.2] needed to convert isovaleric acid to its CoA ester, whereas long-chain acyl-CoA synthetase (acid : CoA ligase (AMP) [EC 6.2.1.3]) has been previously noted in these microsomal preparations (Rock and Snyder, 1974). The chain-length specificity of the later enzyme has been examined in rat liver and the reaction rate with fatty acids shorter than C_6 is negligible (Suzue and Marcel, 1972).

The relative rates of reaction of various short-chain CoA esters are as follows: valerate > isovalerate ~ isobutyrate > caproate ~ butyrate. There was no activity toward acetate and very little with hexanoyl-CoA; long-chain CoA esters were completely inactive. The specificity of the short-chain acyltransferase can account for the exclusion of long-chain acyl-CoAs from the 3-position of the alkyldiacylglycerols in the pink portion of the rabbit harderian gland. However, since the enzyme uses other short-chain acyl-CoAs at least as effectively as isovaleryl-CoA; some mechanism by which isovaleryl-CoA is exclusively presented to the enzyme must be operating.

Tissue slices of the pink portion of the rabbit harderian gland rapidly incorporate labeled leucine into lipid soluble products (Rock and Snyder, 1975a). Thin-layer chromatography revealed that the three isovaleric-containing lipid classes, triacyl-HAG, diacyl-HAG, and alkyldiacylglycerols, were heavily labeled. Gas–liquid chromatography of the 2-chloroethanol esters of the labeled acids revealed >99% of the radioactivity to be associated with isovaleric acid. These data point to leucine as the major source of isovaleric acid in the pink portion of the rabbit harderian gland. Leucine has also been found to be the precursor of isovaleric acid in the porpoise melon tissues (Malins *et al.*, 1972), which possess high levels of glycerides containing isovaleryl moieties (Varanasi and Malins, 1970, 1972; Varanasi *et al.*, 1973). Thus, the harderian gland has a mechanism to present isovaleryl-CoA to the microsomal short-chain acyltransferase.

Tissue slices of the pink portion of the rabbit harderian gland were pulsed for 30 min with L-[U-^{14}C]leucine and chased with 10 mM cold leucine, and the radioactivity in the labeled fraction was then determined at 15-min intervals up to 60 min. Over the time examined, the total radioactivity in the lipids remained relatively constant, as did the radioactivity present in the alkyldiacylglycerols. However, label in the triacyl-HAG increased at the expense of the activity in the diacyl-HAG. This indicates that diacyl-HAG is the precursor of triacyl-HAG. These data also rule out direct hydroxylation of alkyldiacylglycerols followed by acylation as a possible pathway for the formation of triacyl-HAG (Rock and Snyder, 1975a).

IV. Epilogue

The pink portion of the rabbit harderian gland is a rich source of the enzymes involved in the biosynthesis of ether-linked lipids and should continue to serve as a useful model system for the study of these reactions. To date, two of the key enzymes in the ether-lipid pathway have yet to be examined in the harderian gland. Acyl-CoA reductase, the first enzyme in the pathway, has not been studied thoroughly in any tumor or mammalian system. It remains to be determined (1) if the enzyme is coupled to fatty acid synthase, (2) if there is an aldehyde intermediate, (3) if there are two enzymes involved, and (4) the subcellular localization of the enzyme(s). The second enzyme, which synthesizes alkyldihydroxyacetone phosphate from fatty alcohol and acyldihydroxyacetone phosphate, is very active in the harderian gland (Kasama *et al.*, 1973). This is the most mechanistically intriguing enzyme in the pathway, and the harderian gland is a useful source for the solubilization and purification of this enzyme.

With regard to the unusual aspects of harderian gland lipid metabolism, the manner in which the aliphatic moieties are hydroxylated is unknown. However, the placement of the hydroxyl group at the 2-carbon in the white portion and in the middle of the chain in the pink portion would argue for two hydroxylation enzymes having different specificities. In addition, it is not known whether the phospholipids contain hydroxylated aliphatic moieties. The lipid composition of the white portion needs to be examined more thoroughly as it still contains unidentified lipid classes. The occurrence of 2-(*O*-acyl) hydroxy fatty acid esters in the white portion indicates that this organ may also be a good model for the study of these esters and their metabolism.

ACKNOWLEDGMENTS

This work was supported by the Energy Research and Development Administration and the National Cancer Institue (Training Grant No. CA 05287).

References

Björkman, N., Nicander, L., and Schantz, B. 1960. On the histology and ultrastructure of the harderian gland in rabbits. *Z. Zellforsch.* **52**:93–104.

Blank, M. L., Wykle, R. L., Piantadosi, C., and Snyder, F. 1970. The biosynthesis of plasmalogens from labeled *O*-alkyl glycerols in Ehrlich ascites cells. *Biochim. Biophys. Acta* **210**:442–447.

Blank, M. L., Kasama, K., and Snyder, F. 1972. Isolation and identification of an alkyldiacylglycerol containing isovaleric acid. *J. Lipid Res.* **13**:390–395.

Blank, M. L., Wykle, R. L., Alper, S., and Snyder, F. 1974. Microsomal synthesis of the ether analogs of triacylglycerols. Acyl CoA:alkylacylglycerol and acyl CoA:alk-1-enylacylglycerol acyltransferases in tumors and liver. *Biochim. Biophys. Acta* **348**:397–403.

Breckenridge, W. C., and Kuksis, A. 1968. Specific distribution of short-chain fatty acids in molecular distillates of bovine and milk fat. *J. Lipid Res.* **9**:388–393.

Bucana, C. D., and Nadakavukaren, M. J. 1972a. Fine structure of the hamster harderian gland. *Z. Zellforsch.* **129**:178–187.

Bucana, C. D., and Nadakavukaren, M. J. 1972b. Ultrastructural investigation of the postnatal development of the hamster harderian gland. *Z. Zellforsch.* **135**:149–153.

Bucana, C. D., and Nadakavukaren, M. J. 1972c. Innervation of the hamster harderian gland. *Science* **175**:205–206.

Cardinali, D. P., and Wurtman, R. J. 1972. Hydroxyindole-O-methyl transferases in rat pineal, retina and harderian gland. *Endocrinology* **91**:247–252.

Cardinali, D. P., Larin, F., and Wurtman, R. J. 1972. Action spectra for effects of light on hydroxyindole-O-methyl transferases in rat pineal, retina and harderian gland. *Endocrinology* **91**:877–886.

Chae, K., Piantadosi, C., and Snyder, F. 1973. An alternate enzymic route for the synthesis of the alkyl analog of phosphatidic acid involving alkylglycerol. *Biochem. Biophys. Res. Commun.* **51**:119–124.

Chiquoine, A. D. 1958. The identification and electron microscopy of myoepithelial cells in the harderian gland. *Anat. Rec.* **132**:569–577.

Christensen, F., and Dam, H. 1953. A sexual dimorphism of the harderian glands in hamsters. *Acta Physiol. Scand.* **27**:333–336.

Cohn, S. A. 1955. Histochemical observations on the harderian gland of the albino mouse. *J. Histochem. Cytochem.* **71**:342–353.

Grafflin, A. L. 1942. Histological observations upon the porphyrin-excreting harderian gland of the albino rat. *Am. J. Anat.* **55**:43–64.

Hais, I. M., Strych, A., and Chemlar, V. 1968. Preliminary thin-layer chromatographic characterization of the rat harderian gland lipid. *J. Chromatogr.* **35**:179–191.

Hoffman, R. A. 1971. Influence of some endocrine glands, hormones and blinding on the histology and porphyrins of the harderian glands of golden hamsters. *Am. J. Anat.* **132**:463–478.

Itoh, K., and Kasama, K. 1970. Transplantable lipid-producing tumor originating from the harderian gland. *Gann (Jap. J. Cancer Res.)* **61**:271–273.

Jost, U. 1974. 1-Alkyl-2,3-diacyl-*sn*-glycerol, the major lipid in the harderian gland of rabbits. *Hoppe-Seyler's Z. Physiol. Chem.* **355**:422–426.

Kasama, K., Uezumi, N., and Itoh, K. 1970. Characterization and identification of glyceryl ether diesters in harderian gland tumor of mice. *Bichim. Biophys. Acta* **202**:56–66.

Kasama, K., Rainey, W. T., Jr., and Snyder, F. 1973. Chemical identification and enzymatic synthesis of a newly discovered lipid class—hydroxyalkylglycerols. *Arch. Biochem. Biophys.* **154**:648–658.

Kennedy, G. Y. 1970. Harderoporphyrin: A new porphyrin from the harderian glands of the rat. *Comp. Biochem. Physiol.* **36**:21–36.

Kühnel, W. 1966. Enxymhistochemische Untersuchungen an der Harderschen Drüse des Kaninchens. *Histochemie* **7**:230–244.

Kühnel, W. 1971. Struktur und Cytochemie der Harderschen Drüse von Kaninchen. *Z. Zellforsch.* **119**:384–404.

Kühnel, W., and Wrobel, K. H. 1966a. Die Histotopik van Aldolase und Alkohol-Dehydrogenase in der Harderschen Drüse des Kaninchens. *Histochemie* **7**:245–250.

Kühnel, W., and Wrobel, K. H. 1966b. Über die histochemisch farbbare Aktivität der β-D-Glucuronidase und der β-D-Galaktosidase in der Harderian Drüse des Kaninchens. *Albrecht von Graefes Arch. Klin. Exp. Ophthal.* **171**:173–183.

Kühnel, W. and Wrobel, K. H. 1968. Histochemische Studien and der Harderschen Drüse des Kaninchens. *Gegenbaurs Morphol. Jahrb.* **111**:493–500.
Malins, D. C., Robisch, P. A., and Varanasi, U. 1972. Biosynthesis of triacylglycerols containing isovaleric acid. *Biochem. Biophys. Res. Commun.* **48**:314–319.
Margolis, F. L. 1971. Regulation of porphyrin biosynthesis in the harderian gland of inbred mouse strains. *Arch. Biochem. Biophys.* **145**:373–381.
Murawski, U., and Jost, U. 1974. Unsaturated wax esters in the harderian gland of the rat. *Chem. Phys. Lipids* **13**:155–158.
Paltauf, F. 1971. Metabolism of the enantiomeric 1-*O*-alkyl glycerol ethers in the rat intestinal mucosa *in vivo;* incorporation into 1-*O*-alkyl and 1-*O*-alk-l'-enyl glycerol lipids. *Biochim. Biophys. Acta* **239**:38–46.
Paule, W. J., Hayes, E. R., and Marks, B. H. 1955. The harderian gland of the Syrian hamster. *Anat. Rec.* **121**:349–350.
Pitas, R. E., Sampugna, J., and Jensen, R. G. 1967. Triglyceride structure of cow milk fat. I. Preliminary observations on the fatty acid composition of positions 1, 2 and 3. *J. Dairy Sci.* **50**:1332–1336.
Reiter, R. J., and Klein, D. C. 1971. Observations on the pineal gland, the harderian glands, the retina, and the reproductive organs of adult female rats exposed to continuous light. *J. Endocrinol.* **51**:117–125.
Rock, C. O., and Snyder, F. 1974. Biosynthesis of 1-alkyl-*sn*-glycero-3-phosphate via adenosine triphosphate:1-alkyl-*sn*-glycerol phosphotransferase. *J. Biol. Chem.* **249**:5382–5387.
Rock, C. O., and Snyder, F. 1975a. Metabolic interrelation of hydroxy-substituted ether-linked glycerolipids in the pink portion of the rabbit harderian gland. *Arch. Biochem. Biophys.* **171**:631–636.
Rock, C. O., and Snyder, F. 1975b. A short-chain acyl CoA:1-alkyl-2-acyl-*sn*-glycerol acyltransferase from a microsomal fraction of the rabbit harderian gland. *Biochim. Biophys. Acta* **388**:226–230.
Rock, C. O., Fitzgerald, V., Rainey, W. T., Jr., and Snyder, F. 1976. Mass spectral identification of 2-(*O*-acyl) hydroxy fatty acid esters in the white portion of the rabbit harderian gland. *Chem. Phys. Lipids* **17**:207–212.
Schneir, E. S., and Hayes, E. R. 1951. The histochemistry of the harderian gland of the rabbit. *J. Natl. Cancer Inst.* **12**:357–258.
Suzue, G., and Marcel, Y. L. 1972. Kinetic studies on the chain length specificity of long chain acyl coenzyme A synthetase from rat liver microsomes. *J. Biol. Chem.* **247**:6781–6783.
Tashiro, S., Smith, C. C., Badger, E., and Kezur, E. 1940. Chromodacryorrhea, a new criterion for biological assay of acetylcholine. *Proc. Soc. Exp. Biol. Med.* **44**:658–661.
Thompson, G. A., Jr. 1968. The biosynthesis of ether-containing phospholipids in the slug, *Arion aster.* III. Origin of the vinylic ether bond of plasmalogens. *Biochim. Biophys. Acta* **152**:409–411.
Tomio, J. M., and Grinstein, M. 1968. Porphyrin biosynthesis 5. Biosynthesis of protoporphyrin IX in harderian glands. *Eur. J. Biochem.* **6**:80–83.
Tomio, J. M., Tuzman, V., and Grinstein, M. 1968. δ-Aminolevulinate dehydratase from rat harderian gland. Purification and properties. *Eur. J. Biochem.* **6**:84–87.
Tsutsumi, A., Iwata, K., Igawa, K., and Matusura, K. 1966. Histochemical and electron microscopic observations on the harderian gland of the albino rat. *Arch. Histol. Jap.* **27**:553–567.
Varanasi, U., and Malins, D. C. 1970. Ester and ether-linked lipids in the mandibular canal of a propoise (*Phocoena phocoena*). Occurrence of isovaleric acid in glycerolipids. *Biochemistry* **9**:4576–4579.
Varanasi, U., and Malins, D. C. 1972. Triacylglycerols characteristic of propoise acoustic tissues: Molecular structures of diisovaleroylglycerides. *Science* **176**:926–928.

Varanasi, U., Everitt, M., and Malins, D. C. 1973. The isomeric composition of diisovaleroyl-glycerides: A specificity for the biosynthesis of the 1,3-diisovaleroyl structures. *Int. J. Biochem.* **4**:373–378.

Vlahakes, G. J., and Wurtman, R. J. 1972. A Mg^{2+} dependent hydroxyindole *O*-methyltransferase in rat harderian gland. *Biochim. Biophys. Acta* **261**:194–197.

Wetterberg, L. 1972. Increase of harderian gland prophyrin content in castrated male hamsters dependent on light and visual function. *Life Sci.* **11**:541-546.

Wetterberg, L., Geller, G., and Yuwiler, A. 1970a. Harderian gland: An extraretinal photoreceptor influencing the pineal gland in neonatal rats? *Science* **167**:884–885.

Wetterberg, L., Yuwiler, A., Geller, E., and Schapiro, S. 1970b. Harderian gland: Development and influence of early hormonal treatment on porphyrin content. *Science* **168**:996–998.

Wetterberg, L., Yuwiler, A., Ulrich, R., Geller, E., and Wallace, R. 1970c. Harderian gland: Influence on pineal hydroxyindole-*O*-methyltransferase activity in neonatal rats. *Science* **170**:194–196.

Woodhouse, M. A., and Rhodin, J. A. G. 1963. The ultrastructure of the harderian gland of the mouse with particular reference to the formation of its secretory product. *J. Ultrastruct. Res.* **9**:76–98.

Woolley, G. W., and Worley, J. 1954. Sexual dimorphism in the harderian gland of the hamster (*Cricetus auratus*). *Anat. Rec.* **118**:416–417.

Cultured Cells

JOHN M. BAILEY

I. *Introduction*

In this chapter, an exhaustive review of the lipid content and composition of cells in tissue culture will not be attempted. A number of articles are available, particularly those by Rothblat (1969) and Howard and Howard (1974), which cover this topic in an excellent and comprehensive manner. The present article will be concerned mainly with the use of tissue culture to study the mechanisms for regulating lipid biosynthesis by individual cells. Some of the recent developments in the analysis of genetic defects in the regulation of cholesterol metabolism will be described, and the current status of our knowledge concerning the control networks involved in lipid synthesis from glucose and acetate will be reviewed. Finally, the utility of the tissue culture technique for examining the essential fatty acid requirements of cells will be described and the use of metabolic mutants to delineate the pathways of essential fatty acid metabolism will be discussed briefly.

Much biochemical information has been obtained by the use of intact animals and organ or tissue slice techniques. The methods provide information at the level of integration of a number of different cell types which comprise tissues. They can be used to study regulation by control of enzyme synthesis in individual tissues such as the liver. The use of tissue homogenates, various cellular subfractions, and purified enzyme systems is best suited for detailed study of controls at the level of enzyme activity. In contrast, tissue cultures permit one to isolate and study the regulatory mechanisms of lipid metabolism at the cellular level separately from those of the entire organism.

JOHN M. BAILEY • Biochemistry Department, George Washington Medical School, Washington, D.C.

The use of isolated cell types growing under controlled nutritional and environmental conditions makes possible the study of the overall system of control points for the pathways of lipid metabolism. We shall see that cell culture is particularly well suited for measuring systemic type of controls involving interactions of individual cells with the external environment. It is becoming increasingly evident that these controls involve as yet little understood interaction between receptor proteins located in the cell membrane and the biosynthetic machinery for control of enzyme synthesis. It is at this level of interaction that the tissue culture procedure is most useful.

When we use cells in tissue culture to study the control mechanisms, there are several limitations that should be borne in mind. The first of these is that most body cells are in a state of homeostasis. Although considerable turnover of various cellular constituents may occur (with consequent incorporation of radioactive precursors), little net synthesis takes place under normal conditions. In cultured cells, however, there is usually a doubling of cell mass and cell lipids about every 24 hr. Consequently some of the details of the homeostasis mechanisms may be obscured in rapidly growing cells.

The second important point is that the composition of various extracellular fluids *in vivo* may differ considerably from the composition of the serum-supplemented media in which most tissue culture cells are grown. It transpires that such considerations are particularly important for certain constituents of plasma such as the lipoproteins, since some of the individual lipoproteins play a key role in the regulating process.

A number of cell types can be grown in completely chemically defined media, so that the growth requirement for the various lipids found in serum can be studied individually. These synthetic media are especially useful for studying various radioactive-labeled precursors for lipid synthesis such as glucose, acetate, and mevalonic acid. The contribution of these various compounds to the different classes of cell lipids can be measured, and, more importantly, the feedback controls for lipid biosynthesis from these precursors can be studied.

In its early days, there was much mystique surrounding the techniques for growing cells in tissue culture. In particular, it was found necessary to include in the growth media biological components of complex and undefined composition such as serum and embryo extract. Much of the groundwork for the present acceptance of tissue culture as a valid biochemical tool was laid by workers such as Earle, Evans, Fisher, Eagle, Waymouth, Morgan, and others during the 1950s in their painstaking and finally successful search for a complete and chemically defined medium in which cells would grow for indefinite periods in the absence of serum or other biological extracts of unknown composition (for review, see Eagle, 1961). Probably one of the best examples of the manner in which this was accomplished is the development of the National Collection of Type Cultures (NCTC) series of tissue culture media at the National Cancer Institute by Evans, Earle, and others. These efforts culminated with NCTC 109, the first completely successful synthetic medium (Evans *et al.*, 1956). This medium has some 69 different components, including salts, amino acids, vitamins, nucleosides, and coenzymes. The L-2071 strain of mouse fibroblasts has now been growing successfully in this medium for about 20 years, a life-span that is equivalent to over 2000 cell generations.

When cells were grown in serum medium, balance studies of the type indicated in Fig. 1 for mouse lymphoblasts and Table 1 for mouse fibroblasts showed that most of the cell lipids were derived from the serum. In particular, all of the cholesterol found in cells could be accounted for by

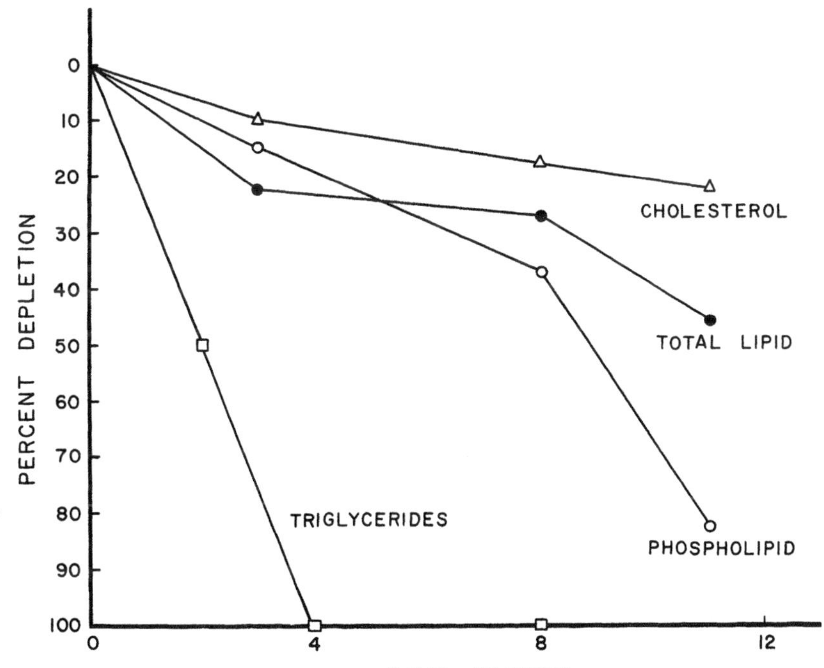

FIGURE 1. Deletion of serum lipids by MBIII cells. The progressive depletion of serum lipids was measured in roller tube cultures over a 10-day growth period in a medium consisting of human placental cord serum diluted with one part of balanced salt solution. In addition, in the 10-day cultures, the glucose concentration was doubled to 0.4%. Initial lipid levels in the cord serum used were as follows: Total lipids 222, cholesterol 86, phospholipid 54, and triglyceride 51 mg%.

TABLE 1
Balance Study on Removal of Lipids from Serum Medium by L-Strain Mouse Fibroblasts[a]

	μg in 10-ml culture medium or in harvested cells			
	Cholesterol	Phospholipid	Triglyceride	Total lipid
Time of growth				
0 days	70	72.9	487	630
2 days	57.3	61.2	265	384
5 days	32.4	21.9	216	270
Lipid utilized (μg)	37.6	51.0	271	360
Lipid found in cells (μg)	40	97	236	383
Serum lipid utilized as percent of cell lipid	94%	53%	115%	94%
Percent contribution to cell dry wt	4.6	11.2	27	44.8

[a] L-strain mouse fibroblasts were grown in NCTC 109 and 10% horse serum. Portions of medium were analyzed at zero time and after 2 and 5 day's growth. Lipid composition of cells harvested at 5 days and lipid depletion in growth medium was measured. Triglycerides were calculated by difference of total lipids and phospholipid-cholesterol values. Lines 1, 2, 3 of table show progressive depletion of each lipid type with days in culture. Line 4 shows total lipids of each type removed from medium over 5-day period. Line 5 shows total lipids of each type recovered from harvested cells. Line 6 expresses depletion in serum lipids as percent of that found in cells. Last line is percent contribution of each lipid type to total dry weight of cell; 94% of cell lipid is potentially derived from serum lipid.

that lost from the medium (Bailey *et al.*, 1959; Bailey, 1966). Experiments with mouse fibroblasts grown in medium NCTC 109, which was free of preformed lipids, showed that the lipid components of the cell including cholesterol, triacylglycerols, and the phospholipids were now derived completely by *de novo* biosynthesis from simple precursors, principally glucose and acetate (Fig. 2) (Bailey, 1966). These two observations indicated that in the presence of serum the *de novo* biosynthetic mechanisms must be inoperative and implied that some type of feedback control existed for regulating lipid biosynthesis by the isolated mammalian cells in response to the lipid composition of the external environment.

The experimental testing of this conclusion has firmly established the tissue culture system as a primary tool in the study of feedback regulation of lipid biosynthesis. The recent extension of these techniques to human diploid skin fibroblast cultures by Goldstein and Brown (1973) has led to important advances in the study of genetic abnormalities in control systems for lipid metabolism in humans. It is this latter aspect—the use of tissue culture cells to study the feedback control mechanisms that regulate the lipid composition of individual cells and thus ultimately affect the lipid composition of both tissue and body fluids—that will be the main topic of this chapter.

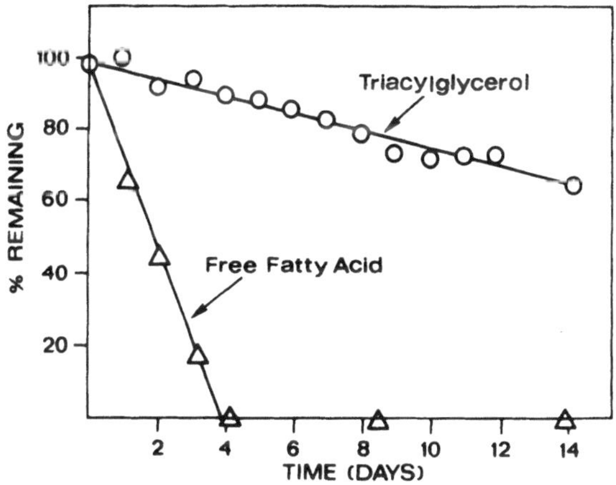

FIGURE 2. Depletion of [^{14}C]triglyceride and [^{14}C]fatty acid by dense cultures of L cells. Replicate culture of L-strain mouse fibroblasts, subcultivated at 1:10 ratios, were grown in medium supplemented with 5% fetal calf serum labeled with either sodium [$^{14}C_1$]palmitate or [carboxy-^{14}C]tripalmitin. Aliquots of medium were removed daily and radioactivity remaining in triglycerides and fatty acids was measured. Total lipid content of the medium initially averaged 0.065 mg per ml, of which about 0.04 mg remained at the end of the experiment.

II. Cellular Lipid Nutrition

The progress of depletion of serum lipids by the MB III strain of mouse lymphoblasts grown in a medium containing human placental cord serum as the principal nutrient source is shown in Fig. 1. Triacylglycerols were most rapidly taken up and were followed in rate by phospholipids, and then cholesterol. This preferential and rapid uptake of triacylglycerols has also been observed for all of the other cell strains we have tested. In later studies it was found that free fatty acids were removed even more rapidly than triacylglycerols (Fig. 2). Experiments in which the medium was supplemented with radioactive fatty acids and triacylglycerols showed that L strain mouse fibroblasts incorporated serum triacylglycerols from the growth medium at rates up to tenfold less than free fatty acids or monoacylglycerols under the same culture conditions. When cells were grown under conditions of limited supply of serum lipids, however, only about 7% of the cell lipid came from serum free fatty acids and up to 28% was provided by the serum triacylglycerols. The major portion of the cell lipid under these conditions was derived by *de novo* synthesis from glucose.

Hydrolysis of exogenous triacylglycerol by lipases in the serum-supplemented tissue culture medium is inadequate to account for the rates of triacylglycerol utilization observed. In addition, heat treatment of serum to inactivate lipases does not significantly affect triacylglycerol uptake. No excretion of lipase activity by cells has been detected.

Incorporation of triacylglycerol labeled in the glycerol portion with ^{3}H and the fatty acid portion with ^{14}C has been measured to determine if the triglyceride was hydrolyzed before uptake. The ^{3}H:^{14}C ratio in the isolated cellular lipids averaged over 70% of that in the original triacylglycerol under all conditions of culture. Furthermore, incorporation of the [^{3}H]glycerol portion of the molecule was not diluted out by addition to the growth medium of a large excess of nonradioactive glycerol or a nonhydrolyzable monoacylglycerol analog. In short-term uptake experiments, most of the radioactivity in cell lipids is found in the triacylglycerol fraction, with only minor amounts in the free fatty acids. In long-term growth experiments the incorporated triacylglycerol is converted extensively (70–80%) to cellular phospholipids with maintenance of the ^{3}H:^{14}C ratio. These results indicate that cells in tissue culture utilize serum triacylglycerols by a mechanism of uptake involving predominantly the intact molecule and requiring no hydrolysis.

By analysis of the harvested cells, it can be shown that most of the decrease in the lipid of serum medium can be accounted for by the lipid found in the cells. Thus over a 4-day period, growth of MB III strain cells in cultures of fairly high (6,000,000 cells/ml) population density resulted in the uptake of all the triacylglycerol (510 μg), 80 μg (15%) of the phospholipid and 70 μg (9%) of the cholesterol. The harvested cells contained 650 μg of total lipid, of which 70 μg (11%) was cholesterol and 360 μg (55%) was phospholipid. Only small amounts of triacylglycerols were found in cells, which suggests conversion of triacylglycerols to phospholipids during or following uptake. The results of complete balance studies such as that given in Table 1 on lipids taken up from the medium and those found in the cultured cells confirm that under the normal conditions of growth in serum medium the bulk of the cell lipids are derived from the serum lipoproteins and that there is very little *de novo* synthesis under these conditions.

III. Factors Influencing Cell Cholesterol Uptake

Cholesterol and its esters are the principal lipids found in advanced atherosclerotic lesions in human arteries and it is perhaps not surprising that much of the effort involving the regulation of lipid metabolism in cultured cells has focused on these compounds. Many factors have been found to influence the cholesterol content of cells in tissue culture. The influence of the type of serum in the medium is of major importance. Cells were grown in different samples of serum having a wide range of total

TABLE 2
Influence of Type of Serum on Cell Cholesterol Content[a]

Type of serum	Number of samples	Mean serum cholesterol (mg%)	Mean cell cholesterol (% of dry wt)
Normal human	20	199	3.60 ± 0.57
Atherosclerotic human	29	229	4.09 ± 0.86
Human placental cord	8	97	2.25 ± 0.23
Normal rabbit	6	86	7.8 ± 2.2
Atherosclerotic rabbit	6	2148	22.2 ± 2.9

[a]Cells (MBIII strain) were grown for 4 days on the different serum samples diluted with one part of balanced saline solution. Cell lipids were extracted and the cholesterol content expressed on the basis of the dry weight of the lipid-extracted cell residue.

cholesterol content (Table 2). These included human sera from normal and atherosclerotic subjects and hyperlipemic serum from rabbits maintained on a diet supplemented with 1 g of cholesterol daily for 3 months. Cells grown on human placental cord serum had the lowest cholesterol content, averaging only 2.25 mg%, and the mean serum level was 97 mg%. This was in marked contrast to cells grown on normal rabbit serum, which has the lowest cholesterol content (86 mg%). Cells grown on sera from both atherosclerotic humans and rabbits contained significantly more cholesterol than cells grown on corresponding normal sera. These early results thus showed that the cholesterol content of the medium *per se* is not the prime factor in determining the cell cholesterol content. We now know that the nature and proportions of the different lipoprotein classes are probably the single most important factor.

The cholesterol content of many cells in culture also appears to be a characteristic of the cell type (Bailey, 1961). In a series of cell lines grown in human placental cord serum, the cholesterol content of the cells varied from 2.8% of the dry weight for the FL strain of human amnion cells to only 0.6% for a strain of rat adrenal cells (Table 3). Since the cell cholesterol is derived exclusively from the serum in these conditions, this implies that the net rate of uptake must be different for each cell type. The possibility of differential breakdown of cholesterol has been ruled out by independent experiments.

A second important control on cholesterol accumulation in cells is a continuous process of cholesterol excretion into the medium that is catalyzed by serum. Cells labeled with radioactive cholesterol either by growth on [^{14}C]cholesterol medium or by being grown on [^{14}C]mevalonic acid as a precursor, continuously excrete labeled cholesterol upon transfer to fresh

TABLE 3
Cholesterol Content of Various Mammalian Cell Strains Grown on the Same Serum Medium[a]

Strain	Cell type	Cholesterol (%)
FL	Human amnion	2.8
D-1 Re	Human chondromyxosarcoma	2.6
ARF	Articular ridges of femur (rat)	2.1
MBIII	Neoplastic mouse lymphoblast	2.0
319	Walker rat sarcoma	1.9
Maben	Human lung carcinoma (Frisch)	1.7
256	Walker rat carcinoma	1.6
72-624	Tumor from subcutaneous areolar tissue (rat)	1.6
HeLa	Epidermoid carcinoma of cervix (human)	1.3
3G29	Rat adrenal (zona fasciculata)	0.6

[a]Cells were harvested from the growth medium after 4 days. Medium in most cases was human placental cord serum diluted with one part of balanced saline. Cholesterol values are expressed as percent of dry weight after extraction of lipid and air drying at 120°C.

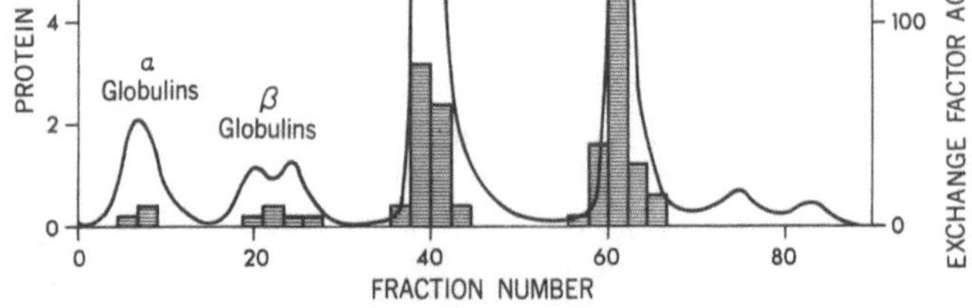

FIGURE 3. Relative ability of serum subfractions to catalyze excretion of [^{14}C]cholesterol from cultured cells. Calf serum was applied to a DEAE cellulose column and eluted with five successive changes in buffer pH and ionic strength. The principal protein components in each peak were identified by paper electrophoresis. The ability of each fraction to catalyze cholesterol excretion from MBIII cells prelabeled with [^{14}C]cholesterol was measured in a standard assay system (Bailey, 1964). The exchange factor activity was associated with the serum α-globulins comprised principally of the high-density lipoprotein class.

nonradioactive medium containing serum. When serum is removed, the excretion process stops and can be reinstated by adding back various serum subfractions, the most effective being the high density lipoproteins or α-globulin fraction of serum (Fig. 3).

Rothblat *et al.* (1967) found that phospholipids also play an important role in cholesterol transport in addition to the serum proteins. The uptake of labeled cholesterol was enhanced by adding serum or delipidized serum proteins and added phospholipids reduced this uptake. Phospholipids also enhanced the excretion of cholesterol from prelabeled cells in the presence of delipidized serum proteins. These effects are reminiscent of the role of phospholipids in the overall structural integrity of the cholesterol–lipoprotein complex.

The correspondence between the ability of the α-globulin fraction to catalyze excretion of cholesterol by cells and the normal function of this serum fraction in binding cholesterol is significant. The fact that intracellularly synthesized cholesterol is exchanged by this same mechanism as readily as is exogenously supplied cholesterol demonstrates the intrinsic metabolic significance of the process in cholesterostasis.

IV. Feedback Regulation of Cholesterol Biosynthesis

In the absence of serum or added lipids the cell cholesterol is derived by *de novo* synthesis from simple precursors, principally from glucose and acetate (Bailey, 1966). When serum is added to the medium, these *de novo* biosynthetic pathways for lipids are almost completely depressed (Fig. 4). Cholesterol biosynthesis is also inhibited by adding cholesterol to the culture medium.

In Fig. 5 the pathways of cholesterol biosynthesis in L strain cells are outlined, together with the location of some of the metabolic blocks involved in the regulation of synthesis. About 20% of the cholesterol in L strain cells grown in chemically defined medium is derived from acetate. It seems that one feedback block when [^{14}C]acetate is used as a precursor is at the 2-carbon level of metabolism, since incorporation of labeled mevalonic acid is not reduced. Also, addition of cholesterol reduces the synthesis of fatty acids from acetate (Bailey, 1966). We have found that this effect is probably mediated via the acetate activating enzyme system, since the levels of this enzyme are reduced 4- to 5-fold when serum lipids are added to the growth medium.

The principal precursor of cholesterol in cells grown in lipid-free medium is glucose, which supplies about 75% of the lipid requirement.

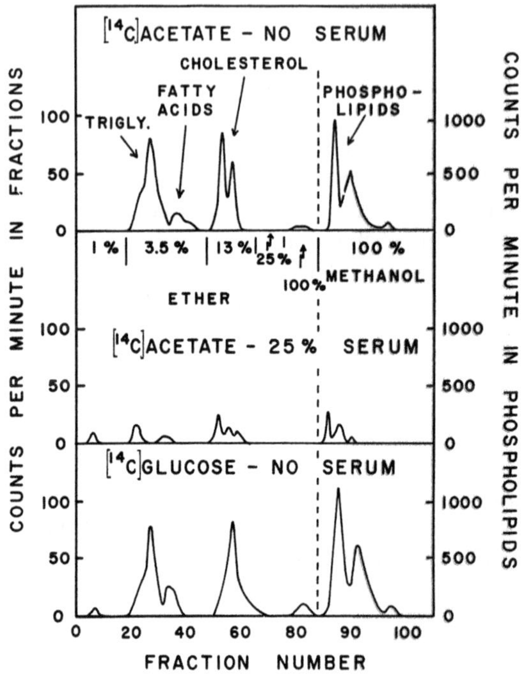

FIGURE 4. Feedback inhibition of [^{14}C]glucose and acetate incorporation into cell lipids. L-Strain mouse fibroblasts were grown for 4 days in chemically defined medium NCTC 109 (serum free) or in the same medium supplemented with 25% human serum. Each culture was supplemented with sodium [1-^{14}C] acetate (0.1 μCi). Cell lipids were extracted and separated into the major fractions shown by chromatography on silicic acid columns. [^{14}C]Acetate incorporation into cell lipids was depressed up to 30-fold by addition of 25% serum. Each lipid class was affected to approximately the same extent. When cells were grown on [^{14}C]glucose-supplemented medium the pattern of incorporation into cell lipids (bottom curve) was similar to that observed with acetate, but the inhibition produced by added serum was not as great.

Incorporation of glucose into total cell lipids is inhibited to a greater extent by serum than by cholesterol. This indicates that the metabolic block in the glucose pathway is probably beyond the point where fatty acid and sterol synthetic systems branch off and suggests that the control point is at the level of the enzyme hydroxy-methyl-glutaryl-CoA (HMG-CoA) reductase in the step immediately preceding the formation of mevalonic acid. This enzyme has been shown to be the major regulatory point for cholesterol synthesis in liver. That additional regulatory controls are also operating at stages beyond mevalonic acid is indicated by the fact that incorporation of labeled mevalonic acid into lipids of L cells is not inhibited by added serum, whereas the incorporation and distribution of radioactivity between cholesterol and other products is changed considerably. These metabolic blocks are of some considerable theoretical interest, since we have found that they are often modified or released during the transformation of normal cells into tumor cells. This is illustrated in Fig. 6 for cholesterol biosynthesis from [^{14}C]mevalonic acid in the normal human diploid fibroblast WI-38 and the tumor cell line WI-38VA-13A derived from it by transformation with oncogenic SV40 virus. Similar release of metabolic blocks has been noted in a number of liver tumors growing *in vivo* (Siperstein *et al.*, 1971).

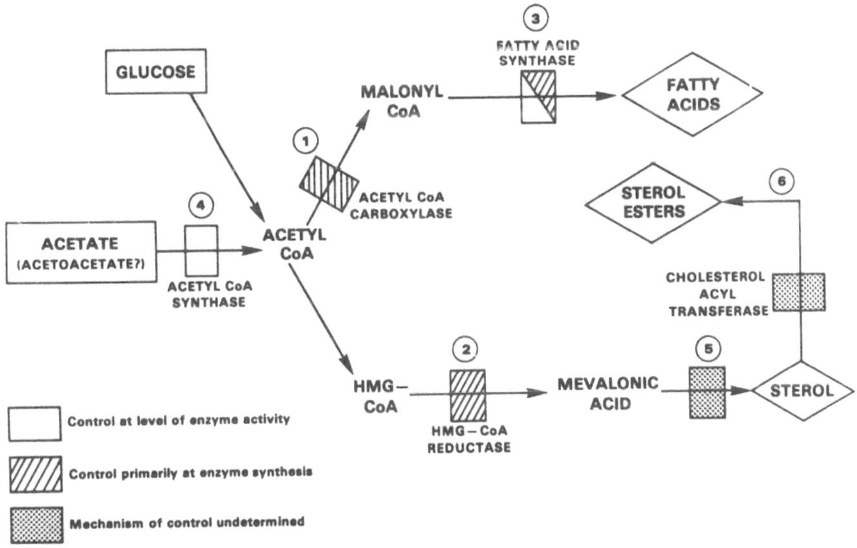

FIGURE 5. Control of lipid biosynthesis in cultured cells.

When cholesterol is added to cells growing in synthetic medium, there is a lag period of 4–8 hr before cholesterol biosynthesis shuts down. Similarly, when cells are transferred from serum into synthetic medium, there is a lag period before synthesis is initiated. Sokoloff and Rothblat (1972) found that the lag periods in the initiation or reduction of synthesis of cholesterol can be attributed in part to delays in the flux of sterol

FIGURE 6. Cholesterol synthesis from [^{14}C]mevalonic acid by normal (WI-38) and tumor-virus-transformed (WI-38VA-13A) human fibroblasts. Cells were grown for 4 days in chemically defined growth medium supplemented with 5% fetal calf serum and 0.1 μCi/ml of [^{14}C]mevalonic acid. Cell lipids were harvested and [^{14}C]cholesterol was isolated by digitonin precipitation and thin-layer chromatography on silicic acid coated plates.

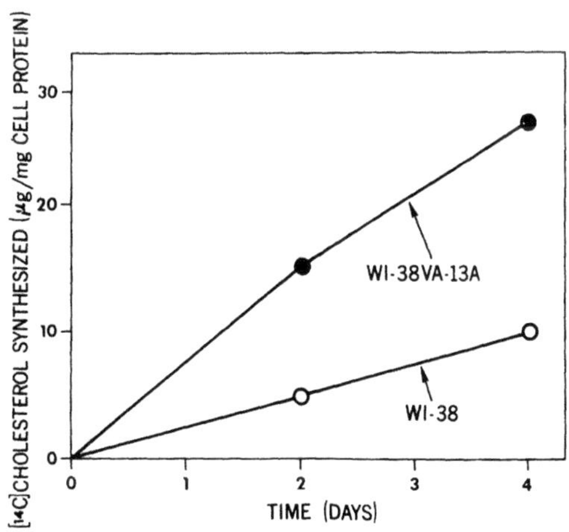

between cells and medium. By adding delipidized serum proteins to the medium they showed that not only is the excretion of cholesterol increased, but that because of the enhanced removal of intracellular cholesterol, the feedback inhibition of synthesis of cholesterol is also released. This stimulation of cholesterol biosynthesis by delipidized serum proteins has also been observed for human skin fibroblasts (Williams and Avigan, 1972) and clearly demonstrates the interaction of transport processes and biosynthetic reactions for sterol synthesis in cultured mammalian cells.

V. Genetic Abnormalities in Regulation of Cholesterol Synthesis

This work has been greatly advanced by the elegant studies of Goldstein and Brown (1973) concerning the metabolic defect known as familial hypercholesterolemia. Skin fibroblasts from individuals homozygous for this disorder have lost feedback control of the enzyme HMG-CoA reductase. Levels of this enzyme in defective fibroblasts grown in serum are some 60-fold higher than those in normal cells. In lipoprotein-free medium there is a rapid induction of the enzyme beginning in a few hours, and the addition of low-density lipoproteins specifically reverses this effect. Abnormal cells show little if any response. However, when free cholesterol is added to the defective cells, an essentially normal reduction in HMG-CoA reductase levels is observed.

By labeling the low-density lipoproteins (LDL) with ^{125}I, Brown and Goldstein (1974) showed that homozygous cells lacked a receptor for the LDL at the cell surface, and, in contrast to normal cells, displayed essentially no specific binding of LDL. Attachment of LDL to normal cells is accompanied by uptake of the cholesterol portion and release of the degraded aproprotein into the medium.

Following addition of LDL to normal cells, there is also a progressive increase in the content of cholesterol acyl-transferase in the cells that accompanies, but does not precisely parallel, the decrease in HMG-CoA reductase activity. There is thus an increase in cholesterol ester synthesis and consequently a large increase in the ratio of ester to free cholesterol in normal cells following addition of LDL.

Many of the effects ascribed to free cholesterol can be induced by much lower concentrations of simple oxygenated derivatives of cholesterol such as 7-keto- and 25-hydroxy-cholesterol (Kandutsch and Chen, 1973; Brown and Goldstein, 1974). There is considerable discussion at present concerning the physiological role of these derivatives, their possible formation by enzymatic conversion of cholesterol inside the cell, and the mechanisms whereby they induce the subsequent feedback inhibition of choles-

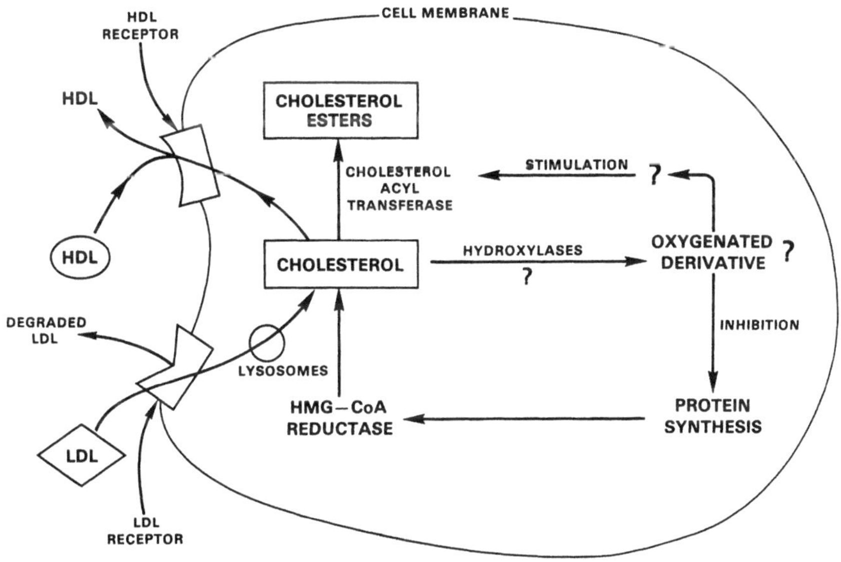

FIGURE 7. Lipoprotein-mediated regulation of cholesterol and cholesterol ester metabolism.

terol biosynthesis. Some of these speculative schemes are incorporated in Fig. 7.

The primary defect in familial hypercholesterolemia from the evidence detailed above thus appears to be the absence of the LDL receptor on the cell membrane. In another genetic variant a defective receptor of much lower affinity than normal is present. These receptor defects result in a relative inability to take up LDL-bound cholesterol. This, in some way not yet completely understood, then leads to failure of the normal feedback control for synthesis of the enzyme HMG-CoA reductase (Fig. 7). There is concomitantly a decrease in the ability to clear low-density lipoproteins from the plasma and the establishment of a new steady state level of LDL in the plasma that is several times the normal level. The defect in a single receptor protein at the cell surface thus becomes manifested as a systemic accumulation of cholesterol in tissues. This leads to the development of generalized atherosclerosis and early death.

VI. Feedback Control of Acetate Incorporation into Cell Lipids

Because of its central role in both lipid and carbohydrate metabolism, acetate has been the precursor of choice for many studies on lipid biosynthesis in tissue culture. These studies including our own have tended to overlook or ignore the fact that free acetate is not a physiologically important component of body fluids. Acetate must first be activated by acetyl-CoA synthase before it can be further utilized (Fig. 5). It seems, however, that the physiological validity of the studies using acetate as a precursor may be somewhat strengthened by the observation (Aas and Bremer, 1968) that acetate is also a good substrate for the aceto-acetyl-CoA synthetase. Acetoacetate, being one of the major ketone bodies, is a common component of serum and other body fluids, particularly during periods when carbohydrate intake is below normal or when body lipid depots are being utilized during periods of fasting. Although the studies to be described here concern mainly the regulation of acetyl-CoA synthetase, it is anticipated that further work will confirm that their physiological significance relates to control of acetoacetyl-CoA formation and the utilization of ketone bodies for lipid synthesis.

When serum lipid is removed from monolayers of mouse fibroblasts, a relatively rapid stimulation of [^{14}C]acetate incorporation into lipid is observed within 2 hr. This stimulation of incorporation is observed in both sterols and fatty acids (Fig. 8). Conversely, an inhibition of [^{14}C]acetate incorporation is observed within 1 to 2 hr when cells cultured in lipid-free medium are transferred to serum-supplemented medium. As with the stimulation, this inhibition is observed both in sterol and fatty acid fractions of the cell lipids (Fig. 9). There is thus a coordination of fatty acid and cholesterol metabolism in the cells. Inhibition of [^{14}C]acetate incorporation

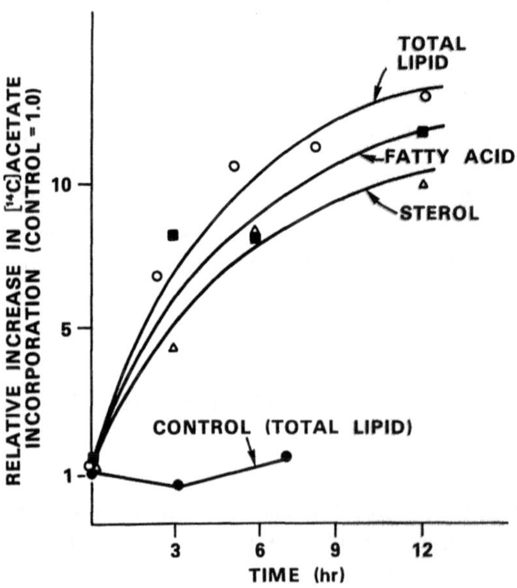

FIGURE 8. Time-course of stimulation of [^{14}C]acetate incorporation into lipids of L cells after removal of exogenous lipid. Replicate cultures of L-929 cells were grown to confluency in medium supplemented with 20% fetal bovine serum. At zero time monolayers were washed three times with balanced salts solution and transferred to medium supplemented with delipidized fetal calf serum. At the indicated time [^{14}C]acetate was added to the medium and the cells pulsed for 1 hr. Monolayers then were rinsed three times with CBSS, the cells were harvested by scraping, cell lipid was extracted and saponified, and radioactivity was determined. Control represents cultures where medium was unchanged at zero time (Howard and Bailey, 1974).

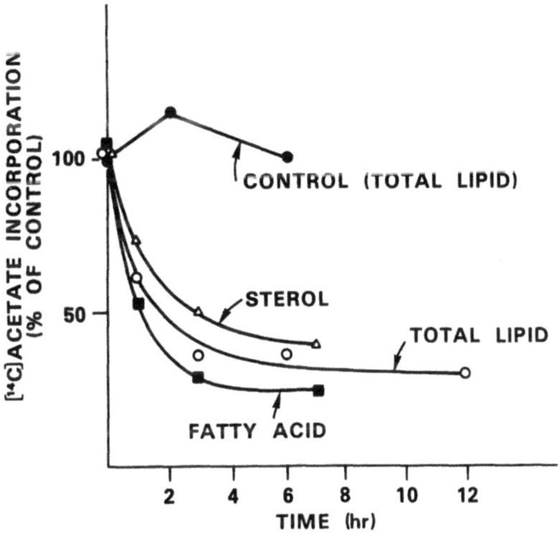

FIGURE 9. Time-course of inhibition of [^{14}C]acetate incorporation into lipids of L cells after addition of serum to the medium. Replicate cultures of L-929 cells were grown to confluency in medium supplemented with delipidized serum protein. At zero time monolayers were washed three times with balanced salts solution and transferred to medium supplemented with 20% fetal bovine serum. At the indicated time [^{14}C]acetate was added to the medium and the cells were pulsed for 1 hr. Cell lipid was extracted and saponified and radioactivity was determined. Values for experimental cultures are expressed as percent of control, which represents 1-hr pulses in cultures where the medium was unchanged at zero time (Howard and Bailey, 1974).

is observed in sterol as well as free fatty acid and glycerolipid fractions in cultures grown in lipid-free medium and then transferred to fatty acid-supplemented medium. Similarly, cultures transferred to media supplemented with cholesterol show inhibition of acetate incorporation not only into sterols, but also into the fatty acid and glycerolipid fractions. Cycloheximide, actinomycin D, and mitomycin C do not block the early stimulation or inhibition of [^{14}C]acetate incorporation into total lipid. Consequently, these effects are not dependent on nucleic acid or protein synthesis.

A five-fold decrease in acetyl-CoA synthetase activity was observed in homogenates of cultures grown in the presence of exogenous lipid. When cells grown in serum-supplemented medium are transferred to serum-free medium, a stimulation of enzyme activity occurs within 2–3 hr, which reaches a maximum by 6 hr. Conversely, when cells are cultured in lipid-free medium and transferred to serum-supplemented medium inhibition of enzyme activity occurs within a similar time frame.

Cycloheximide, actinomycin D, and mitomycin C have no influence on the stimulation or inhibition of enzyme activity observed in response to changes in exogenous lipids (Fig. 10). Inhibition of enzyme activity also is observed in cells cultured in lipid-free medium that were transferred to medium containing either fatty acid or cholesterol, indicating that fatty acid and cholesterol might be part of the components of serum lipids that can influence the activity of acetyl-CoA synthetase. However, these com-

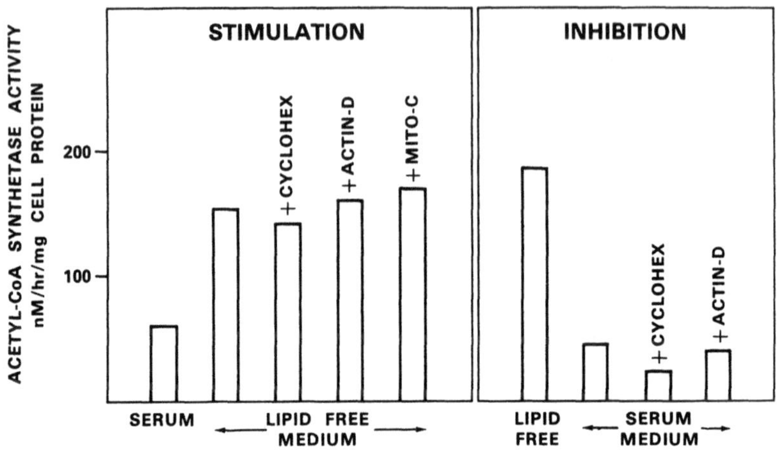

FIGURE 10. Failure of inhibitors of protein and nucleic acid synthesis to inhibit induction and repression of acetyl-CoA synthetase activity in cell cultures. To determine the effect on induction of enzyme activity, replicate cultures of strain L929 mouse fibroblasts were grown in medium supplemented with serum lipids (20% fetal bovine serum) and transferred to lipid-free medium at zero time. To study their effects on repression of enzyme activity the cells were cultivated in medium supplemented with delipidized serum protein and transferred to serum-supplemented medium at zero time. In each case, acetyl-CoA synthetase was assayed 6 hr after the change in medium was made (Howard and Bailey, 1974).

pounds have no direct effect on the enzyme when added to the reaction mixture *in vitro*.

The control points for lipid synthesis identified and partially characterized in cultured cells are illustrated schematically in Fig. 5. From the available evidence it seems that two of the major control points are at the first committed steps in the biosynthesis of fatty acids and sterols (Fig. 5, steps 1 and 2). It appears that these steps are regulated via the induction and repression of the appropriate enzymes acetyl-CoA carboxylase (Jacobs *et al.,* 1973) and HMG-CoA reductase (Brown *et al.,* 1974). Another control point in fatty acid synthesis via enzyme induction and repression occurs at the level of fatty acid synthase (Fig. 5, step 3) (Alberts *et al.,* 1975). A fourth point of control that would influence both pathways is via regulation of the acetate or acetoacetate activating enzyme (Fig. 5, step 4); the mechanism, which is so far unidentified, appears to be at the level of enzyme activity rather than induction and repression of synthesis. Figure 5 also indicates that other control points may exist between mevalonic acid and cholesterol (Fig. 5, step 5) and between cholesterol and its esters (Fig. 5, step 6).

It is difficult, however, to determine which control point is rate-limiting and most important physiologically. It is possible that, although a coordinate effect between fatty acid and sterol synthesis is observed, separate points of regulation in fatty acid and sterol pathways might be the rate-limiting ones. It is clear in fatty acid synthesis that the time-course of regulation of lipid synthesis from acetate cannot be accounted for by the kinetics of regulation of acetyl-CoA carboxylase or fatty acid synthase alone. The definition of the relative role of various control points in sterol synthesis is more difficult since the induction and repression of HMG-CoA reductase occurs within 4–6 hr of a change in the exogenous lipid. Therefore, a stimulation of [^{14}C]acetate incorporation into sterol would reflect either changes in HMG-CoA reductase or regulation at the level of acetate activation or both. Similarly, inhibition of [^{14}C]acetate incorporation into sterol could reflect changes in either HMG-CoA reductase or acetate activation and in addition be subjected to dilution effects in the oxidation of fatty acids to acetyl-CoA. Some indications of the relative importance of acetyl-CoA synthetase and HMG-CoA reductase in regulation of sterol synthesis comes from data of Bates and Rothblat (1974). In comparing the inhibition of sterol synthesis from [^{14}C]acetate and [^{3}H]H_2O, they observed a greater inhibition of synthesis in L cells using acetate as a precursor, the difference presumably reflecting controls at the level of 2-carbon metabolism.

From experiments of this type an overall picture of the control points and their relative importance to the regulation of lipid synthesis from these various endogenous precursors is beginning to emerge. In contrast, however, to the considerable amount of information accumulated on the control of acetate (or acetoacetate) incorporation, there is much less known about the controls on the utilization of the physiologically and quantitatively much more important compound—glucose. As indicated in Fig. 5, there is no evidence yet for control points between glucose and acetyl-CoA; the implication is that the entire regulation of entry of glucose into pathways for lipid biosynthesis is at the level of the acetyl-CoA carboxylase for fatty acids and HMG-CoA reductase for sterol synthesis. Since such a loose control beyond the first committed step is unusual, it seems that further developments in this area are possible.

VII. *Defective Control of Lipid Metabolism in Tumors and Transformed Cells*

A number of abnormalities in lipid metabolism have been observed in both tumor cells and cultured cells following transformation by oncogenic viruses. The relation of the changes observed to the oncogenic process itself is not clear at present, but a direct relation (i.e., at the cause and effect level) seems unlikely for a number of reasons. Perhaps the most significant is that the transformed state in tissue culture is usually associated with major changes in the number, appearance, and qualitative pattern of the chromosomes from the diploid to the heteroploid state. It seems reasonable to assume that alterations of this magnitude in the chromosomal complement could lead to significant changes in enzyme patterns and control networks purely as a coincidental accompaniment of transformation. Despite the possibility that these alterations in lipid metabolism may have very little to do with the oncogenic state per se, the enzyme deletions and defective regulatory systems that are found make these transformed cells useful tools for the study of the regulatory mechanisms themselves.

One of the most frequent occurrences is a partial loss of control of feedback inhibition for HMG-CoA reductase (Fig. 5, step 2). This relates to the original observations by Siperstein *et al.* (1971) on a similar defect in hepatomas *in vivo*. Consequently, lipid synthesis from acetate proceeds at a higher level in the W1-38Va13A cell line that has been transformed by the oncogenic SV40 virus than in the normal diploid W138 human fibroblasts from which it was derived (Fig. 11).

A second loss of control in sterol biosynthesis is observed in the transformation of mevalonic acid to cholesterol (Fig. 5, step 5). Therefore, in a large number of tumor and transformed cell lines, cholesterol synthesis proceeds at a much more rapid rate than in the normal diploid cells W138 and W126 (Fig. 12) (Bailey *et al.,* 1972).

The lack of control of the enzyme HMG-CoA reductase is not related to loss of the surface receptor for LDL since the transformed cells bind

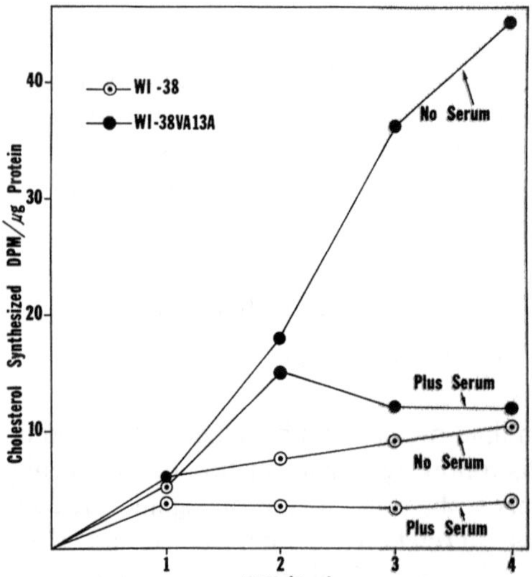

FIGURE 11. Cholesterol synthesis from [^{14}C]acetate by normal diploid WI-38 and virus transformed WI-38VA13A cells. Approximately 10^7 cells/ml were suspended in medium containing sodium [1-^{14}C]acetate of known specific activity and incubated at 37°C. At intervals samples were removed and washed. Lipids were extracted and the cholesterol was isolated by thin-layer chromatography.

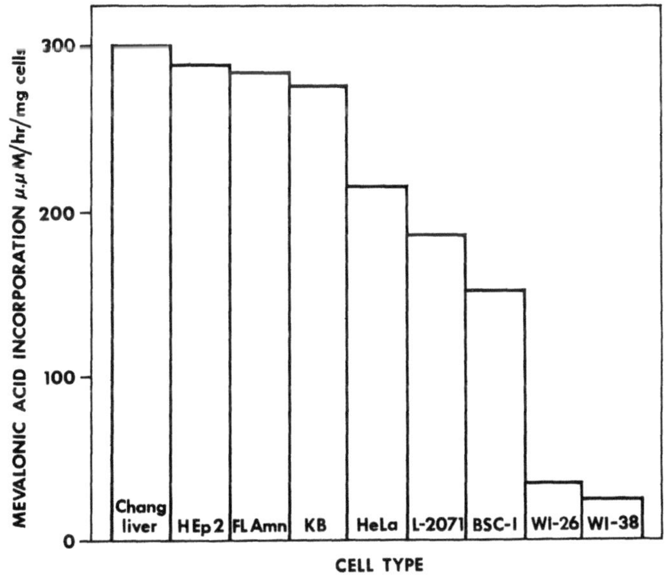

FIGURE 12. Sterol biosynthesis from [2-^{14}C]mevalonic acid by diploid and heteroploid cell lines. H Ep 2, Hela, and KB lines were derived from tumors; FL-Amnion and Chang Liver are lines derived from human tissues; L-2071 and BSC-1 arose similarly from mouse and monkey, respectively. WI-26 and WI-38 are human diploid cell lines. Replicate cultures of each cell type were grown to confluency in medium supplemented with 10% serum. The monolayers of cells were then washed and exposed to serum-free medium containing [^{14}C]-mevalonic acid for 96 hr. Cells were washed and harvested by scraping and the lipids extracted. Lipid classes were separated by silicic acid column chromatography. Cholesterol was eluted with 17.5% diethyl ether in hexane.

LDL to about the same degree as normal cells (Bailey *et al.,* 1976). Consequently, the defect must lie in the system subsequent to the entry of cholesterol into the cell whereby the internal cholesterol pool is coupled in some way to the control of synthesis of HMG-CoA reductase.

Another example of an enzyme defect, which is frequently found in transformed and tumor cells, lies in the pathways for metabolic interconversion of the essential fatty acids. Conversion of linoleic acid (18:2,*n*-6) to arachidonic acid (20:4, *n*-6), which is the immediate precursor of the PG2 series of prostaglandins, requires the sequential action of three enzymes (Fig. 13). The first of these, $\Delta 6$ reductase, introduces the third double bond into the molecule. This enzyme is lost in many tumor cells. Consequently, a cell such as the Ehrlich ascites tumor, when grown in tissue culture in the presence of ^{14}C-labeled linoleic acid, is unable to synthesize arachidonic acid (Fig. 14).

A third defect, which is frequently observed in lipid metabolism in

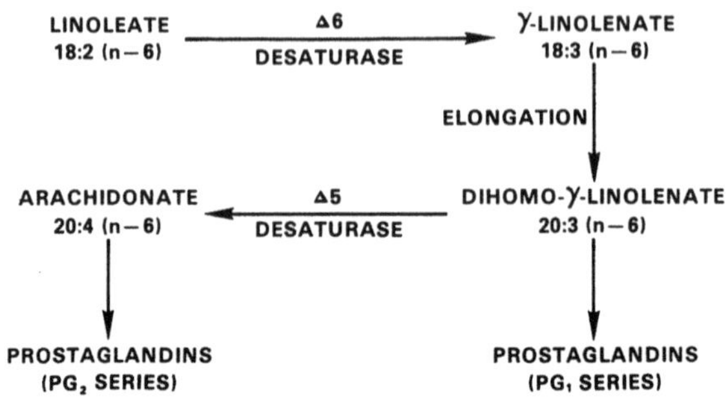

FIGURE 13. Pathways of essential fatty acid interconversion in mammalian cells. The polyunsaturated fatty acid linoleic acid (18:2,*n*-6) can give rise to prostaglandins of both the PG1 and the PG2 series in cultured cells by the sequence of reactions shown (Dunbar and Bailey, 1975). Three successive reactions a Δ6 desaturation followed by an elongation and a Δ5 desaturation are required for the complete conversion of linoleate to arachidonate. Many heteroploid or transformed cells cannot desaturate linoleic acid but readily convert dihomo γ linoleate to arachidonate (Tables 5 and 6). These observations demonstrate that separate Δ5 and Δ6 desaturases are involved and that loss of the latter enzyme is a frequent accompaniment of cell transformation.

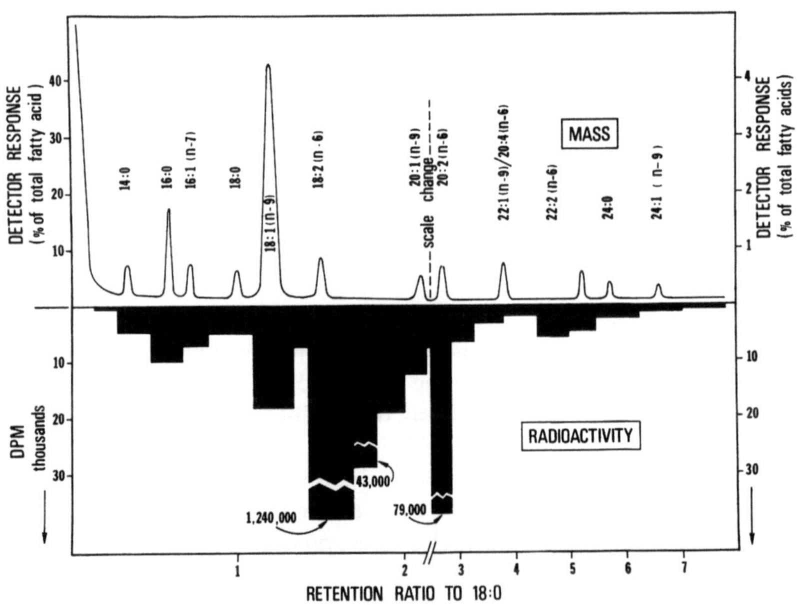

FIGURE 14. Enzymatic conversions of [1-^{14}C]linoleic acid by Ehrlich ascites cells growing in lipid-free medium. Lipid extracts of cells were converted to the fatty acid methyl ester and chromatographed using a stream splitter to recover a portion for analysis of radioactivity. The elution patterns of mass and radioactivity are plotted on the same horizontal axis to show correspondence of peaks. There was no evidence of desaturation of the carbon chain, indicating that the Δ6 desaturase is missing.

tumors, is the accumulation of a class of compounds known as the ether lipids. Their accumulation could be related to the loss of *sn*-glycerol-3-P dehydrogenase. The production of *sn*-glycerol-3-P from glucose, which is used for the synthesis of lipid esters, is decreased and the resulting increase in availability of dihydroxyacetone-P (the glycerol source of ether lipids) is consistent with the elevated alkyl and alk-1-enyl lipid ethers. It is of particular interest that this phenomenon is also related to the abnormally high aerobic glycolytic capacities of most tumor cells. The loss of ability to transfer reducing equivalents in the form of NADH from the cytoplasm to the mitochondria via the *sn*-glycerol-3-P-dehydrogenase "shuttle" results in overproduction of lactic acid even in the presence of adequate supplies of oxygen (Howard *et al.*, 1972).

VIII. Essential Fatty Acid Requirements and Metabolism

It has been recognized for many years that polyunsaturated fatty acids of the linoleic acid type are necessary for the nutrition of whole animals. Some of the early information on the lipid requirements of cells in culture was obtained by Evans *et al.* (1956) during development of the NCTC series of protein-free chemically defined media. Although the methyl esters of linoleate, linolenate, and arachidonate were added to the earlier media of this series, it was later found that these compounds were not required. The essential fatty acids could be omitted from the chemically defined media with no deleterious effects on cell growth (McQuilkin *et al.*, 1957). Similar conclusions can be drawn from the growth of cells in the chemically defined media developed by Morgan *et al.* (1950), Eagle (1959), and Waymouth (1959).

Gerschenson *et al.* (1967a) studied the effects of essential fatty acid (EFA) deficiency on the growth and mitochondrial function of HeLa cells in culture. These cells failed to grow in fat-deficient medium and did not exhibit an increase in eicosatrienoic acid. However, a decrease in the relative amount of 18:2(*n*-6) and 20:4(*n*-6) and an increase in 18:1(*n*-9) were observed. When compared with serum-fed controls, homogenates of cells that were cultured in lipid-free medium consistently showed an uncoupling of oxidative phosphorylation and loss of respiratory control with succinate, α-ketoglutarate, or β-hydroxybutyrate as a substrate. In a similar study with beating heart cells from rats, Gerschenson *et al.* (1967b) found that EFA deficiency affected the phosphorylation coupled to the oxidation of α-ketoglutarate and β-hydroxybutyrate, but not succinate. The fact that addition of either linoleate or arachidonate prevented these changes was interpreted to indicate a requirement for one or the other of these acids for maintaining the physiological integrity of the mitochondria.

In studies of clonal growth of Chinese hamster strain CHD-3 cells in synthetic medium, Ham (1963) found that an apparent requirement for serum albumin could be replaced by linoleic acid. Plating efficiency was markedly improved by the addition of this acid.

In contrast to these findings, the L-2071 subline of mouse fibroblast has been cultured in chemically defined, lipid-free medium since 1956 by Evans *et al.* Bailey and Menter (1967) examined the essential fatty acid content of L-2071 strain mouse fibroblasts growing in synthetic lipid-free medium (NCTC 135). A persistent residual level of linoleic acid was found, suggesting an efficient conservation mechanism for traces of linoleic acid present as a contaminant in the culture environment. Geyer (1967) showed that when L-cells were growing rapidly in NCTC 135, polyunsaturated fatty acids including linoleic acid could not be detected. These findings have been confirmed by Anderson *et al.* (1969). Although polyunsaturated fatty acids were absent, the cells had a rapid generation time and a tightly coupled mitochondrial function.

Mitochondria of L-929 mouse fibroblasts cultured in lipid- and protein-free medium for 9 years have been studied by Kagawa *et al.* (1970). Essential fatty acids were not found in these cultures either and the monounsaturated fatty acids, 16:1(n-7) and 18:1(n-9) were present in cell lipids in increased proportions. Mitochondrial functions were unimpaired with respect to phospholipid content, cytochromes, ATPase, electron transport, and oxidative phosphorylation *in situ*. Oxidative phosphorylation in isolated mitochondria, however, was found to be unstable. The same workers have recently reported the successful culture of ten mammalian cell lines in lipid and protein-free medium (Takaoka and Katsuta, 1971).

It thus seems fairly well established that a number of cell lines growing and metabolizing normally contain no detectable amounts of essential fatty acids in the cell lipids. This conclusion should be qualified by the observation that all culture environments may contain traces of essential fatty acids present as contaminants on the glassware and in the chemicals used to make up the synthetic growth media or introduced during the filtration and sterilization procedures. Significant amounts of linoleic acid and other lipids were consistently found in samples of commercially obtained chemically defined "lipid-free" culture media (Bailey *et al.*, 1972). The tissue culture experiments, therefore, do not completely eliminate the possibility that some metabolic product of the essential fatty acids is present in the cells, although the results do indicate fairly conclusively that the acids *per se* are not required for the structural and metabolic functions of the cell. Cell lines that have now been grown in the various media containing no essential fatty acids are included in Table 4.

TABLE 4
Essential Fatty Acid Requirements of Various Cell Types in Tissue Culture

Cell line	Essential fatty acid requirement	References
Mouse fibroblasts (L-929 and L-2071)	None	Evans *et al.* (1956); Geyer (1967); Bailey and Menter (1967); Anderson *et al.* (1969)
Chinese hamster (C3HD)	Improved plating efficiency	Ham (1963)
Mouse embryo fibroblasts (NCTC 3681)	None	Evans *et al.* (1964)
Mouse lymphoma (NCTC 3749)	None	Evans *et al.* (1964)
Chinese hamster sarcoma (NCTC 4206)	None	Evans *et al.* (1964)
Mouse parotid cells (NCTC 4075)	None	Evans *et al,* (1964)
Mouse kidney cells (NCTC 3526)	None	Evans *et al.* (1964)
Human skin epithelium (NCTC 3075)	None	Evans *et al.* (1964)
Human fetal cells (NCTC 2627)	None	Evans *et al.* (1964)
Human cervical carcinoma HeLa-S3 (NCTC 3952)	None	Evans *et al.* (1964)
Mouse liver parenchymal cells (NCTC 4067)	None	Evans *et al.* (1964)
Human cervical carcinoma (HeLa-S3)	Improved growth and mitochondrial function	Gerschenson *et al.* (1967a)
Embryonic rat heart cells	Improved mitochondrial function	Gerschenson *et al.* (1967b)
Mouse Ehrlich ascites	None	Dunbar *et al.* (1969)
Mouse sarcoma 180 ascitic form	None	Bailey and Dunbar (1971)
Rat liver parenchymal cells (JCT-21P3)	None	Takaoka and Katsuta (1971)
Rat thymus reticulum cells (JCT-20P3)	None	Takaoka and Katsuta (1971)
Rat ascites hepatoma (JCT-16P3)	None	Takaoka and Katsuta (1971)

Cell growth rates and morphology in most of these deficient cell lines are undistinguishable from controls. In addition, mitochondrial function in the deficient cells is essentially unchanged. As indicated in Fig. 15, cells showed normal respiratory characteristics when malonate was the substrate. Similar basal rates of oxygen consumption and a similar degree of stimulation by ADP and DNP indicates that NADH respiration is unimpaired in cells having no polyunsaturated fatty acids.

The requirement for essential fatty acids in the intact animal may, therefore, represent some "supracellular" requirement related to the integration of cell activities in a multicellular organism. The complete deletion of polyunsaturated lipids could possibly result in unfavorable modification of the physical and biological properties of living tissues. For example, decreased plating efficiency of cultured cells and the increased permeability of EFA-deficient tissues to water may relate to changes in cellular adhesion properties. The replacement of essential fatty acids by unsaturated acids of the eicosatrienoic type in EFA-deficient animals may represent some compensatory mechanism at the tissue level. Accumulation of the unusual polyunsaturates in the absence of linoleic acid might be a consequence of

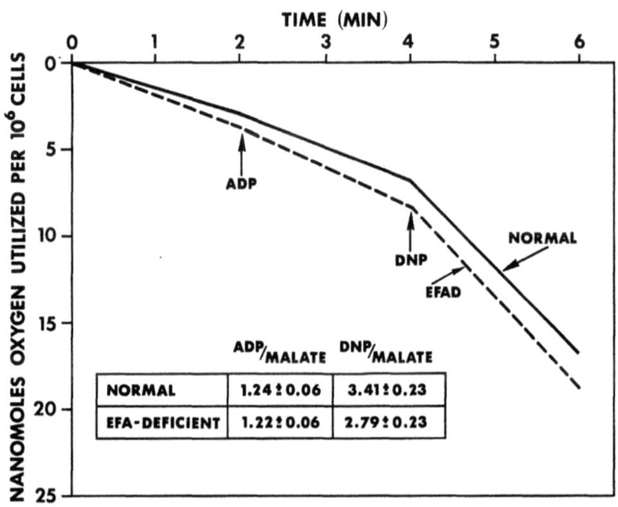

FIGURE 15. Malate respiration in normal and essential fatty-acid-deficient Ehrlich ascites cells. Respiration of cells in the presence of 15 mM malate was measured at 25°C using an oxygen electrode. Results are expressed as nanomoles of oxygen utilized per million cells per minute. ADP (140 μM) and DNP (50 μM) were added to the reaction chamber at the times indicated to determine the degree of mitochondrial coupling and the rate of uncoupled respiration. The similar basic rate of oxygen consumption and similar degree of stimulation by ADP (the respiratory control index) and DNP indicate that NADH-linked respiration is unimpaired by the absence of polyunsaturates.

TABLE 5
Distribution of Radioactive Fatty Acids in Cells Incubated with [1-^{14}C]Linoleic Acid

	Radioactivity in given fatty acid (%)[a]					
	Diploid cells[b]			Transformed cells[b]		
Fatty acid	WI-38	Ric Mil	Lala	HSDM	L929	L2071
18:2(*n*-6)	50.6	59.8	54.1	99.2	97.6	98.8
20:2(*n*-6)	2.8	5.0	1.5	0.8	2.4	1.2
20:3(*n*-6)	9.4	11.5	7.7	0	0	0
20:4(*n*-6)	34.6	21.3	36.7	0	0	0
Indicated metabolic capabilities (see Fig. 13)						
Δ6 Desaturase	+	+	+	−	−	−
Elongation enzymes	+	+	+	+	+	+
Δ5 Desaturase	+	+	+			

[a]The average incorporation of linoleic acid (18:2) for the six cell lines was 54.5 ± 7.1 μg per mg of cell protein over the 4-day growth period. This represented approximately 34% of that initially present in the original growth medium. There were no significant differences in linoleic acid uptake between normal and transformed cells.
[b]WI-38, human embryonic lung fibroblasts; Ric Mil and Lala, human skin fibroblasts; HSDM, mouse fibrosarcoma; L929 and L2071, transformed mouse fibroblasts (Dunbar and Bailey, 1975).

the action of the desaturating enzymes on alternative substrates. This is supported by the evidence that the absence of polyunsaturated acids in tumor cells grown in lipid-free media is related to loss of the necessary desaturating enzymes (Tables 5 and 6).

A question of particular pertinence is whether the known function of the essential fatty acids as the precursors of prostaglandins may explain

TABLE 6
Distribution of Radioactivity in Fatty Acids from Cells Incubated with Dihomo-γ-[1-^{14}C]Linolenic Acid

	Radioactivity in given fatty acid (%)[a]			
		Transformed cells		
Fatty acid	Diploid cells (W-138)	L929	HSDM	L2071
20:3(*n*-6)	20.0	47.8	36.6	62.6
20:4(*n*-6)	77.6	45.4	63.4	37.4

[a] The average incorporation of dihomo-γ-[1-^{14}C]linolenic acid (20:3) was 1.8 ± 0.2 μg per mg of cell protein during the 4-day growth period, which represented approximately 64% of that originally present in the growth medium (Dunbar and Bailey, 1975). Δ5 Desaturase as present in all cell lines. See Fig. 13 for metabolic pathways.

their "essential" biological role. The absence of other known functions for the essential fatty acids and the apparent inability of other fatty acids to replace them in synthesis of biologically active prostaglandins suggests that prostaglandin formation possibly represents the only absolute requirement for these acids. While the preponderance of the evidence from the tissue culture studies is not inconsistent with this conclusion, the possibility that the essential fatty acids fulfill some additional essential function in intact tissues has not been eliminated.

IX. Summary and Conclusions

The tissue culture procedure has become a powerful tool for studying complex feedback mechanisms for control of lipid metabolism in mammalian cells. The development of chemically defined lipid-free media led to the discovery that most cells can satisfy their lipid requirements by biosynthesis from simple components in the medium, principally glucose and acetate. When serum lipoproteins are added, these biosynthetic mechanisms are repressed and the cell lipids are obtained from the serum. Although delineation of the feedback networks is far from complete, the major control points have now been identified and a number of new concepts concerning the mechanisms of coupling to the external lipid environment are beginning to emerge. During lipid synthesis from glucose, the feedback control points are acetyl-CoA carboxylase for fatty acids and HMG-CoA reductase for cholesterol. Both operate primarily by repressing enzyme synthesis. Lipid synthesis from acetate is regulated via the enzyme acetyl-CoA synthetase. It is believed that *in vivo* the physiologically more abundant ketone body, acetoacetate, is the natural substrate. Control of acetyl-CoA synthetase operates at the level of enzyme activity. Fatty acid synthesis is also regulated via fatty acid synthase by a dual mechanism involving both repression or induction of enzyme synthesis and regulation via interconversion of the apo- and holoenzymes.

Considerable advances have been made recently in delineating the molecular mechanisms whereby cholesterol synthesis in human diploid skin fibroblasts is regulated by the serum cholesterol concentration. The principal feedback inhibitor is the low-density lipoprotein class of buoyant density 1.019–1.063.

The sequence of events during inhibition of cholesterol synthesis by low-density lipoprotein appears to be as follows: The intact lipoproteins attach to specific LDL receptors at the membrane and enter the cell continuously via pinocytosis and autophagy. The apoprotein portion is degraded, probably in the lysosomes, and excreted. The cholesterol moiety induces inhibition of cholesterol synthesis by repressing formation of

HMG-CoA reductase. This may involve earlier conversion to some oxygenated derivative, since 7-keto- and 25-hydroxy-cholesterol are about 100 times more effective and markedly reduce the lag period associated with repression by pure cholesterol.

Intracellularly synthesized cholesterol is excreted from the cell mainly in combination with the high-density lipoproteins. These combine with high-density lipoprotein-specific membrane receptor sites and under normal conditions serve to catalyze a rapid two-way flux of cholesterol between cells and medium. Concomitantly with the repression of HMG-CoA reductase is an activation of cholesterol acyltransferase that enhances accumulation of cholesterol esters. This is not blocked by inhibitors of protein synthesis such as cycloheximide and probably depends upon modification of pre-existing enzyme protein. When LDL is removed from the medium, most of these processes are reversed. This control network is defective in skin fibroblasts of individuals homozygous for the disorder familial hypercholesterolemia because their cells lack the membrane receptor for low-density lipoproteins (Goldstein *et al.,* 1974). Consequently repression of cholesterol synthesis and removal of low-density lipoproteins operates at suboptimal levels. The resulting elevation in serum low-density lipoproteins and accumulation of cholesterol in tissues leads to generalized atherosclerosis and early death.

Tissue culture has been used to study the essential fatty acid requirements of mammalian cells. Numerous cell lines have now been grown for extended periods amounting to hundreds of generations in the complete absence of essential fatty acids. Growth rates and morphological characteristics are unaffected by the deficiency. Mutants, which have lost certain of the enzymes for interconversion of polyunsaturated fatty acids, are proving useful in the delineation of the pathways of essential fatty acid metabolism. The main conclusion from these studies is that the requirement for essential fatty acids by the intact organism probably reflects their role as precursors of the prostaglandins.

A number of abnormalities in lipid metabolism have been observed both in tumor cells and in cultured cells following transformation by oncogenic viruses. These include partial loss of control of feedback inhibition of HMG-CoA reductase by a mechanism that does not involve loss of the membrane receptor and loss of control in converting mevalonic acid to cholesterol. A second example lies in the metabolic pathway for conversion of linoleic acid to prostaglandins in which a $\Delta 6$ reductase is lost in many tumors. A third defect frequently observed in tumors is the accumulation of ether lipids. This could be related to the decreased activity of cytoplasmic *sn*-glycerol-3-P dehydrogenase, a phenomenon that is also related to the abnormally high glycolytic capacity of tumor cells.

It is particularly impressive that, with the exception of those in tumor cells noted above, all of the cellular control mechanisms for lipid biosynthesis remain intact during hundreds of generations of cultivation in chemically defined lipid-free media. They are reactivated within a few hours of transfer back to serum-containing medium. It is apparent that further elucidation of the molecular mechanisms of these control systems will involve some fundamental advances in our understanding of the control of gene expression in mammalian cells.

References

Aas, M., and Bremer, J. 1968. Enzymatic formation of acetoacetyl coenzyme A. *Biochim. Biophys. Acta* **164**:157–166.

Alberts, A. W., Fergusion, K., Hennesy, S., and Vagelos, P. R. 1974. Feedback control of fatty acid synthetase. *J. Biol. Chem.* **249**:5241–5249.

Alberts, A. W., Strauss, A. W., Hennesy, S., and Vagelos, P. R. 1975. Regulation of synthesis of fatty acid synthetase. *Fed. Proc.* **34**:213.

Anderson, R. E., Cummins, R. B., Walton, M., and Snyder, F. 1969. Lipid metabolism of cells grown in tissue culture 0-alkyl, 0-alk-1-enyl and acyl moieties of LM cells. *Biochim. Biophys. Acta* **176**:491–501.

Bailey, J. M. 1961. Lipid metabolism in cultured cells I. Factors affecting cholesterol uptake. *Proc. Soc. Exp. Biol. Med.* **107**:30–34.

Bailey, J. M., 1964. Lipid metabolism in cultured cells III. The cholesterol excretion process. *Am. J. Physiol.* **207**:1221–1225.

Bailey, J. M. 1966. Lipid metabolism in cultured cells VI: Lipid biosynthesis in serum and synthetic growth media. *Biochim. Biophys. Acta* **125**:226–236.

Bailey, J. M., and Dunbar, L. M. 1971. Lipid metabolism in cultured cells: Growth of tumor cells deficient in essential fatty acids. *Cancer Res.* **31**:91–97.

Bailey, J. M., and Menter, J. 1967. Lipid metabolsim in cultured cells VIII: linoleic acid content of cells grown in lipid-free synthetic medium. *Proc. Soc. Exp. Biol. Med.* **125**:101–105.

Bailey, J. M., Gey, G. O., and Gey, M. K. 1959. Utilization of serum lipids by cultured mammalian cells. *Proc. Soc. Exp. Biol. Med.* **100**:686–692.

Bailey, J. M., Howard, B. V., Dunbar, L. M., and Tillman, S. 1972. Control of lipid metabolism in cultured cells. *Lipids* **7**:125–134.

Bailey, J. M., Allan, T., Butler, E. J., and Wu, J-D. 1976. Detective regulation of cholesterol biosynthesis in tumor-virus and hypercholesterolemic human skin fibroblasts: A comparative study. VIII Annual Miami Winter Symposium, 136.

Bates, S. R., and Rothblat, G. 1974. Regulation of cellular sterol flux and synthesis by human serum lipoproteins. *Biochim. Biophys. Acta* **360**:38–55.

Brown, M. S., and Goldstein, J. L. 1974. Familial hypercholesterolemia: Defective binding of lipoproteins to cultured fibroblasts associated with impaired regulation of HMG CoA reductase. *Proc. Natl. Acad. Sci. USA* **71**:788–792.

Brown, M. S., Dana, S. E., and Goldstein, J. L. 1974. Defective regulation of HMG CoA reductase in human skin fibroblast cultures. *J. Biol. Chem.* **249**:789–796.

Dunbar, L. M., and Bailey, J. M. 1975. Enzyme deletions and essential fatty acid metabolism in cultured cells. *J. Biol. Chem.* **250**:1152–1154.

Dunbar, L. M., Viener, R. S., and Bailey, J. M. 1969. Growth of ascites tumors deficient in essential fatty acids. *Fed. Proc.* **28**:557.

Eagle, H. 1959. Amino acid metabolism in mammalian cell cultures. *Science* **130**:432–437.
Eagle, H. 1961. Biochemistry of cultured mammalian cells. *Annu. Rev. Biochem.* **30**:605–625.
Evans, V. J., Bryant, J. C., Fioramonti, M. C., McQuilkin, W. T., Sanford, K. K., and Earle, W. R. 1956. Studies on nutrient media for tissue cells *in vitro* I: A protein-free chemically defined medium for strain L cells. *Cancer Res.* **16**:77–94.
Evans, V. J., Bryant, J. C., Kerr, H. A., and Schilling, E. L. 1964. Chemically defined media for mammalian cells. *Exp. Cell Res.* **36**:439–474.
Gerschenson, L. E., Mead, J. G., Harary, I., and Haggerty, D. F. 1967a. Studies on the effects of essential fatty-acids on growth rate, fatty-acid composition, oxidative phosphorylation and respiratory control of HeLa cells in culture. *Biochim. Biophys. Acta* **131**:42–49.
Gerschenson, L. E., Mead, J. G., Harary, I., and Haggerty, D. F. 1967b. Studies *in vitro* on single beating rat-heart cells. X. The effect of linoleic and palmitic acids on beating and mitochondrial phosphorylation. *Biochim. Biophys. Acta* **131**:50–58.
Geyer, R. P. 1967. Uptake and retention of fatty acids, p. 33. *In* G. H. Rothblat and D. Kritchevsky (eds.). Lipid Metabolism in Tissue Culture Cells. The Wistar Institute Press, Philadelphia.
Goldstein, J. L., and Brown, M. S. 1973. Familial hypercholesterolemia: Identification of a defect in the regulation of HMG CoA reductase activity associated with overproduction of cholesterol. *Proc. Natl. Acad. Sci.* **70**:2804–2808.
Goldstein, J. L., Dana, S. E., and Brown, M. S. 1974. Esterification of low density lipoprotein cholesterol in human fibroblasts and its absence in homozygous familial hypercholesterolemia. *Proc. Natl. Acad. Sci. USA* **71**:4288–4292.
Ham, R. G. 1963. Albumin replacement by fatty-acids in clonal growth of mammalian cells. *Science* **140**:802–803.
Howard, B. V., and Bailey, J. M. 1974. Acetyl coenzyme A synthetase and the regulation of lipid synthesis from acetate in cultured cells. *J. Biol. Chem.* **249**:7912–7921.
Howard, V. B., and Howard, W. J. 1974. Lipid metabolism in cultured cells. *Adv. Lipid Res.* **12**:51–96.
Howard, B. W., Morris, H. P., and Bailey, J. M. 1972. α-glycerol phosphate dehydrogenase levels and growth rate in tumors and cultured cells. *Cancer Res.* **32**:1533–1542.
Jacobs, R. A., Sly, W. S., and Majerus, P. W. 1973. Feedback regulation of acetyl coenzyme A carboxylase in cultured cells. *J. Biol. Chem.* **248**:1268–1276.
Kagawa, Y., Takaoka, T., and Katsuta, H. 1970. Absence of essential fatty acids in mammalian cell strains cultured in lipid- and protein-free chemically defined synthetic media. *J. Biochem.* **68**:133–136.
Kandutsch, A. A., and Chen, H. W. 1973. Inhibition of sterol synthesis in cultured mouse cells by 7α-hydroxy-cholesterol, 7β-hydroxy-cholesterol and 7-keto-cholesterol. *J. Biol. Chem.* **248**:8408–8417.
McQuilkin, W. P., Evans, V. J., and Earle, W. R. 1957. Adaptation of additional lines of NCTC 929 cells to chemically defined protein-free medium NCTC 109. *J. Natl. Cancer Inst.* **19**:886–892.
Morgan, J. F., and Parker, R. C. 1950. Nutrition of cells in tissue culture I. Initial studies in a synthetic medium. *Proc. Soc. Exp. Biol. Med.* **73**:1–8.
Rothblat, G. H. 1969. Lipid metabolism in cultured cells. *Adv. Lipid Res.* **7**:135–145.
Rothblat, G. H., Hartzell, R., Mialhe, H., and Kritchevsky, D. 1967. Cholesterol metabolism in tissue culture cells, pp. 129–146. *In* G. H. Rothblat and D. Kritchevsky (eds.). Lipid Metabolism in Tissue Culture Cells. The Wistar Institute Press, Philadelphia.
Siperstein, M., Gyde, A., and Morris, H. 1971. Loss of feedback control of HMG CoA reductase in hepatomas. *Proc. Natl. Acad. Sci. USA* **68**:315–322.
Sokoloff, L., and Rothblat, G. H. 1972. Regulation of sterol synthesis in L cells: Steady-state and transitional responses. *Biochim. Biophys. Acta* **172**:172–181.

Takaoka, T., and Katsuta, H. 1971. Long-term cultivation of mammalian cell strains in protein- and lipid-free chemically-defined synthetic media. *Exp. Cell Res.* **67**:295–300,

Waymouth, C. 1959. Rapid proliferation of sublines of NCTC clone 929 (strain L) mouse cells in a simple chemically-defined medium (MB752/1). *J. Natl. Cancer Inst.* **22**:1003–1015.

Williams, C. D., and Avigan, J. 1972. *In vitro* effects of serum proteins and lipids on lipid synthesis in human skin fibroblasts and leukocytes grown in culture. *Biochim. Biophys. Acta* **260**:413–423.

Lipid Changes in Membranes during Growth and Development

C. A. Pasternak

I. Introduction

It must be stated at the outset that this is not a comprehensive review. Instead, a few systems have been picked out for detailed analysis. The choice depends on the fact that they appear, at least to the author, to provide fairly clearcut examples of growth and development that are also amenable to experimental attack. Several have been studied in his own laboratory and consequently this chapter presents a somewhat personalized approach to the study of lipid changes in membranes during growth and development.

A. Growth Versus Development

Growth and development are overlapping concepts. The growth of a fertilized egg into an adult organism involves many developmental stages; the development of a particular tissue such as liver or limb is accompanied by growth from some tens or hundreds of cells to some 10^8 or more. For the purposes of this chapter, the terms will be distinguished as follows: Growth is defined as the proliferation of cells without structural or functional change; development as a change in the structure and function of cells. Growth and development do not necessarily occur in concert. On the

C. A. Pasternak • Department of Biochemistry, St. George's Hospital Medical School, Blackshaw Road, Tooting, London SW17 OQT.

contrary, they are often alternatives in cellular behavior. The production of a clone of antibody-producing cells involves growth without development. The later stages of erythropoiesis involve development without growth; whether the trigger for such development, i.e., the commitment to an altered expression of certain genes, involves cell division or not, is another matter.

The most convenient systems for studying growth in the absence of development are established lines of cells in culture. Provided that the external milieu of a stable clone of cells is kept constant, daughter cells deviate minimally from their parents. One can then investigate the extent to which constancy of cellular structure and function is superimposed on cyclical changes as cells progress from mitosis to interphase and on to mitosis again. It may be argued that successive stages of the cell cycle are themselves developmental events (Mitchison, 1971). But if one accepts the more common definition of development, i.e., differentiation, as the stable alteration of structure and function (Pasternak, 1970), successive stages of the cell cycle are not developmental events. In any case, it would appear convenient to keep cell cycle-dependent changes distinct from developmental ones involved in differentiation, if only because growth and differentiation are sometimes mutually exclusive events.

In this chapter attention will be focused on membranes. This encompasses most of the cellular phospholipids and much of its cholesterol. Neutral lipid droplets (mainly triglyceride) are thus excluded. Although triglycerides constitute a metabolic pool of phospholipid precursors, as well as being an important intracellular energy reservoir, changes in their composition or amount are due less to passage through a particular phase of the cell cycle or to a specific developmental stimulus, than to alterations in food supply or other environmental factor. In other words it is adaptive changes rather than developmental ones that predominantly affect the triglyceride metabolism of cells (e.g., Pasternak, 1970).

The plasma membrane will be discussed in greatest detail. Many developmental changes are expressed at the cell surface: The appearance of specific receptors, their reaction with hormones or immunological factors, resultant changes in membrane permeability or in the activity of membrane-embedded enzymes, are examples. Of course, other membranes also show changes, often of a grosser type; the proliferation of endoplasmic reticulum in response to hormones (Tata, 1967a,b) may be cited in this regard. The nuclear membrane is of particular importance with respect to the cell cycle in view of its apparent disappearance during the mitotic phase. Unfortunately few studies on its assembly or composition have yet been made (Kay and Johnston, 1973). Mitochondrial membranes, about which much more is known (Getz, 1970, 1972), probably do not alter markedly during growth and development, since their function remains the same; the transient alterations underlying the mechanism of oxidative phosphorylation fall outside the scope of this review.

B. Role of Membrane Lipids

In order to assess the relevance of lipid changes in membranes during growth and development, the role of membrane lipids has to be understood. It is clear that in every type of cell—microbial, plant, or animal—the internal environment is separated from the external by a predominantly lipid membrane; the major function of the membrane appears to be to limit the movement of water-soluble substances into and out of cells. The presence of carbohydrate-containing walls around such membranes in microbes and plants has more to do with the elaboration of shape and rigidity than with the maintenance of correct internal environment; equally the carbohydrate-rich glycocalyx covering some animal cells probably has little effect on their permeability properties. Lipid membranes also encase subcellular organelles and certain viruses. An additional function in these instances may be the ability to fuse with the plasma membrane of cells.

The reason why the major lipid components of biological membranes are phospholipids is less obvious. Suffice it to say that a membrane consisting of a hydrophobic interior sandwiched between two hydrophilic layers combines the necessary function of chemical insulation with that of thermodynamic stability, and that a double layer of phospholipids is an effective means of achieving it. Further probing into the reason for the chemical makeup of phospholipids—why they are based predominantly on a glycerol moiety, for example—is as teleological a question as that seeking to answer why particular L-amino acids or D-sugars have evolved as biological molecules. Some type of anionic grouping seems essential, and while phosphate has evolved as the most usual, it appears to be partially replaceable by carboxyl during phosphate starvation in certain bacterial cells (Tempest *et al.*, 1968).

An important property of membranes is their fluidity (Singer and Nicolson, 1972); that is the ability of the component molecules to move in a lateral direction independently of each other. Membrane fluidity is determined largely by its lipid composition, the important parameters being the length and degree of unsaturation of the phospholipid acyl chains, and the ratio of cholesterol to phospholipid. Changes in fluidity have been postulated to occur during the cell cycle and to accompany such developmental events as the stimulation of lymphocytes or the emergence of cancer cells. They are discussed under the appropriate headings (sections II.B, III.B, and VI.B).

Superimposed on these major roles of membrane lipids are minor functions related to more specific structures: glycolipid antigens of the ABO and related serological systems (Watkins, 1974), glycolipid receptors for cholera (Hollenberg *et al.,* 1974) and tetanus (Van Heyningen, 1974) toxins, particular phospholipids associated with hormonal responsiveness (Blecher 1968, 1969; Rodbell *et al.,* 1968), and with membrane-embedded enzymes (Coleman, 1973; Fourcans and Jain, 1974) of which calcium-ATPase (Warren *et al.,* 1974a,b) is a well-studied example. These functions of membrane lipids will not be discussed in detail; full accounts are given in the references cited.

C. *Compositional Changes: Synthesis, Insertion, and Turnover*

The lipid composition of a membrane is determined by two main factors: (i) the synthesis of the components, generally at the endoplasmic reticulum, and (ii) their insertion into the membrane. While many details of the synthesis of phospholipids, glycolipids, and cholesterol have been clarified (see Vol. 1, the chapter Introduction and the chapter Liver), little is known of the insertion process. For example, it is not yet established whether molecules are inserted independently of each other by some kind of transfer molecule (e.g., Wirtz, 1974) or whether intact pieces of endoplasmic reticulum "flow" to plasma membranes (Hirano *et al.,* 1972) or other sites. All that is clear is that the size of newly inserted pieces in animal (Leskes *et al.,* 1971) as well as in bacterial (Fox, 1972) membranes is likely to be rather small. As a result it is not known how specificity of insertion (for example cholesterol and glycolipids predominantly into the plasma membranes of cells) is achieved. The same problem of course applies to membrane-destined proteins, also.

Superimposed on these two factors, which account for the net synthesis of membranes, are factors that operate to modulate the composition of existing membranes: (iii) excision, perhaps by the opposite mechanism to that for insertion, and (iv) degradation, generally by lysozomal enzymes. In addition, (v) partial turnover of phospholipids *in situ* occurs; degradation of one of the acyl chains is followed by reacylation ("Lands" cycle, Hill and Lands, 1970; Numa and Yamashita, 1974); by insertion of a fatty acid different from the one removed, chemical modification of existing phospholipids is achieved. With phosphatidylinositol, degradation of the hydrophilic portion also occurs (Dawson, 1966); neither the route for resynthesis nor the significance of this type of turnover is clear, except that a relation with many "short-acting" hormones seems indicated (Michell, 1975).

Phospholipids in general have rather high rates of turnover [achieved by a combination of mechanisms (i)–(iv) as well as by mechanism (v)], so that opportunities for chemical modification exist. An important part of this review is to assess the extent to which such modification accompanies specific phases of growth and development.

II. *The Cell Cycle*

A. *Methodology*

Two methods are available for studying mammalian cells at different stages of the cell cycle. In the first, synchronous growth is induced by some technique, and cells are sampled sequentially over the period of one complete cycle. In the second, exponentially growing cells are physically separated into those in mitosis, those in G_1, those in S, etc. (Fig. 1). Both methods have been extensively reviewed (Stubblefield, 1968; Mitchison, 1971; Nias and Fox, 1971; Schindler and Schaer, 1972; Pasternak, 1973a; Shall, 1973).

Because mammalian cells are generally easier to grow in monolayer than in suspension cultures, most studies have been carried out with cells attached to petri dishes made of glass or plastic. Another reason is that a common method currently used to synchronize cells is the addition of fresh serum to cultures halted at confluency, though in fact the method works equally well with suspensions (Knox and Pasternak, 1975, 1977). Use of monolayer cells for membrane studies has two drawbacks. First the cells have to be detached from the vessel surface by one means or another, and this may result in components being left behind (e.g., Rosenberg, 1960; Weiss, 1961; Takeichi, 1971; Knox and Pasternak, 1975). Second, the surface

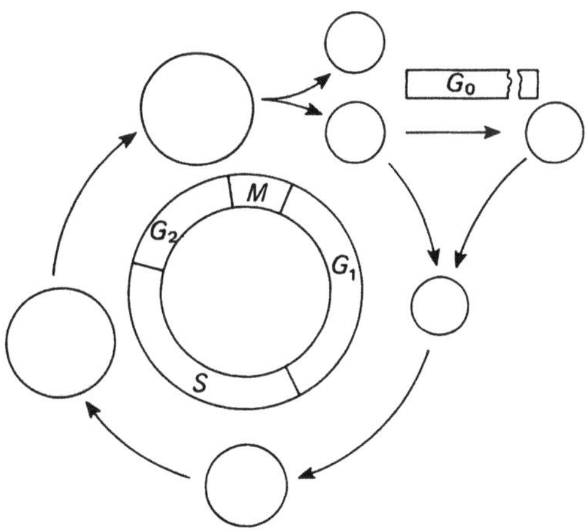

FIGURE 1. The cell cycle. The conventional abbreviations are used: M—mitosis; G_1 and G_2—gap periods 1 and 2; S—DNA synthetic period; G_0—period characteristic of quiescent cells (see section III).

membrane may itself show changes as a result of alterations not directly connected with the cell cycle; attachment to the culture vessel and to neighboring cells, the breaking of such contacts, and the movement over vessel and surrounding cells (Ambrose, 1975). Suspension cells are therefore preferable for an analysis of cycle-dependent surface changes.

B. Surface Membranes

As stable clones of cultured cells procede from one mitosis to the next, each constituent—be it DNA, glycolytic enzyme or membrane lipid—is exactly doubled. It is known that with DNA the synthetic period is restricted to only a part of the cell cycle, the S phase. With RNA and protein, synthesis is continuous throughout interphase and the amount of new material made exceeds twofold to compensate for any turnover that occurs; the same is true for total phospholipids (Pasternak, 1974). In many instances individual molecular species of protein or RNA are synthesized at discrete times and several enzymes show fluctuations of activity (Klevecz and Ruddle, 1968; Mitchison, 1971). In order to derive meaningful data for phospholipids, fractionation into subcellular organelles is a prerequisite. In any case the surface membrane is of special interest for the following reason (Pasternak, 1976).

When suspension cells, which are generally spherical, progress through interphase, their volume doubles between G_1 and G_2; since most constituents also double between G_1 and G_2, the density remains approximately constant. The surface area of a spherical cell, however, increases by only $2^{2/3}$ or 1.6-fold as volume doubles, the remaining 0.4-fold being achieved during cytokinesis. Does the "density" of the plasma membrane also remain constant, in which case appreciable synthesis and insertion of its components must occur during mitosis as well as during interphase? Or are synthesis and insertion restricted to G_2 and mitosis [as suggested by preliminary results (Bosmann and Winston, 1970) with L5178Y cells], or are they in concert with other cellular constituents, to interphase? Isolation of plasma membranes at different stages of the cell cycle is clearly an important goal.

This goal has been achieved with NIL hamster and P815Y mouse cells (each of which remains spherical throughout the cell cycle) by separating plasma membrane of cells at different stages of the cell cycle in a zonal rotor. Since the percentage of the total cellular protein that is in the plasma membrane remains constant, one may conclude that plasma membrane proteins double during interphase, and do not increase during mitosis (Graham *et al.*, 1973). The same is true of plasma membrane phospholipids (Graham *et al.*, 1973) and of cholesterol (Table 1). The implication of this result is that the "density" of the plasma membrane does *not* remain

TABLE 1
Plasma Membrane Cholesterol and Fatty Acids during the Cell Cycle[a]

Fraction number	Cholesterol[b] (μg)	Fatty acid (μg)[b] 16:0	16:1	18:0	18:1	18:2	20:4
1 ↑	680	124	36	180	108	220	50
2	620	137	34	176	112	231	46
3 G_1	640	127	34	186	110	230	48
4 ↓ ↑	670	128	36	179	108	222	45
5 S	620	136	36	187	117	227	45
6 ↑ ↓	680	130	33	178	110	229	50
7 G_2 ↓	630	127	36	180	112	240	51

[a] P815Y mouse mastocytoma cells were separated by zonal centrifugation according to position in the cell cycle and plasma membrane isolated (Graham *et al.*, 1973). The content of cholesterol (Crawford, 1958) and of phospholipid acyl chains, separated and analyzed by gas–liquid chromatography, was measured. Reproduced with permission from Sumner (1974).
[b] Values correspond to 1 mg of plasma membrane phospholipid.

constant. But how is the surfeit of plasma membrane components in late interphase accommodated?

The thickness of the plasma membrane does not increase during the cell cycle (Knutton, 1976). The density of intramembranous particles is said to change dramatically during the cell cycle of mouse L cells in suspension and of Chinese hamster cells in monolayer (Scott *et al.*, 1971), but this has not been substantiated with other suspension [P815Y] or monolayer [mouse 3T3, (Knutton, unpublished results) or hamster BHK (Torpier *et al.*, 1975)] cells (Knutton, 1976). In any case it is difficult to account for a concerted increase in protein, phospholipid, and cholesterol by an increased density of intramembranous particles. The answer appears to lie in the extent to which the cell surface is elaborated into microvilli. Using scanning electron microscopy as well as freeze-fracture techniques to assess the number of microvilli per cell, it has been found that cells in early G_1 have (on average) 3% of their surface elaborated into microvilli, whereas G_2 cells have 6% (Knutton *et al.*, 1975). The net result is that while the *apparent* surface area increases only 1.6-fold between G_1 and G_2, the *actual* surface area increases twofold, in concert with the insertion of its components (Fig. 2). During cytokinesis, the surface area generated by the microvilli is unfolded to provide the surface necessary for the formation of two daughter cells.

Changes in the density of intramembranous particles in the plasma membrane have been interpreted as indicating an altered membrane fluidity (Barnett *et al.*, 1974a). With P815Y cells in suspension culture, neither the cholesterol content nor the chain length or degree of unsaturation of the constituent phospholipid molecules (Table 1) appears to change signifi-

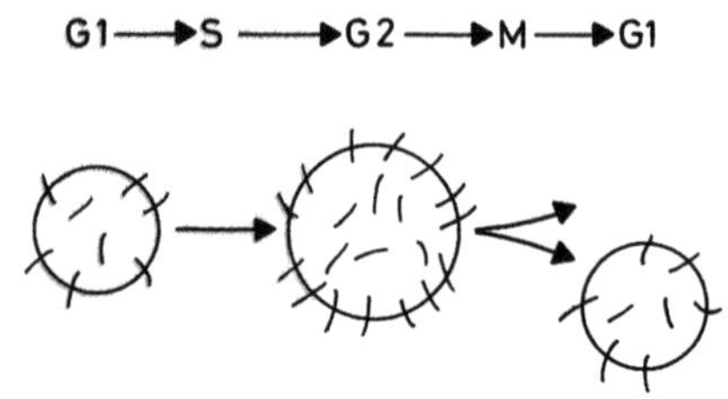

Volume	1	2	2x1
Apparent surface area	1	1·6	2x1
Actual surface area	1	2	2x1

FIGURE 2. Surface changes during the cell cycle. A cell that remains spherical throughout the cycle is illustrated. Volume and surface area of a newly formed G_1 cell are arbitrarily taken as 1. Calculation of actual surface area takes into account the number and mean size of the microvilli, illustrated diagrammatically.

cantly, nor is the distribution of phospholipid classes* (PC, PE, SM, or PI) altered. If changes in lipid fluidity of the plasma membrane occur during the cell cycle, they must be small and/or localized in a few discrete areas of the plasma membrane. Certainly newly synthesized phospholipids have the same composition as pre-existing ones. Whether newly synthesized components are inserted directly in the form of microvilli, or whether microvilli are elaborated only after insertion, remains to be elucidated. The former mechanism might be an attractive one for membrane assembly.

So far as glycolipids are concerned, both neutral and acidic (gangliosides) species follow the general pattern of gradual insertion during interphase, largely in G_1 (Table 2). A similar conclusion has been reached with regard to Forssman antigen and other glycolipids (Wolf and Robbins, 1974). Fluctuations in plasma membrane sialic acid, fucose, or other sugar components (Pasternak *et al.*, 1974) must therefore reflect the predominant carbohydrate-containing species, namely, glycoproteins (the plasma membrane of P815Y cells has over 4 times as much sugar in glycoprotein as in glycolipid). It is glycoproteins that provide the major contribution to surface charge (as carboxyl groups) and that may be involved in changes of electrophoretic mobility during mitosis (Brent and Forrester, 1967; Mayhew, 1967).

*PC—phosphatidylcholine; PE—phosphatidylethanolamine; SM—sphingomyelin; PI—phosphatidylinositol.

C. Other Membranes

Few data are on hand concerning membrane lipids in other subcellular organelles during the cell cycle. Certainly the rate of incorporation of phospholipid precursors into the nucleus (Pegorano *et al.*, 1972) and other organelles (Warmsley *et al.*, 1970) increases at the same rate as incorporation into whole cells, but the action of phospholipid exchange proteins (Dawson, 1973; Wirtz, 1974) in such cells (Pasternak and Bergeron, 1970: Table 4) prevents definitive interpretation of this result. So far as slowly growing compared with rapidly growing cells are concerned, the results of Bergelson *et al.* (1974) are relevant: No significant differences in phospholipid composition, of mitochondria or microsomes, between normal and regenerating rat liver, were found.

Several mitochondrial enzymes have been measured. Succinate–cytochrome *c* reductase (EC 1.3.99.1) and cytochrome *c* oxidase (EC 1.9.3.1), both of which are membrane-bound enzymes requiring a specific lipid environment (Coleman, 1973), appear to double gradually during interphase (Warmsley *et al.*, 1970; though note Bosmann, 1971). So does the succinate-mediated increase in fluorescence caused by the binding of 1-anilinonaphthalene-8-sulfonate (Warmsely *et al.*, 1970) to a supposedly hydro-

TABLE 2
Gangliosides during the Cell Cycle[a]

Fraction number		Ganglioside content[b]	Percentage in band			
			1	2	3	4
1	↑	48	7	22	35	36
2			53	6	24	34	36
3	G_1	60	9	24	31	36
4			59	8	25	33	34
5	↓ ↑	68	7	24	33	36
6	S	75	7	21	35	37
7	↑ ↓	79	8	25	33	34
8	G_2 ↓	85	6	23	35	36

[a] P815Y mastocytoma cells were separated by zonal centrifugation according to position in the cell cycle (Graham *et al.*, 1973). A chloroform:methanol extract was washed with water, the aqueous layer concentrated and total ganglioside measured (Warren, 1959). Individual components were separated by thin-layer chromatography in chloroform:methanol:2.5 N ammonium hydroxide (60:40:9), detected by spraying with resorcinol reagent, scraped off the plate and analyzed spectrophotometrically at 580 nm. The fastest-moving band (1) has the mobility of *N*-acetyl-galactosaminylmonosialolactose-ceramide, the next (band 2) of galactosyl-*N*-acetyl-galactosaminylmonosialolactose-ceramide; two slower-moving bands (3 and 4) have not been identified. Reproduced with permission from Sumner (1974).

[b] nmoles of ganglioside sialic acid/10^8 cells.

TABLE 3
Fractionation of Labeled Sea Urchin Embryos[a]

Cell fraction		Phospholipids $[^3H]$ (dpm × 10^{-3})	Phospholipids Amount (μg)	Phospholipids Specific action (dpm × 10^{-3}/μg)	Protein $[^{14}C]$ (dpm × 10^{-2})	Protein Amount (μg)	Protein Specific action (dpm × 10^{-3}/μg)	Na/K-ATPase Specific action (μmole/mg prot/hr)
Homogenate		780	155	5.0	3300	111	30	N.D.
Supernatant		380	56	6.8	1320	42	31	N.D.
Fraction	1	50	12	4.2	48	9	5	<0.1
	2	9	2	4.5	123	11	11	<0.1
	3	<5	4	—	60	1.4	43	<0.1
	4	<6	<1	—	45	1.1	41	<0.1
	5	<4	<1	—	39	1.1	35	0.84
	6	<4	<1	—	18	0.4	45	2.64
	7	15	7	2.5	190	2.5	76	0.25
	8	40	15	2.7	180	1.8	100	0.23
Pellet		150	63	2.4	1830	59	31	N.D.

[a] *S. purpuratus* eggs were fertilized, exposed to [Me-^{3}H]choline (8 μCi/ml) and [^{14}C]valine (0.08 μCi/ml) for 15 min, washed and allowed to develop for 22 hr, by which time all had reached the stage of hatched blastula (Pasternak, 1973c). Embryos were dissociated in 0.4 M glucose, homogenized and spun (1100 × *g* for 10 min) to yield a supernatant and pellet (largely nuclear) fraction. The supernatant was further fractionated by spinning through a gradient (10–60%) of sucrose. Radioactivity, phospholipid phosphorus (Bartlett, 1959), protein (Pasternak, 1973c), and Na/K-ATPase (Avruch and Wallach, 1971) were measured. Unpublished experiments in collaboration with E. G. Baker, D. Epel, and J. M. Graham. N.D. = not determined.

phobic membrane component (Brocklehurst *et al.*, 1970). This fits with a gradual increase in mitochondrial protein (Warmsley *et al.*, 1970); mitochondrial DNA appears to be duplicated during S and early G_2 (Bosmann, 1971), the actual *number* of mitochondria presumably doubling thereafter. Studies should now be aimed at elucidating the fate of the nuclear membrane during mitosis, so that the mechanism of its assembly and dispersion, as well as possible changes in lipid components, may be assessed.

III. Stimulation of Quiescent Cells

When cells in monolayer or suspension culture reach confluency or saturation density, they generally become arrested between mitosis and the next round of DNA synthesis. This period has been termed G_0 in order to distinguish it from the G_1 phase of cycling cells (Fig. 1). Although the concept of a G_0 phase distinct from an abnormally long G_1 phase has been questioned (Burns and Tannock, 1970; Smith and Martin, 1973), it is agreed that when such cells are stimulated to resume growth—by addition of extra serum or of hormones, by wounding, or by other treatment (e.g., Knox and Pasternak, 1977), they progress through the remainder of G_1 and subsequent periods in a normal manner. As mentioned, serum stimulation has been used to synchronize cultured cells.

However, some of the changes, such as an immediate stimulation of nutrient uptake and a fall in the concentration of intracellular cyclic AMP, do not occur in this manner in the G_1 period of cycling cells, and in that sense the phenomenon has attributes of development as well as of growth. In fact, there is a resemblance to the progression of events that takes place when lymphocytes are stimulated *in vivo* or *in vitro* by antigen or mitogen such as phytohemagglutinin or other lectins. Since circulating lymphocytes do not normally divide, and since in this instance dramatic morphological changes accompany the stimulation (Ling, 1971), such initiation of growth is clearly a developmental event. Moreover the activation of unfertilized eggs, which are also arrested in G_1, by sperm or other agents has similarities to the stimulation of cultured cells and lymphocytes; a change in membrane potential and a movement of calcium ion may be cited (Mazia, 1974). The latter event is certainly a developmental one. Having drawn attention to these similarities, the three situations will now be discussed, the last-mentioned in section IV.

A. Cultured Cells

The turnover of phospholipids is increased when mouse 3T3 cells (Cunningham, 1972), chick embryo fibroblasts (Pasternak, 1972), rat hepatoma (HTC) cells (Pasternak and Knox, 1973), or rat embryo fibroblasts (Ristow *et al.*, 1973) are exposed to additional serum. Phosphatidylinositol is particularly affected. Turnover also appears to be greater in subconfluent than in confluent mouse embryo fibroblasts (Diringer and Koch, 1973). It is not known whether extra synthesis and degradation are exactly matched, or whether the phospholipid content changes. Nor is it known which membranes are involved, though it is likely that the plasma membrane is initially affected. This is because a loss of phospholipids (Knox and Pasternak, 1973; Pasternak and Knox, 1973; Ristow *et al.*, 1973) occurs at the same time as the component(s) of serum becomes bound to the cell surface (Knox and Pasternak, 1975, 1977). Although the loss is small and has been observed as yet only with prelabeled cells, it is not due to exchange with lipids present in serum; dilute trypsin, for example, has the same effect (P. Knox, unpublished observations). Whether the phospholipids that are released are located in the bilayer proper, or are bound to a peripheral protein, has to be determined. Also, the relation of phospholipid release to the change in membrane potential (Hulser and Frank, 1971) needs to be assessed.

An increase in the content and turnover of surface glycolipids takes place when various cell lines reach confluency (Hakomori, 1975a); this may be related to the intercellular transglycosylation mechanism that has been postulated (Roseman, 1971; Roth and White, 1972) to underlie intercellular adhesion. However, no loss of glycolipid or a component thereof has so far been observed to occur when quiescent cells are stimulated by serum. It is therefore premature to correlate this change with the release of phospholipids described above.

B. Lymphocytes

Addition of phytohemagglutinin or other lectins to lymphocytes causes a stimulation of phospholipid turnover (Fisher and Mueller, 1968; Pasternak and Friedrichs, 1970); as in serum stimulation of cultured cells, phosphatidylinositol is predominantly affected. Again, it is not known whether the effect is primarily at the cell surface or whether the net content of phosphatidylinositol changes. Degradation is largely via the phosphatidylinositol-specific phospholipase C that results in the formation of cyclic phosphorylinositol (Michell, 1975). This enzyme, which appears to be associated with microtubular subunits (Quinn, 1973), presumably acts from within the cell and may be activated in response to a more immediate stimulus at the cell surface.

Concomitant with this stimulation of turnover at the polar moiety of phosphatidyl inositol, an increased turnover of acyl chains by Lands cycle enzymes (transacylases) takes place (Ferber and Resch, 1973). Acyl chain turnover is probably several-fold faster than phospholipase-C-mediated turnover (Resch and Ferber, 1972). In this instance, there is good evidence that much of the activity is at the plasma membrane (Ferber *et al.*, 1972). What is particularly interesting is the fact that the nature of the acyl chains becomes modified; by substituting highly unsaturated fatty acids for less unsaturated ones, the actual fluidity of the plasma membrane is increased (Ferber *et al.*, 1974; note also Barnett *et al.*, 1974b). The situation, which occurs also during antigenic stimulation of lymphocytes, is therefore a rather clearcut example of a developmental change involving membrane lipids; similar modification of phospholipid acyl side chains by the activity (Hill and Lands, 1970) may occur during formation of pulmonary surfactant (dipalmitoylphosphatidylcholine) and in the aging of red cells. The exact relationship between the change in lymphocytes and subsequent developmental events is not yet clear. On the one hand calcium entry, with which antigenic stimulation may be associated, is alone able to stimulate lymphocytes (Maino *et al.*, 1974). On the other hand much of the turnover occurs after calcium entry has taken place (Freedman *et al.*, 1975). Moreover, constituents *other* than membrane lipids, such as microtubules and microfilaments (Berlin *et al.*, 1974; Yahara and Edelman, 1975), appear to contribute to the increased fluidity of stimulated lymphocytes.

IV. Embryonic Development

The most striking example of development is undoubtedly that of the embryo. From a single germ cell are produced some thousands of cells, without an overall increase in size or mass, arranged in definitive clusters or primitive organs each of which already displays some of the biochemical functions characteristic of the adult animal. The process occurs in several stages (cleavage, formation of blastula, of gastrula, and so on), the earliest of which are remarkably similar in most animal species. Since methods for the study of mammalian embryos are only now being developed (e.g., Herbert and Graham, 1974), investigators have turned in the past to more accessible systems; Echinoderms (especially sea urchins) and Amphibia (mainly *Bufo arenarum* and *Xenopus laevis*) have proved to be the most fruitful, and will accordingly be described in this review.

A. Sea Urchins

Cleavage, the successive division of a single cell, has many of the characteristics of growth defined in section I (see also section III). Little overt differentiation occurs and the process is essentially one of nuclear multiplication and plasma membrane assembly; changes in the amount of endoplasmic reticulum or mitochondria are minimal.

On the other hand cells decrease in size at every cleavage and are obviously not identical replicas of their parent. Moreover, the initial plasma membrane is covered by distinct layers which become modified on fertilization and eventually disappear altogether. Since the total surface area of a sea urchin embryo such as *Strongylocentrotus purpuratus* increases some 6- to 7-fold between fertilization and the formation of a blastula (containing 400–500 cells), plasma membrane is the predominant species of newly synthesized membranes (Pasternak, 1973b).

Unlike the situation in cultured mammalian cells (section II) membrane formation is not necessarily the result of *de novo* phospholipid synthesis, since the yolk of unfertilized eggs is rich in most lipid species. The question that therefore arises is: Are the phospholipids of yolk utilized intact, or are they degraded and resynthesized, perhaps into different molecules, before insertion? Attempts to detect changes in composition must await successful isolation of purified plasma and other membranes [a possible approach is indicated below (Table 3)]. For while the overall lipid composition does not change significantly during early development (Fagerlund, 1969; Pasternak, 1973b), the amount of phospholipids in yolk, mitochondria, and endoplasmic reticulum would mask any changes occurring in the plasma membrane alone. Recourse has therefore been made to the use of isotopically labeled phospholipid precursors. Not unexpectedly, perhaps, this has yielded conflicting results.

Following the initial observation that incorporation of labeled acetate and glycerol into the lipids of *Paracentrotus lividus* is stimulated by fertilization (Mohri, 1964), it became clear that increased incorporation into phospholipids is partly the result of an elevated uptake of labeled precursors (Pasternak, 1973b; Schmell and Lennarz, 1974). Uptake of amino acids and other compounds is likewise stimulated by fertilization (Epel *et al.*, 1969; Pasternak, 1973b). Even allowing for an increase in specific activity of the precursor pool, a small but significant increase in incorporation (over a low rate in the unfertilized egg) into phosphatidylcholine (Pasternak, 1973b) and phosphatidylinositol (Schmell and Lennarz, 1974) yet occurs. Moreover, inhibitors of DNA synthesis like hydroxyurea, that arrest nuclear multiplication, inhibit phospholipid synthesis (Pasternak, 1973b). Other investigators, however, have not observed any difference between fertilized and unfertilized eggs, and instead find considerable incorporation into phosphatidylcholine (Byrd, 1975) or phosphatidylethanolamine (Schmell and Lennarz, 1974) by each.

Attempts to detect changes in the activity of enzymes involved in phospholipid synthesis have not revealed any stimulation following fertilization. Choline kinase (EC 2.7.1.32) is apparently unchanged (Pasternak, 1973b; Byrd, 1975) and choline phosphotransferase (EC 2.7.8.2) actually decreases (Ewing, 1973). Definitive evidence of a link between nuclear activity and phospholipid synthesis would be achieved if nucleated and enucleated egg fragments (Tyler and Tyler, 1966) were to respond differently to activation. Technical difficulties have so far dogged this approach (Pasternak, unpublished results).

It may be concluded that some phospholipid synthesis does take place during early cleavage. Whether this is a stimulation over the basal level occurring in unfertilized eggs is not clear. In order to assess the amount of newly synthesized phospholipid that becomes inserted specifically into the plasma membrane, autoradiographic visualization or biochemical isolation of plasma membrane from eggs exposed to a pulse of labeled precursor is necessary. The first approach has not proved successful so far (Pasternak, unpublished), and the second has yielded inconclusive results as follows.

The low rate of incorporation into phospholipids means that insufficient label enters any membrane to allow detection following a short pulse of isotope. When embryos are incubated for a longer time following a pulse of, for example, [^{3}H]choline, isotope appears in all membrane fractions. While a fraction enriched in Na/K-ATPase certainly contains much labeled phospholipid, the specific activity is no higher than that in any other fraction (Table 3). That this result is not an artifact of the fractionation procedure is shown by the fact that the specific activity of protein (labeled by exposure to [^{14}C]valine), does vary by more than 10-fold among the different fractions (Table 3). Equilibration of newly synthesized phospholipid among all cellular membranes might result from the action of phospholipid exchange proteins, previously demonstrated in other animal species (Dawson, 1973; Wirtz, 1974). Sea urchin embryos indeed contain phospholipid exchange protein (Table 4; note that unlike viable cells, mammalian erythrocytes are devoid of activity). Therefore, it is not possible to draw any definitive conclusions regarding the insertion of newly synthesized phospholipids in membranes, especially as exchange proteins do not appear to catalyze a net transfer of phospholipids. Perhaps the most useful conclusion to be drawn is that isotopically labeled phospholipid precursors are as unlikely to yield information regarding sites of synthesis and insertion in sea urchin embryos as they are in other animal cells (Quinn and Pasternak, 1976).

B. *Frogs*

With amphibian embryos, results are not much more conclusive. Since amphibian eggs are rather impermeable to added precursors, even after fertilization, tracers are generally injected into a fertile female together with pituitary extract or gonadotrophin to induce ovulation. Oocytes removed from gonadal tissue, or eggs excreted spontaneously, contain labeled phospholipids (Pechen *et al.,* 1974; Table 5). What is not clear is the extent to which incorporation is mediated via the liver, or whether the incorporation is due to net synthesis during oocyte maturation or to turnover.

TABLE 4

Phospholipid Exchange Proteins in Cell Extracts[a]

Sea urchin embryos		P815Y Cells		Lettree cells		Rat liver		Human erythrocytes	
Amount added (mg protein)	Exchange activity[b]	Amount added (mg protein)	Exchange activity[b]	Amount added (mg protein)	Exchange activity[b]	Amount added (mg protein)	Exchange activity[b]	Amount added (mg protein)	Exchange activity[b]
1.1	0.01	1.5	0.12	1.4	0.05	1.7	0.04	1.6	<0.01
2.7	0.06	3.0	0.14	3.4	0.11	3.4	0.20	3.9	0.08
6.4	0.22	7.4	0.32	6.8	0.24	6.8	0.38	9.4	<0.01

[a] *S. purpuratus* embryos (hatched blastulae) were homogenized (4 vols. of 0.25 M sucrose/1 mM EDTA//10 mM Tris pH 7.4), spun (105,000 × *g* for 1 hr) and the supernatant fraction incubated (30 min at 25°C) with ^{32}P-labeled rat liver mitochondria (3.8 mg; 5.3 × 10^6 dpm/μmole phospholipid phosphorus) and rat liver microsomes (1.1 mg) according to Wirtz *et al.* (1972). The mixture was spun (15,000 × *g* for 10 min) and radioactivity and phospholipid phosphorus (Bartlett, 1959) in the supernatant (containing labeled microsomes) measured. The exchange activity of mouse P815Y tumor cells grown *in vitro*, of Lettree cells grown intraperitoneally in mice, of rat liver and of human erythrocytes, assayed in an identical manner, is shown for comparison. Unpublished results of R. M. Abra, reproduced with permission.

[b] Final specific activity of microsomes (plus supernatant) in dpm × 10^{-6}/μmole phospholipid phosphorus; the value in the absence of added cell extract has been subtracted.

TABLE 5
[^{32}P]*Phospholipids in Frog Embryos*[a]

Developmental stage	% of total ^{32}P		
	Phospholipids	Aqueous ext.	Pellet
Unfertilized[b]	7	31	61
1st cleavage[c]	3	20	77
8-cell stage	3	20	76
128-cell stage	3	18	78
Early neurula	5	30	65
Neurula	9	27	63

[a] Female *Xenopus laevis* adults were injected intraperitoneally with human chorionic gonadotrophin and 8 hr later with a similar amount together with 2 mCi of [^{32}P]inorganic phosphate (Gurdon, 1974). Eggs, which began to be laid approx 8 hr later, were collected in 0.11 M saline, fertilized with a testicular macerate, and allowed to develop at room temperature; the embryos were washed, stripped of jelly coat (Gurdon, 1974), and homogenized in 20 volumes of chloroform:methanol (2:1). The homogenate was spun to yield an insoluble pellet (largely nucleic acids and protein); the supernatant was extracted with 0.2 volume of water and the upper (aqueous) and lower (lipid) layers separated, washed, and analyzed. Unpublished experiments in collaboration with C. F. Graham and K. J. Micklem.
[b] Mean of 3 separate determinations.
[c] Mean of 2 separate determinations.

When labeled eggs are fertilized, little additional incorporation takes place prior to formation of blastula (Pechen *et al.*, 1974; Table 5), though it increases thereafter (Pechen and Bazan, 1974). The incorporation of water-soluble ^{32}P compounds taking place initially, is predominantly into DNA, not phospholipids (Table 5).

Incorporation of labeled precursors by egg homogenates is likewise negligible, at least prior to midgastrula (Pechen and Bazan, 1976). Neither the total amount, nor the composition of phospholipid classes, changes between fertilization and the stage of gill circulation (Barassi and Bazan, 1974). It is therefore likely that phospholipids synthesized before fertilization are inserted intact into growing membranes. A similar conclusion has been reached with regard to chick embryos (Siek and Newburgh, 1965). Moreover, a redistribution of labeled phospholipids among subcellular organelles of frog embryos has been demonstrated (Pechen and Bazan, 1974). However, the participation of phospholipid turnover cannot be ruled out, since it is difficult to "chase" labeled precursors out of the impermeable eggs.

In summary, the distinction between insertion into new plasma membranes of yolk phospholipids intact, and insertion of phospholipids derived from yolk by degradation and resynthesis, is not easy to make, either for sea urchins or for frogs. This is due partly to the existence of phospholipid exchange proteins and partly to turnover occurring even before fertilization. This situation presumably operates also in mammalian embryos, so that use of isotopic precursors of phospholipids is unlikely to reveal useful data regarding the assembly of lipids in new plasma membranes. It would seem more profitable to examine other, albeit minor, components, such as glycolipids, that are located specifically at the cell surface.

V. Other Developmental Situations

A. Muscle: Membrane Fusion

The fusion of membranes, whether between cells or between subcellular organelles, is often part of an event in development or growth. Expulsion of the nucleus by fusion with the plasma membrane during the maturation of mammalian erythrocytes, the fusion of a row of myoblast cells into a myotubule during the development of muscle fibers, the membrane fusion occurring during hormonally triggered endocytosis or exocytosis, are all examples associated with specific developmental events (Poste and Allison, 1973). The resealing of the surface of two daughter cells at the end of cytokinesis exemplifies an event of growth. Since the demonstration that lysophospholipids are particularly potent inducers of intercellular fusion (Poole *et al.*, 1970; Ahkong *et al.*, 1972), attention has focused on the possibility that lysophospholipids, generated by operation of transacylases (see section I.C), might be involved in membrane fusion.

A suitable system for study is intercellular fusion caused by Sendai and similar viruses (Harris, 1970; Poste, 1972), despite the fact that the event is one of abnormal (section VI) rather than of normal development. All attempts to detect changes in the turnover of phospholipids, by transacylases or by phospholipase C, or in the level of lysophospholipids, have so far failed (Falke *et al.*, 1967; Elsbach *et al.*, 1969; Pasternak and Micklem, 1974). Of course, the amount of lysophospholipids necessary to trigger membrane fusion may be rather small, and a transient increase at a localized site might escape detection. On the other hand, lysophosphatidylcholine added to developing myoblasts actually inhibits fusion (Reporter and Norris, 1973), and the presence of large amounts of lysophosphatidylcholine in chromaffin granules (Blaschko *et al.*, 1967) does not in itself lead to fusion. An intriguing alternative is that a concerted action of phospholipid exchange proteins on two apposing membranes brings about fusion, and that this is the major physiological role of such proteins.

In the case of myotubule formation, the possibility that the myoblast surface becomes gradually more prone to fusion as a result of an altered fluidity has been examined. Measurement of the lipid composition of plasma membrane isolated from cultured chick myoblasts before, during, and after fusion, has revealed no significant change in the degree of unsaturation or of chain length of the constituent phospholipids, nor in the cholesterol:phospholipid ratio (Kent *et al.*, 1974). Although there may be differences between the lipid composition of cultured chick cells and that of chick or mammalian muscles developing *in vivo*, compositional changes do not appear to underlie the step of membrane fusion.

B. Testis

There are reasons for supposing that the metabolism of essential fatty acids changes with age. Two tissues that are likely to reveal resultant differences in the phospholipid composition are liver and testis. When these were examined in rats aged 2–20 months, no significant difference in content of polyunsaturated fatty acids was detected (Burdett *et al.*, 1971). When the rats were subjected to a fat-free diet, on the other hand, considerable differences became apparent. Such experiments, therefore, amplify the point made in section I.A that changes in membrane lipids during growth and development are often due more to dietary and other environmental factors than to endogenously programmed alterations.

In contrast to phospholipids, testis of several mammalian species contains a glycolipid, the appearance of which is correlated with a precise stage in spermatogenesis. It is sulfoglyceroglactolipid, and the activity of galactolipid sulfotransferase increases dramatically before its accumulation (Kornblatt *et al.*, 1974).

C. Brain

With brain, the lipid composition certainly does change with age (Rouser *et al.*, 1972). Partly this is due to the development of specific cell types, such as neurons, glial cells, and capillaries, each of which has a specific function. Since the distribution of subcellular organelles varies among these cells, it is perhaps not surprising that the lipid content varies also. The difficulty of obtaining pure preparations of specific cells has hampered studies to date. As new techniques for cell separation and propagation *in vitro* are developed, however, it is likely that significant correlations between structure and function will emerge. For of all membranes, those of the nervous tissue are probably the most dependent on the nature of their constituent lipids.

VI. Abnormal Development

A. Sphingolipidoses

The best documented examples of the effects of an alteration in membrane lipids has proved, somewhat paradoxically, to come from an examination of certain pathological states. The sphingolipidoses have been known for some time to be genetic diseases of greater or lesser severity, affecting different organs such as liver, spleen, or brain; the common defect is an abnormal accumulation of sphingolipids. It is this accumulation, generally in the form of intracellular lipid droplets, that leads to toxic symptoms. The reason for the accumulation has proved to be not an increased synthesis, but a decreased degradation, of sphingolipids [except in one instance where a synthetic enzyme is at fault (Fishman *et al.*, 1975)]. In other words, sphingolipids normally turn over at significant rates and failure to maintain this turnover has deleterious consequences.

The enzymes that are at fault have been recognized and found to be absent in the diseased state. Examples are sphingomyelinase (a phospholipase-C type of enzyme) in Niemann–Pick disease, leading to an accumulation of sphingomyelin; hexosaminidase A in Tay–Sachs disease, leading to an accumulation of *N*-acetylgalactosaminyl-(sialosyl) lactosyl ceramide (G_{M2}); a glucosidase in Gaucher's disease, leading to an accumulation of glucocerebroside; and so forth. The relevant literature has been extensively reviewed (Neufeld *et al.*, 1975; Volk and Schneck 1975), and here attention is drawn only to the following point: The absence of a particular enzyme has been shown to be a genetic defect, but accumulation of lipid is much more severe in some tissues than in others. In Tay–Sachs disease, for example, G_{M2} accumulates predominantly in nervous tissue, whereas in Gaucher's disease glucocerebroside accumulates predominantly in spleen. One must therefore conclude *either* that several genetically distinct tissue-specific isoenzymes exist (as with phosphorylase of muscle and liver) *or* that the rate of turnover of different sphingolipids varies greatly from tissue to tissue. The latter is the more likely. Absence of hexosaminidase A in Tay–Sachs disease, for example, can be demonstrated in serum and in tissues other than those that show significant accumulation of G_{M2} (O'Brien, 1973). In other words, it is the combination of a high rate of synthesis in *specific* tissues, coupled with a *generalized* loss of degradative enzyme, that leads to sphingolipidoses. This explains why lipids such as sphingomyelin, that in the main turnover relatively slowly, yet accumulate in certain situations.

B. Cancer

As in the sphingolipidoses, the basic defect in cancer cells is genetic. What is not known is whether the fault lies in the loss (or acquisition) of particular genes, or merely in a heritable change in their expression. More important, it is not at all clear what the nature of the phenotypic outcome of the altered genes is, and how this is related to malignant behavior. What does seem to have emerged is the realization that a major defect of cancer cells is expressed at the cell surface. Unlike some hormonally triggered developmental changes in which specific surface proteins have been recognized to play a part, no protein has yet been identified as being specifically altered in cancer cells. The role of membrane lipids is even less clear (e.g., Bergelson, 1972; Wood, 1973). Two changes in particular have been suggested to be relevant. The first concerns certain glycolipids and the second the fluidity of the surface membrane.

The systems that have revealed these changes have been the use of virally transformed cells in culture. The realization that certain viruses are capable of causing a stable, heritable transformation of cells such that the growth properties change to those typical of cancer cells has led to their widespread use. In particular, transformed cells reach a higher saturation density in monolayer than their untransformed counterparts, owing to some defect in density-dependent inhibition of growth (Dulbecco, 1969). Related—it is generally presumed causally—is the enhanced ability of transformed cells to give rise to tumors when injected into animals.

When the glycolipids of transformed cells are examined, it is found that there is a loss of gangliosides (Brady and Fishman, 1974) and neutral fucolipids (Hakomori, 1975b). At the same time shorter-chain precursors accumulate, owing to the absence of specific glycosyl transferases (Cumar *et al.,* 1970; Kijimoto and Hakomori, 1971); in other words the opposite effect to that in the sphingolipidoses. Comparison of tumorigenicity (that is, the ability to induce tumors when injected into animals), of several transformed and untransformed cell lines with their glycolipid composition, however, has failed to demonstrate a correlation (Sakiyama and Robbins, 1973). While the glycolipid composition of cells does appear to be related to the degree of confluency (section III.A), the relation is clearly not a fundamental part of the difference between normal and malignant cells.

Differences in the fluidity of the surface membrane of normal and transformed cells have been proposed (Barnett *et al.,* 1974a,b) to account for the following observations. Transformed cells require a much lower concentration of lectins, such as wheat germ agglutinin or concanavalin A, to cause agglutination than do normal cells (Burger, 1971). The difference is not due to an alteration in binding capacity, but rather to a temperature-dependent ability of the receptor sites to "cluster" in the plane of the plasma membrane (Nicolson, 1972). Since clustering of intramembranous particles appears to be related to the lipid fluidity of membranes, a connection between fluidity and cancer seems plausible (Barnett *et al.,* 1974a). Moreover, leukemic lymphocytes have less cholesterol (presumed to be located predominantly in the plasma membrane) relative to phospholipids than do normal lymphocytes (Inbar and Shinitzky, 1974a; Vlodavsky and Sachs, 1974), and their lipid fluidity, as indicated by measurement of their microviscosity with a fluorescent probe, is greater (Shinitzky and Inbar, 1974).

However, (i) the lipid fluidity of transformed cells, determined with a spin-labeled probe, is not different from that of the untransformed parent (Gaffney *et al.*, 1974; Gaffney, 1975); (ii) the lipid composition (degree of unsaturation, average chain length or cholesterol content) of normal and transformed cell plasma membrane is the same (Micklem *et al.*, 1976); and (iii) clustering of lectin receptors is not necessarily related to clustering of intramembranous particles anyway (Karnovsky and Unanue, 1973; Pinto da Silva and Martinez-Palomo, 1975; Micklem *et al.*, 1976). What, then, determines agglutinability by lectins? Although the answer is not at present clear, it is likely to lie more with the correct positioning of intracellular elements such as microfilaments and microtubules (Berlin *et al.*, 1974; Yahara and Edelman, 1975) than with the lipid composition of the cell surface. Certainly the concept of membrane and lipid fluidity requires better definition. In any case, while lectin-mediated agglutination clearly is altered in transformed cells, the change does not appear to be an integral property of the malignant state (Kao and Harris, 1975).

With regard to the cholesterol:phospholipid ratio, the situation is unclear. The plasma membrane of some rat and mouse hepatomas has more cholesterol than that of normal liver (van Hoeven and Emmelot, 1972). In rats the increase is accompanied by a decrease in polyunsaturation of phospholipids, whereas in mice it is not (van Hoeven *et al.*, 1975). In transformed fibroblasts, the cholesterol:phospholipid ratio of plasma membranes is the same as in the untransformed counterpart (Micklem *et al.*, 1976). The decreased level of cholesterol in leukemic lymphocytes, despite its apparent association with malignancy (Inbar and Shinitzky, 1974a,b), is therefore not a general phenomenon of cancer cells. One can only conclude that cell-specific differences are a more important determinant of cholesterol content than the state of malignancy.

VII. Conclusions

It is apparent that major compositional changes in membrane lipids are not a prerequisite of growth or development. This is in marked contrast to membrane proteins, which do change, not only during development (of which an altered protein composition is the very basis), but also during the cell cycle (Fox and Pardee, 1971; Mitchison, 1971). Of course the same degree of change is hardly to be expected, since the heterogeneity of proteins is considerably greater than that of phospholipids, even allowing for extensive variation of acyl chain composition. Moreover, the specificity of protein function allows detection of very minor amounts. In the case of glycolipids, the total quantity of any one type that is present in a membrane is also small, and some clearcut changes are indeed observed (e.g., sections

III.A and V.B). Nevertheless, the constancy of phospholipid composition during growth and development is somewhat surprising and one may seek a reason.

Perhaps the mistake has been to regard phospholipids as cellular components akin to macromolecules such as proteins, nucleic acids, or polysaccharides. It is true that, like these macromolecules, phospholipids are the end product of a biosynthetic sequence. But phospholipids are not macromolecules, either in mass (molecular weight less than 1,000) or in size (volume less than 4 nm^3). They should be compared rather with the *constituents* of macromolecules, like amino acids, nucleotides or sugars. The equivalent "macromolecule" in the case of phospholipids is a multimolecular vesicle or portion of a membrane. It so happens that in this instance the constituents are joined not by covalent bonds, but by hydrophobic and other interactions. Single phospholipid molecules do not exist as discrete entities in membranes.

Seen in this context, it becomes easier to understand why the phospholipid composition of yolk in an unfertilized egg, for example, is similar to that of a developing embryo. Just as the amino acid content, but not the protein composition, of yolk and embryo is similar, so the composition of phospholipids is similar also. It is the assembly of a relatively small number of structurally distinct phospholipid "building-blocks" into multimolecular structures, that changes as yolk is used to make embryonic membranes.

Another aspect that may become clarified by this concept is the continual turnover of phospholipids, which is apparent even in quiescent cells prior to fertilization. For the turnover should be compared not to the degradation and resynthesis of protein (which is minimal in such cells), but to the turnover of metabolically labile amino acids like glutamate or serine. Just as other amino acids, like valine or leucine, do not turn over so extensively, nor do some phospholipids like sphingomyelin. To push the analogy a little further, the acyl exchange cycle referred to as partial turnover (section I.C), which is the major component of the rapid turnover of phospholipids, is a kind of "metabolic flux." This links phospholipids to other lipids and to the basal metabolism of the cell, in the way that the "flux" through glutamate $\rightleftharpoons$ α-ketoglutarate, or through glucose 1-P $\rightleftharpoons$ glucose 6-P $\leftrightharpoons$ fructose 6-P, links amino acids, sugars, and other basic metabolic pathways. The latter reactions occur in any cell that is viable, whether quiescent or not, for maintenance of viability depends on energy supply, and it is through these reactions that the supply is achieved.

One difference between phospholipids and metabolic intermediates like amino acids or sugar phosphates lies in their location. Intracellular phospholipids are entirely in membranes; water-soluble intermediates are in the cytoplasm. As a result, the turnover of phospholipids causes membranes themselves to acquire a "metabolic fluidity" they would not otherwise possess (Pasternak, 1973c).

Like metabolic intermediates, phospholipids assembled into multimolecular structures have other functions. The correct positioning of a dioleoyl type of phospatidylcholine molecule rather than one of dipalmitoyl type, in the calcium ATPase complex (Warren *et al.*, 1974a), is as crucial for biological function as is the presence of a serine residue rather than an aspartate residue at position 189 in chymotrypsin (Porter, 1971) or of a *N*-acetyl galactosamine residue rather than a galactose residue in the terminal position of blood group substance A of red cells (Watkins, 1974). It is such functions of phospholipids that are involved in growth and development. The apparent disappearance and reassembly of the nuclear membrane during mitosis, the insertion of lipids into a growing cell in the form of microvilli and their subsequent unfolding during cytokinesis, the dispersion of yolk droplets and insertion of the constituent lipids into growing membranes during cleavage, the displacement of phospholipids from the cell surface during serum stimulation, the rearrangement of surface lipids during membrane fusion or during differentiation as new hormone receptors are synthesized and inserted, the reshaping of the cell surface as cells become malignant and invasive—these are all examples of an alteration in the *assembly* of membrane lipids. In short, it is *spatial* changes in membrane lipids, not *chemical* ones, that are the more important in relation to growth and development. The physical probes of membrane structure that are currently being developed are ideally suited for investigating such architectural changes, and a future review of this topic is likely to be rich in new results of this nature.

Acknowledgments

The author is grateful to colleagues for unpublished material and to the Cancer Research Campaign for financial support.

References

Ahkong, Q. F., Cramp, F. C., Fisher, D., Howell, J. I., and Lucy, J. A. 1972. Studies on chemically induced cell fusion. *J. Cell Sci.* **10**:769–787.

Ambrose, E. J. 1975. The surface properties of tumour cells, pp. 27–57. *In* E. J. Ambrose and F. J. C. Roe (eds). Biology of Cancer, 2nd ed. Ellis Horwood, Chichester.

Avruch, J., and Wallach, D. F. H. 1971. Preparation and properties of plasma membrane and endoplasmic reticulum fragments from isolated rat fat cells. *Biochim. Biophys. Acta* **233**:334–347.

Barassi, C. A., and Bazan, N. G. 1974. Metabolic heterogeneity of phosphoglyceride classes and subfractions during cell cleavage and early embryogenesis: Model for cell membrane biogenesis. *J. Cell. Physiol.* **84**:101–114.

Barnett, R. E., Furcht, L. T., and Scott, R. E. 1974a. Differences in membrane fluidity and structure in contact-inhibited and transformed cells. *Proc. Natl. Acad. Sci. USA* **71**:1992–1994.

Barnett, R. E., Scott, R. E., Furcht, L. T., and Kersey, J. H. 1974b. Evidence that mitogenic lectins induce changes in lymphocyte membrane fluidity. *Nature* **249**:465–466.

Bartlett, G. R. 1959. Phosphorus assay in column chromatography. *J. Biol. Chem.* **234**:466–468.

Bergelson, L. D. 1972. Tumor lipids. *Progr. Chem. Fats Lipids* **13**:1–59.

Bergelson, L. D., Dyatlovitskaya, E. V., Sorokina, I. B., and Gorkova, H. P. 1974. Phospholipid composition of mitchondria and microsomes from regenerating rat liver and hepatomas of different growth rate. *Biochim. Biophys. Acta* **360**:361–365.

Berlin, R. D., Oliver, J. M., Ukena, T. E., and Yin, H. H. 1974. Control of cell surface topography. *Nature* **247**:45–46.

Blaschko, H., Firemark, H., Smith, A. D., and Winkler, H. 1967. Lipids of the adrenal medulla. Lysolecithin, a characteristic constituent of chromaffin granules. *Biochem. J.* **104**:545–549.

Blecher, M. 1968. Action of insulin on a glucose transport mechanism in the plasma membrane of the isolated adipose cell. Participation of membrane phospholipids and cyclic adenosine monophosphate in the transport process, pp. 145–161. *In* Gunma Symposia on Endocrinology, Vol. 5. Gunma University, Maebashi, Japan.

Blecher, M. 1969. Insulin-like antilipolytic actions of phospholipase A in isolated rat adipose cells. *Biochim. Biophys. Acta* **187**:380–384.

Bosmann, H. B. 1971. Mitochondrial biochemical events in a synchronized mammalian cell population. *J. Biol. Chem.* **246**:3817–3823.

Bosmann, H. B., and Winston, R. A. 1970. Synthesis of glycoprotein, glycolipid, protein and lipid in synchronized L5178Y cells. *J. Cell Biol.* **45**:23–33.

Brady, R. O., and Fishman, P. M., 1974. Biosynthesis of glycolipids in virus-transformed cells. *Biochim. Biophys. Acta* **355**:121–148.

Brent, T. P., and Forrester, J. A. 1967. Changes on surface charge of Hela cells during the cell cycle. *Nature* **215**:92–93.

Brocklehurst, J. R., Freedman, R. B., Hancock, D. J., and Radda, G. K. 1970. Membrane studies with polarity-dependent and excimer-forming fluorescent probes. *Biochem. J.* **116**:721–731.

Burdett, P., Davies, T., Griffin, H., and Hall, D. A. 1971. The effect of age on metabolism of poly-unsaturated fatty acids, pp. 69–76. Symp. on Nutrition in Old Age, Vol. X. Swedish Nutrition Foundation.

Burger, M. M. 1971. Cell surfaces in neoplastic transformation. *Curr. Top. Cell Regul.* **3**:135–193.

Burns, F. J., and Tannock, I. F. 1970. On the existence of a Go phase in the cell cycle. *Cell Tissue Kinet.* **3**:321–334.

Byrd, E. W. 1975. Phospholipid metabolism following fertilization in sea urchin eggs and embryos. *Dev. Biol.* **146**:309–316.

Coleman, R. 1973. Membrane-bound enzymes and membrane ultrastructure. *Biochim. Biophys. Acta* **300**:1–30.

Crawford, N. 1958. An improved method of determination of free and total cholesterol using the ferric chloride reaction. *Clin. Chim. Acta* **3**:357–367.

Cumar, F. A., Brady, R. O., Kolodny, E. H., McFarland, V. W., and Mora, P. T. 1970. Enzymatic block in the synthesis of gangliosides in DNA virus-transformed tumorigenic mouse 11 lines. *Proc. Natl. Acad. Sci. USA* **67**:757–764.

Cunningham, D. D. 1972. Changes in phospholipid turnover following growth of 3T3 mouse cells to confluency. *J. Biol. Chem.* **247**:2464–2470.

Dawson, R. M. C. 1966. The metabolism of animal phospholipids and their turnover in cell membranes, pp. 69–115. *In* P. N. Campbell and G. D. Greville (eds.). Essays in Biochemistry, Vol. 2. Academic Press, London.

Dawson, R. M. C. 1973. The exchange of phospholipids between cell membranes. *Sub-Cell. Biochem.* **2**:69–89.

Diringer, H., and Koch, M. A. 1973. Differences in the metabolism of phospholipids depending on cell population density. *Biochem. Biophys. Res. Commun.* **51**:967–971.

Dulbecco, R. 1969. Cell transformation by viruses. *Science* **166**:962–968.

Elsbach, P., Holmes, K. V., and Choppin, P. W. 1969. Metabolism of lecithin and virus-induced cell fusion. *Proc. Soc. Exp. Biol. Med.* **130**:903–908.

Epel, D., Pressman, B. C., Elsaesser, S., and Weaver, A. M. 1969. The program of structural and metabolic changes following fertilization of sea urchin eggs, pp. 279–298. *In* G. M. Padilla, G. L. Whitson, and I. L. Cameron (eds). The Cell Cycle: Gene–Enzyme Interactions. Academic Press, New York.

Ewing, R. D., 1973. Cholinephosphotransferase activity during early development of the sea urchin, *Arbacia punctulata*. *Dev. Biol.* **31**:234–241.

Fagerlund, U. H. M. 1969. Lipid metabolism, pp. 123–134. *In* M. Florkin and B. T. Scheer (eds.) Chemical Zoology, Vol 3. Academic Press, New York.

Falke, D., Schiefer, H. G., and Stoffel, W. 1967. Lipoid-Analysen bei Riesenzellbildung durch Herpes virus Hominis. *Z. Naturforsch* **22b**:1360–1362.

Ferber, E., and Resch, K. 1973. Phospholipid metabolism of stimulated lymphocytes: Activation of acyl–CoA: lysolecithin acyltransferases in microsomal membranes. *Biochim. Biophys. Acta* **296**:335–349.

Ferber, E., Resch, K., Wallach, D. F. H., and Imm, W. 1972. Isolation and characterization of lymphocyte plasma membranes. *Biochim. Biophys. Acta* **266**:494–504.

Ferber, E., Reilly, C. E., de Pasquale, G., and Resch, K. 1974. Lymphocyte stimulation by mitogens: Increase in membrane fluidity caused by changes of fatty acid moities of phospholipids, pp. 529–534. *In* Proceedings of the 8th Leucocyte Culture Conference, Academic Press, New York.

Fisher, D. B., and Mueller, G. C. 1968. An early alteration in the phospholipid metabolism of lymphocytes by phytohemagglutinin. *Proc. Natl. Acad. Sci. USA* **60**:1396–1402.

Fishman, P. H., Max. S. R., Tallman, J. F., Brady, R. O., Maclaren, N. K., and Cornblath, M. 1975. Deficient ganglioside biosynthesis: A novel human sphingolipidosis. *Science* **187**:68–70.

Fourcans, B., and Jain, M. K. 1974. Role of phospholipids in transport and enzymic reactions. *Adv. Lipid Res.* **12**:148–226.

Fox, C. F. 1972. Membrane assembly, pp. 342–385. *In* C. F. Fox and A. D. Keith (eds.). Membrane Molecular Biology. Sinauer Assoc. Inc., Stanford, Connecticut.

Fox, T. O., and Pardee, A. B. 1971. Proteins made in the mammalian cell cycle. *J. Biol. Chem.* **246**:6159–6165.

Freedman, M. H., Raff, M. C., and Gomperts, B. 1975. Induction of increased calcium uptake in mouse T lymphocytes by concanavalin A and its modulation by cyclic nucleotides. *Nature* **255**:378–382.

Gaffney, B. J. 1975. Fatty acid chain flexibility in the membrane of normal and transformed fibroblasts. *Proc. Natl. Acad. Sci. USA* **72**:664–668.

Gaffney, B. J., Branton, P. E., Wickus, G. G., and Hirschberg, C. B. 1974. Fluid lipid regions in normal and Rous sarcoma virus transformed chick embryo fibroblasts, pp. 97–114. *In* A. S. Kaplan (ed.). Viral Transformation and Endogenous Viruses. Academic Press, New York.

Getz., G. S. 1970. Lipids in membrane development. *Adv. Lipid Res.* **8**:175–223.

Getz, G. S. 1972. Organelle Biogenesis, pp. 386–438. *In* C. F. Fox and A. D. Keith (eds.). Membrane Molecular Biology. Sinauer Assoc. Inc., Stamford, Connecticut.

Graham, J. M., Sumner, M. C. B., Curtis, D. H., and Pasternak, C. A. 1973. Sequence of events in plasma membrane assembly during the cell cycle. *Nature* **246**:291–295.
Gurdon, J. B. 1974. The content of gene expression in animal development. (Appendix C3). Clarendon Press, Oxford. 160 pp.
Hakomori, S. 1975a. Structures and organisation of cell surface glycolipids dependency on cell growth and malignant transformation. *Biochim. Biophys. Acta* **417**:55–89.
Hakomori, S. 1975b. Fucolipids and blood group glycolipids in normal and tumor tissue. *Progr. Biochem. Pharmacol.* **10**:167–196.
Harris, H. 1970. Cell Fusion. Clarendon Press, Oxford. 108 pp.
Herbert, M. C., and Graham, C. F. 1974. Cell determination and biochemical differentiation of the early mammalian embryo. *Curr. Top. Dev. Biol.* **8**:151–178.
Hill, E. E., and Lands, W. E. M., 1970. Phospholipid metabolism, pp. 185–277. *In* S. J. Wakil (ed.). Lipid Metabolism, Academic Press, New York.
Hirano, H., Parkhouse, B., Nicolson, G. L., Lennox, E. S., and Singer, S. J. 1972. Distribution of saccharide residues on membrane fragments from a myeloma-cell homogenate: Its implications for membrane biogenesis. *Proc. Natl. Acad. Sci. USA* **69**:2945–2949.
Hollenberg, M. D., Fishman, P. H., Bennett, V., and Cuatrecasas, P. 1974. Cholera toxin and cell growth: Role of membrane gangliosides. *Proc. Natl. Acad. Sci. USA* **71**:4224–4228.
Hulser, D. F., and Frank, W. 1971. Stimulation of embryonic rat cells in culture by a protein isolated from fetal calf serum. I. Electrophysiological measurements at the cell surface membranes. *Z. Naturforsch.* **26b**:1045–1048.
Inbar, M., and Shinitzky, M. 1974a. Cholesterol as a bioregulator in the development and inhibition of leukemia. *Proc. Natl. Acad. Sci. USA* **71**:4229–4231.
Inbar, M., and Shinitzky, M. 1974b. Increase of cholesterol level in the surface membrane of lymphoma cells and its inhibitory effect on ascites tumor development. *Proc. Natl. Acad. Sci. USA* **71**:2128–2130.
Kao, F. T., and Harris, H. 1975. Lack of correlation between malignancy and sensitivity to killing by concanavalin A. *J. Nat. Cancer Inst.* **54**:767–768.
Karnovsky, M. J., and Unanue, E. R. 1973. Mapping and migration of lymphocyte surface macromolecules. *Fed. Proc.* **32**:55–59.
Kay, R. R., and Johnston, I. R. 1973. The nuclear envelope: Current problems of structure and of function. *Sub-Cell. Biochem.* **2**:127–166.
Kent, C., Schimmel, S. D., and Vagelos, P. R. 1974. Lipid composition of plasma membranes from developing chick muscle cells in culture. *Biochim. Biophys. Acta* **360**:312–321.
Kijimoto, S., and Hakomori, S. 1971. Enhanced glycolipid: α-galactosyl-transferase activity in contact-inhibited hamster cells, and loss of this response in polyoma transformants. *Biochem. Biophis. Res. Commun.* **44**:557–563.
Klevecz, R. R., and Ruddle, F. H. 1968. Cyclic changes in synchronized mammalian cell cultures. *Science* **159**:634–636.
Knox, P., and Pasternak, C. A. 1973. Serum-mediated membrane changes. *Biochem. Soc. Trans.* **1**:430–431.
Knox, P., and Pasternak, C. A. 1975. The binding of serum components to cultured cells, pp. 289–295. *In* C. A. Pasternak (ed.). Radioimmunoassay in Clinical Biochemistry. Heyden and Son, London.
Knox, P., and Pasternak, C. A. 1977. The cell surface and growth in vitro, pp. 195–218. *In* C. A. Jamieson and D. M. Robinson (eds.). Mammalian Cell Membranes, Vol. 5. Butterworths, London.
Knutton, S. 1976. Structural changes in the plasma membrane of synchronized P815Y mastocytoma cells. *Exp. Cell Res.* **102**:109–116.
Knutton, S., Sumner, M. C. B., and Pasternak, C. A., 1975. The role of microvilli in surface changes of synchronized P815T mastocytoma cells. *J. Cell Biol.* **66**:568–576.
Kornblatt, M. J., Knapp, A., Levine, M., Schachter, H., and Murray, R. K. 1974. Studies on

the structure and formation during spermatogenesis of the sulfoglycerogalactolipid of rat testis. *Canad. J. Biochem.* **52**:689–697.

Leskes, A., Siekevitz, P., and Palade, G. E. 1971. Differentiation of endoplasmic reticulum in hepatocytes. *J. Cell Biol.* **49**:288–302.

Ling, N. R. 1971. Lymphocyte stimulation. North-Holland Publishing Co., Amsterdam, 290 pp.

Maino, V. C., Green, N. M., and Crumpton, M. J. 1974. The role of calcium ions in initiating transformation of lymphocytes. *Nature* **251**:324–327.

Mayhew, E. C. 1967. Effect of ribonuclease and neuraminidase on the electrophoetic mobility of tissue culture cells in parasynchronous growth. *J. Cell Physiol.* **60**:305–309.

Mazia, D. 1974. The chromosome cycle in the cell cycle, pp. 265–272. *In* G. M. Padilla, I. L. Cameron and A. Zimmerman (eds.). Cell Cycle Controls, Academic Press, New York.

Michell, R. H. 1975. Inositol phopholipids and cell surface receptor function. *Biochim. Biophys. Acta* **415**:81–147.

Micklem, K. J., Abra, R. M., Knutton, S., Graham, J. M., and Pasternak, C. A. 1976. The fluidity of normal and transformed cell plasma membrane. *Biochem. J.* **154**:561–566.

Mitchison, J. M. 1971. The biology of the cell cycle. Cambridge University Press, Cambridge. 313 pp.

Mohri, H. 1964. Utilization of ^{14}C-labelled acetate and glycerol for lipid synthesis during the early development of sea urchin embryos. *Biol. Bull.* **126**:440–455.

Neufeld, E. F., Lim, T. W., and Shapiro, L. J. 1975. Inherited disorders of lysozomal metabolism. *Annu. Rev. Biochem.* **44**:357–376.

Nias, A. H. W., and Fox, M. 1971. Synchronization of mammalian cells with respect to the mitotic cycle. *Cell Tissue Kinet.* **4**:375–398.

Nicolson, G. L. 1972. Topography of membrane concanavalin A sites modified by proteolysis. *Nature New Biol. (London)* **239**:193–197.

Numa, S., and Yamashita, S. 1974. Regulation of lipogenesis in animal tissues. *Curr. Top. Cell Reg.* **8**:197–246.

O'Brien, J. S. 1973. Tay–Sachs disease: From enzyme to prevention. *Fed. Proc.* **32**:191–199.

Pasternak, C. A. 1970. Biochemistry of Differentiation. Wiley-Interscience, London. 189 pp.

Pasternak, C. A. 1972. Relief from contact inhibition. Early increase in phospholipid turnover. *J. Cell Biol.* **53**:231–234.

Pasternak, C. A. 1973a. Synchronization of mammalian cells by size separation, pp. 247–261. *In* A. I. Laskin and J. A. Last (eds.). Nucleic Acid Biosynthesis, Marcel Dekker, New York.

Pasternak, C. A. 1973b. Phospholipid synthesis in cleaving sea urchin eggs: Model for specific membrane assembly. *Dev. Biol.* **30**:403–410.

Pasternak, C. A. 1973c. Significance of phospholipid turnover. *Biochem. Soc. Trans.* **1**:333–336.

Pasternak, C. A. 1974. Biochemical aspects of the cell cycle, pp. 399–414. *In* A. T. Bull, J. R. Lagnado, J. O. Thomas, and K. F. Tipton (eds.). Companion to Biochemistry, Longman, London.

Pasternak, C. A. 1976. The cell surface in relation to the growth cycle. *J. Theoret. Biol.* **58**:365–382.

Pasternak, C. A., and Bergeron, J. J. M. 1970. Turnover of mammalian phospholipids. Stable and unstable components in neoplastic mast cells. *Biochem. J.* **119**:473–480.

Pasternak, C. A., and Friedrichs, B. 1970. Turnover of mammalian phospholipids. Rates of turnover and metabolic heterogeneity in cultured human lymphocytes and in tissues of healthy starved and vitamin A-deficient rats. *Biochem. J.* **119**:481–488.

Pasternak, C. A., and Knox, P. 1973. Phospholipid turnover as a determinant of membrane function, pp. 171–179. *In* P. W. Kent (ed.). Membrane Mediated Information, Vol. 1, Medical and Technical Publishing Co., Lancaster, England.

Pasternak, C. A., and Micklem, K. J. 1974. The biochemistry of virus-induced cell fusion. Changes in membrane integrity. *Biochem. J.* **140**:405–411.

Pasternak, C. A., Sumner, M. C. B., and Collin, R. C. L. S. 1974. Surface changes during the cell cycle, pp. 117–124. *In* G. M. Padilla, I. L. Cameron, and A. Zimmerman (eds.). Cell Cycle Controls. Academic Press, New York.

Pechen, A. M., and Bazan, N. G. 1974. Membrane ^{32}P-phospholipid labelling in early developing toad embryos. *Exp. Cell Res.* **88**:432–435.

Pechen, A. M., and Bazan, N. G. 1976. Lipid metabolism during early development using label precursors incorporated during oogenesis and in cell-free embryo homogenates. *Lipids* in press.

Pechen, A. M., Bonini, I. C., and Bazan, N. G. 1974. Distributional changes of ^{32}P labelled acid-soluble phosphates and phospholipids among subcellular fractions during early vertebrate embryonic development. *Biochim. Biophys. Acta* **372**:388–399.

Pegoraro, L., Galanti, N., Stein, G., and Baserga, R. 1972. The synthesis of phospholipids in the nucleus and nuclear membrane of synchronized He La cells. *Cell Tissue Kinet.* **15**:65–77.

Pinto Da Silva, P., and Martinez-Palomo, A. 1975. Distribution of membrane particles and gap junctions in normal and transformed 3T3 cells studied *in situ*, in suspension, and treated with concanavalin A. *Proc. Natl. Acad. Sci. USA* **72**:572–576.

Poole, A. R., Howell, J. I., and Lucy, J. A. 1970. Lysolicithin and cell fusion. *Nature* **227**:810–813.

Porter, R. R. 1971. The combining sites of antibodies. Harvey Lectures **65**:157–174.

Poste, G. 1972. Mechanisms of virus-induced cell fusion. *Int. Rev. Cytol.* **33**:157–252.

Poste, G., and Allison, A. C. 1973. Membrane fusion. *Biochim. Biophys. Acta* **300**:421–465.

Quinn, P. J. 1973. The association between phosphatidylinositol phosphodiesterase activity and a specific sub-unit of microtubular protein in rat brain. *Biochem. J.* **133**:273–281.

Quinn, P. J., and Pasternak, C. A. 1977. Properties of normal cell membranes: Lipids. *In* H. A. Blough and J. M. Tiffany (eds.). Cell Membranes and Viral Envelopes, Academic Press, London, in press.

Reporter, M., and Norris, G. 1973. Reversible effects of lysolecithin on fusion of cultured rat muscle cells. *Differentiation* **1**:83–95.

Resch, K., and Ferber, E. 1972. Phospholipid metabolism of stimulated lymphocytes. *Eur. J. Biochem.* **27**:153–161.

Ristow, H. J., Frank, W., and Fronlich, M. 1973. Stimulation of embryonic rat cells by calf serum. V. Metabolism of inositol and choline phospholipids. *Z. Naturforsch.* **28c**:188–197.

Rodbell, M., Jones, A. B., Chiappe de Cingolani, G. E., and Birnbaumer, L. 1968. The actions of insulin and catabolic hormones on the plasma membrane of the fat cells. *Rec. Progr. Horm. Res.* **24**:215–254.

Roseman, S. 1971. The synthesis of complex carbohydrates by multiglycosyl transferase systems and their potential function in intercellular adhesion. *Chem. Phys. Lipids* **5**:270–297.

Rosenberg, M. D. 1960. Microexudates from cells grown in tissue culture. *Biophys. J.* **1**:137–159.

Roth, S., and White, D. 1972. Intercellular contact and cell-surface galactosyl transferase activity. *Proc. Natl. Acad. Sci. USA* **69**:485–489.

Rouser, G., Kritchevsky, G., Yamamoto, A., and Baxter, C. F. 1972. Lipids in the nervous system of different species as a function of age: Brain, spinal cord, peripheral nerve, purified whole cell preparations, and subcellular particulates: Regulatory mechanisms and membrane structure. *Adv. Lipid Res.* **10**:261–360.

Sakiyama, H., and Robbins, P. W. 1973. Glycolipid synthesis and tumorigenicity of clones isolated from the NIL 2 line of hamster embryo fibroblasts. *Fed. Proc.* **32**:86–90.

Schindler, R., and Schaer, J. C. 1972. Preparation of synchronous cell cultures from early interphase cells obtained by sucrose gradient-centrifugation, pp. 43–65. *In* D. M. Prescott (ed.). Methods in Cell Biology, Vol. 6. Academic Press, N.Y.

Schmell, E., and Lennarz, W. J. 1974. Phospholipid metabolism in the eggs and embryos of the sea urchin *Arbacia punctulata. Biochemistry* **13**:4114–4121.

Scott, R. E., Carter, R. L., and Kidwell, W. R. 1971. Structural changes in membranes of synchronized cells demonstrated by freeze cleavage. *Nature New Biol. (London)* **233**:219–220.

Shall, S. 1973. Selection synchronization by velocity sedimentation separation of mouse fibroblast cells grown in suspension culture, pp. 269–285. *In* D. M. Prescott (ed.). Methods in Cell Biology, Vol 7. Academic Press, New York.

Shinitzky, M., and Inbar, M. 1974. Difference in microviscosity induced by different cholesterol levels in the surface membrane lipid layer of normal lymphocytes and malignant lymphoma cells. *J. Mol. Biol.* **85**:603–615.

Siek, T. J., and Newburgh, R. W. 1965. Origin of phospholipids in the chick embryo during development. *J. Lipid. Res.* **6**:556–564.

Singer, S. J., and Nicolson, G. L., 1972. The fluid mosaic model of the structure of cell membranes. *Science* **175**:720–731.

Smith, J. A., and Martin, L. 1973. Do cells cycle? *Proc. Natl. Acad. Sci. USA* **70**:1263–1267.

Stubblefield, E. 1968. Synchronization methods for mammalian cell cultures, pp. 25–42. *In* D. M. Prescott (ed.). Methods in Cell Physiology, Vol. 3, Academic Press, N.Y.

Sumner, M. C. B. 1974. Biogenesis of the plasma membrane. D. Phil. thesis. University of Oxford, England. 139 pp.

Takeichi, M. 1971. Changes in the properties of cell-substrate adhesion during cultivation of chicken fibroblasts *in vitro* in a serum-free medium. *Exp. Cell Res.* **68**:88–96.

Tata, J. R. 1967a. The formation and distribution of ribosomes during hormone-induced growth and development. *Biochem. J.* **104**:1–16.

Tata, J. R. 1967b. The formation, distribution and function of ribosomes and microsmal membranes during induced amphibian metamorphsis. *Biochem. J.* **105**:783–801.

Tempest, D. W., Dicks, J. W., and Ellwood, D. C. 1968. Influence of growth condition on the concentration of potassium in *Bacillus subtilis* var. *niger* and its possible relationship to cellular ribonucleic acid, teichonic acid and teichuronic acid. *Biochem. J.* **106**:237–243.

Torpier, G., Montagnier, L., Bignard, J. -M., and Vigier, P. 1975. A structural change of the plasma membrane induced by oncogenic viruses: quantitative studies with the freeze-fracture technique. *Proc. Natl. Acad. Sci. USA* **72**:1675–1698.

Tyler, A., and Tyler, B. S. 1966. Physiology of fertilization and early development, pp. 683–741. *In* R. A. Boolootian (ed.). Physiology of Echinodermata, John Wiley, New York.

Van Heyningen, W. E. 1974. Gangliosides as membrane receptors for tetanus toxin, cholera toxin and serotonin. *Nature* **249**:415–417.

Van Hoeven, R. P., and Emmelot, P. 1972. Studies on plasma membranes. XVIII. Lipid class composition of plasma membranes isolated from rat and mouse liver and hepatomas. *J. Membrane Biol.* **9**:105–126.

Van Hoeven, R. P., Emmelot, P., Krol, J. H., and Oomen-Meulemans, E. P. M. 1975. Studies on plasma membranes. XXII. Fatty acid profiles of lipid classes in plasma membranes of rat and mouse livers and hepatomas. *Biochim. Biophys. Acta* **380**:1–11.

Vlodavsky, I., and Sachs, L. 1974. Difference in the cellular cholesterol to phospholipid ratio in normal lymphocytes and in lymphocytic leukemic cells. *Nature* **250**:67–68.

Volk, B. W., and Schneck, L. (eds.) 1975. The gangliosidoses. Plenum, New York. 270 pp.

Warmsley, A. M. H., Phillips, B., and Pasternak, C. A. 1970. The use of zonal centrifugation to study membrane formation during the life cycle of mammalian cells. *Biochem. J.* **120**:683–688.

Warren, G. B., Toon, P. A., Birdsall, N. J. M., Lee, A. G., and Metcalfe, J. C. 1974a.

Reversible lipid titrations of the activity of pure adenosine triphosphatase–lipid complexes. *Biochemistry* **13**:5501–5507.

Warren, G. B., Toon, P. A., Birdsall, N. J. M., Lee, A. G., and Metcalfe, J. C. 1974b. Reconstitution of a calcium pump using defined membrane components. *Proc. Natl. Acad. Sci. USA* **71**:622–626.

Warren, L. 1959. The thiobarbituric acid assay of sialic acids. *J. Biol. Chem.* **234**:1971–1975.

Watkins, W. M. 1974. Blood-group substances: their nature and genetics, pp. 293–360. *In* D. M. Surgenor (ed.). The Red Blood Cell, Vol. 1, 2nd ed. Academic Press, New York.

Weiss, L. 1961. Studies on cellular adhesion in tissue culture. IV. The alteration of substrata by cell surfaces. *Exp. Cell Res.* **25**:504–517.

Wirtz, K. W. A. 1974. Transfer of phospholipids between membranes. *Biochim. Biophys. Acta* **344**:95–117.

Wirtz, K. W. A., Kamp, H. H., and Van Deenen, L. L. M. 1972. Isolation of a protein from beef liver which specifically stimulates the exchange of phosphatidylcholine. *Biochim. Biophys. Acta* **274**:606–617.

Wolf, B. A., and Robbins, P. W. 1974. Cell mitotic cycle synthesis of NIL hamster glycolipids including the Forssman antigen. *J. Cell Biol.* **61**:676–687.

Wood, R. (ed.) 1973. Tumor lipids: Biochemistry and Metabolism. American Oil Chemists Society, Champaign, Illinois. 324 pp.

Yahara, I., and Edelman, G. M. 1975. Modulation of lymphocyte receptor mobility by locally bound concanavalin A. *Proc. Natl. Acad. Sci. USA* **72**:1579–1583.

Index

MIX
Papier aus verantwortungsvollen Quellen
Paper from responsible sources
FSC® C105338

If you have any concerns about our products,
you can contact us on
ProductSafety@springernature.com

In case Publisher is established outside the EU,
the EU authorized representative is:
Springer Nature Customer Service Center GmbH
Europaplatz 3, 69115 Heidelberg, Germany

Printed by Libri Plureos GmbH
in Hamburg, Germany